James H. Capshew

# SCIENCE IN SOCIETY
## Classical and Contemporary Readings

# SCIENCE IN SOCIETY

## Classical and Contemporary Readings

*Edited by*
Byron E. Wall
*Associate of the Institute for the History and Philosophy of Science and Technology, University of Toronto*

**WALL & THOMPSON**
*Toronto*

**Canadian Cataloguing in Publication Data**

Main entry under title:

Science in society

ISBN 0-921332-25-4

1. Science – Social aspects.
2. Science and civilization.
3. Technology – Social aspects.
4. Technology and civilization.
I. Wall, Byron Emerson, 1943– .

Q175.5.S25 1989 500 C89-09
4892-5

ISBN 0–921332–25–4
Printed in Canada.
1 2 3 4 5 93 92 91 90 89

# Table of Contents

## Unit 3: Humans in the Environment / 201

## Unit 4: Current Issues in Science / 315

**Photograph following the next page:**
Scientific equipment in the Swiss Polytechnic in Zurich where Einstein studied from 1896 to 1899.

# PREFACE

This book is a collection of readings suitable as source material in courses in science, technology, and society, or the history and philosophy of science and technology, and similar related courses in colleges and universities and at the final year of high school. The book was created specifically for the Ontario Academic Course "Science in Society"; the topics covered and the sequence in which they are presented follow the curriculum guideline for that course. However, the book is generally applicable to and should be useful in a variety of courses in which these topics are examined.

American and other non-Canadian readers will find a few more references to Canadian situations and sources than they might otherwise expect, but the issues discussed really have no national boundaries and are just as relevant in the United States and other English-speaking countries as in Canada.

This is a long book, made deliberately so in order to give a wide choice of readings on each topic in order to accommodate different approaches to the subject matter, different emphases within the course topics, and different interests and abilities among the students. The selections are a combination of classical documents or commentaries from the history of science and technology and recent pieces, chosen for their readability and interest. In general each topic has some short, easier articles, and some longer, more demanding articles. Articles marked in the table of contents and on their first page with an asterisk (*) are more challenging and are intended for the better prepared students. Alternatively, these articles might be assigned as special projects or as starting points for essays.

Each article is followed by questions and topics for discussion and writing. There is considerable variation in the depth and complexity of the questions, generally with the easier questions first and the more demanding ones further on down the list. Many of these questions and topics, would make suitable essay or research paper topics.

This volume can be used as the sole text for a course, in which case the introductory overview for each section will help draw the separate articles together. Instructors should be aware that the articles have been selected in order that they stand on their own and are interesting by themselves, rather than that they "cover" the topic fully. Therefore many issues that might have been included are skipped entirely, or mentioned only in passing. For example, in the unit on the nature of technology, there are several articles about the history of technology, but nothing *per se* on the industrial revolution; in the unit on the nature of science, there is very little about the development of chemistry, and so on.

Those instructors who wish to ensure that all the major topics are given some consideration may wish to use this book in conjunction with a survey text. A good choice would be *Children of Prometheus: A History of Science and Technology* by James MacLachlan (Toronto: Wall & Thompson, 1989). Another approach is to supplement this book with research projects and further readings that can be found in libraries. A companion volume to this book is *Science in Society: An Annotated Guide to Resources* compiled by the Institute for the History and Philosophy of Science and Technology of the University of Toronto (Toronto: Wall & Thompson, 1989), which is intended to help anyone interested in science in society find the more useful sources. It follows the same organization of topics as in this book.

*Byron E. Wall*
*Toronto, July 24, 1989*

AMPERE

*Unit 1*

# The Nature of Science

- The Nature of Scientific Knowledge
- The Development of the Physical Sciences
- The Development of the Life Sciences

At the first Solvay Congress in Brussels in 1911. Most of Europe's top physicists of the early twentieth century are in the picture, including Marie Curie, seated, and Albert Einstein, standing, second from right.

*Issue*

# The Nature of Scientific Knowledge

It would seem to be a good idea to begin a book that looks at the context of science with a definition of science itself. But defining something as complex and far-reaching as science is no easy task. A whole branch of philosophy—the philosophy of science—is devoted to analysing the nature of scientific knowledge and its reasoning processes. The literature of the philosophy of science is vast, difficult, and full of areas of disagreement. Yet ordinary people speak of science frequently and at least *think* they are making themselves understood.

The articles in this section are intended to help clarify some of the major issues in defining the nature of science without getting too far afield in some of the more subtle points. William Hanna's article, "The Nature of Scientific Reasoning," lays out some of the main issues that are discussed in greater detail in other articles in this unit and which are referred to throughout the book. Isaac Asimov's "The Eureka Phenomenon" examines the creative process in science, particularly as it applies to finding new relationships and hypotheses. Berton Roueché illustrates deductive reasoning in "The Orange Man," where established scientific knowledge is used to diagnose an illness. Carl Sagan writes about science as a special kind of thinking about the world and considers the limits of human knowledge in "Can We Know the Universe?" Thomas Henry Huxley's lecture, "We All Use the Scientific Method," shows how the processes of observation, induction, hypothesis formation and evaluation that are considered hallmarks of the scientific method are also everyday tools of common sense thinking. An example of the careful analysis of scientific thinking by philosophers of science is Carl G. Hempel's "Scientific Inquiry: Invention and Test." Hempel shows how carefully chosen discriminating observations and experiments can enable a scientist to posit a likely hypothesis and then test it by testing its logical consequences. Finally, to illustrate the real scientist at work, Marie Curie's lecture, "The Discovery of Radium," describes her work investigating the nature of radioactivity and the methods she employed. However, more than the methods, what she really conveys is the joy of scientific work and the beauty of scientific knowledge.

# The Nature of Scientific Reasoning

## William R. Hanna

Just what *is* science anyway? If we want to talk sensibly about science in society, we had better have some clear idea what we mean by "science" and what makes a subject or statement "scientific." This article, by William R. Hanna of Humber College in Toronto, explores the basic issues.

What is known as modern science came into existence only a few hundred years ago, yet it has had a profound impact on almost every facet of our lives. The western world is inundated with the practical applications of scientific knowledge. These applications have led to improvements in farming, manufacturing, transportation, health—the list goes on and on. But the technical advances based on scientific knowledge have not only led to an easier and more comfortable life. There is also a negative side represented by such things as: the possibility of nuclear destruction, global pollution, changes to the climate, etc. In addition to improvements and threats, there are whole new fields being developed that open new possibilities and present unknown consequences. Computer technology, genetic engineering and molecular biology are all new fields that have only recently emerged. The impact of science and technology on our daily existence is of such magnitude that it would be folly to remain ignorant of both new developments and of the nature of scientific reasoning and the scientific process.

Given the way in which these advances, problems and benefits impinge upon our consciousness through the products of applied science, it is easy to confuse technological achievements with "science" itself. When not bewildered by the confusing array of technological "toys," public perception tends to focus on the complexity of scientific procedures and on the gulf separating the "real" world from the world in which scientists work. We tend to forget that science is a very human activity—a way of thinking. Over the last three hundred years, however, this way of thinking has dramatically expanded the range of human knowledge. It has both created new problems and offered tantalizing new visions. To understand the problems and to share in the visions, it is necessary to look more closely at the nature and limits of scientific reasoning.

* * *

When we talk about science, we normally refer to the activity of investigating nature. This activity must be faithful to the evidence gathered through careful observation and experimentation and it must follow certain rules of logic. Science studies particular aspects of our knowledge upon which, in principle, there can be universal agreement. It is perceived as seeking, and getting, maximum consensus.

The word itself refers to both a body of knowledge and the set of rules by which this knowledge is gathered. A dictionary definition of science might state that science is: "Knowledge of the world ascertained by observation, critically examined and classified systematically under general principles." This knowledge is also presumed to explain or encompass past discoveries and to predict future developments or occurrences. It is measurable insofar as the relation between hypotheses and their implications may be evaluated. A scientist, therefore, generates ideas which can then be tested through observation. The emphasis is not on the generation of ideas, but on their verification through testing.

Although there is room in science for luck and

guesses and "bolts from the blue," science is usually perceived as disciplined investigation. It is a series of questions followed by an attempt to answer these questions through systematic steps. Facts are observed, hypotheses are formed, and hypotheses are tested through controlled experimentation. Verified results of such methodology become the generalizations known as scientific theories or laws.

While the practical applications of science have certainly stimulated its development, curiosity plays an important role as well. There is a quest for general truths—truths that are illustrated by particular facts. These facts may be known by direct observation. A ball falls when released. The moon changes position in the sky. These are matters open to inspection. Science is not a simple record of such phenomena: it is an attempt to understand them. It is an effort to formulate general laws that describe underlying patterns and their interrelation. When a scientist observes an event, he looks for the natural law which governs that event and for the underlying principle that explains how it fits into the greater order of things. The role of theories is to explain facts. Galileo (1564–1642) formulated the laws of terrestrial mechanics to explain the motion of bodies on or near the earth. Kepler (1571–1630) formulated the laws of celestial mechanics to explain planetary motion. It was up to Newton (1642–1727) to unify and explain them through his Theory of Universal Gravitation and the Laws of Motion.

Strictly speaking, then, scientific method connects present observation to past experience. This allows for the generation of a hypothesis from which predictions can be made. Findings or evidence are then subject to the verdict of experience. The merit of the hypothesis must be assessed. The basic idea is straightforward. Starting with some knowledge, we infer that, if X then Y. Then, given some figures, we predict the likely result as exactly as possible. The prediction is put to the test of experience and is either upheld or refuted. If refuted, we either discard the hypothesis or try an amended version.

As mentioned above, hypotheses must be relevant to the facts they are intended to explain. Conversely, the fact in question must be deducible from the proposed hypothesis. Hypotheses must also be testable—there must be the possibility of making observations that either confirm or disprove the hypothesis. They must also be compatible and consistent with other, "established" hypotheses. Sometimes, there may be two or more hypotheses which are relevant, compatible and subject to verification. Scientists then look to predictive power and simplicity to choose among conflicting theories. The greater the range of observable facts which can be deduced from a hypothesis and the simpler the explanation, the more likely it is that a given hypothesis will be accepted.

* * *

It is important to remember that scientific laws, theories and hypotheses are not certainties. Invariability in science is only invariable until proven otherwise. Laws of substance, for example, describe such things as elements and compounds. At one time, the freezing and boiling temperatures of water were thought to be invariable. It was discovered, however, that the application of pressure may effect a significant change on these. The law that explains these properties of water had to be modified. Scientists can never be certain that important factors may not have been considered. Generally speaking, laws state invariable relations and regularities about nature which are subject to empirical confirmation. Theories attempt to explain laws. The history of science is littered with the corpses of theories that could not explain newly discovered laws. Theories are necessary but fallible inventions of the mind, constructed in the effort to make sense of the observable world. They are a valuable tool of inquiry, but they are never certainties.

We have made reference to both deduction and induction. These are both elements of logic, and logic plays a key role in scientific method. In essence, logic is the study of the principles of correct reasoning. It analyses arguments and the clarity of their expression. Logic does not determine the quality of evidence (facts) nor the quality of conclusions. Its role is to determine the quality of the relation between evidence and conclusion—arguments. Scientific theories are built on arguments, and the success of a theory is based, in part, upon the soundness of its logic.

Science is often considered to be very methodical and pedestrian: observation of data, determination of relations and regularities, divination of organizing and underlying patterns. This is a misrepresentation of scientific method. While the method is indeed rigorous, and the testing is perhaps routine, the very fact that hypotheses and theories are not certainties indicates their conjectural nature. They are educated guesses. There is conjecture involved in every use of evidence. Underlying patterns are not written on every event or experience. The act of discovery is the art of informed speculation; the proposing of answers to questions beginning "What if..."

Speculation is one of the distinguishing features of humankind. In this sense, science is a tool used more or less rigorously by everyone. It has limits and limitations. Its products are both a curse and a boon. Offering both a pathway to knowledge and the challenge of discovery, science is not likely to be discarded because of the consequences of its applications. The technological manifestations of science are its products. If products malfunction, create hazards, etc., you don't blame the tools used in production—you look to the manufacturer.

## Questions and Topics for Discussion and Writing

1. Give a definition of science and a definition of another kind of knowledge which you would not classify as science.
2. Make lists of subject statements, and theories in three categories: (1) those which are clearly scientific, (2) those which are clearly not scientific, and (3) those which could be either scientific or not.
3. Discuss some of the items selected in category (3) in question 2 above. Explain how each item could be called scientific and how it could be called non-scientific.

*William R. Hanna*

# The Eureka Phenomenon

## Isaac Asimov

Discussions of the nature of science usually concentrate on aspects of science which *differ* from non-science and which have to do with verification and reasoning procedures. But while science does have special means by which facts become established and accepted, there would be no body of scientific knowledge at all without the creative process that conceived the scientific propositions to begin with. In this article prolific science writer Isaac Asimov discusses the creative process in science as revealed in some famous examples.

In the old days, when I was writing a great deal of fiction, there would come, once in a while, moments when I was stymied. Suddenly, I would find I had written myself into a hole and could see no way out. To take care of that, I developed a technique which invariably worked.

It was simply this—I went to the movies. Not just any movie. I had to pick a movie which was loaded with action but which made no demands on the intellect. As I watched, I did my best to avoid any conscious thinking concerning my problem, and when I came out of the movie I knew exactly what I would have to do to put the story back on the track.

It never failed.

In fact, when I was working on my doctoral dissertation, too many years ago, I suddenly came across a flaw in my logic that I had not noticed before and that knocked out everything I had done. In utter panic, I made my way to a Bob Hope movie—and came out with the necessary changes in point of view.

It is my belief, you see, that thinking is a double phenomenon like breathing.

You can control breathing by deliberate voluntary action: you can breathe deeply and quickly, or you can hold your breath altogether, regardless of the body's needs at the time. This, however, doesn't work well for very long. Your chest muscles grow tired, your body clamors for more oxygen, or less, and you relax. The automatic involuntary control of breathing takes over, adjusts it to the body's needs and unless you have some respiratory disorder, you can forget about the whole thing.

Well, you can think by deliberate voluntary action, too, and I don't think it is much more efficient on the whole than voluntary breath control is. You can deliberately force your mind through channels of deductions and associations in search of a solution to some problem and before long you have dug mental furrows for yourself and find yourself circling round and round the same limited pathways. If those pathways yield no solution, no amount of further conscious thought will help.

On the other hand, if you let go, then the thinking process comes under automatic involuntary control and is more apt to take new pathways and make erratic associations you would not think of consciously. The solution will then come while you *think* you are *not* thinking.

The trouble is, though, that conscious thought involves no muscular action and so there is no sensation of physical weariness that would force you to quit. What's more, the panic of necessity tends to force you to go on uselessly, with each added bit of useless effort adding to the panic in a vicious cycle.

It is my feeling that it helps to relax, deliberately, by subjecting your mind to material complicated enough to occupy the voluntary faculty of thought, but superficial enough not to engage the deeper involuntary one. In my case, it is an action movie; in your case, it might be something else.

I suspect it is the involuntary faculty of thought that gives rise to what we call "a flash of intuition," something that I imagine must be merely the result of unnoticed thinking.

Perhaps the most famous flash of intuition in the

history of science took place in the city of Syracuse in third-century B.C. Sicily. Bear with me and I will tell you the story.

About 250 B.C., the city of Syracuse was experiencing a kind of Golden Age. It was under the protection of the rising power of Rome, but it retained a king of its own and considerable self-government; it was prosperous; and it had a flourishing intellectual life.

The king was Hieron II, and he had commissioned a new golden crown from a goldsmith, to whom he had given an ingot of gold as raw material. Hieron, being a practical man, had carefully weighed the ingot and then weighed the crown he received back. The two weights were precisely equal. Good deal!

But then he sat and thought for a while. Suppose the goldsmith had subtracted a little bit of the gold, not too much, and had substituted an equal weight of the considerably less valuable copper. The resulting alloy would still have the appearance of pure gold, but the goldsmith would be plus a quantity of gold over and above his fee. He would be buying gold with copper, so to speak, and Hieron would be neatly cheated.

Hieron didn't like the thought of being cheated any more than you or I would, but he didn't know how to find out for sure if he had been. He could scarcely punish the goldsmith on mere suspicion. What to do?

Fortunately, Hieron had an advantage few rulers in the history of the world could boast. He had a relative of considerable talent. The relative was named Archimedes and he probably had the greatest intellect the world was to see prior to the birth of Newton.

Archimedes was called in and was posed the problem. He had to determine whether the crown Hieron showed him was pure gold, or was gold to which a small but significant quantity of copper had been added.

If we were to reconstruct Archimedes' reasoning, it might go as follows. Gold was the densest known substance (at that time). Its density in modern terms is 19.3 grams per cubic centimeter. This means that a given weight of gold takes up less volume than the same weight of anything else! In fact, a given weight of pure gold takes up less volume than the same weight of *any* kind of impure gold.

The density of copper is 8.92 grams per cubic centimeter, just about half that of gold. If we consider 100 grams of pure gold, for instance, it is easy to calculate it to have a volume of 5.18 cubic centimeters. But suppose that 100 grams of what looked like pure gold was really only 90 grams of gold and 10 grams of copper. The 90 grams of gold would have a volume of 4.66 cubic centimeters, while the 10 grams of copper would have a volume of 1.12 cubic centimeters; for a total value of 5.78 cubic centimeters.

The difference between 5.18 cubic centimeters and 5.78 cubic centimeters is quite a noticeable one, and would instantly tell if the crown were of pure gold, or if it contained 10 per cent copper (with the missing 10 per cent of gold tucked neatly in the goldsmith's strongbox).

All one had to do, then, was measure the volume of the crown and compare it with the volume of the same weight of pure gold.

The mathematics of the time made it easy to measure the volume of many simple shapes: a cube, a sphere, a cone, a cylinder, any flattened object of simple regular shape and known thickness, and so on.

We can imagine Archimedes saying, "All that is necessary, sire, is to pound that crown flat, shape it into a square of uniform thickness, and then I can have the answer for you in a moment."

Whereupon Hieron must certainly have snatched the crown away and said, "No such thing. I can do that much without you; I've studied the principles of mathematics, too. This crown is a highly satisfactory work of art and I won't have it damaged. Just calculate its volume without in any way altering it."

But Greek mathematics had no way of determining the volume of anything with a shape as irregular as the crown, since integral calculus had not yet been invented (and wouldn't be for two thousand years, almost). Archimedes would have had to say, "There is no known way, sire, to carry through a non-destructive determination of volume."

"Then think of one," said Hieron testily.

And Archimedes must have set about thinking of one, and gotten nowhere. Nobody knows how long he thought, or how hard, or what hypotheses he considered and discarded, or any of the details.

What we do know is that, worn out with thinking,

Archimedes decided to visit the public baths and relax. I think we are quite safe in saying that Archimedes had no intention of taking his problem to the baths with him. It would be ridiculous to imagine he would, for the public baths of a Greek metropolis weren't intended for that sort of thing.

The Greek baths were a place for relaxation. Half the social aristocracy of the town would be there and there was a great deal more to do than wash. One steamed one's self, got a massage, exercised, and engaged in general socializing. We can be sure that Archimedes intended to forget the stupid crown for a while.

One can envisage him engaging in light talk, discussing the latest news from Alexandria and Carthage, the latest scandals in town, the latest funny jokes at the expense of the country-squire Romans—and then he lowered himself into a nice hot bath which some bumbling attendant had filled too full.

The water in the bath slopped over as Archimedes got in. Did Archimedes notice that at once, or did he sigh, sink back, and paddle his feet awhile before noting the water-slop? I guess the latter. But, whether soon or late, he noticed, and that one fact, added to all the chains of reasoning his brain had been working on during the period of relaxation when it was unhampered by the comparative stupidities (even in Archimedes) of voluntary thought, gave Archimedes his answer in one blinding flash of insight.

Jumping out of the bath, he proceeded to run home at top speed through the streets of Syracuse. He did *not* bother to put on his clothes. The thought of Archimedes running naked through Syracuse has titillated dozens of generations of youngsters who have heard this story, but I must explain that the ancient Greeks were quite lighthearted in their attitude toward nudity. They thought no more of seeing a naked man on the streets of Syracuse, than we would on the Broadway stage.

And as he ran, Archimedes shouted over and over, "I've got it! I've got it! " Of course, knowing no English, he was compelled to shout it in Greek, so it came out, "Eureka! Eureka! "

Archimedes' solution was so simple that anyone could understand it—once Archimedes explained it.

If an object that is not affected by water in any way, is immersed in water, it is bound to displace an amount of water equal to its own volume, since two objects cannot occupy the same space at the same time.

Suppose, then, you had a vessel large enough to hold the crown and suppose it had a small overflow spout set into the middle of its side. And suppose further that the vessel was filled with water exactly to the spout, so that if the water level were raised a bit higher, however slightly, some would overflow.

Next, suppose that you carefully lower the crown into the water. The water level would rise by an amount equal to the volume of the crown, and that volume of water would pour out the overflow and be caught in a small vessel. Next, a lump of gold, known to be pure and exactly equal in weight to the crown, is also immersed in the water and again the level rises and the overflow is caught in a second vessel.

If the crown were pure gold, the overflow would be exactly the same in each case, and the volume of water caught in the two small vessels would be equal. If, however, the crown were of alloy, it would produce a larger overflow than the pure gold would and this would be easily noticeable.

What's more, the crown would in no way be harmed, defaced, or even as much as scratched. More important, Archimedes had discovered the "principle of buoyancy."

And was the crown pure gold? I've heard that it turned out to be alloy and that the goldsmith was executed, but I wouldn't swear to it.

How often does this "Eureka phenomenon" happen? How often is there this flash of deep insight during a moment of relaxation, this triumphant cry of "I've got it! I've got it! " which must surely be a moment of the purest ecstasy this sorry world can afford?

I wish there were some way we could tell. I suspect that in the history of science it happens *often*; I suspect that very few significant discoveries are made by the pure technique of voluntary thought; I suspect that voluntary thought may possibly prepare the ground (if even that), but that the final touch, the real inspiration, comes when thinking is under involuntary control.

But the world is in a conspiracy to hide the fact. Scientists are wedded to reason, to the meticulous working out of consequences from assumptions to the careful organization of experiments designed to check those consequences. If a certain line of experiments ends nowhere, it is omitted from the final report. If an inspired guess turns out to be correct, it is *not* reported as an inspired guess. Instead, a solid line of voluntary thought is invented after the fact to lead up to the thought, and that is what is inserted in the final report.

The result is that anyone reading scientific papers would swear that *nothing* took place but voluntary thought maintaining a steady clumping stride from origin to destination, and that just can't be true.

It's such a shame. Not only does it deprive science of much of its glamour (how much of the dramatic story in Watson's *Double Helix* do you suppose got into the final reports announcing the great discovery of the structure of DNA? [1]), but it hands over the important process of "insight," "inspiration," "revelation" to the mystic.

The scientist actually becomes ashamed of having what we might call a revelation, as though to have one is to betray reason—when actually what we call revelation in a man who has devoted his life to reasoned thought, is after all merely reasoned thought that is not under voluntary control.

Only once in a while in modern times do we ever get a glimpse into the workings of involuntary reasoning, and when we do, it is always fascinating. Consider, for instance, the case of Friedrich August Kekule von Stradonitz.

In Kekule's time, a century and a quarter ago, a subject of great interest to chemists was the structure of organic molecules (those associated with living tissue). Inorganic molecules were generally simple in the sense that they were made up of few atoms. Water molecules, for instance, are made up of two atoms of hydrogen and one of oxygen ($H_2O$). Molecules of ordinary salt are made up of one atom of sodium and one of chlorine (NaCl), and so on.

Organic molecules, on the other hand, often contained a large number of atoms. Ethyl alcohol molecules have two carbon atoms, six hydrogen atoms, and an oxygen atom ($C_2H_6O$); the molecule of ordinary cane sugar is $C_{12}H_{22}O_{11}$, and other molecules are even more complex.

Then, too, it is sufficient, in the case of inorganic molecules generally, merely to know the kinds and numbers of atoms in the molecule; in organic molecules, more is necessary. Thus, dimethyl ether has the formula $C_2H_6O$, just as ethyl alcohol does, and yet the two are quite different in properties. Apparently, the atoms are arranged differently within the molecules—but how to determine the arrangements?

In 1852, an English chemist, Edward Frankland, had noticed that the atoms of a particular element tended to combine with a fixed number of other atoms. This combining number was called "valence." Kekule in 1858 reduced this notion to a system. The carbon atom, he decided (on the basis of plenty of chemical evidence) had a valence of four; the hydrogen atom, a valence of one; and the oxygen atom, a valence of two (and so on).

Why not represent the atoms as their symbols plus a number of attached dashes, that number being equal to the valence? Such atoms could then be put together as though they were so many Tinker Toy units and "structural formulas" could be built up.

It was possible to reason out that the structural formula of ethyl alcohol was

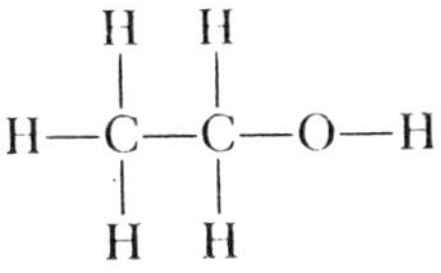

while that of dimethyl ether was

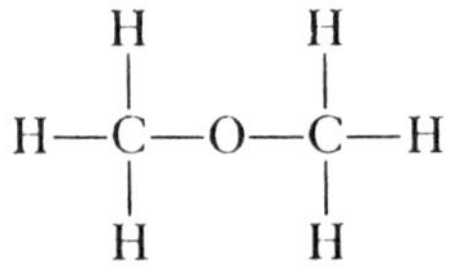

[1] I'll tell you, in case you're curious. None! [Asimov's note.]

In each case, there were two carbon atoms, each with four dashes attached; six hydrogen atoms, each with one dash attached; and an oxygen atom with two ashes attached. The molecules were built up of the same components, but in different arrangements.

Kekule's theory worked beautifully. It has been immensely deepened and elaborated since his day, but you can still find structures very much like Kekule's Tinker Toy formulas in any modern chemical textbook. They represent oversimplifications of the true situation, but they remain extremely useful in practice even so.

The Kekule structures were applied to many organic molecules in the years after 1858 and the similarities and contrasts in the structures neatly matched similarities and contrasts in properties. The key to the rationalization of organic chemistry had, it seemed, been found.

Yet there was one disturbing fact. The well-known chemical benzene wouldn't fit. It was known to have a molecule made up of equal numbers of carbon and hydrogen atoms. Its molecular weight was known to be 78 and a single carbon-hydrogen combination had a weight of 13. Therefore, the benzene molecule had to contain six carbon-hydrogen combinations and its formula had to be $C_6H_6$.

But that meant trouble. By the Kekule formulas, the hydrocarbons (molecules made up of carbon and hydrogen atoms only) could easily be envisioned as chains of carbon atoms with hydrogen atoms attached. If all the valences of the carbon atoms were filled with hydrogen atoms, as in "hexane," whose molecule looks like this—

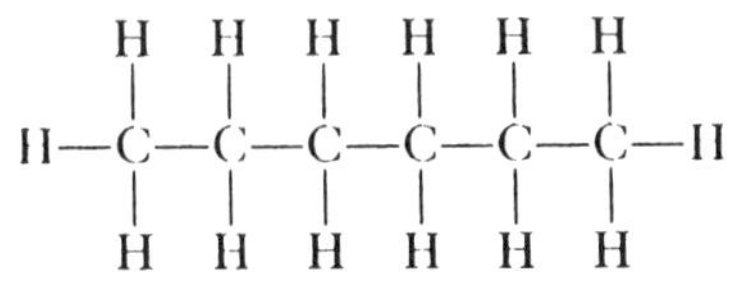

the compound is said to be saturated. Such saturated hydrocarbons were found to have very little tendency to react with other substances.

If some of the valences were not filled, unused bonds were added to those connecting the carbon atoms. Double bonds were formed as in "hexene"—

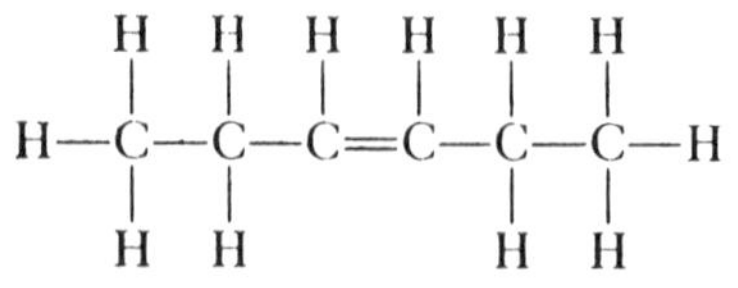

Hexene is unsaturated, for that double bond has a tendency to open up and add other atoms. Hexene is chemically active.

When six carbons are present in a molecule, it takes fourteen hydrogen atoms to occupy all the valence bonds and make it inert—as in hexane. In hexene, on the other hand, there are only twelve hydrogens. If there were still fewer hydrogen atoms, there would be more than one double bond; there might even be triple bonds, and the compound would be still more active than hexene.

Yet benzene, which is $C_6H_6$ and has eight fewer hydrogen atoms than hexane, is *less* active than hexene, which has only two fewer hydrogen atoms than hexane. In fact, benzene is even less active than hexane itself. The six hydrogen atoms in the benzene molecule seem to satisfy the six carbon atoms to a greater extent than do the fourteen hydrogen atoms in hexane.

For heaven's sake, why?

This might seem unimportant. The Kekule formulas were so beautifully suitable in the case of so many compounds that one might simply dismiss benzene as an exception to the general rule.

Science, however, is not English grammar. You can't just categorize something as an exception. If the exception doesn't fit into the general system, then the general system must be wrong.

Or, take the more positive approach. An exception can often be made to fit into a general system, provided the general system is broadened. Such broadening generally represents a great advance and for this reason, exceptions ought to be paid great attention.

For some seven years, Kekule faced the problem of benzene and tried to puzzle out how a chain of six carbon atoms could be completely satisfied with as few as six hydrogen atoms in benzene and yet be left unsatisfied with twelve hydrogen atoms in hexane.

Nothing came to him!

And then one day in 1865 (he tells the story himself) he was in Ghent, Belgium, and in order to get to some destination, he boarded a public bus. He was tired and, undoubtedly, the droning beat of the horses' hooves on the cobblestones lulled him. He fell into a comatose half-sleep.

In that sleep, he seemed to see a vision of atoms attaching themselves to each other in chains that moved about. (Why not? It was the sort of thing that constantly occupied his waking thoughts.) But then one chain twisted in such a way that head and tail joined, forming a ring—and Kekule woke with a start.

To himself, he must surely have shouted "Eureka," for indeed he had it. The six carbon atoms of benzene formed a ring and not a chain, so that the structural formula looked like this:

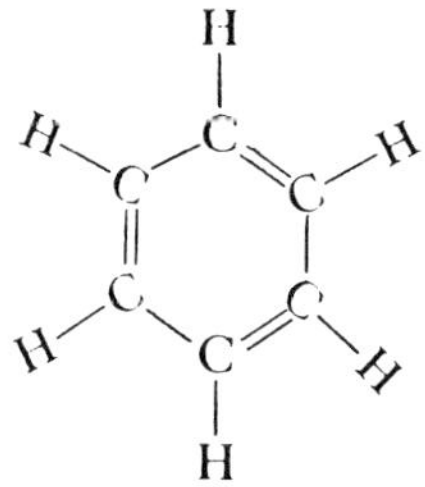

To be sure, there were still three double bonds, so you might think the molecule had to be very active—but now there was a difference. Atoms in a ring might be expected to have different properties from those in a chain and double bonds in one case might not have the properties of those in the other. At least, chemists could work on that assumption and see if it involved them in contradictions.

It didn't. The assumption worked excellently well. It turned out that organic molecules could be divided into two groups: aromatic and aliphatic. The former had the benzene ring (or certain other similar rings) as part of the structure and the latter did not. Allowing for different properties within each group, the Kekule structures worked very well.

For nearly seventy years, Kekule's vision held good in the hard field of actual chemical techniques, guiding the chemist through the jungle of reactions that led to the synthesis of more and more molecules. Then, in 1932, Linus Pauling applied quantum mechanics to chemical structure with sufficient subtlety to explain just why the benzene ring was so special and what had proven correct in practice proved correct in theory as well.

* * *

Other cases? Certainly.

In 1764, the Scottish engineer James Watt was working as an instrument maker for the University of Glasgow. The university gave him a model of a Newcomen steam engine, which didn't work well, and asked him to fix it. Watt fixed it without trouble, but even when it worked perfectly, it didn't work well. It was far too inefficient and consumed incredible quantities of fuel. Was there a way to improve that?

Thought didn't help, but a peaceful, relaxed walk on a Sunday afternoon did. Watt returned with the key notion in mind of using two separate chambers, one for steam only and one for cold water only, so that the same chamber did not have to be constantly cooled and reheated to the infinite waste of fuel.

The Irish mathematician William Rowan Hamilton worked up a theory of "quaternions" in 1843 but couldn't complete that theory until he grasped the fact that there were conditions under which $p \times q$ was *not* equal to $q \times p$. The necessary thought came to him in a flash one time when he was walking to town with his wife.

The German physiologist Otto Loewi was working on the mechanism of nerve action, in particular, on the chemicals produced by nerve endings. He awoke at 3 A.M. one night in 1921 with a perfectly clear notion of the type of experiment he would have to run to settle a key point that was puzzling him. He wrote it down and went back to sleep. When he woke in the morning, he found he couldn't remember what his inspiration had been. He remembered he had written it down, but he couldn't read his writing.

The next night, he woke again at 3 A.M. with the clear thought once more in mind. This time, he didn't fool around. He got up, dressed himself, went straight to the laboratory and began work. By 5 A.M. he had proved his point and the consequences of his findings became important enough in later years so that in 1936 he received a share in the Nobel prize in medicine and physiology.

*Isaac Asimov*

How very often this sort of thing must happen, and what a shame that scientists are so devoted to their belief in conscious thought that they so consistently obscure the actual methods by which they obtain their results.

From Isaac Asimov, *The Left Hand of the Election (Garden City, N.Y.: Doubleday, 1971).*

## Questions and Topics for Discussion and Writing

1. Why does Asimov want the reader to understand the role of the "Eureka Phenomenon" in science? What is the conception of scientific reasoning that he contrasts it to?
2. Why is it important that the scientist *not* be concentrating on the problem to be solved to make the breakthrough insight possible?
3. Asimov contrasts science to English grammar, which can accommodate exceptions to rules. Why can science not have such exceptions?

# The Orange Man

## Berton Roueché

An important part of scientific thinking is reasoning by deduction. Deduction is a part of every stage of science, but it is particularly characteristic of applications of a body of scientific knowledge to investigations. The practice of medicine provides especially good examples.

How does a doctor diagnose an illness? What are the means by which the patient's symptoms guide the doctor to the identification of the disease at hand and to the proper treatment? This article by science and medical writer Berton Roueché describes the diagnostic process for a rather bizarre case.

Around eleven-thirty on the morning of December 15, 1960, Dr. Richard L. Wooten, an internist and an assistant professor of internal medicine at the University of Tennessee College of Medicine, in Memphis, was informed by the receptionist in the office he shares with several associates that a patient named (I'll say) Elmo Turner was waiting to see him. Dr. Wooten remembered Turner, but not much about him. He asked the receptionist to fetch him Turner's folder, and then, when she had done so, to send Turner right on in. The folder refreshed his memory. Turner was fifty-three years old, married, and a plumber by trade, and over the past ten years, Dr. Wooten had seen him through an attack of pneumonia and referred him along for treatment of a variety of troubles, including a fractured wrist and hip-joint condition. There were footsteps in the hall. Dr. Wooten closed the folder. The door opened, and Turner—a short, thick, muscular man—came in. Dr. Wooten had risen to greet him, but for a moment he could only stand and stare. Turner's face was orange—a golden, pumpkin orange. So were both his hands.

Dr. Wooten found his voice. He gave Turner a friendly good morning, asked him to sit down, and remarked that it had been a couple of years since their last meeting. Turner agreed that it had. He had been away. He had been working up in Alaska—in Fairbanks. He and his wife were back in Memphis only on a matter of family business. But, being in town, he though he ought to pay Dr. Wooten a visit. There was something that kind of bothered him. Dr. Wooten listened with half an ear. His mind was searching through the spectrum of pathological skin discolorations. There were many diseases with pigmentary manifestations. There was the paper pallor of pituitary disease. There was the cyanotic blue of congenital heart disease. There was the deep Florida tan of thyroid dysfunction. There was the jaundice yellow of liver damage. There was the bronze of hemochromatosis. As far as he knew, however, there was no disease that colored its victims orange. Turner's voice recalled him. In fact, he was saying, he was worried. Dr. Wooten nodded. Just what seemed to be the trouble? Turner touched his abdomen with a bright-orange hand. He had a pain down there. His abdomen had been sore off and on for over a year, but now it was more than sore. It hurt. Dr. Wooten gave an encouraging grunt, and waited. He waited for Turner to say something about his extraordinary color. But Turner had finished. He had said all he had to say. Apparently, it was only his abdomen that worried him.

Dr. Wooten stood up. He asked Turner to come along down the hall to the examining room. His color, however bizarre, could wait. A chronic abdominal pain came first. And not only that. The cause of Turner's pain was probably also the cause of his color. That seemed, at least, a reasonable assumption. They entered the examining room. Dr. Wooten switched on the light above the examination table and turned and looked at Turner. The light in his office had been an ordinary electric light, and ordinary electric light has a faintly yellow tinge. The examining room had a true-color daylight light.

But Turner's color owed nothing to tricks of light. His skin was still an unearthly golden orange. Turner stripped to the waist and got up on the table and stretched out on his back. His torso was as orange as his face. Dr. Wooten began his examination. He found the painful abdominal area, and carefully pressed. There was something there. He could feel an abnormality—a deep-seated mass about the size of an apple. It was below and behind the stomach, and he thought it might be sited at the liver. He pressed again. It wasn't the liver. It was positioned too near the center of the stomach for that. It was the pancreas.

Dr. Wooten moved away from the table. He had learned all he could from manual exploration. He waited for Turner to dress, and then led the way back to his office. He told Turner what he had found. He said he couldn't identify the mass he had felt, and he wouldn't attempt to guess. Its nature could be determined only by a series of X-ray examinations. That, he was sorry to say, would require a couple of days in the hospital. The pancreas was seated too deep to be accessible to direct X-ray examination, and an indirect examination took time and special preparation. Turner listened, and shrugged. He was willing to do whatever had to be done. Dr. Wooten swivelled around in his chair and picked up the telephone. He put in a call to the admitting office of Baptist Memorial Hospital, an affiliate of the medical school, and had a few words with the reservations clerk. He swivelled back to Turner. It was all arranged. Turner would be expected at Baptist Memorial at three o'clock that afternoon. Turner nodded, and got up to go. Dr. Wooten waved him back into his chair. There was one more thing. It was about the color of his skin. How long had it been like that? Turner looked blank. Color? What color? What was wrong with the color of his skin? Dr. Wooten hesitated. He was startled. There was no mistaking Turner's reaction. He was genuinely confused. He didn't know about his color—he really didn't know. And that was an interesting thought. It was, in fact, instructive. It clearly meant that Turner's change of color was not a sudden development. It had come on slowly, insidiously, imperceptibly. He realized that Turner was waiting, that his question had to be answered. Dr. Wooten answered it. Turner looked even blanker. He gazed at his hands, and then at Dr. Wooten. He didn't see anything unusual about his color. His skin was naturally ruddy. It always looked this way.

Dr. Wooten let it go at that. There was no point in pressing the matter any further right then. It would only worry Turner, and he was worried enough already. The matter would keep until the afternoon, until the next day, until he had a little more information to work with. He leaned back and lighted a cigarette, and changed the subject. Or seemed to. Had Turner ever met the senior associate here? That was Dr. Hughes—Dr. John D. Hughes. No? Well, in that case...Dr. Wooten reached for the telephone. Dr. Hughes's office was just next door, and he arrived a moment later. He walked into the room and glanced at Turner, and stopped—and stared. Dr. Wooten introduced them. He described the reason for Turner's visit and the mass he had found in the region of the pancreas. Dr. Hughes subdued his stare to a look of polite attention. They talked for several minutes. When Turner got up again to go, Dr. Wooten saw him to the door. He came back to his desk and sat down. Well, what did Dr. Hughes make of that? Had he ever seen or read or even heard of a man that color before? Dr. Hughes said no. And he didn't know what to think. He was completely flabbergasted. He was rather uneasy, too. That, Dr. Wooten said, made two of them.

* * *

Turner was admitted to Baptist Memorial Hospital for observation that afternoon at a few minutes after three. He was given the usual admission examination and assigned a bed in a ward. An hour or two later, Dr. Wooten, in the course of his regular hospital rounds, stopped by Turner's bed for the ritual visit of welcome and reassurance. Turner appeared to be no more than reasonably nervous, and Dr. Wooten found that satisfactory. He then turned his attention to Turner's chart and the results of the admission examination. They were, as expected, unrevealing. Turner's temperature was normal. So were his pulse rate (seventy-eight beats a minute), his respiration rate (sixteen respirations a minute), and his blood pressure (a hundred and ten systolic, eighty diastolic). The results of the urinalysis and of an electrocardiographic examination were also normal. Before resuming his rounds, Dr. Wooten satisfied himself that the really important examinations had been scheduled. These were comprehensive X-ray studies of the chest, upper gastrointestinal tract, and colon. the first two examinations were down for the following morning.

They were made at about eight o'clock. When Dr. Wooten reached the hospital on a midmorning tour, the radiologist's report was in and waiting. It more than confirmed Dr. Wooten's impression of the location of the mass. It defined its nature as well. The report read, "Lung fields are clear. Heart is normal. Barium readily traversed the esophagus and entered the stomach. In certain positions, supine projections, an apparent defect was seen on the stomach. However, this was extrinsic to the stomach. It may well represent a pseudocyst of the pancreas. No lesions of the stomach itself were demonstrated. Duodenal bulb and loop appeared normal. Stomach was emptying in a satisfactory manner." Dr. Wooten put down the report with a shiver of relief. A pancreatic cyst—even a pseudocyst—is not a trifling affliction, but he welcomed that diagnosis. The mass on Turner's pancreas just might have been a tumor. It hadn't been a likely possibility—the mass was too large and the symptoms were too mild—but it had been a possibility.

Dr. Wooten went up to Turner's ward. He told Turner what the X-ray examination had shown and what the findings meant. A cyst was a sac retaining a liquid normally excreted by the body. A pseudocyst was an empty sac—a mere dilation of space. The only known treatment of a pancreatic cyst was surgical, and surgery involving the pancreas was difficult and dangerous. Surgery was difficult because of the remote location of the pancreas, and dangerous because of the delicacy of the organs surrounding the pancreas (the stomach, the spleen, the duodenum) and the delicacy of the functions of the pancreas (the production of enzymes essential to digestion and the secretion of insulin). Fortunately, however, treatment was seldom necessary. Most cysts—particularly pseudocysts—had a way of disappearing as mysteriously as they had come. It was his belief that this was such a cyst. In that case, there was nothing much to do but be patient. And careful. Turner was to guard his belly from sudden bumps or strains. A blow or a wrench could cause a lot of trouble.

Nevertheless, Dr. Wooten went on, he wanted Turner to remain in the hospital for at least another day. There was a final X-ray of the colon to be made, and several other tests. In view of this morning's findings, the examination was, he admitted, very largely a matter of form. The cause of Turner's abdominal pain was definitely a pseudocyst of the pancreas. But prudence required an X-ray, and it would probably be done the next day. It was usual, for technical reasons, to let a day elapse between an upper-gastrointestinal study and a colon examination. Two of the other tests were indicated by the X-ray findings. One was a test for diabetes—the glucose-tolerance test. Diabetes was a possible complication of a cyst of the pancreas. Pressure from the cyst could produce diabetes by disrupting the production of insulin in the pancreatic islets of Langerhans. Such pressure could also cause another complication—a blockage of the common bile duct. The diagnostic test for that was a chemical analysis of the blood serum for the presence of the bile pigment known as bilirubin. Dr. Wooten paused. The time had come to reopen the subject that he had tactfully dropped the day before. He reopened it. It was possible, he said, that the bilirubin test might help explain the unusual color of Turner's skin. And Turner's skin *was* a most unusual color. He held up an adamant hand. No, Turner was mistaken. His color *had* changed in the past year or two. It wasn't a natural ruddiness. It was a highly unnatural orange. It was a sign that something was wrong, and he intended to find out what. That was the reason for a third test he had ordered. It was a diagnostic blood test for a condition called hemochromatosis. Hemochromatosis was a disturbance of iron metabolism that deposited iron in the skin and stained it the color of bronze. To be frank, he didn't hope for much from either of the pigmentation tests. Turner's color wasn't the bronze of hemochromatosis, and it wasn't the yellow of jaundice. the possibility of jaundice was particularly remote. The whites of Turner's eyes were still white, and that was usually where jaundice made its first appearance. But he had to carry out the tests. He had to be sure. The process of elimination was always an instructive process. And they didn't have long to wait. The results of the tests would be ready sometime that afternoon. He would be back to see Turner then.

Dr. Wooten spent the next few hours at the hospital and his office. He had other patients to see, other problems to consider, other decisions to make. But Turner remained on his mind. His first impression, like so many first impressions, had been mistaken. It now seemed practically certain that Turner's color had no connection with Turner's pancreatic cyst. They were two quite different complaints. And that returned him to the question he had asked himself when Turner walked into his office. What did an orange skin signify? What disease had the power to turn its victims orange? The answer, as before, was none. But perhaps this wasn't in the usual sense a disease. Perhaps it was a drug-induced reaction. Many chemicals in common therapeutic and diagnostic use were capable of producing conspicuous skin discolorations. Or it might be related to diet.

The question hung in Dr. Wooten's mind all day. It was still hanging there when he headed back to Turner's ward. On the way, he picked up the results of the tests he had ordered that morning, and they did nothing to resolve it. Turner's total bilirubin level was 0.9 milligrams per hundred milliliters, or normal. The total iron-binding capacity was also normal—286 micrograms per hundred milliliters. And he didn't have diabetes. When Dr. Wooten came into the ward, he found Turner's wife at his bedside and Turner in a somewhat altered state of mind. He said he had begun to think that maybe Dr. Wooten was right about the color of his skin. There must be something peculiar about it. There had been a parade of doctors and nurses past his bed ever since early morning. Mrs. Turner looked bewildered. She hadn't noticed anything unusual about her husband's color. She hadn't thought about it—the question had never come up. But now that it had, she had to admit that he did look kind of different. He did look kind of orange. But what was the reason? What in the world could cause a thing like that? Dr. Wooten said he didn't know. The most he could say at the moment was that certain possibilities had been eliminated. He summarized the results of the three diagnostic tests. Another possible cause, he then went on to say, was drugs. Medicinal drugs. Certain medicines incorporated dyes or chemicals with pigmentary properties. Turner shook his head. Maybe so, he said, but that was out. It had been months since he had taken any kind of drug except aspirin.

Dr. Wooten was glad to believe him. Drugs had been a rather farfetched possibility. The color changes they produced were generally dramatically sudden and almost never lasting. He turned to another area—to diet. What did Turner like to eat? What, for example, did he usually have for breakfast? That was no problem, Turner said. His breakfast was almost always the same—orange juice, bacon and eggs, toast, coffee. And what about lunch? Well, that didn't change much, either. He ate a lot of vegetables—carrots, rutabagas, squash, beans, spinach, turnips, things like that. Mrs. Turner laughed. That, she said, was putting it mildly. He ate carrots the way some people eat candy. Dr. Wooten sat erect. Carrots, he was abruptly aware, were rich in carotene. So were eggs, oranges, rutabagas, squash, beans, spinach, and turnips. And carotene was a powerful yellow pigment. What, he asked Mrs. Turner, did she mean about the way her husband ate carrots? Mrs. Turner laughed again. She meant just what she said. Elmo was always eating carrots. Eating carrots and drinking tomato juice. Tomato juice was his favorite drink. And carrots were his favorite snack. He ate raw carrots all day long. He ate four or five of them a day. Why, driving down home from Alaska last week, he kept her busy just scraping and slicing and feeding him carrots. Turner gave an embarrassed grin. His wife was right. He reckoned he did eat a lot of carrots. But he had his reasons. You needed extra vitamins when you lived in Alaska. You had to make up for the long, dark winters—the lack of sunlight up there. Dr. Wooten stood up to go. What the Turners had told him was extremely interesting. He was sure, he said, that Turner's appetite for carrots was a clue to the cause of his color. It was also, as it happened, misguided. The so-called "sunshine vitamin" was vitamin D. The vitamin with which carrots and other yellow vegetables were abundantly endowed was Vitamin A.

There was a telephone just down the hall from Turner's ward. Dr. Wooten stopped and put in a call to the hospital laboratory. He arranged with the technician who took the call for a sample of Turner's blood to be tested for an abnormal concentration of carotene. Then he left the hospital and cut across the campus to the Mooney Memorial Library. He asked the librarian to let him see what she could find in the way of clinical literature of carotenemia and any related nutritional skin discolorations. He was elated by what he had learned from the Turners, but he knew that it wasn't enough. He had seen several cases of carotenemia. An excessive intake of carotene was a not uncommon condition among health-bar habitués and other amateur nutritionists. But carotene didn't color people orange. It colored them yellow. Or such had been his experience.

The librarian reported that papers on carotenemia were scarce. She had, however, found three clinical studies that looked as though they might be useful. Here was one of them. She handed Dr. Wooten a bound volume of the *Journal of the American Medical Association* for 1919, and indicated the relevant article. It was a report by two New York city investigators—Alfred F. Hess and Victor C. Myers—entitled "Carotenemia: A New Clinical Picture." Dr. Wooten knew their report, at least by reputation. It was the original study in the field. The opening descriptive paragraphs refreshed his memory and confirmed his judgment. they read:

> About a year ago one of us (A.F.H.) observed that two children in a ward containing about twenty-five infants, from a year to a year and a half in age, were developing a yellowish complexion. This coloration was not confined to the face, but involved, to a less extent, the entire body, being most evident on the palms of the hands.... For a time, we were at a loss to account for this peculiar phenomenon, when our attention was directed to the fact that these two children, and only these two, were receiving a daily ration of carrots in addition to their milk and cereal. For some time we had been testing the food value of dehydrated vegetables, and when the change in color was noted, had given these babies the equivalent of 2 tablespoonfuls of fresh carrots for a period of six weeks.
>
> It seemed as if this mild jaundiced hue might well be the result of the introduction into the body of a pigment rather than the manifestation of a pathologic condition. Attention was accordingly directed to the carrots, and the same amount of this vegetable was added to the dietary of two other children of about the same age. In the one instance, after an interval of about five weeks, a yellowish tinge of the skin was noted, and about two weeks later the other baby had become somewhat yellow. There was a decided difference in the intensity of color of the four infants, indicating probably that the alteration was in part governed by individual idiosyncrasy. On omission of the carrots from the dietary, the skin gradually lost its yellow color, and in the course of some weeks regained its normal tint.

The librarian returned to Dr. Wooten's table with the other references. Both were contributions to the *New England Journal of Medicine*. One was entitled "Skin Changes of Nutritional Origin," and had been written by Harold Jeghers, an associate professor of medicine at the Boston University School of Medicine, in 1943. The other was the work of three faculty members of the Harvard Medical School—Peter Reich, Harry Shwachman, and John M. Craig—and was entitled "Lycopenemia: A Variant of Carotenemia." It had appeared in 1960. Dr. Wooten looked first at "Skin Changes of Nutritional Origin." It was a comprehensive survey, and it read, in part:

> The carotenoid group of pigments color the serum and fix themselves to the fat of the dermis and subcutaneous tissues, to which they impart the yellow tint.... Edwards and Duntley showed by means of spectrophotometric analysis of skin color in human beings that carotene is present in every normal skin and is one of the five basic pigments that determine the skin color of every living person. Clinically, therefore, carotenemia refers to the presence of an excess over normal of carotene in the skin and serum.... In most cases carotenemia results simply from excess use of foods rich in the carotenoid pigments. Individuals probably vary in the ease with which carotenemia develops, which is evidenced by the fact that many vegetarians do not develop it. It is said to develop more readily in those who sweat profusely. Except for the yellow color produced, it appears to be harmless, even though present for months. It eventually disappears over several weeks to months when the carotene consumption is reduced.

Dr. Wooten moved on to the third report. ("This investigation concerns a middle-aged woman whose prolonged and excessive consumption of tomato juice led to the discoloration of their skin.") He read it slowly through from beginning to end, and then turned back and reread certain passages:

> Although carotenemia due to the ingestion of foods containing a high concentration of beta carotene is a commonly described disorder, a similar condition secondary to the ingestion of tomatoes and associated with high serum levels of lycopene has not previously been reported.... Lycopene is a common carotenoid pigment widely distributed through nature. It is most familiar as the red pigment of

> tomatoes, but has been detected in many animals and vegetables....It is also frequently found in human serum and liver, especially when tomatoes are eaten. But lycopene is not well known medically because, unlike beta carotene, it is physiologically inert and has not been involved in any form of illness.

Dr. Wooten closed the volume. Turner was not only a heavy eater of carrots. He was also a heavy drinker of tomato juice. Carrots are rich in carotene and tomatoes are rich in lycopene. Carotene is a yellow pigment and lycopene is red. And yellow and red make orange.

Dr. Wooten completed his record of the case with a double diagnosis: pseudocyst of the pancreas and carotenemia-lycopenemia. The results of the X-ray examination of Turner's colon were normal ("Terminal ileum was visualized. No pathology was demonstrated in the colon"), and the carotene test showed a high concentration of serum carotenoids (495 micrometers per hundred milliliters, compared to a normal concentration of 50 to 350 micrometers per hundred milliliters). The diagnosis of lycopenemia was made from the clinical evidence. Turner was discharged from the hospital on December 17. His instructions were to avoid abdominal blows, carrots (and other yellow vegetables), and tomatoes in any form. Four months later, on April 16, 1961, he reported to Dr. Wooten that his skin had recovered its normal ruddiness. Two years later, in 1963, he returned again to Memphis and dropped in on Dr. Wooten for a visit. His abdominal symptoms had long since disappeared, and a comprehensive examination showed no sign of the pseudocyst.

* * *

Elmo Turner was the first recorded victim of the condition known as carotenemia-lycopenemia. He is not, however, the only one now on record. Another victim turned up in 1964. She was a woman of thirty-five, a resident of Memphis, and a patient of Dr. Wooten's. He had been treating her for a mild diabetes since 1962, and had put her at that time on the eighteen-hundred-calorie diet recommended by the American Diabetes Association. She had faithfully followed the diet, but in order to do so had eaten heavily of low-calorie vegetables, and (as she confirmed, with some surprise, when questioned) the vegetables she ate most heavily were carrots and tomatoes. She ate at least two cups of carrots and at least two whole tomatoes every day. Dr. Wooten was unaware of this until she walked into his office one October day in 1964 for her semi-annual consultation. He greeted her as calmly as he could, and asked her to sit down. He would be back in just a moment. He stepped along the hall to Dr. Hughes's office and looked in. Dr. Hughes was alone.

"Have you got a minute, John?" Dr. Wooten said.

"Sure, " Dr. Hughes said. "What is it? "

"I'd like you to come into my office," Dr. Wooten said. "I'd like to show you something."

*Berton Roueché*

---

From *The Medical Detectives* (Harold Ober Associates, Inc., 1967).

---

## Questions and Topics for Discussion and Writing

1. Outline the deductive process used by Dr. Wooten to diagnose Elmo Turner's condition. Indicate the possibilities considered by Dr. Wooten and how he confirmed or rejected each one.
2. To what extent did Dr. Wooten's diagnosis depend on his prior knowledge and training and to what extent did it depend on information gained after Elmo Turner came to see him?
3. Discuss the role of the different sources of information used by Dr. Wooten: Elmo Turner's own report, the physical examination, the laboratory tests and procedures, research in the medical literature, Mrs. Turner's comments.

# Can We Know the Universe?

## Carl Sagan

In this article, astronomer, science writer, and TV host, Carl Sagan discusses science as a way of thinking and scientific knowledge as a special kind of knowledge. He also considers the limits of possible human knowledge.

> Nothing is rich but the inexhaustible wealth of nature. She shows us only surfaces, but she is a million fathoms deep.
>
> *Ralph Waldo Emerson*

Science is a way of thinking much more than it is a body of knowledge. Its goal is to find out how the world works, to seek what regularities there may be, to penetrate to the connections of things—from subnuclear particles, which may be the constituents of all matter, to living organisms, the human social community, and thence to the cosmos as a whole. Our intuition is by no means an infallible guide. Our perceptions may be distorted by training and prejudice or merely because of the limitations of our sense organs, which, of course, perceive directly but a small fraction of the phenomena of the world. Even so straightforward a question as whether in the absence of friction a pound of lead falls faster than a gram of fluff was answered incorrectly by Aristotle and almost everyone else before the time of Galileo. Science is based on experiment, on a willingness to challenge old dogma, on an openness to see the universe as it really is. Accordingly, science sometimes requires courage—at the very least the courage to question the conventional wisdom.

Beyond this the main trick of science is to *really* think of something; the shape of clouds and their occasional sharp bottom edges at the same altitude everywhere in the sky; the formation of a dewdrop on a leaf; the origin of a name or a word—Shakespeare, say, or "philanthropic"; the reason for human social customs—the incest taboo, for example; how it is that a lens in sunlight can make paper burn; how a "walking stick" got to look so much like a twig; why the Moon seems to follow us as we walk; what prevents us from digging a hole down to the center of the Earth; what the definition is of "down" on a spherical Earth; how it is possible for the body to convert yesterday's lunch into today's muscle and sinew; or how far is up—does the universe go on forever, or if it does not, is there any meaning to the question of what lies on the other side? Some of these questions are pretty easy. Others, especially the last, are mysteries to which no one even today knows the answer. They are natural questions to ask. Every culture has posed such questions in one way or another. Almost always the proposed answers are in the nature of "Just So Stories," attempted explanations divorced from experiment, or even from careful comparative observations.

But the scientific cast of mind examines the world critically as if many alternative worlds might exist, as if other things might be here which are not. Then we are forced to ask why what we see is present and not something else. Why are the Sun and the Moon and the planets spheres? Why not pyramids, or cubes, or dodecahedra? Why not irregular, jumbly shapes? Why so symmetrical, worlds? If you spend any time spinning hypotheses, checking to see whether they make sense, whether they conform to what else we know, thinking of tests you can pose to substantiate or deflate your hypotheses, you will find yourself doing science. And as you come to practice this habit of thought more and more you will get better and better at it. To penetrate into the heart of the thing—even a little thing, a blade of grass, as Walt Whitman said—is to experience a kind of exhilaration that, it may be, only human beings of all the beings on this planet can feel. We are an intelligent species and the use of our intelligence quite properly gives us pleasure. In this respect the brain is like a muscle. When we think well, we feel good. Understanding is a kind of ecstasy.

But to what extent can we really know the universe around us? Sometimes this question is posed by people who hope the answer will be in the negative, who are fearful of a universe in which everything might one day be known. And sometimes we hear pronouncements from scientists who confidently state that everything worth knowing will soon be known—or even is already known—and who paint pictures of a Dionysian or Polynesian age in which the zest for intellectual discovery has withered, to be replaced by a kind of subdued languor, the lotus eaters drinking fermented coconut milk or some other mild hallucinogen. In addition to maligning both the Polynesians, who were intrepid explorers (and whose brief respite in paradise is now sadly ending), as well as the inducements to intellectual discovery provided by some hallucinogens, this contention turns out to be trivially mistaken.

Let us approach a much more modest question: not whether we can know the universe or the Milky Way Galaxy or a star or a world. Can we know, ultimately and in detail, a grain of salt? Consider one microgram of table salt, a speck just barely large enough for someone with keen eyesight to make out without a microscope. In that grain of salt there are about $10^{16}$ sodium and chlorine atoms. This is a 1 followed by 16 zeros, 10 million billion atoms. If we wish to know a grain of salt, we must know at least the three-dimensional positions of each of these atoms. (In fact, there is much more to be known—for example, the nature of the forces between the atoms—but we are making only a modest calculation.) Now, is this number more or less than the number of things which the brain can know?

How much *can* the brain know? There are perhaps $10^{11}$ neurons in the brain, the circuit elements and switches that are responsible in their electrical and chemical activity for the functioning of our minds. A typical brain neuron has perhaps a thousand little wires, called dendrites, which connect it with its fellows. If, as seems likely, every bit of information in the brain corresponds to one of these connections, the total number of things knowable by the brain is no more than $10^{14}$, one hundred trillion. But this number is only one percent of the number of atoms in our speck of salt.

So in this sense the universe is intractable, astonishingly immune to any human attempt at full knowledge, We cannot on this level understand a grain of salt, much less the universe.

But let us look a little more deeply at our microgram of salt. Salt happens to be a crystal in which, except for defects in the structure of the crystal lattice, the position of every sodium and chlorine atom is predetermined. If we could shrink ourselves into this crystalline world, we would see rank upon rank of atoms in an ordered array, a regularly alternating structure—sodium, chlorine, sodium, chlorine, specifying the sheet of atoms we are standing on and all the sheets above us and below us. An absolutely pure crystal of salt could have the position of every atom specified by something like 10 bits of information.[1] This would not strain the information-carrying capacity of the brain.

If the universe had natural laws that governed its behavior to the same degree of regularity that determines a crystal of salt, then, of course, the universe would be knowable. Even if there were many such laws, each of considerable complexity, human beings might have the capability to understand them all. Even if such knowledge exceeded the information-carrying capacity of the brain, we might store the additional information outside our bodies—in books, for example, or in computer memories—and still, in some sense, know the universe.

Human beings are, understandably, highly motivated to find regularities, natural laws. The search for rules, the only possible way to understand such a vast and complex universe, is called science. The universe forces those who live in it to understand it. Those creatures who find everyday experience a muddled jumble of events with no predictability, no

[1] Chlorine is a deadly poison gas employed on European battlefields in World War I. Sodium is a corrosive metal which burns upon contact with water. Together they make a placid and unpoisonous material, table salt. Why each of these substances has the properties it does is a subject called chemistry, which requires more than 10 bits of information to understand.

regularity, are in grave peril. The universe belongs to those who, at least to some degree, have figured it out.

It is an astonishing fact that there *are* laws of nature, rules that summarize conveniently—not just qualitatively but quantitatively—how the world works. We might imagine a universe in which there are no such laws, in which the $10^{80}$ elementary particles that make up a universe like our own behave with utter and uncompromising abandon. To understand such a universe we would need a brain at least as massive as the universe. It seems unlikely that such a universe could have life and intelligence, because beings and brains require some degree of internal stability and order. But even if in a much more random universe there were such beings with an intelligence much greater than our own, there could not be much knowledge, passion or joy.

Fortunately for us, we live in a universe that has at least important parts that are knowable. Our common-sense experience and our evolutionary history have prepared us to understand something of the workaday world. When we go into other realms, however, common sense and ordinary intuition turn out to be highly unreliable guides. It is stunning that as we go close to the speed of light our mass increases indefinitely, we shrink toward zero thickness in the direction of motion, and time for us comes as near to stopping as we would like. Many people think that this is silly, and every week or two I get a letter from someone who complains to me about it. But it is a virtually certain consequence not just of experiment but also of Albert Einstein's brilliant analysis of space and time called the Special Theory of Relativity. It does not matter that these effects seem unreasonable to us. We are not in the habit of traveling close to the speed of light. The testimony of our common sense is suspect at high velocities.

Or consider an isolated molecule composed of two atoms shaped something like a dumbbell—a molecule of salt, it might be. Such a molecule rotates about an axis through the line connecting the two atoms. But in the world of quantum mechanics, the realm of the very small, not all orientations of our dumbbell molecule are possible. It might be that the molecule could be oriented in a horizontal position, say, or in a vertical position, but not at many angles in between. Some rotational positions are forbidden. Forbidden by what? By the laws of nature. The universe is built in such a way as to limit, or quantize, rotation. We do not experience this directly in everyday life; we would find it startling as well as awkward in sitting-up exercises, to find arms outstretched from the sides or pointed up to the skies permitted but many intermediate positions forbidden. We do not live in the world of the small, on the scale of $10^{-13}$ centimeters, in the realm where there are twelve zeros between the decimal place and the one. Our common-sense intuitions do not count. What does count is experiment—in this case observations from the far infrared spectra of molecules. They show molecular rotation to be quantized.

The idea that the world places restrictions on what humans might do is frustrating. Why *shouldn't* we be able to have intermediate rotational positions? Why *can't* we travel faster than the speed of light? But so far as we can tell, this is the way the universe is constructed. Such prohibitions not only press us toward a little humility; they also make the world more knowable. Every restriction corresponds to a law of nature, a regularization of the universe. The more restrictions there are on what matter and energy can do, the more knowledge human beings can attain. Whether in some sense the universe is ultimately knowable depends not only on how many natural laws there are that encompass widely divergent phenomena, but also on whether we have the openness and the intellectual capacity to understand such laws. Our formulations of the regularities of nature are surely dependent on how the brain is built, but also, and to a significant degree, on how the universe is built.

For myself, I like a universe that includes much that is unknown and, at the same time, much that is knowable. A universe in which everything is known would be static and dull, as boring as the heaven of some weak-minded theologians. A universe that is unknowable is no fit place for a thinking being. The ideal universe for us is one very much like the universe we inhabit. And I would guess that this is not really much of a coincidence.

---

From Carl Sagan, *Broca's Brain* (New York: Random House, 1979).

*Carl Sagan*

## Questions and Topics for Discussion and Writing

1. Summarize Sagan's conception of what science is. What would he view as non-science?
2. In what way, according to Sagan, can we know the universe and in what way can we not know it?
3. Why are some of the implications of science seemingly unreasonable? Should they be reasonable to us for us to accept them?
4. What relationship does Sagan see between order in the universe and our ability to know and understand it?

*Carl Sagan*

# We All Use the Scientific Method

## Thomas Henry Huxley

Thousands of science textbooks begin with a short discussion of "the scientific method," encapsulated in a few pat sentences. These are dutifully memorized and spat forth by zillions of students on their first tests and then, whether the sentences are forgotten or not, they are at least forgotten about. But, if there *is* a method to science, what is it and how does it relate to the real world? This lecture by T. H. Huxley tries to answer these questions by characterizing scientific method in everyday life situations.

Huxley, known best as the dynamic popularizer and persuader for the theory of evolution in the nineteenth century—they called him "Darwin's Bulldog"—was a tireless writer and speaker for science in general. This lecture was one that Huxley gave to "workingmen" to help advance the cause of technical education.

The method of scientific investigation is nothing but the expression of the necessary mode of working of the human mind. It is simply the mode at which all phenomena are reasoned about, rendered precise and exact. There is no more difference, but there is just the same kind of difference, between the mental operations of a man of science and those of an ordinary person, as there is between the operations and methods of a baker or of a butcher weighing out his goods in common scales, and the operations of a chemist in performing a difficult and complex analysis by means of his balance and finely-graduated weights. It is not that the action of the scales in the one case, and the balance in the other, differ in the principles of their construction or manner of working; but the beam of one is set on an infinitely finer axis than the other, and of course turns by the addition of a much smaller weight.

You will understand this better, perhaps, if I give you some familiar example. You have all heard it repeated, I dare say, that men of science work by means of induction and deduction, and that by the help of these operations, they, in a sort of sense, wring from Nature certain other things, which are called natural laws, and causes, and that out of these, by some cunning skill of their own, they build up hypotheses and theories. And it is imagined by many, that the operations of the common mind can be by no means compared with these processes, and that they have to be acquired by a sort of special apprenticeship to the craft. To hear all these large words, you would think that the mind of a man of science must be constituted differently from that of his fellow men; but if you will not be frightened by terms, you will discover that you are quite wrong, and that all these terrible apparatus are being used by yourselves every day and every hour of your lives.

There is a well-known incident in one of Moliere's plays, where the author makes the hero express unbounded delight on being told that he had been talking prose during the whole of his life. In the same way, I trust that you will take comfort, and be delighted with yourselves, on the discovery that you have been acting on the principles of inductive and deductive philosophy during the same period. Probably there is not one here who has not in the course of the day had occasion to set in motion a complex train of reasoning, of the very same kind, though differing of course in degree, as that which a scientific man goes through in tracing the causes of natural phenomena.

A very trivial circumstance will serve to exemplify this. Suppose you go into a fruiterer's shop, wanting an apple,—you take up one, and, on biting it, you find it is sour; you look at it, and see that it is hard and green. You take up another one, and that too is hard, green, and sour. The shopman offers you a third; but, before biting it, you examine it, and find that it is hard and green, and you immediately say that you will not have it, as it must be sour, like those that you have already tried.

Nothing can be more simple than that, you think; but if you will take the trouble to analyse and trace out into its logical elements what has been done by the mind, you will be greatly surprised. In the first place, you have performed the operation of induction. You found that, in two experiences, hardness and greenness in apples went together with sourness. It was so in the first case, and it was confirmed by the second. True, it is a very small basis, but still it is enough to make an induction from; you generalise the facts, and you expect to find sourness in apples where you get hardness and greenness. You found upon that a general law, that all hard and green apples are sour; and that, so far as it goes, is a perfect induction. Well, having got your natural law in this way, when you are offered another apple which you find is hard and green, you say, "All hard and green apples are sour; this apple is hard and green, therefore this apple is sour." That train of reasoning is what logicians call a syllogism, and has all its various parts and terms—its major premise, its minor premise, and its conclusion. And, by the help of further reasoning, which, if drawn out, would have to be exhibited in two or three other syllogisms, you arrive at your final determination, "I will not have that apple." So that, you see, you have, in the first place, established a law by induction, and upon that you have founded a deduction, and reasoned out the special conclusion of the particular case. Well now, suppose, having got your law, that at some time afterwards, you are discussing the qualities of apples with a friend: you will say to him, "It is a very curious thing—but I find that all hard and green apples are sour!" Your friend says to you, "But how do you know that?" You at once reply, "Oh, because I have tried them over and over again, and have always found them to be so." Well, if we were talking science instead of common sense, we should call that an experimental verification. And, if still opposed, you go further, and say, "I have heard from the people in Somersetshire and Devonshire, where a large number of apples are grown, that they have observed the same thing. It is also found to be the case in Normandy, and in North America. In short, I find it to be the universal experience of mankind wherever attention has been directed to the subject." Whereupon, your friend, unless he is a very unreasonable man, agrees with you, and is convinced that you are quite right in the conclusion you have drawn. He believes, although perhaps he does not know he believes it, that the more extensive verifications are,—that the more frequently experiments have been made, and results of the same kind arrived at,—that the more varied the conditions under which the same results are attained, the more certain is the ultimate conclusion, and he disputes the question no further. He sees that the experiment has been tried under all sorts of conditions, as to time, place, and people, with the same result; and he says with you, therefore, that the law you have laid down must be a good one, and we must believe it.

In science we do the same thing;—the philosopher exercises precisely the same faculties, though in a much more delicate manner. In scientific inquiry it becomes a matter of duty to expose a supposed law to every possible kind of verification, and to take care, moreover, that this is done intentionally, and not left to a mere accident, as in the case of the apples. And in science, as in common life, our confidence in a law is in exact proportion to the absence of variation in the result of our experimental verifications. For instance, if you let go your grasp of an article you may have in your hand, it will immediately fall to the ground. That is a very common verification of one of the best established laws of nature—that of gravitation. The method by which men of science establish the existence of that law is exactly the same as that by which we have established the trivial proposition about the sourness of hard and green apples. But we believe it in such an extensive, thorough, and unhesitating manner because the universal experience of mankind verifies it, and we can verify it ourselves at any time; and that is the strongest possible foundation on which any natural law can rest.

So much, then, by way of proof that the method of establishing laws in science is exactly the same as that pursued in common life. Let us now turn to another matter (though really it is but another phase of the same question), and that is, the method by which, from the relations of certain phenomena, we prove that some stand in the position of causes towards the others.

I want to put the case clearly before you, and I will therefore show you what I mean by another familiar example. I will suppose that one of you, on coming down in the morning to the parlour of your house, finds that a tea-pot and some spoons which had been left in the room on the previous evening are gone,—the window is open, and you observe the mark of a dirty hand on the window-frame, and perhaps, in addition to that, you notice the impress of a hob-nailed shoe on the gravel outside. All these phenomena have struck your attention instantly, and before two seconds have passed you say, "Oh, somebody has broken open the window, entered the room, and run off with the spoons and the tea-pot!" That speech is out of your mouth in a moment. And you will probably add, "I know there has; I am quite sure of it!" You mean to say exactly what you know; but in reality you are giving expression to what is, in all essential particulars, an hypothesis. You do not *know* it at all; it is nothing but an hypothesis rapidly framed in your own mind. And it is an hypothesis founded on a long train of inductions and deductions.

What are those inductions and deductions, and how have you got at this hypothesis? You have observed, in the first place, that the window is open; but by a train of reasoning involving many inductions and deductions, you have probably arrived long before at the general law—and a very good one it is—that windows do not open of themselves; and you therefore conclude that something has opened the window. A second general law that you have arrived at in the same way is, that tea-pots and spoons do not go out of a window spontaneously, and you are satisfied that, as they are not now where you left them, they have been removed. In the third place, you look at the marks on the window-sill, and the shoe-marks outside, and you say that in all previous experience the former kind of mark has never been produced by anything else but the hand of a human being; and the same experience shows that no other animal but man at present wears shoes with hob-nails in them such as would produce the marks in the gravel. I do not know, even if we could discover any of those "missing links" that are talked about, that they would help us to any other conclusion! At any rate the law which states our present experience is strong enough for my present purpose. You next reach the conclusion, that as these kinds of marks have not been left by any other animals than men, or are liable to be formed in any other way than by a man's hand and shoe, the marks in question have been formed by a man in that way. You have, further, a general law, founded on observation and experience, and that, too, is, I am sorry to say, a very universal and unimpeachable one,—that some men are thieves; and you assume at once from all these premises—and that is what constitutes your hypothesis—that the man who made the marks outside and on the window-sill, opened the window, got into the room, and stole your tea-pot and spoons. You have now arrived at a *vera causa*;—you have assumed a cause which, it is plain, is competent to produce all the phenomena you have observed. You can explain all these phenomena only by the hypothesis of a thief. But that is a hypothetical conclusion, of the justice of which you have no absolute proof at all; it is only rendered highly probable by a series of inductive and deductive reasonings.

I suppose your first action, assuming that you are a man of ordinary common sense, and that you have established this hypothesis to your own satisfaction, will very likely be to go off for the police, and set them on the track of the burglar, with the view to the recovery of your property. But just as you are starting with this object, some person comes in, and on learning what you are about, says, "My good friend, you are going on a great deal too fast. How do you know that the man who really made the marks took the spoons? It might have been a monkey that took them, and the man may have merely looked in afterwards." You would probably reply, "Well, that is all very well, but you see it is contrary to all experience of the way tea-pots and spoons are abstracted; so that, at any rate, your hypothesis is less probable than mine." While you are talking the thing over in this way, another friend arrives, one of that good kind of people that I was talking of a little while ago. And he might say, "Oh, my dear sir, you are certainly going on a great deal too fast. You are most presumptuous. You admit that all these occurrences took place when you were fast asleep, at a time when you could not possibly have known anything about what was taking place. How do you know that the laws of Nature are not suspended during the night? It may be that there has been some kind of supernatural interference in this case." In point of fact, he declares that your hypothesis is one of which you cannot at all demonstrate the truth, and that you are by no means sure that the laws of Nature are the same when you are asleep as when you are awake.

Well, now, you cannot at the moment answer that kind of reasoning. You feel that your worthy friend has you somewhat at a disadvantage. You will feel perfectly convinced in your own mind, however, that you are quite right, and you say to him, "My good friend, I can only be guided by the natural probabilities of the case, and if you will be kind enough to stand aside and permit me to pass, I will go and fetch the police." Well, we will suppose that your journey is successful, and that by good luck you meet with a policeman; that eventually the burglar is found with your property on his person, and the marks correspond to his hand and to his boots. Probably any jury would consider those facts a very good experimental verification of your hypothesis, touching the cause of the abnormal phenomena observed in your parlour, and would act accordingly.

Now, in this suppositious case, I have taken phenomena of a very common kind, in order that you might see what are the different steps in an ordinary process of reasoning, if you will only take the trouble to analyse it carefully. All the operations I have described, you will see, are involved in the mind of any man of sense in leading him to a conclusion as to the course he should take in order to make good a robbery and punish the offender. I say that you are led, in that case, to your conclusion by exactly the same train of reasoning as that which a man of science pursues when he is endeavouring to discover the origin and laws of the most occult phenomena. The process is, and always must be, the same; and precisely the same mode of reasoning was employed by Newton and Laplace in their endeavours to discover and define the causes of the movements of the heavenly bodies, as you, with your own common sense, would employ to detect a burglar. The only difference is, that the nature of the inquiry being more abstruse, every step has to be most carefully watched, so that there may not be a single crack or flaw in your hypothesis. A flaw or crack in many of the hypotheses of daily life may be of little or no moment as affecting the general correctness of the conclusions at which we may arrive; but, in a scientific inquiry, a fallacy, great or small, is always of importance, and is sure to be in the long run constantly productive of mischievous, if not fatal results.

Do not allow yourselves to be misled by the common notion that an hypothesis is untrustworthy simply because it is an hypothesis. It is often urged, in respect to some scientific conclusion, that, after all, it is only an hypothesis. But what more have we to guide us in nine-tenths of the most important affairs of daily life than hypotheses, and often very ill-based ones? So that in science, where the evidence of an hypothesis is subjected to the most rigid examination, we may rightly pursue the same course. You may have hypotheses and hypotheses. A man may say, if he likes, that the moon is made of green cheese: that is an hypothesis. But another man, who has devoted a great deal of time and attention to the subject, and availed himself of the most powerful telescopes and the results of the observations of others, declares that in his opinion it is probably composed of materials very similar to those of which our own earth is made up: and that is also only an hypothesis. But I need not tell you that there is an enormous difference in the value of the two hypotheses. That one which is based on sound scientific knowledge is sure to have a corresponding value; and that which is a mere hasty random guess is likely to have but little value. Every great step in our progress in discovering causes has been made in exactly the same way as that which I have detailed to you. A person observing the occurrence of certain facts and phenomena asks, naturally enough, what process, what kind of operation known to occur in Nature applied to the particular case, will unravel and explain the mystery? Hence you have the scientific hypothesis; and its value will be proportionate to the care and completeness with which its basis had been tested and verified. It is in these matters as in the commonest affairs of practical life: the guess of the fool will be folly, while the guess of the wise man will contain wisdom. In all cases, you see that the value of the result depends on the patience and faithfulness with which the investigator applies to his hypothesis every possible kind of verification.

*Thomas Henry Huxley*

## Questions and Topics for Discussion and Writing

1. Describe the process of induction and the formation of a hypothesis illustrated by the hard, green apples.
2. Discuss the objections raised against the conclusion that a thief has stolen the spoons and the teapot. What is the role of past experience in the evaluation of the hypothesis versus the objections to it?
3. Why does Huxley say that the two hypotheses about the composition of the moon have great differences in value? What makes the hypothesis that the moon is similar to the earth more valuable that the hypothesis that it is made of green cheese?

# * Scientific Inquiry

## Carl G. Hempel

If there is a methodology to scientific work, then there must be a logic to it, a logic that can be examined and analysed by philosophers. Philosophers who devote themselves to the analysis of the logic and methodology of science call themselves philosophers of science. One of the most prominent of these is Professor Carl G. Hempel of Princeton University. This excerpt from his book, *Philosophy of Natural Science,* explores the logic of forming and testing hypotheses. Hempel illustrates his analysis with a truly fascinating case study from the history of medicine.

### A Case History as an Example

As a simple illustration of some important aspects of scientific inquiry let us consider Semmelweis' work on childbed fever. Ignaz Semmelweis, a physician of Hungarian birth, did this work during the years from 1844 to 1848 at the Vienna General Hospital. As a member of the medical staff of the First Maternity Division in the hospital, Semmelweis was distressed to find that a large proportion of the women who were delivered of their babies in that division contracted a serious and often fatal illness known as puerperal fever or childbed fever. In 1844, as many as 260 out of 3,157 mothers in the First Division, or 8.2 per cent, died of the disease; for 1845, the death rate was 6.8 per cent, and for 1846, it was 11.4 per cent. These figures were all the more alarming because in the adjacent Second Maternity Division of the same hospital, which accommodated almost as many women as the First, the death toll from childbed fever was much lower: 2.3, 2.0, and 2.7 per cent for the same years. In a book that he wrote later on the causation and the prevention of childbed fever, Semmelweis describes his efforts to resolve the dreadful puzzle.[1]

He began by considering various explanations that were current at the time; some of these he rejected out of hand as incompatible with well-established facts; others he subjected to specific tests.

One widely accepted view attributed the ravages of puerperal fever to "epidemic influences," which were vaguely described as "atmospheric-cosmic-telluric changes" spreading over whole districts and causing childbed fever in women in confinement. But how, Semmelweis reasons, could such influences have plagued the First Division for years and yet spared the Second? And how could this view he reconciled with the fact that while the fever was raging in the hospital, hardly a case occurred in the city of Vienna or in its surroundings: a genuine epidemic, such as cholera, would not be so selective. Finally, Semmelweis notes that some of the women admitted to the First Division, living far from the hospital, had been overcome by labor on their way and had given birth in the street: yet despite these adverse conditions, the death rate from childbed fever among these cases of "street birth" was lower than the average for the First Division.

On another view, overcrowding was a cause of mortality in the First Division. But Semmelweis points out that in fact the crowding was heavier in the Second Division, partly as a result of the desperate efforts of patients to avoid assignment to the

[1] The story of Semmelweis' work and of the difficulties he encountered forms a fascinating page in the history of medicine. A detailed account, which includes translations and paraphrases of large portions of Semmelweis' writings, is given in W. J. Sinclair, *Semmelweis: His Life and His Doctrine* (Manchester, England: Manchester University Press, 1909). Brief quoted phrases in this chapter are taken from this work. The highlights of Semmelweis' career are recounted in the first chapter of P. de Kruif, *Men Against Death* (New York: Harcourt, Brace & World, Inc., 1932).

notorious First Division. He also rejects two similar conjectures that were current, by noting that there were no differences between the two Divisions in regard to diet or general care of the patients.

In 1846, a commission that had been appointed to investigate the matter attributed the prevalence of illness in the First Division to injuries resulting from rough examination by the medical students, all of whom received their obstetrical training in the First Division. Semmelweis notes in refutation of this view that (a) the injuries resulting naturally from the process of birth are much more extensive than those that might be caused by rough examination; (b) the midwives who received their training in the Second Division examined their patients in much the same manner but without the same ill effects; (c) when, in response to the commission's report, the number of medical students was halved and their examinations of the women were reduced to a minimum, the mortality, after a brief decline, rose to higher levels than ever before.

Various psychological explanations were attempted. One of them noted that the First Division was so arranged that a priest bearing the last sacrament to a dying woman had to pass through five wards before reaching the sickroom beyond: the appearance of the priest, preceded by an attendant ringing a bell, was held to have a terrifying and debilitating effect upon the patients in the wards and thus to make them more likely victims of childbed fever. In the Second Division, this adverse factor was absent, since the priest had direct access to the sickroom. Semmelweis decided to test this conjecture. He persuaded the priest to come by a roundabout route and without ringing of the bell, in order to reach the sick chamber silently and unobserved. But the mortality in the First Division did not decrease.

A new idea was suggested to Semmelweis by the observation that in the First Division the women were delivered lying on their backs; in the Second Division, on their sides. Though he thought it unlikely, he decided "like a drowning man clutching at a straw," to test whether this difference in procedure was significant. He introduced the use of the lateral position in the First Division, but again, the mortality remained unaffected.

At last, early in 1847, an accident gave Semmelweis the decisive clue for his solution of the problem. A colleague of his, Kolletschka, received a puncture wound in the finger, from the scalpel of a student with whom he was performing an autopsy, and died after an agonizing illness during which he displayed the same symptoms that Semmelweis had observed in the victims of childbed fever. Although the role of microorganisms in such infections had not yet been recognized at the time, Semmelweis realized that "cadaveric matter" which the student's scalpel had introduced into Kolletschka's blood stream had caused his colleague's fatal illness. And the similarities between the course of Kolletschka's disease and that of the women in his clinic led Semmelweis to the conclusion that his patients had died of the same kind of blood poisoning: he, his colleagues, and the medical students had been the carriers of the infectious material, for he and his associates used to come to the wards directly from performing dissections in the autopsy room, and examine the women in labor after only superficially washing their hands, which often retained a characteristic foul odor.

Again, Semmelweis put his idea to a test. He reasoned that if he were right, then childbed fever could be prevented by chemically destroying the infectious material adhering to the hands. He therefore issued an order requiring all medical students to wash their bands in a solution of chlorinated lime before making an examination. The mortality from childbed fever promptly began to decrease, and for the year 1848 it fell to 1.27 per cent in the First Division, compared to 1.33 in the Second.

In further support of his idea, or of his *hypothesis*, as we will also say, Semmelweis notes that it accounts for the fact that the mortality in the Second Division consistently was so much lower: the patients there were attended by midwives, whose training did not include anatomical instruction by dissection of cadavers.

The hypothesis also explained the lower mortality among "street births": women who arrived with babies in arms were rarely examined after admission and thus had a better chance of escaping infection.

Similarly, the hypothesis accounted for the fact that the victims of childbed fever among the newborn babies were all among those whose mothers had contracted the disease during labor; for then the infection could be transmitted to the baby before birth, through the common bloodstream of mother and child, whereas this was impossible when the mother remained healthy.

Further clinical experiences soon led Semmelweis to broaden his hypothesis. On one occasion, for example, he and his associates, having carefully disinfected their hands, examined first a woman in labor who was suffering from a festering cervical cancer; then they proceeded to examine twelve other women in the same room, after only routine washing without renewed disinfection. Eleven of the twelve patients died of puerperal fever. Semmelweis concluded that childbed fever can be caused not only by cadaveric material, but also by "putrid matter derived from living organisms."

### Basic Steps in Testing a Hypothesis

We have seen how, in his search for the cause of childbed fever, Semmelweis examined various hypotheses that had been suggested as possible answers. How such hypotheses are arrived at in the first place is an intriguing question which we will consider later. First, however, let us examine how a hypothesis, once proposed, is tested.

Sometimes, the procedure is quite direct. Consider the conjectures that differences in crowding, or in diet, or in general care account for the difference in mortality between the two divisions. As Semmelweis points out, these conflict with readily observable facts. There are no such differences between the divisions; the hypotheses are therefore rejected as false.

But usually the test will be less simple and straightforward. Take the hypothesis attributing the high mortality in the First Division to the dread evoked by the appearance of the priest with his attendant. The intensity of that dread, and especially its effect upon childbed fever, are not as directly ascertainable as are differences in crowding or in diet, and Semmelweis uses an indirect method of testing. He asks himself: Are there any readily observable effects that should occur if the hypothesis were true? And he reasons: *If* the hypothesis were true, *then* an appropriate change in the priest's procedure should be followed by a decline in fatalities. He checks this implication by a simple experiment and finds it false, and he therefore rejects the hypothesis.

Similarly, to test his conjecture about the position of the women during delivery, he reasons: *If* this conjecture should be true, *then* adoption of the lateral position in the First Division will reduce the mortality. Again, the implication is shown false by his experiment, and the conjecture is discarded.

In the last two cases, the test is based on an argument to the effect that *if* the contemplated hypothesis, say H, is true, then certain observable events (e.g., decline in mortality) should occur under specified circumstances (e.g., if the priest refrains from walking through the wards, or if the women are delivered in lateral position); or briefly, if H is true, then so is I, where I is a statement describing the observable occurrences to be expected. For convenience, let us say that I is inferred from, or implied by, H; and let us call I a *test implication of the hypothesis H.*

In our last two examples, experiments show the test implication to be false, and the hypothesis is accordingly rejected. The reasoning that leads to the rejection may be schematized as follows:

*If H is true, then so is I.*
(*a*) *But (as the evidence shows) I is not true.*
______________
*H is not true.*

Any argument of this form, called *modus tollens* in logic, is deductively valid; that is, if its premisses (the sentences above the horizontal line) are true, then its conclusion (the sentence below the horizontal line) is unfailingly true as well. Hence, if the premisses of *(a)* are properly established, the hypothesis H that is being tested must indeed be rejected.

Next, let us consider the case where observation or experiment bears out the test implication I. From his hypothesis that childbed fever is blood poisoning produced by cadaveric matter, Semmelweis infers that suitable antiseptic measures will reduce fatalities from the disease. This time, experiment shows the test implication to be true. But this favorable outcome does not conclusively prove the hypothesis true, for the underlying argument would have the form

*If H is true, then so is I.*
(b) *(As the evidence shows) I is true.*
———————
*H is true.*

And this mode of reasoning, which is referred to as the *fallacy of affirming the consequent*, is deductively invalid, that is, its conclusion may be false even if its premisses are true. This is in fact illustrated by Semmelweis' own experience. The initial version of his account of childbed fever as a form of blood poisoning presented infection with cadaveric matter essentially as the one and only source of the disease; and he was right in reasoning that if this hypothesis should be true, then destruction of cadaveric particles by antiseptic washing should reduce the mortality. Furthermore, his experiment did show the test implication to be true. Hence, in this case, the premisses of *(b)* were both true. Yet, his hypothesis was false, for as he later discovered, putrid material from living organisms, too, could produce childbed fever.

Thus, the favorable outcome of a test, i.e., the fact that a test implication inferred from a hypothesis is found to be true, does not prove the hypothesis to be true. Even if many implications of a hypothesis have been borne out by careful tests, the hypothesis may still be false. The following argument still commits the fallacy of affirming the consequent:

*If H is true, then so are* $I_1, I_2, ..., I_n$.
(c) *(As the evidence shows* $I_1, I_2, ..., I_n$ *are all true.*
———————
*H is true.*

This, too, can be illustrated by reference to Semmelweis' final hypothesis in its first version. As we noted earlier, his hypothesis also yields the test implications that among cases of street births admitted to the First Division, mortality from puerperal fever should be below the average for the Division, and that infants of mothers who escape the illness do not contract childbed fever; and these implications, too, were borne out by the evidence—even though the first version of the final hypothesis was false.

But the observation that a favorable outcome of however many tests does not afford conclusive proof for a hypothesis should not lead us to think that if we have subjected a hypothesis to a number of tests and all of them have had a favorable outcome, we are no better off than if we had not tested the hypothesis at all. For each of our tests might conceivably have had an unfavorable outcome and might have led to the rejection of the hypothesis. A set of favorable results obtained by testing different test implications, $I_1, I_2, ..., I_n$ of a hypothesis, shows that as far as these particular implications are concerned, the hypothesis has been borne out; and while this result does not afford a complete proof of the hypothesis, it provides at least some support, some partial corroboration or confirmation for it. The extent of this support will depend on various aspects of the hypothesis and of the test data ...

Let us now consider another example,[1] which will also bring to our attention some further aspects of scientific inquiry.

As was known at Galileo's time, and probably much earlier, a simple suction pump, which draws water from a well by means of a piston that can be raised in the pump barrel, will lift water no higher than about 34 feet above the surface of the well. Galileo was intrigued by this limitation and suggested an explanation for it, which was, however, unsound. After Galileo's death, his pupil Torricelli advanced a new answer. He argued that the earth is surrounded by a sea of air, which, by reason of its weight exerts pressure upon the surface below, and that this pressure upon the surface of the well forces water up the pump barrel when the piston is raised. The maximum length of 34 feet for the water column in the barrel thus reflects simply the total pressure of the atmosphere upon the surface of the well.

It is evidently impossible to determine by direct inspection or observation whether this account is correct, and Torricelli tested it indirectly. He reasoned that *if* his conjecture were true, *then* the pressure of the atmosphere should also be capable

[1] The reader will find a fuller account of this example in Chap. 4 of J. B. Conant's fascinating book, *Science and Common Sense* (New Haven: Yale University Press, 1951). A letter by Torricelli setting forth his hypothesis and his test of it, and an eyewitness report on the Puy-de-Dôme experiment are reprinted in W. F. Magie, *A Source Book in Physics* (Cambridge: Harvard University Press, 1963), pp. 70–75.

of supporting a proportionately shorter column of mercury; indeed, since the specific gravity of mercury is about 14 times that of water, the length of the mercury column should be about $\frac{34}{14}$ feet, or slightly less than 2 ½ feet. He checked this test implication by means of an ingeniously simple device, which was, in effect, the mercury barometer. The well of water is replaced by an open vessel, containing mercury; the barrel of the suction pump is replaced by a glass tube sealed off at one end. The tube is completely filled with mercury and closed by placing the thumb tightly over the open end. It is then inverted, the open end is submerged in the mercury well, and the thumb is withdrawn; whereupon the mercury column in the tube drops until its length is about 30 inches—just as predicted by Torricelli's hypothesis.

A further test implication of that hypothesis was noted by Pascal, who reasoned that if the mercury in Torricelli's barometer is counterbalanced by the pressure of the air above the open mercury well, then its length should decrease with increasing altitude, since the weight of the air overhead becomes smaller. At Pascal's request, this implication was checked by his brother-in-law, Périer, who measured the length of the mercury column in the Torricelli barometer at the foot of the Puy-de-Dôme, a mountain some 4,800 feet high, and then carefully carried the apparatus to the top and repeated the measurement there while a control barometer was left at the bottom under the supervision of an assistant. Périer found the mercury column at the top of the mountain more than three inches shorter than at the bottom, whereas the length of the column in the control barometer had remained unchanged throughout the day.

## The Role of Induction in Scientific Inquiry

We have considered some scientific investigations in which a problem was tackled by proposing tentative answers in the form of hypotheses that were then tested by deriving from them suitable test implications and checking these by observation or experiment.

But how are suitable hypotheses arrived at in the first place?

It is sometimes held that they are inferred from antecedently collected data by means of a procedure called *inductive inference*, as contradistinguished from deductive inference, from which it differs in important respects.

In a deductively valid argument, the conclusion is related to the premisses in such a way that if the premisses are true then the conclusion cannot fail to be true as well. This requirement is satisfied, for example, by any argument of the following general form:

> If *p*, then *q*.
> It is not the case that *q*.
> ———————
> It is not the case that *p*.

Brief reflection shows that no matter what particular statements may stand at the places marked by the letters "*p*" and "*q*," the conclusion will certainly be true if the premisses are. In fact, our schema represents the argument form called *modus tollens*, to which we referred earlier.

Another type of deductively valid inference is illustrated by this example:

> Any sodium salt, when put into the flame of a Bunsen burner, turns the flame yellow.
> This piece of rock salt is a sodium salt.
> ———————
> This piece of rock salt, when put into the flame of a Bunsen burner, will turn the flame yellow.

Arguments of the latter kind are often said to lead from the general (here, the premiss about all sodium salts) to the particular (a conclusion about the particular piece of rock salt). Inductive inferences, by contrast, are sometimes described as leading from premisses about particular cases to a conclusion that has the character of a general law or principle. For example, from premisses to the effect that each of the particular samples of various sodium salts that have so far been subjected to the Bunsen flame test did turn the flame yellow, inductive inference supposedly leads to the general conclusion that all sodium salts, when put into the flame of a Bunsen burner, turn the flame yellow. But in this case, the truth of the premisses obviously does *not* guarantee the truth of the conclusion; for even if it is the case that all samples of sodium salts examined so far did turn the Bunsen flame yellow, it remains quite possible that new kinds of sodium salt might yet be found that do not conform to this generalization. Indeed, even some kinds of sodium salt that have already been tested with positive result might conceivably fail to satisfy the generalization under special physical conditions (such as very strong magnetic fields or the like) in which they have not yet been examined. For this reason, the premisses of an inductive inference are often said to imply the conclusion only with more or less high probability, whereas the premisses of a deductive inference imply the conclusion with certainty.

The idea that in scientific inquiry, inductive inference from antecedently collected data leads to appropriate general principles is clearly embodied in the following account of how a scientist would ideally proceed:

> If we try to imagine how a mind of superhuman power and reach, but normal so far as the logical processes of its thought are concerned, ... would use the scientific method, the process would be as follows: First, all facts would be observed and recorded, *without selection* or *a priori* guess as to their relative importance. Secondly, the observed and recorded facts would be analyzed, compared, and classified, *without hypothesis or postulates* other than those necessarily involved in the logic of thought. Third, from this analysis of the facts generalizations would be inductively drawn as to the relations, classificatory or causal, between them. Fourth, further research would be deductive as well as inductive, employing inferences from previously established generalizations.[1]

This passage distinguishes four stages in an ideal scientific inquiry: (1) observation and recording of all facts, (2) analysis and classification of these facts, (3) inductive derivation of generalizations from them, and (4) further testing of the generalizations. The first two of these stages are specifically assumed not to make use of any guesses or hypotheses as to how the observed facts might be interconnected; this restriction seems to have been imposed in the belief that such preconceived ideas would introduce a bias and would jeopardize the scientific objectivity of the investigation.

But the view expressed in the quoted passage—I will call it *the narrow inductivist conception of scientific inquiry*—is untenable, for several reasons. A brief survey of these can serve to amplify and to supplement our earlier remarks on scientific procedure.

First, a scientific investigation as here envisaged could never get off the ground. Even its first phase could never be carried out, for a collection of *all* the facts would have to await the end of the world, so to speak; and even all the facts *up to now* cannot be collected, since there are an infinite number and variety of them. Are we to examine, for, example, all the grains of sand in all the deserts and on all the beaches, and are we to record their shapes, their weights, their chemical composition, their distances from each other, their constantly changing temperature, and their equally changing distance from the center of the moon? Are we to record the floating thoughts that cross our minds in the tedious process? The shapes of the clouds overhead, the changing color of the sky? The construction and the trade name of our writing equipment? Our own life histories and those of our fellow investigators? All these and untold other things, are, after all, among "all the facts up to now."

Perhaps, then, all that should be required in the first phase is that all the *relevant* facts be collected. But relevant to what? Though the author does not mention this, let us suppose that the inquiry is concerned with a specified *problem*. Should we not then begin by collecting all the facts—or better, all available data—relevant to that problem? This notion still makes no clear sense. Semmelweis sought to solve one specific problem, yet he collected quite different kinds of data at different stages of his inquiry. And rightly so; for what particular sorts of data it is reasonable to collect is not determined by the problem under study, but by a tentative answer to it that the investigator entertains in the form of a conjecture or hypothesis. Given the conjecture that mortality from childbed fever was increased by the terrifying appearance of the priest and his attendant with the death bell, it was relevant to collect data on the consequences of having the priest change his routine; but it would have been totally irrelevant to check what would happen if doctors and students disinfected their hands before examining their patients. With respect to Semmelweis' eventual contamination hypothesis, data of the latter kind were

---

[1] A. B. Wolfe, "Functional Economics," in *The Trend of Economics*, ed. R. G. Tugwell, 450 (New York: Alfred A. Knopf, 1924) [italics are quoted].

clearly relevant, and those of the former kind totally irrelevant.

Empirical "facts" or findings, therefore, can be qualified as logically relevant or irrelevant only in reference to a given hypothesis, but not in reference to a given problem.

Suppose now that a hypothesis *H* bas been advanced as a tentative answer to a research problem: what kinds of data would be relevant to *H*? Our earlier examples suggest an answer: A finding is relevant to *H* if either its occurrence or its nonoccurrence can be inferred from *H*. Take Torricelli's hypothesis, for example. As we saw, Pascal inferred from it that the mercury column in a barometer should grow shorter if the barometer were carried up a mountain. Therefore, any finding to the effect that this did indeed happen in a particular case is relevant to the hypotheses; but so would be the finding that the length of the mercury column had remained unchanged or that it had decreased and then increased during the ascent, for such findings would refute Pascal's test implication and would thus disconfirm Torricelli's hypothesis. Data of the former kind may be called positively, or favorably, relevant to the hypothesis; those of the latter kind negatively, or unfavorably, relevant.

In sum, the maxim that data should be gathered without guidance by antecedent hypotheses about the connections among the facts under study is self-defeating, and it is certainly not followed in scientific inquiry. On the contrary, tentative hypotheses are needed to give direction to a scientific investigation. Such hypotheses determine, among other things, what data should be collected at a given point in a scientific investigation.

It is of interest to note that social scientists trying to check a hypothesis by reference to the vast store of facts recorded by the U.S. Bureau of the Census, or by other data-gathering organizations, sometimes find to their disappointment that the values of some variable that plays a central role in the hypothesis have nowhere been systematically recorded. This remark is not, of course, intended as a criticism of data gathering: those engaged in the process no doubt try to select facts that might prove relevant to future hypotheses; the observation is simply meant to illustrate the impossibility of collecting "all the relevant data" without knowledge of the hypotheses to which the data are to have relevance.

The second stage envisaged in our quoted passage is open to similar criticism. A set of empirical "facts" can be analyzed and classified in many different ways, most of which will be unilluminating for the purposes of a given inquiry. Semmelweis could have classified the women in the maternity wards according to criteria such as age, place of residence, marital status, dietary habits, and so forth; but information on these would have provided no clue to a patient's prospects of becoming a victim of childbed fever. What Semmelweis sought were criteria that would be significantly connected with those prospects; and for this purpose, as he eventually found, it was illuminating to single out those women who were attended by medical personnel with contaminated hands; for it was with this characteristic, or with the corresponding class of patients, that high mortality from childbed fever was associated.

Thus, if a particular way of analyzing and classifying empirical findings is to lead to an explanation of the phenomena concerned, then it must be based on hypotheses about how those phenomena are connected; without such hypotheses, analysis and classification are blind.

Our critical reflections on the first two stages of inquiry as envisaged in the quoted passage also undercut the notion that hypotheses are introduced only in the third stage, by inductive inference from antecedently collected data. But some further remarks on the subject should be added here.

Induction is sometimes conceived as a method that leads, by means of mechanically applicable rules, from observed facts to corresponding general principles. In this case, the rules of inductive inference would provide effective canons of scientific discovery; induction would be a mechanical procedure analogous to the familiar routine for the multiplication of integers, which leads, in a finite number of predetermined and mechanically performable steps, to the corresponding product. Actually, however, no such general and mechanical induction procedure is available at present; otherwise, the much studied problem of the causation of cancer, for example, would hardly have remained unsolved to this day. Nor can the discovery of such a procedure ever be expected. For—to mention one reason—scientific hypotheses and theories are usually couched in terms that do not occur at all in the description of the empirical findings on which they rest, and which they serve to explain. For example, theories about the atomic and subatomic structure of matter contain terms such as "atom," "electron," "proton," "neutron," "psi-function," etc.; yet they are based on laboratory findings about the spectra of various gases, tracks in cloud and bubble chambers, quantitative aspects of chemical reactions, and so forth—all of which can be described without the use of those "theoretical terms." Induction rules of the kind here envisaged would therefore have to provide a mechanical routine for constructing, on the basis of the given data, a hypothesis or theory stated in terms of some quite novel concepts, which are nowhere used in the description of the data themselves. Surely, no general mechanical rule of procedure can be expected to achieve this. Could there be a general rule, for example, which, when applied to the data available to Galileo concerning the limited effectiveness of suction pumps, would, by a mechanical routine, produce a hypothesis based on the concept of a sea of air?

To be sure, mechanical procedures for inductively "inferring" a hypothesis on the basis of given data may be specifiable for situations of special, and relatively simple, kinds. For example, if the length of a copper rod bas been measured at several different temperatures, the resulting pairs of associated values for temperature and length may be represented by points in a plane coordinate system, and a curve may be drawn through them in accordance with some particular rule of curve fitting. The curve then graphically represents a general quantitative hypothesis that expresses the length of the rod as a specific function of its temperature. But note that this hypothesis contains no novel terms; it is expressible in terms of the concepts of temperature and length, which are used also in describing the data. Moreover, the choice of "associated" values of temperature and length as data already presupposes a guiding hypothesis; namely that with each value of the temperature, exactly one value of the length of the copper rod is associated, so that its length is indeed a function of its temperature alone. The mechanical curve-fitting routine then serves only to select a particular function as the appropriate one. This point is important; for suppose that instead of a copper rod, we examine a body of nitrogen gas enclosed in a cylindrical container with a movable piston as a lid, and that we measure its volume at several different temperatures. If we were to use this procedure in an effort to obtain from our data a *general* hypothesis representing the volume of the gas as a function of its temperature, we would fail, because the volume of a gas is a function both of its temperature and of the pressure exerted upon it, so that at the same temperature, the given gas may assume different volumes.

Thus, even in these simple cases, the mechanical procedures for the construction of a hypothesis do only part of the job, for they presuppose an antecedent, less specific hypothesis (i.e., that a certain physical variable is a function of one single other variable), which is not obtainable by the same procedure.

There are, then, no generally applicable "rules of induction," by which hypotheses or theories can be mechanically derived or inferred from empirical data. The transition from data to theory requires creative imagination. Scientific hypotheses and theories are not *derived* from observed facts, but *invented* in order to account for them. They constitute guesses at the connections that might obtain between the phenomena under study, at uniformities and patterns that might underlie their occurrence. "Happy guesses" of this kind require great ingenuity, especially if they involve a radical departure from current modes of scientific thinking, as did, for example, the theory of relativity and quantum theory. The inventive effort required in scientific research will benefit from a thorough familiarity with current knowledge in the field. A complete novice will hardly make an important scientific discovery, for the ideas that may occur to him are likely to duplicate what has been tried before or to run afoul of well-established facts or theories of which he is not aware …

Scientific knowledge, as we have seen, is not arrived at by applying some inductive inference procedure to antecedently collected data, but rather by what is often called "the method of hypothesis," i.e. by inventing hypotheses as tentative answers to a problem under study, and then subjecting these to empirical test. It will be part of such test to see whether the hypothesis is borne out by whatever relevant findings may have been gathered before its formulation; an acceptable hypothesis will have to fit the available relevant data. Another part of the test will consist in deriving new test implications from the hypothesis and checking these by suitable observations or experiments. As we noted earlier, even extensive testing with entirely favorable results does not establish a hypothesis conclusively, but provides only more or less strong support for it. Hence, while scientific inquiry is certainly not inductive in the narrow sense we have examined in some detail, it may be said to be *inductive in a wider sense*, inasmuch as it involves the acceptance of hypotheses on the basis of data that afford no deductively conclusive evidence for it, but lend it only more or less strong "inductive support" or confirmation. And any "rules of induction" will have to be conceived, in analogy to the rules of deduction, as canons of validation rather than of discovery. Far from generating a hypothesis that accounts for given empirical findings, such rules will presuppose that both the empirical data forming the "premisses" of the "inductive argument" and a tentative hypothesis forming its "conclusion" are *given*. The rules of induction would then state criteria for the soundness of the argument. According to some theories of induction, the rules would determine the strength of the support that the data lend to the hypothesis, and they might express such support in terms of probabilities.

From Carl G. Hempel, *Philosophy of Natural Science* (Englewood Cliffs, NJ: Prentice-Hall, 1966).

## Questions and Topics for Discussion and Writing

1. Why did Hempel choose to illustrate the process of scientific inquiry by using the case of Semmelweis's discovery of the causes of childbed fever? Show how each of the steps of Semmelweis's investigation illustrate different kinds of reasoning.
2. What is the role of testing in this process? How are effective tests of Semmelweis's hypotheses constructed? Is this a straightforward mechanical process or does it require an inventiveness comparable to making the hypothesis in the first place?
3. Why is the "ideal scientific inquiry" sketched by Hempel an impossibility? Discuss how Galileo might have attempted to explain the limitation on the effectiveness of a suction pump using the method of "ideal scientific inquiry."

# The Discovery of Radium

## Marie Curie

The following short article is from an address by one of the world's greatest experimental physicists, Marie Curie, to an audience at Vassar College in Massachusetts acknowledging a gift by American women to her of a gram of radium for use in her research. Radium was (and is) extremely expensive and difficult to produce.

Marie Curie (1867–1934) was the first person ever to receive a second Nobel Prize. The first, the Nobel Prize in physics in 1903, she shared with her husband, Pierre Curie, and Henri Becquerel. The second, the Nobel Prize in chemistry in 1911, she received alone. Her life was one of tremendous hardships and strenuous work. After Pierre Curie's tragic death in a traffic accident, Marie Curie took over his teaching position and thereby became the first woman professor at the Sorbonne in its 650-year history.

In this article she describes the general course of her work on radiation.

I could tell you many things about radium and radioactivity and it would take a long time. But as we cannot do that, I shall give you only a short account of my early work about radium. Radium is no more a baby; it is more than twenty years old, but the conditions of the discovery were somewhat peculiar, and so it is always of interest to remember them and to explain them.

We must go back to the year 1897. Professor Curie and I worked at that time in the laboratory of the School of Physics and Chemistry where Professor Curie held his lectures. I was engaged in some work on uranium rays which had been discovered two years before by Professor Becquerel. I shall tell you how these uranium rays may be detected. If you take a photographic plate and wrap it in black paper and then on this plate, protected from ordinary light, put some uranium salt and leave it a day, and the next day the plate is developed, you notice on the plate a black spot at the place where the uranium salt was. This spot has been made by special rays which are given out by the uranium and are able to make an impression on the plate in the same way as ordinary light. You can also test those rays in another way, by placing them on an electroscope. You know what an electroscope is. If you charge it, you can keep it charged several hours and more, unless uranium salts are placed near to it. But if this is the case the electroscope loses its charge and the gold or aluminum leaf falls gradually in a progressive way. The speed with which the leaf moves may be used as a measure of the intensity of the rays; the greater the speed, the greater the intensity.

I spent some time in studying the way of making good measurements of the uranium rays, and then I wanted to know if there were other elements, giving out rays of the same kind. So I took up a work about all known elements and their compounds and found that uranium compounds are active and also all thorium compounds, but other elements were not found active, nor were their compounds. As for the uranium and thorium compounds, I found that they were active in proportion to their uranium or thorium content. The more uranium or thorium, the greater the activity, the activity being an atomic property of the elements, uranium and thorium.

Then I took up measurements of minerals and I found that several of those which contain uranium or thorium or both were active. But then the activity was not what I would expect; it was greater than for uranium or thorium compounds, like the oxides which are almost entirely composed of these elements. Then I thought that there should be in the minerals some unknown element having a much greater radioactivity than uranium or thorium. And I wanted to find and to separate that element, and I settled to that work with Professor Curie. We thought it would be done in several weeks or months, but it was not so. It took many years of hard work to finish that task. There was not *one* element; there were several of them. But the most important is radium, which could be separated in a pure state.

All the tests for the separation were done by the method of electrical measurements with some kind of electroscope. We just had to make chemical separations and to examine all products obtained, with respect to their activity. The product which retained the radioactivity was considered as that one which had kept the new element; and, and as the radioactivity was more strong in some products, we knew that we had succeeded in concentrating the new element. The radioactivity was used in the same was as a spectroscopical test.

The difficulty was that there is not much radium in a mineral; this we did not know at the beginning. But we now know that there is not even one part of radium in a million parts of good ore. And, too, to get a small quantity of pure radium salt, one is obliged to work up a huge quantity of ore. And that was very hard in a laboratory.

We had not even a good laboratory at that time. We worked in a hangar where there were no improvements, no good chemical arrangements. We had no help, no money. And because of that, the work could not go on as it would have done under better conditions. I did myself the numerous crystallizations which were wanted to get the radium salt separated from the barium salt, with which it is obtained, out of the ore. And in 1902 I finally succeeded in getting pure radium chloride and determining the atomic weight of the new element, radium, which is 226, while that of barium is only 137.

Later I could also separate the metal radium, but that was a very difficult work; and, as it is not necessary for the use of radium to have it in this state, it is not generally prepared that way.

Now, the special interest of radium is in the intensity of its rays, which is several million times greater than the uranium rays. And the effects of the rays make the radium so important. If we take a practical point of view, then the most important property of the rays is the production of physiological effects on the cells of the human organism. These effects may be used for the cure of several diseases. Good results have been obtained in many cases. What is considered particularly important is the treatment of cancer. The medical utilization of radium makes it necessary to get that element in sufficient quantities. And so a factory of radium was started, to begin with, in France, and later in America, where a big quantity of ore named carnotite is available. America does not produce many grams of radium every year but the price is still very high because the quantity of radium contained in the ore is so small. The radium is more than a hundred thousand times dearer than gold.

But we must not forget that when radium was discovered no one knew that it would prove useful in hospitals. The work was one of pure science. And this is a proof that scientific work must not be considered from the point of view of the direct usefulness of it. It must be done for itself, for the beauty of science, and then there is always the chance that a scientific discovery may become, like the radium, a benefit for humanity.

But science is not rich; it does not dispose of important means; it does not generally meet recognition before the material usefulness of it has been proved. The factories produce many grams of radium every year, but the laboratories have very small quantities. It is the same for my laboratory, and I am very grateful to the American women who wish me to have more of radium, and give me the opportunity of doing more work with it.

The scientific history of radium is beautiful. The properties of the rays have been studied very closely. We know that particles are expelled from radium with a very great velocity, near to that of light. We know that the atoms of radium are destroyed by expulsion of these particles, some of which are atoms of helium. And in that way it has been proved that the radioactive elements are constantly disintegrating, and that they produce, at the end, ordinary elements, principally helium and lead. That is, as you see, a theory of transformation of atoms, which are not stable, as was believed before, but may undergo spontaneous changes.

Radium is not alone in having these properties. Many having other radioelements are known already: the polonium, the mesothorium, the radiothorium, the actinium. We know also radioactive gases, named emanations. There is a great variety of substances and effects in radioactivity. There is always a vast field left to experimentation and I hope that we may have some beautiful progress in the following years. It is my earnest desire that some of you should carry on this scientific work, and keep for your ambition the determination to make a permanent contribution to science.

---

A lecture given by Mme. Curie at Vassar College in the 1920s in appreciation of the gift from American women of a gram of radium.

## Questions and Topics for Discussion and Writing

1. Why was Marie Curie so interested in radioactivity that she would devote so many years of strenuous labour to isolating and measuring radium?

2. Describe the "beauty of science" that Curie refers to. What are some of the "beautiful" aspects of her work, as she might have seen them?

3. Curie died of leukemia, doubtless caused by overexposure to radioactive substances. Before her work, the dangers of exposure to radiation were not known. Comment on the risks of such scientific investigation to the investigator. Is the knowledge gained (and the "beauty" beheld) worth the danger encountered?

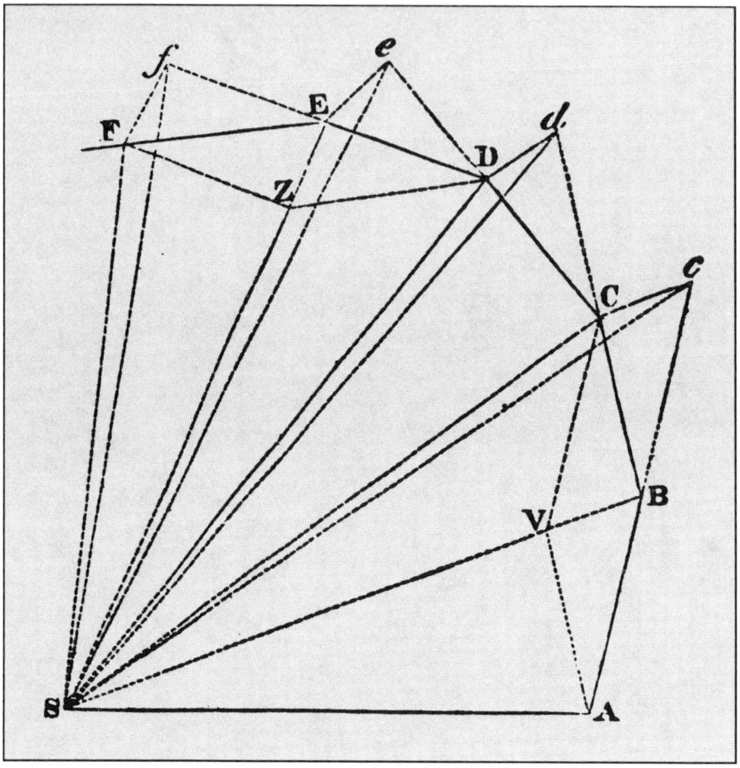


Newton's diagram of the effect of a centripetal force upon a uniformly moving body.

*Issue*

# The Development of the Physical Sciences

The great corpus of science can be examined more closely by focusing on only a part of science. All of what is science can be divided into more manageable parts using any of a number of criteria—method of investigation, or logical structure of theories, for instance—but the most usual division is by the subject matter studied. A common hierarchy, which will also be followed here, is to divide all science into two groups: the natural sciences and the social sciences. The natural sciences are those which study the phenomena of nature, while the social sciences study the organizations and structures of human culture.

While there is much overlap of methods of study and subject matter studied, it is also true that there are vast differences between the natural sciences and the social sciences, differences due to the fact that in the social sciences human interactions present a level of unpredictability that must be allowed for and accounted for throughout all social theories. It is therefore a common practice to consider social science as something very different from the natural sciences, and to use the term *science* to refer to the natural sciences alone. That convention will also be followed here, and this book will concern itself almost exclusively with natural science and its interaction with society.

The natural sciences divide most simply and obviously into two great sections: the physical sciences and the life sciences. The life sciences are concerned with the phenomena of nature that are in some sense alive while the physical sciences are concerned with everything else. The physical sciences include astronomy, physics, chemistry, and geology, among others, and, for convenience, we shall here also consider mathematics (though many people consider mathematics not to be a science at all).

The history of the physical sciences shows a definite sequence of development of the different component sciences, a sequence that says a great deal about the structure of these sciences. Physical sciences develop rapidly and become fruitful areas of inquiry when reliable physical laws can be stated and tested, and used as the basis for discovering other laws. Most often, this has required a precise mathematical formulation. Therefore what is required is the discovery of relationships that are simple enough to be stated in mathematical terms, and mathematics that is sophisticated enough to express those relationships.

The order in which the physical sciences "came of age" is: mathematics, astronomy, physics, chemistry, and geology. Mathematics was made rigorous and thereby reliable and fruitful in antiquity, notably by the Greeks. Astronomy was a precise observational science in antiquity, but moved forward rapidly after mathematically precise and simple descriptions of the planetary motions were found in the seventeenth century. Physics was a subject of speculation, but not of measurement nor reliable mathematical formulation until the seventeenth century. Chemistry followed in the eighteenth century; geology in the nineteenth.

The articles in this section focus primarily on developments in physics, with some attention to astronomy and a mention of mathematics. Also the period of greatest attention is the sixteenth and seventeenth centuries—the period known as the Scientific Revolution. Byron Wall's essay, "What the Copernican Revolution is All About," discusses the profound shift in thinking brought about by the theory of Copernicus that it was the earth that travelled around the sun, and not vice-versa. Wall's "Proposition I.47 of Euclid: The Pythagorean Theorem" illustrates the rigor and beauty of Greek mathematics which formed the backbone of scientific theories in all the physical sciences. Thomas H. Leith sketches the life and personality of the recluse who set the Scientific Revolution in motion in "Copernicus: The Man."

Quite the opposite kind of personality was the argumentative and outspoken Galileo, who did far more to push the Copernican Revolution than did Copernicus himself. Galileo's "Starry Messenger" is one of Galileo's monographs that supported the Copernican view with observations of heavenly phenomena seen for the first time with the aid of the newly invented telescope. Pushing Copernicanism was not just a matter of arguing a point of view for the sake of knowledge alone. Galileo's advocacy flew in the face of the established dogma of the Catholic Church and brought him the wrath of the Inquisition. "The Sentence of Galileo by the Inquisition" is the decree of the Inquisition which placed Galileo under house arrest for the rest of his life, giving the reasons for its actions.

The Scientific Revolution began with the Copernican Revolution and ended with the birth or rebirth of all the major modern physical sciences. The person who drew together the various theories and discoveries of the Scientific Revolution into one great synthesis of physical theory was Isaac Newton. Thomas H. Leith's "Newton: The Man" is a brief sketch of Newton's life and very peculiar personality. Newton's synthesis survived as the foundation of physical theory until it was finally supplanted by developments in physics in the early twentieth century. The most notable physicist of this period was Albert Einstein, whose theory of relativity made physics more exact in the realm of very large distances and very fast speeds. James A. Newman's "Einstein" is an appreciation of Einstein and his work, but also an insightful essay on the workings of the scientific mind.

# What the Copernican Revolution is All About

**Byron E. Wall**

The Copernican "Revolution," was a revolution in two senses. By attributing the "revolution" to Copernicus, reference is made to Copernicus' chief piece of writing, his book *On the Revolutions of the Heavenly Spheres*. Hence, the "revolution" is that of the earth's annual motion and those of other planets around the sun, or alternately, of the daily rotation of the earth on its own axis.

On the other hand, the more political sense of "revolution" is also intended, because the Copernican Revolution represented a turning of mankind's allegiances away from one system to another. In this case the systems were not of government, but of beliefs and modes of thought. Perhaps it would be better to go with the metaphor farther and say that the Copernican Revolution *was* an overthrow of one system of government for another, where the powers of these governments rested not in brute forces but in systems of ideas that exercised controls, first, on the thoughts of men, and thereby, indirectly, on their actions.

### Ancient Astronomy

Astronomy is a very ancient science. By the standards we now use to determine what is or is not a science, astronomy was the first subject to become a science. Astronomy today is an inordinately complex field, involving assimilating data from huge and expensive optical and radio telescopes, subjecting these data to intricate mathematical analysis, interpreting the results in terms of both the modern physical theories that apply to vast distances and speeds and the theories applicable to sub-atomic particles. From this research comes information about the size and nature of the universe, its history, and perhaps its future, and in a more immediately practical form, information that will assist the development of aerospace technology. Ancient astronomy was of course very different in its specific manifestations, but in a general way its aims and even its methods have much in common with astronomy today.

Using a very inclusive definition of what constitutes astronomy, it is obvious that all early cultures must have had some kind of astronomical theory, even if only the bare observation that the sun rises and sets every day. But there is evidence that virtually all ancient civilizations went a lot farther than that.

The ancient Mayans in Central America had an advanced mathematics, an essential prerequisite for the development of astronomical theory, but their astronomy confined itself mainly to the regulation of the calendar. It's not that calendars are not important—they predict the seasons and hence the proper times for sowing and harvesting crops, and they determine the proper dates and times for religious rituals. It was indeed concern with the calendar that gave great impetus to European astronomy in the time of Copernicus. What the Mayans didn't do was apply their sophisticated mathematics to an analysis of the complex motions of the heavenly bodies.

In European astronomy, calculations of this sort were the central concerns of the subject. The accuracy of some of the calculations of planetary positions made in ancient Greece, using cumbersome mathematics and naked-eye observations, is staggering. Careful observation and painstaking calculation makes the ancient astronomy of the Mediterranean civilizations look very modern. One important difference between ancient and modern, though, was the choice of heavenly bodies studied.

Today's astronomers are primarily concerned with very distant stars. The ancients treated these as a more or less fixed background against which the objects of their attention moved.

The general idea was that all the stars, or what *we* would call stars, were equidistant from the earth, and were fixed, as it were, on a huge sphere which turned around on its own axis once every twenty-four hours, carrying everything with it. The ancients were not able to detect any movement of these stars with respect to each other, only the daily rotation of the whole troupe. What fascinated them were seven exceptions to this principle. They called them "wandering stars". We still call them by the Greek word for "wanderer"—planet. There were seven planets in ancient astronomy—Mercury, Venus, Mars, Jupiter, Saturn, and also the Sun and the Moon. They all had these characteristics in common: they made a cycle through the sky every twenty-four hours, just like the fixed stars, but they did not exactly keep pace with them. They had, it seemed, a motion of their own, in addition to partaking of the motion of the sphere of the fixed stars. Moreover, the paths of their motions relative to their fixed counterparts were all in a narrow band that cut diagonally across the large stellar sphere. That narrow band was where all the action took place. The stars that were fixed in it became the landmarks by which the wanderings of the planets could be detected. Since the sun took one year to complete its journey around the band, the band could be divided into twelve sections, so that the sun spent a month, more or less, in each one. In each of the twelve sections, the brightest stars formed a recognizable pattern, or constellation, each of which was given the name of the animal figure, or in Greek, *zoidion*, that the constellation was purported to resemble.

Thus, what *ancient* astronomy, at any rate, was all about, was the movements of seven spots of light through the signs of the *zodiac*. Of what value was this arcane study other than the opportunity it provided for mental gymnastics? If we confine the subject to the rotation of the sphere of the fixed stars, including the very slight deviation or wobble of the sphere in its daily motion, which was noticed by the ancients and carefully recorded, plus the study of the motions of the sun and moon as they move against the fixed stars, then we have an adequate basis for regulating the calendar, making agricultural decisions that depend on the seasons, for navigating land and sea journeys by the stars, and even for accurately predicting eclipses. But the most complicated studies were those devoted to accounting for the motions of Mercury, Venus, Mars, Jupiter and Saturn. It is not easy to find a utilitarian function for the results of the study of these planets. At least, it is not easy if we insist on evaluating their work from our own worldview. The fact that the objects of their study were planetary motions through the signs of the zodiac alerts us to the significance of their work for astrological usages. If the positions of the planets among the signs of the zodiac were held to influence the destiny of individuals on earth, then the precise determination of those positions in the past and future could be considered extremely important. Indeed, the terms astrologer and astronomer were virtually interchangeable throughout the ancient and medieval world. Casting horoscopes provided a considerable source of income for astronomers in pre-modern times.

Helpful as the horoscope industry may have been in providing livelihoods for astronomers, the concerns of astrology were not major factors in the Copernican revolution. The justification for ancient planetary astronomy that *was* crucial later in the Copernican era had to do with the overall cosmology of Greek philosophy.

### Aristotle's Cosmology

The dominant feature of Greek philosophy was its tidiness—everything in its place in a coherent order. This was nowhere more evident than in cosmology. The cosmological theory that had the greatest acceptance was Aristotle's. According to Aristotle, the world was divided into two neat compartments. The inner compartment was the world of generation and corruption. It was the world we lived in, where things grew and died and moved around in a seemingly complex disorder. This apparent disorder was put in philosophical order by analyzing material things into combinations of earth, air, fire and water, and attributing natural tendencies to each of these elements. Things composed of earth and water moved downward, that is, inward toward the centre of the world, because of their inherent heaviness. Fire and air moved up-

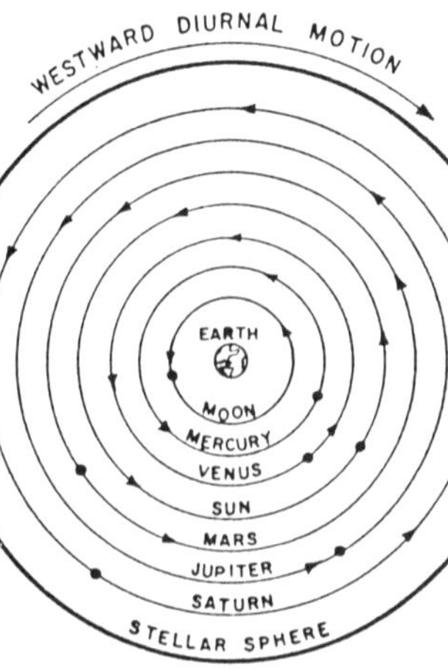


FIGURE 1.1: Aristotle's universe of concentric spheres.

ward, that is, away from the centre because of their inherent lightness. Forces which interfered with these natural motions were attributed to external, violent causes. A good example is that of a stone thrown upward: its original upward motion is due to the force of the agent that threw it. When the force is expended, the stone resumes the motion inherent in its heavy nature and falls to the ground. By such arguments was every physical event in the corruptible world accounted for. The corruptible world included the earth itself and everything between it and the moon, which was believed to be the closest of the planets.

But from the moon outward to the fixed stars, everything was different. Out there things did not grow and die, nor were their motions random. Things neither fell toward the earth nor floated farther and farther away from it. They just went round and round, day after day and year after year. By the way, meteors, shooting stars, comets and other such items involved in unique occurrences were believed to be all below the moon. The physics that explained motions in the world around us just would not do for motions in the heavens. Hence for that outer world, a completely different physics was expounded.

Up beyond the moon, there was no earth, air, fire or water, but a fifth element, or quintessence,—a crystalline substance that filled all space. To this fifth element, only one kind of motion was natural: circular. The motions were circular for two reasons: (1) It was what they seemed to be, in general. The stars went round in circles, night after night; the sun came up in the east, set in the west, then came up in the east the next day, and so on. (2) Circular motions have eternity built into their definitions. Motions in straight lines require beginnings and endings, generations and corruptions. Circular motions require no starting and stopping. So we get a nice tidy system of celestial mechanics if everything up there goes obediently around in circles.

## The Problem of the Planets

But here we come smack up against the problem of the planets. These mysterious wanderers threaten to upset the whole system. Not only do they not go nicely around in simple circles in step with the other stars, not only do they run a course in approximately the opposite direction, they actually seem to start and stop, then turn around and go in the other direction for brief periods of time. Sometimes they actually seem to do loops in the sky. The only saving grace is that they seem to perform these antics over and over again, so that there must be some regular order to their motions.

The task for astronomers, then, was to find out how to account for the apparent motions in terms of circular motions, perhaps as the result of several circular motions compounded together. This is the task that astronomers were charged with by philosophers. Plato referred to it as "saving the phenomena"; it might have been more appropriately called "saving the philosophy."

Aristotle proposed his own solution to the problem of the planets, adapted from an idea of another Greek thinker. In Aristotle's version, each planet was on a different spherical shell, all shells being concentric with the earth. Between the shells on which a planet was stuck were intermediate shells. Except for the planets, which were each on different shells, and the fixed stars, which were all on the nearly outermost shell, the rest of the substance of the shells was invisible.

The outermost shell was the prime mover, which somehow communicated its motion to the next shell, but the next shell rotated on its own in a slightly different direction, so that anything visible

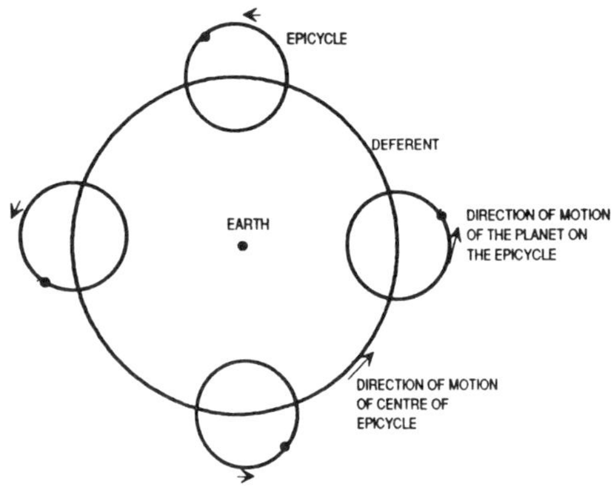


FIGURE 1.2: A planet travelling on an epicycle, which is fixed on a rotating deferent.

on it would move with the combined motions of the outer shell in one direction and its own shell in another direction. That shell in turn carried with it the next shell, which also rotated in a different direction, and so on. Thus visible points on the surfaces of these shells, such as planets, would appear to be moving quite irregularly, but actually their motions were merely the combined effects of numerous circular motions from the rotations of other shells.

The idea was ingenious in principle, but unfortunately it could never be made to work very well. It just did not seem possible to think up an arrangement of shells that could move in such a way as to specify the planetary motions that had been observed and recorded.

### The System of Ptolemy

By the end of the second century A.D., another conception for accounting for the planetary motions had been worked out. This was the system perfected by Claudius Ptolemy, a Greek astronomer whose work we have come to know by its Arabic title, *Almagest.* It was the general system of Ptolemy that was the accepted basis for astronomy in the time of Copernicus, and it was in particular Ptolemy's work, *Almagest*, that Copernicus directed his work against.

Ptolemy's system accepted the basic Aristotelian conception for the "inner" world between the moon and the earth, and also the general idea of the sphere of the fixed stars that rotated every twenty-four hours. He was also able to account for the motions of the planets with combinations of circular motions. But instead of postulating a set of concentric spherical shells, like onion skins revolving every which way, his way of accounting for the irregular planetary motions involved circles which turned on circles. Arthur Koestler has called it the Ferris Wheel universe. For each planet, one big circle called the deferent described a path through the zodiac that represented the general direction of the planet's motion. Somewhere on the edge of the deferent was the centre of a second, generally smaller circle called the epicycle. At the edge of the epicycle was the actual planet. Both the deferent and the epicycle turned, at rates specified by Ptolemy for each planet, which resulted in the planet moving in what appeared from the vantage point of the earth to be the irregular combinations of forward, backward, starting and stopping, and looping motions that had so troubled Greek philosophy. But of course in Ptolemy's system these irregular appearances were actually to be just the visual result of combined smooth, circular, and of course, eternal, motions.

The great advantage of Ptolemy's system over Aristotle's was that it worked—it could indeed account for the planetary motions within a reasonable margin of error. It was pretty complicated though. The concept of deferents and epicycles had been around for a long time before Ptolemy had worked it into a reliable system. In getting the bugs out, he had to introduce a number of inelegant details. For example, the centre of the deferents were not necessarily the centre of the earth; they could be taken to be eccentric to the earth, which would explain why a planet would seem to complete one-half of its course in a shorter time than it would take to travel through the other half. For similar reasons, Ptolemy did not require that the epicycles turn at uniform speeds around their respective centres, as long as they turned uniformly around some other chosen point within their circumferences.

But, after all the necessary adjustments, Ptolemy did come up with a way of accounting for the planetary positions that had been observed and

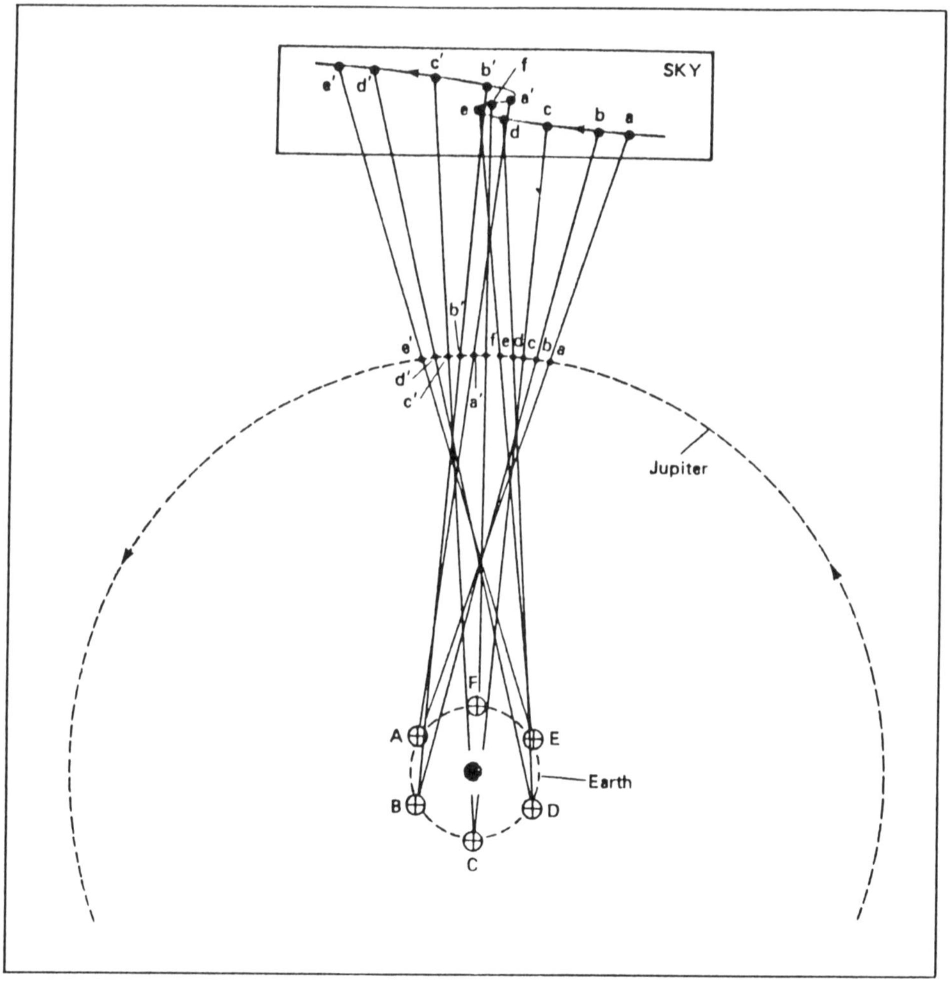


FIGURE 1.3: The apparent "retrograde" motion of Jupiter against the backdrop of the "fixed" stars can be seen to be an illusion caused by the combinations of the motions of the Earth and Jupiter around the Sun.

predicting future positions of the planets as they would appear against the stars in the zodiac. The philosophical requirements of the perfect heavens had been fulfilled. The planets were seen to move in never-ending circular pathways at what could be at least construed as uniform, that is, unchanging, speeds. It was regular enough that it was possible to construct a mechanical clock-work system that would show the positions of the planets as the gears turned round. In the Middle Ages, just such a clock was built along Ptolemaic principles.

The real fault of Ptolemy's system was that despite meeting the philosophical requirements that had been specified, the system just didn't make philosophical sense. It's all fine to specify precise combinations of circular paths and speeds for the planets and to show that they don't really start and stop, but even in Ptolemy's plan, the planets are supposed to be really doing loops in the sky as they pass through the zodiac. Why should they be doing that?

In Aristotle's onion-skin system, the planets were carried around on solid pieces of stuff. The way they were held up and moved around had some logic to it, even if the motions themselves didn't. Ptolemy had the motions, but no explanation for them—no explanation for what was moving the planets around and no explanation for why they should be moving in this peculiar fashion.

The problem of what was moving the planets around was a subject of much study in the Middle Ages and Renaissance. It isn't just one "clock" on which all the planets are shown, but seven different clocks, one for each planet. The motions of each planet are conceived differently. There is no principle to unify them.

### The Copernican Viewpoint

It is the work of Copernicus that provides the unifying principle. By taking the sun, or rather a point near the sun, as the main centre of the motion of the planets, he could dispense with the epicycles, for their main function. By taking the sun as fixed, and making the earth into a wanderer like the other planets, the epicycles that Ptolemy needed can be seen to be just an artifact of our perception due to the combination of the motions of ourselves and of the planets all around the sun. The planets don't do loops and other dances in the sky, they just appear to, because as we watch their motions around the sun, we see them from different angles due to our own motion. The vast sphere of the fixed stars doesn't spin around at enormous velocity every day, instead we revolve on our own earthly axis, and that makes the stars appear to move. Moreover, all that is special about the stars in the zodiac is that they happen to be the stars that fall in the line of vision between ourselves and the planets.

But now we come up against the more political sense of revolution in the Copernican revolution—the overthrow of a complete system of ideas. In the Aristotelian worldview, even as amended by Ptolemy, we lived in the centre of generation and corruption, of living and dying, stopping and going, in short, in an imperfect, temporal world. Out there, beyond the moon, everything is perfect and eternal and changeless. But if *we* are also eternally on a celestial body, does that mean that we are also eternal, perfect, incorruptible, and, for that matter, weightless? No, that can't be, so it must then be that there is no sharp distinction between the perfect and imperfect worlds. Maybe it's all like it is here on earth. But then why do the planets move eternally around the sun? Why is it that when we let go of something heavy it doesn't fly away to the sun if the sun is the centre? Why don't birds and clouds disappear toward the west if we are spinning around toward the east? If the heavens that we see aren't perfect, where is perfection to be found? Where is heaven?

These questions put to the Aristotelian worldview were not just idle analyses of the thought of some ancient philosopher. Aristotle's system was taken over by the Church during the Middle Ages and made into its own. Scripture had been interpreted in an Aristotelian framework. God's world was to be understood with the help of the heathen philosopher. The heavens were supposed to be telling the glory of God, and reminding us of our imperfections down here on earth. Now what were we supposed to think?

Copernicus was not unaware of some of these difficulties. He offered some possible explanations for the more paradoxical questions about terrestrial physics on a moving earth which was not at the centre of the world, and gave a rather poetic justification for the sun being in the centre. But thanks to an overzealous colleague, the Lutheran theologian Osiander, these more shattering questions and the answers they lead to could be quietly put off for a while. When Copernicus' book appeared, Osiander had anonymously added a preface saying that the sun-centered system given in the book was to be viewed as merely a convenience for purposes of astronomical *calculations*, and need not be taken literally. And that is precisely what was done with the Copernican system for the next half-century or more, with some notable exceptions including Giordano Bruno, who dared to take the system literally and draw too many embarrassing implications from it, and was burnt at the stake in consequence.

Sooner or later the issue had to be faced; it had to be laid at the feet of the general educated public and be discussed and analyzed in detail. Until then *astronomers* could use the Copernican system in their calculations and ignore the consequences of taking it literally. Non-astronomers who could not appreciate the technical reasons favouring the Copernican system could easily dismiss the notion of the earth as a moving planet as utter nonsense, and did just that.

Because the Aristotelian-Ptolemaic conception of the world was so entrenched, it was accepted as just an elaboration of common sense—which in a way it was, of course. It was therefore accepted not just as a good theory, but as the simple truth. So self-evident did the immobility of the earth seem, that for the general public to abandon that view, a complete revolution of their normal thinking would have to take place. In the sense of revolution here intended, the monumental work of Copernicus was hardly a suitable vehicle for bringing it about, despite the appearance of the word "revolution" in the title of Copernicus' book.

*On the Revolutions* is, except to those with a taste for tedious mathematical calculation, a very dull, stupefying piece of writing with very little to capture the imagination, much less launch a revolution in human thought by the boldness of its polemic. This is a point that has been repeatedly emphasized to me by my students whenever I have assigned them to read sections of it. Even granting the adventurous climate of the Renaissance and the newfound respect for ideas that permeated the revival of learning, it is hard to see how the Copernican view could begin to get serious consideration outside of mathematically astute circles—unless it was pressed very vigorously by a very strong personality.

**Galileo, the Revolutionary Leader**

As it turns out, that is exactly what happened in Italy in the first decades of the seventeenth century. The strong personality was Galileo. Galileo was astute, bold, fearless—one might even say reckless. He used all the tricks known to debating societies, encyclopedia salesmen and movie stars to sell his product, the Copernican system, which he happened to believe in. And by and large, he succeeded.

He took apart the arguments of his opponents, the Aristotelian scholars, and made laughing stock of them. He produced new and totally unsuspected evidence that defied assimilation into the Aristotelian conception, but which fit nicely into *his* view. He managed to get the book in which he set out his strongest arguments banned, which of course only increased the demand for it. He even employed the technique for becoming famous, which we in the post-Watergate era now know to be flawless, of getting himself convicted for a major crime.

The new evidence came from the telescope, then recently invented and much improved by Galileo. With it he discovered many new facts about the heavens which undermined the Aristotelian conception. The moon, for example, he could see to be rocky and uneven. It had mountains, which Galileo calculated to be as large as those on earth; it had craters; it had dark spots which Galileo mistakenly thought to be seas. It was therefore much like the earth in its irregular configuration. In other words, it was not smooth and perfect as fit a heavenly body according to the Aristotelians.

He also noticed that besides the illumination of the moon by the sun, the moon had a secondary light that must have come via reflection of the sun's light off the earth. And of course, the earth has a rough and irregular surface. This answered the objection that if the heavenly bodies were not smooth and polished, they could not reflect light.

The visible phases of the moon were not important for the Copernican system, as the moon was held to go around the earth in any case. But phases of the *planets* would be another matter. With his telescope, Galileo was able to see that Venus had phases like the moon's. These could *only* be explained if Venus travelled around the sun. Yet in the Ptolemaic system, Venus was placed between the sun and the earth at all times.

The placement of the moon in the Copernican system was an anomaly. If all the planets were supposed to go around the sun, why did the moon go around the earth instead? Galileo was able to forestall this objection by finding through his telescope that the moon was not an exception. Jupiter also had *moons* that circled it—four of them that he could see.

Galileo published these and other findings in a short book called *The Starry Messenger*. In Galileo's book the observations are carefully reported, and the reader is told how to construct a telescope so he can go and see for himself. The book caused a stir and prompted some "refutations" to the effect that the observations were impossible and mistaken or were caused by distortions in the telescope itself. A breach had been made, but there were still plenty of defences available.

Galileo answered his opponents, published other findings, invited people to use his telescope under his guidance, and eventually began to address himself to issues that were not strictly astronomical, but were concerned with implications in other areas, for example whether there was a necessary connection between Aristotelian cosmology and Christian theology. This was dangerous ground. Galileo gained the reputation of a trouble-maker. He was warned by the Vatican. For a time he kept quiet, but during this time he worked out his next arguments and how they would be presented.

Eventually he published the book which got him in deep trouble, *The Dialogue Concerning the Two Chief World Systems*, by which is meant the Copernican versus the Aristotelian-Ptolemaic systems. The *Dialogue* is a lengthy book, written in the style of a Plato dialogue, with three characters: a mathematician-scientist who espouses the Copernican system, an Aristotelian scholar who defends the status quo, and a neutral interlocutor who is open to being convinced one way or the other. Galileo wrote the *Dialogue* in Italian, not in Latin, which meant that he was deliberately playing to the galleries, that is, to the ordinary educated citizen rather than to the scholars, who he viewed as too entrenched in their fixed ideas.

As one would expect, the mathematician-scientist has no trouble destroying all the objections of the Aristotelian while making a flawless case for Copernicus. The Aristotelian, who is called Simplicio, is revealed as a true simpleton who again and again argues himself into absurdities, fails to see the point and blithely continues to maintain his original position. The neutral observer slowly is won over to the Copernican view after having made numerous reasonable objections which he eventually sees are satisfactorily answered within the Copernican system.

In 1632 the *Dialogue* was published. Five months later it was banned. Galileo was called before the Inquisition, tried and pronounced to be vehemently suspected of heresy, ordered to recant his false opinions, and was placed under house arrest for the remainder of his life.

### The Basis of Knowledge: Experience versus Authority

Galileo lost his battle for Copernicus, but of course the war was eventually won, and Galileo's fate at the hands of the Inquisition *promoted*, rather than retarded his cause. It did so because it pinpointed the crucial issue that makes the word revolution in the Copernican Revolution refer to more than planetary motions. That issue had to do with whether knowledge, at least knowledge about the external world, rests on a basis of sense experience or on the basis of authority. In this sense, the motions of the earth and the planets merely provided the test case.

Aristotle's system *was* based on sense experience, the sense experience of ancient Greece—a point Galileo repeatedly emphasized. But what was true for Aristotle was not true for the Scholastic Aristotelians, for whom the basis of knowledge was the set of books, including Aristotle's, that established the framework into which all experience had to fit. As only a few had the requisite learning to have mastered these texts, it was their lot to explain the universe to those who did not enjoy this direct access to the truth. Those who gained more of their knowledge through their senses in the daily experience of living were placed in the position of having to disregard their experience or to interpret it according to well worked out principles understood only by the scholars. That is why it was revolutionary for Galileo to address himself to the larger audience of laymen by writing in Italian instead of Latin.

Galileo's *Dialogue*, the work that we take to be the most influential in convincing the public of the truth of the Copernican world-view, was actually not pro-Copernican so much as *anti*-anti-Copernican. There are two reasons for this. One was that Galileo had received specific instructions from the Vatican not to hold or defend the Copernican system in print. Galileo interpreted this to mean that he was *not* forbidden to attack the position of those who opposed Copernicus. As we know, the Inquisition did not agree with Galileo on this fine distinction.

The second reason provides us with a glimpse of the motive that led Galileo to go to all this trouble at all. Galileo defended the theory of the long-dead Copernicus because he believed it true. Granting that, we can take it a step further and say that Galileo concerned himself with the theory of Copernicus because, to the extent that he could show the Copernican system to be true, he could thereby show the Aristotelian theory to be false. If the Aristotelian world-view could be seen to be false, then the whole authority of the ancients collapsed.

What Galileo wished to do was break the stranglehold of Aristotelianism, and it suited his purposes to use Copernicus to do it. That *this* was his major interest becomes clear when we examine Galileo's own contributions to astronomy. With the telescope, Galileo made a number of observations that lent support to the general idea that the earth moves around a stationary sun. But Galileo did not concern himself with detailed, precise measurements of heavenly motions the way that astronomers of his own era or even in antiquity did. Similarly, Galileo made no attempt to advance Copernican theory by working out a mathematical account of his new observations. In contrast, say, to his contemporary, Kepler, who advanced Copernican astronomy by discovering that the planets moved not in circles, but ellipses, Galileo expounded a simplified version of the Copernican theory in the *Dialogue* that did *not* fit the available observational data at all. Copernicus had hoped to show that his system was more accurate than Ptolemy's. The version that Galileo presented to his readers would have been much worse. But none of that mattered if all Galileo wished to accomplish was to show that the ancient theory was built on sand.

Galileo was particularly incensed about grand, sweeping theories that were unthinkingly accepted by his contemporaries. More than anything else, he wished to show that these theories are sustained only by ignorance. In a passage in the *Dialogue* the supposedly impartial observer remarks:

> There is not a single effect in nature, even the least that exists, such that the most ingenious theorists can arrive at a complete understanding of it. This vain presumption of understanding everything can have no other basis than never understanding anything. For anyone who had experienced just once the perfect understanding of one single thing, and had truly tasted how knowledge is accomplished, would recognize that of the infinity of other truths he understands nothing.

For Galileo the way to knowledge was to build it up bit by bit from observing how nature works, not by imagining great schemes. The major thrust of Galileo's scientific work that was not concerned with the Copernican view is to show how this is accomplished. In the *Dialogue*, there is enough of an indication of Galileo's conception of the road to knowledge for the reader to get the general idea. In his later book, *Two New Sciences*, that Galileo wrote while under house arrest and then smuggled out to be published in a Protestant country, that pathway is shown much more explicitly.

The way is through controlled experiment and testing of hypotheses. It relies on the data of sense perception, but with a *refinement* that made the Copernican system an excellent illustration for Galileo. One learns ultimately through the senses, but one must not rely on *common* sense for understanding. For the principles of nature to be grasped, one must experiment and reason very carefully to avoid being misled by appearances. It is *common* sense that supports the Aristotelian world-view, and only reasoning carefully from the observations that supports Copernicus.

### The Triumph of Empiricism

The trial of Galileo had its effect. Italy had been a great centre of scientific activity. After the trial, science in Italy was severely retarded, and has never quite recovered. But elsewhere in the world, the trial showed those who had nothing to fear from the Inquisition that reliance on the authority of ancient philosophy leads nowhere. The new philosophy that took its place was empiricism. Its scientific basis was experiment and reason, and its watchword was to be wary of unwarranted generalizations.

Galileo had a comrade in arms for his battle for empiricism against authority in his English contemporary, Sir Francis Bacon. They had at least this much in common, but they differed on many particulars, belief in the Copernican system being one of them. More generally, Bacon distrusted the use of mathematics and logic to advance from direct observation, fearing that it would lead to the same kind of grand systems that both he and Galileo despised. But Bacon was eloquent on the value of collecting empirical data and, by a process of induction, discerning true scientific principles. Bacon's writings became a manifesto for the kind of science that Galileo had fought for with the Copernican theory as a weapon.

The end of the Copernican revolution comes with Sir Isaac Newton in England in the late seventeenth century. It was Newton who finally made physical sense of Copernicanism by showing how the new laws of physics developed by Galileo for terrestrial phenomena account for the motions of the planets as they had been described by Kepler's improvements on Copernican theory. Newton's unified physics vindicated Copernicus' celestial revolutions and established a firm basis for the new philosophy of empirical science.

That is the end of the story of what the Copernican revolution is all about, but there is one final irony. Galileo used Copernicus as a lever to overthrow the stifling authority of Aristotle. But Newton used Galileo to establish Copernicus and to lay down a new great system that engulfed all science under its authoritative and comprehensive laws. Bacon, arguing the sterility of ancient Greek science, was equally ready to dispense with the best developed part of Greek scientific thought, mathematics. But Newton, his own countryman, completed the rout of ancient science by creating a *mathematical*, deductive basis for the new science. In all revolutions, it is easier for the reins of power to change hands than it is to change the reins of power.

## Questions and Topics for Discussion and Writing

1. In what way did the Aristotelian world view make more "sense" than the Copernican system?
2. Why did Aristotle have one set of rules (one *physics*) for the world "below the moon" and a different set for the world beyond the moon?
3. In what way was the system of Ptolemy an improvement over Aristotle's, and in what way was it not? In what way was the system of Copernicus an improvement over Ptolemy's, and in what way not? In what way was the system of Copernicus *not* an improvement over Aristotle's system?
4. What is the role of Galileo's own astronomical observations in establishing the Copernican system?
5. Discuss the last sentence of the article with respect to the Copernican revolution. What are the "reins of power" that have changed hands? How have the "reins" changed and how have they not changed?

*Byron E. Wall*

# Proposition I.47 of Euclid: The Pythagorean Theorem

## Byron E. Wall

The crowning achievement of ancient Greek science was the development of rigorous mathematics. The mathematics known to ancient Greeks is astonishing in its complexity, subtlety and, above all, its logical rigor. The form in which most of what we know as Greek mathematics came down to us is the thirteen books of Euclid's *Elements*, written about 300 B.C. Euclid attempted to encompass most of the mathematics known in his time in a single multi-volume treatise which began from simple axioms and developed each successive mathematical theorem, or "Proposition," from the axioms and previous propositions, using only logical inference. The approach had its limitations, not the least of which arose from Euclid's characterizing all mathematics, including number theory, as geometry, but overall the *Elements* was so successful that it remained the standard, indeed the only, textbook of geometry used, in various translations, for two thousand years. Many of the world's greatest scientists and mathematicians can trace their decision to pursue a career in science to their studying Euclid's *Elements* and becoming enthralled by the beauty and elegance of exacting mathematics.

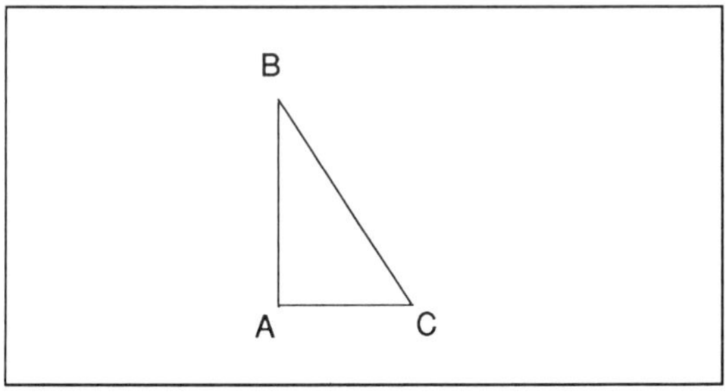


**FIGURE 1.4:** A right triangle.

By far, the most famous result of Greek mathematics is the theorem we ascribe to Pythagoras and which appears in Euclid as Proposition 47 in Book I. For hundreds of years this "Pythagorean" theorem was known simply as Proposition I.47. It states the familiar result, used constantly in modern science, that the area of a square constructed on the hypotenuse of a right triangle (that is, on the side opposite the right angle) is equal to the sum of the areas of squares constructed on the other two sides.

To prove this, Euclid first establishes several prior results (in Propositions I.1 to I.46), including such notions, familiar to anyone who has studied plane geometry, as (1) that triangles are congruent if one triangle has two sides and the angle between the sides equal in size to corresponding sides and

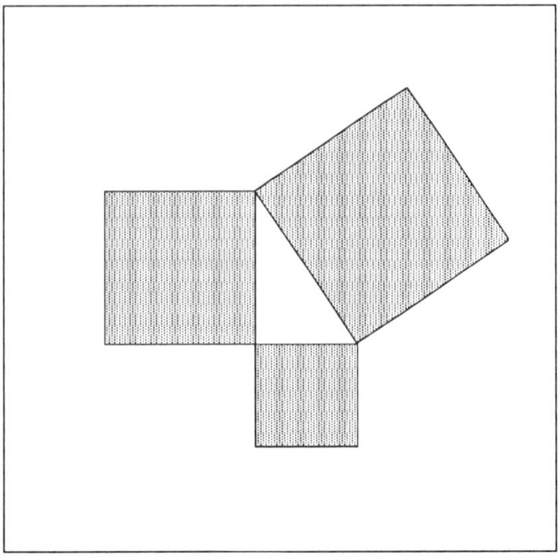

**FIGURE 1.5:** The squares on the sides of a right triangle.

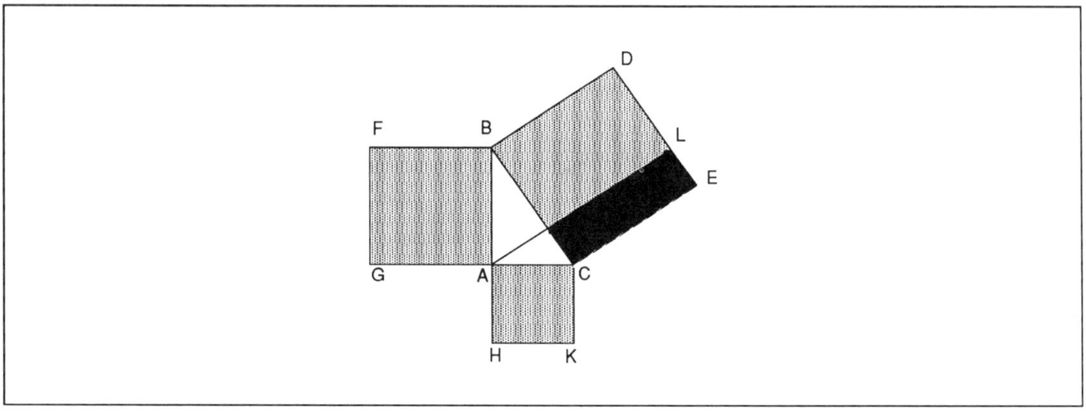


FIGURE 1.6: The square on the hypotenuse divided into two rectangles by a line from the vertex of the right angle, parallel to the sides of the square.

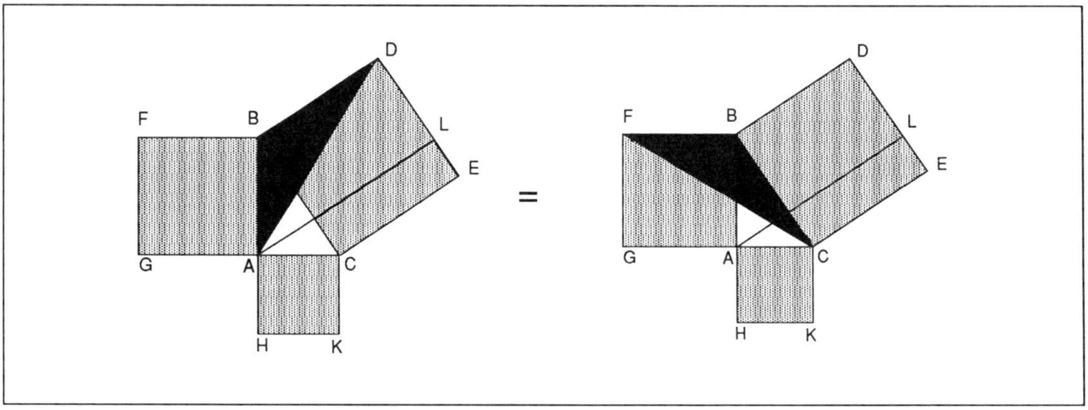


FIGURE 1.7: Triangle ABD is congruent to triangle FBC.

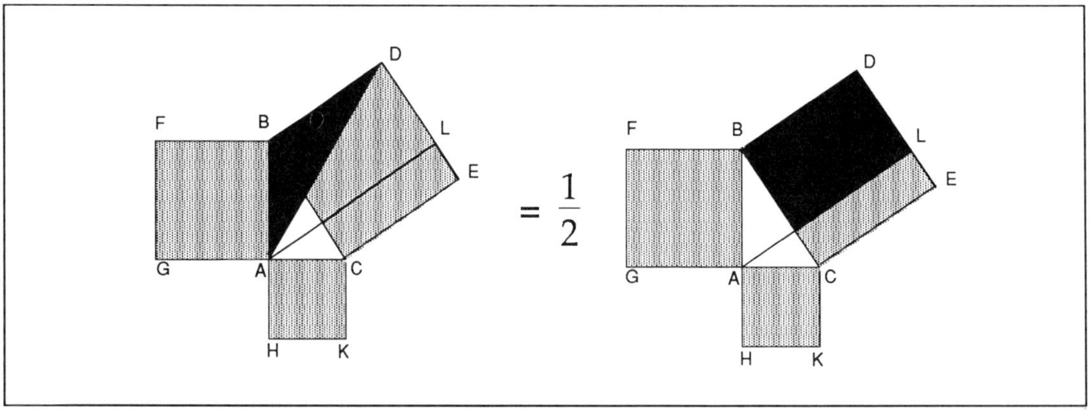


FIGURE 1.8: Triangle ABD is equal in area to ½ of rectangle BL.

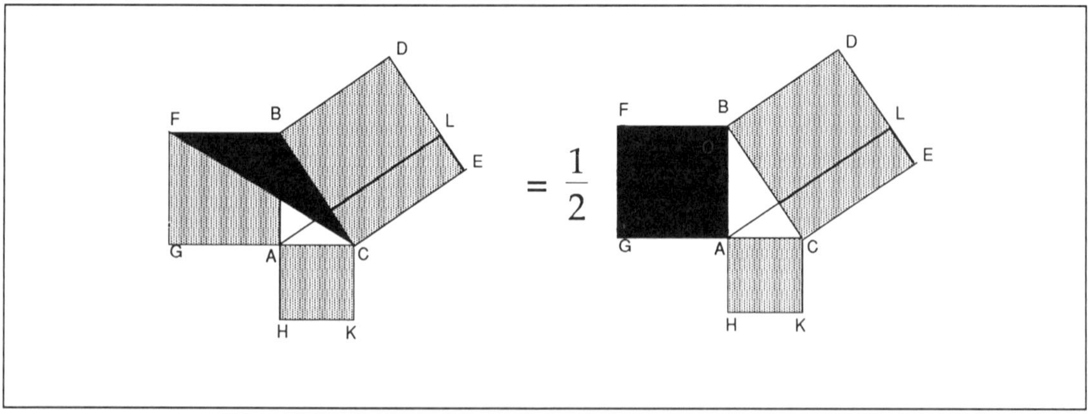


FIGURE 1.9: Triangle FBC is equal in area to ½ of square FA.

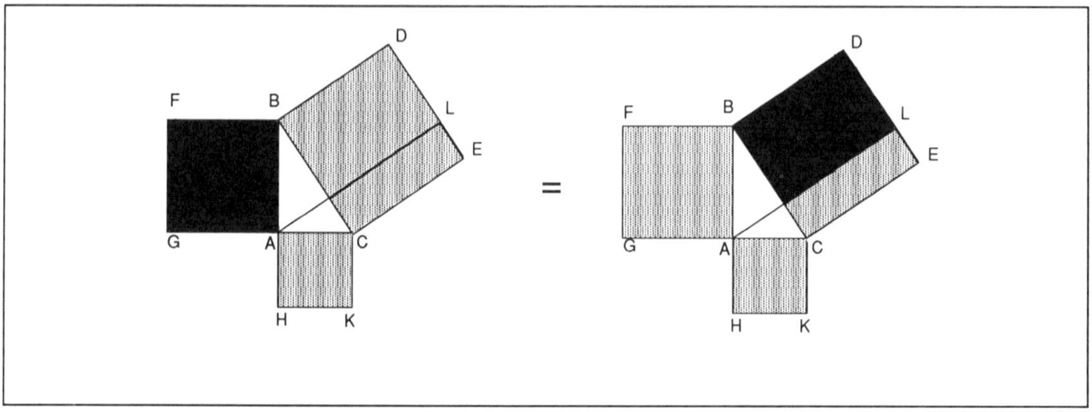


FIGURE 1.10: Square GB is equal in area to rectangle BL.

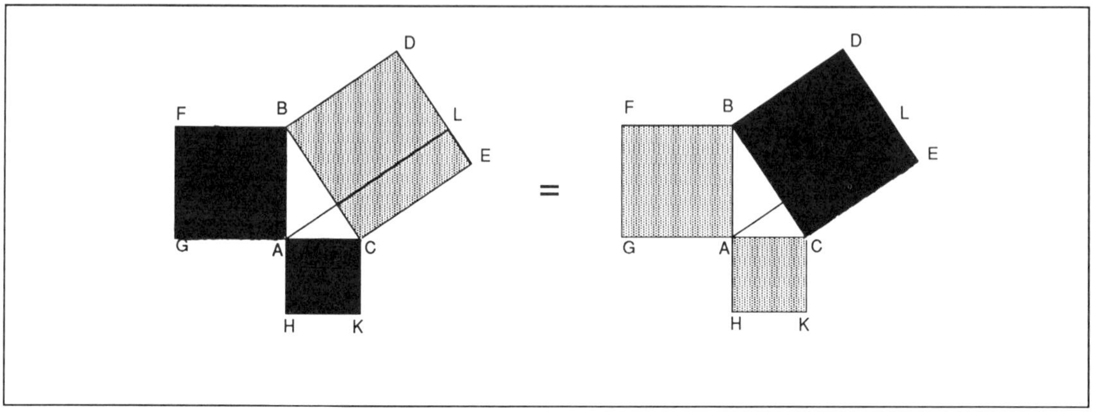


FIGURE 1.11: Squares GB and HC together are equal to square BE.

angle of the other triangle, and (2) that the area of a triangle is exactly one half that of a parallelogram with the same base and height.

His proof is very general and applies to *any* right triangle, but it is easier to see by visualizing an example. (See Figure 1.4.) First he constructs the squares on each side (Figure 1.5), then draws a line from the vertex of the right angle parallel to the sides of the square on the hypotenuse. Thus he divides the square on the hypotenuse into two rectangles. (Figures 1.6.)

The thrust of the proof is to show that each of the *other* two squares is equal in area to one of the rectangles which together make the larger square. In the most interesting and ingenious part of the proof, Euclid draws lines connecting vertices of the original right triangle to the far vertices of the squares opposite them, thus creating triangles which have bases and heights equal, respectively, to those of the rectangles which form the larger square and those of the other two squares. Euclid then shows that pairs of these (new) triangles are congruent to each other (and therefore have equal area). (Figure 1.7.)

Since squares and rectangles are also parallelograms, the area of each triangle is exactly half of the area of one of the squares or rectangles. (Figures 1.8 and 1.9) Therefore each smaller square is equal in area to one of the rectangles, (Figure 1.10) therefore the sum of the areas of the smaller squares is equal to the area of the larger square. (Figure 1.11.)

The result is magical and of immense usefulness. The proof is elegant but not at all obvious until it has been pointed out. Here is how it was expressed by Euclid, in the translation by Sir Thomas L. Heath:

### PROPOSITION I.47

*In right-angled triangles the square on the side subtending the right angle is equal to the squares on the sides containing the right angle.*

Let *ABC* be a right-angled triangle having the angle *BAC* right; I say that the square on *BC* is equal to the squares on *BA*, *AC*.

For let there be described on *BC* the square *BDEC*, and on *BA*, *AC* the squares *GB*, *HC*; through *A* let *AL* be drawn parallel to either *BD* or *CE*, and let *AD*, *FC* be joined.

Then, since each of the angles *BAC*, *BAG* is right, it follows that with a straight line *BA*, and at the point *A* on it, the two straight lines *AC*, *AG* not lying on the same side make the adjacent angles equal to two right angles; therefore *CA* is in a straight line with *AG*.

For the same reason *BA* is also in a straight line with *AH*. And, since the angle *DBC* is equal to the angle *FBA*: for each is right: let the angle *ABC* be added to each; therefore the whole angle *DBA* is equal to the whole angle *FBC*. And, since *DB* is equal to *BC*, and *FB* to *BA*, the two sides *AB*, *BD* are equal to the two sides *FB*, *BC* respectively; and the angle *ABD* is equal to the angle *FBC*; therefore the base *AD* is equal to the base *FC*, and the triangle *ABD* is equal to the triangle *FBC*.

Now the parallelogram *BL* is double of the triangle *ABD*, for they have the same base *BD* and are in the same parallels *BD*, *AL*.

And the square *GB* is double of the triangle *FBC*, for they again have the same base *FB* and are in the same parallels *FB*, *GC*.

[But the doubles of equals are equal to one another.]

Therefore the parallelogram *BL* is also equal to the square *GB*.

Similarly, if *AE*, *BK* be joined, the parallelogram *CL* can also be proved equal to the square *HC*; therefore the whole square *BDEC* is equal to the two squares *GB*, *HC*.

And the square *BDEC* is described on *BC*, and the squares *GB*, *HC* on *BA*, *AC*. Therefore the square on the side *BC* is equal to the squares on the sides *BA*, *AC*.

Therefore etc. Q.E.D.[1]

[1] All Euclid's proofs end with "Therefore etc.," meaning that one can now repeat the proposition which was stated at the beginning, this time not as a proposition, but as a proved theorem. The traditional initials at the very end, "Q.E.D.," stand for "*quod erat demonstrandum,*" "that which was to be proved."

# Copernicus: The Man


## Thomas H. Leith


In antiquity the followers of Pythagoras formed a secret cult, members of which swore not to discuss their beliefs with non-members. As much of their philosophy was concerned with abstruse numerical relationships it would have been difficult for them to be understood by outsiders anyway, and they could easily have been ridiculed by those who could not understand the basis of their thoughts. Though he lived about two thousand years after the Pythagoreans, Nicolas Copernicus would have been comfortable as a Pythagorean. Only after much pleading and cajoling by colleagues did Copernicus allow the publication of his *Revolutions of the Heavenly Spheres*, the work that launched the Copernican Revolution. Here are some of the main points of his life, as sketched by Professor Thomas H. Leith of York University:

He was born in 1473 in Torun on the Vistula River which flows into the Baltic. His father, a merchant, died when Nicolas was 10. His uncle, a Bishop, took charge of Nicolas and his sisters and brothers. At 18 Nicholas went to the University of Cracow (he was there while Columbus discovered America) but did not take the degree since his uncle called him home in order to place him as a Canon in the Frauenburg Cathedral. With the economic decline of the whole area, and especially Torun, this was to give him security. But the Canon he was to replace died a few days too soon, in an uneven month, and the custom was for the Pope to appoint on these alternate months. Copernicus missed, but 2 years later another Canon needed replacement and the uncle got him the post.

For these two years and for the next eight years Copernicus studied in Bologna and Padua at which places his interest in astronomy grew. He could stay at the University as he had no duties at the Cathedral. At 30 he took a Doctor's degree in Canon law from the University of Ferrara which was possible in those days even though he hadn't studied there. In 1506, at 33, he went to his uncle's residence castle at Heilsberg for the next 6 years, ostensibly as house physician (he had studied some medicine in Padua). While there he wrote in the *Commentary* an outline of his astronomical ideas but it was not published until the late 19th century.

In 1512 the uncle died and Copernicus took up his duties at the Frauenburg Cathedral. Here he had considerable leisure and a good income. About 1530 he finished the manuscript of the *Revolutions of the Heavenly Spheres* (early drafts began about 1515) and locked it away, making only occasional corrections later. In his 30 years there he wrote little else of import but did publish a translation of a Byzantine poet. Much of his time was involved in practicing medicine and in State affairs. The *Commentary* had become known as it was privately circulated and his views were widely known despite his failure to publish or teach. It earned him a request to help the Lateran Council reform the calendar (it occurred in 1582, but he refused, saying the motion of the Sun and Moon weren't then well enough known). We know too that the personal secretary of the pope lectured on his system and that a cardinal requested earnestly that he publish.

Copernicus hesitated to publish, however, for a variety of reasons:

1. Most of the data which he was using were known to be in error or even inconsistent. He introduced about 27 revisions of his own but even these were cruder than the observations of the Alexandrians. Consequently, the model built to fit the data was, in detail, likely to be overly complex and unsatisfactory.
2. The Reformation was bringing increasing Catholic reaction and a fear of any further departure from traditional belief made it increasingly difficult to argue that his system

was, at least in its broad principles, truer than any older system.

3. He was quite tolerant of Protestants and had friends among them. This made his own position rather difficult and his views would be, as a consequence, somewhat suspect.
4. He had an intense fear of ridicule by the uninformed and wished only the learned to debate his ideas. Indeed a play was produced near Frauenburg which did include such ridicule. There are signs that Copernicus was rather Pythagorean, that is, that he believed in keeping technical questions within the "cult" able to analyze them on their merits.
5. He almost surely believed that his views were true in principle even if dubious in detail. Consequently, while some found the views useful in improving the calendar and found the model an improved hypothesis over older ones, the difficulty with calling it true was another matter. If it were true, what then of the literal interpretation of certain Biblical texts held traditionally by most theologians? The great Lutheran scholar Melanchthon pointed this out in a book published six years after Copernicus died. Luther is also quoted as having criticized Copernicus. Catholics likewise recognized this difficulty and the *Revolutions* was placed on the Index of Forbidden Books 73 years after the book was published.

Then a young admirer, a Lutheran mathematician by the name of Rheticus, saw the manuscript of the Revolution and he and Copernicus' bishop in 1537 together tried to talk him into printing it. They failed, but Rheticus was allowed to publish the theme of the manuscript without using Copernicus' name. It repeatedly takes care to point out that Copernicus (called 'my master') goes against the ancients only when facts force him to: likely less to reassure Luther and Melanchton than to preserve Copernicus' peace of mind. The book was a masterpiece of abstracting masses of tables and complex charts and took only 10 weeks. It attracted some attention and Copernicus was urged from all sides to publish the original manuscript. Rheticus left for an important post in Leipzig in 1542, but by then he had convinced Copernicus that if he didn't publish he'd be ridiculed far worse now than if he did. Rheticus had copied the entire manuscript correcting some errors and obtained permission to publish it in Lutheran Nuremberg—so that the added weight of the Lutheran Duke of Prussia there could add his support to it in the face of Luther's and Melanchton's opposition. The book was still in press when Rheticus left for Leipzig, and Osiander, a leading Lutheran theologian, was asked to write the anonymous preface to placate the theologians and the Aristotelians whose ridicule Copernicus feared. It claimed the ideas were hypotheses (as in Ptolemy they were merely a calculating device); certainly the Copernican scheme was as good as or better than the ancient ones, but astronomy couldn't give certainty. Kepler (66 years later) reported its true author but many thought that Copernicus had written it.

Copernicus died in his tower, where he had pretty well hidden for so long, after a cerebral haemorrhage five months earlier, on May 24, 1543.

---

Thomas H. Leith, *The Nature and Growth of Science*, 64-66 (Toronto: Wall & Thompson, 1989).

---

## Questions and Topics for Discussion and Writing

1. Was Copernicus justified in withholding publication for so long?
2. Comment on the breadth of Copernicus's education.

*Thomas H. Leith*

# The Starry Messenger

## Galileo Galilei

In the following excerpt from *The Starry Messenger*, Galileo announces the discoveries he made about the surface of the moon with the aid of the newly invented telescope. Elsewhere in *The Starry Messenger,* Galileo reported that the mysterious "Milky Way" was indeed a vast conglomeration of stars which could not be distinguished with the naked eye, and he reported his discovery of four satellites of Jupiter. These and other astronomical discoveries buttressed Galileo's arguments that the heavens were not unlike the Earth and that the Earth could quite logically move around the sun.

Great indeed are the things which in this brief treatise I propose for observation and consideration by all students of nature. I say great, because of the excellence of the subject itself, the entirely unexpected and novel character of these things, and finally because of the instrument by means of which they have been revealed to our senses.

Surely it is a great thing to increase the numerous host of fixed stars previously visible to the unaided vision, adding countless more which have never before been seen, exposing these plainly to the eye in numbers ten times exceeding the old and familiar stars.

It is a very beautiful thing, and most gratifying to the sight, to behold the body of the moon, distant from us almost sixty earthly radii, as if it were no farther away than two such measures—so that its diameter appears almost thirty times larger, its surface nearly nine hundred times, and its volume twenty-seven thousand times as large as when viewed with the naked eye. In this way one may learn with all the certainty of sense evidence that the moon is not robed in a smooth and polished surface but is in fact rough and uneven, covered everywhere, just like the earth's surface, with huge prominences, deep valleys, and chasms.

Again, it seems to me a matter of no small importance to have ended the dispute about the Milky Way by making its nature manifest to the very senses as well as to the intellect. Similarly it will be a pleasant and elegant thing to demonstrate that the nature of those stars which astronomers have previously called "nebulous" is far different from what has been believed hitherto. But what surpasses all wonders by far, and what particularly moves us to seek the attention of all astronomers and philosophers, is the discovery of four wandering stars not known or observed by any man before us. Like Venus and Mercury, which have their own periods about the sun, these have theirs about a certain star that is conspicuous among those already known, which they sometimes precede and sometimes follow, without ever departing from it beyond certain limits. All these facts were discovered and observed by me not many days ago with the aid of a spyglass which I devised, after first being illuminated by divine grace. Perhaps other things, still more remarkable, will in time be discovered by me or by other observers with the aid of such an instrument, the form and construction of which I shall first briefly explain, as well as the occasion of its having been devised. Afterwards I shall relate the story of the observations I have made.

About ten months ago a report reached my ears that a certain Fleming had constructed a spyglass by means of which visible objects, though very distant from the eye of the observer, were distinctly seen as if nearby. Of this truly remarkable effect several experiences were related, to which some persons gave credence while others denied them. A few days later the report was confirmed to me in a letter from a noble Frenchman at Paris, Jacques Badovere, which caused me to apply myself wholeheartedly to inquire into the means by which I might arrive at the invention of a similar instrument. This

I did shortly afterwards, my basis being the theory of refraction. First I prepared a tube of lead, at the ends of which I fitted two glass lenses, both plane on one side while on the other side one was spherically convex and the other concave. Then placing my eye near the concave lens I perceived objects satisfactorily large and near, for they appeared three times closer and nine times larger than when seen with the naked eye alone. Next I constructed another one, more accurate, which represented objects as enlarged more than sixty times. finally, sparing neither labor nor expense, I succeeded in constructing for myself so excellent an instrument that objects seen by means of it appeared nearly one thousand times larger and over thirty times closer than when regarded with our natural vision.

It would be superfluous to enumerate the number and importance of the advantages of such an instrument at sea as well as on land. But forsaking terrestrial observations, I turned to celestial ones, and first I saw the moon from as near at hand as if it were scarcely two terrestrial radii away. After that I observed often with wondering delight both the planets and the fixed stars, and since I saw these latter to be very crowded, I began to seek (and eventually found) a method by which I might measure their distances apart....

Now let us review the observations made during the past two months, once more inviting the attention of all who are eager for true philosophy to the first steps of such important contemplations. Let us speak first of that surface of the moon which faces us. For greater clarity I distinguish two parts of this surface, a lighter and a darker; the lighter part seems to surround and to pervade the whole hemisphere, while the darker part discolors the moon's surface like a kind of cloud, and makes it appear covered with spots. Now those spots which are fairly dark and rather large are plain to everyone and have been seen throughout the ages; these I shall call the "large" or "ancient" spots, distinguishing them from others that are smaller in size but so numerous as to occur all over the lunar surface, and especially the lighter part. The latter spots had never been seen by anyone before me. From observations of these spots repeated many times I have been led to the opinion and conviction that the surface of the moon is not smooth, uniform, and precisely spherical as a great number of philosophers believe it (and the other heavenly bodies) to be, but is uneven, rough, and full of cavities and prominences, being not unlike the face of the earth, relieved by chains of mountains and deep valleys. The things I have seen by which I was enabled to draw this conclusion are as follows.

On the fourth or fifth day after new moon, when the moon is seen with brilliant horns, the boundary which divides the dark part from the light does not extend uniformly in an oval line as would happen on a perfectly spherical solid, but traces out an uneven, rough, and very wavy line as shown in the figure below. Indeed, many luminous excrescences extend beyond the boundary into the darker portion, while on the other hand some dark patches invade the illuminated part. Moreover a great quantity of small blackish spots, entirely separated from the dark region, are scattered almost all over the area illuminated by the sun with the exception only of that part which is occupied by the large and ancient spots. Let us note, however, that the said small spots always agree in having their blackened parts directed toward the sun, while on the side opposite the sun they are crowned with bright contours, like shining summits. there is a similar sight on earth about sunrise, when we behold the valleys not yet flooded with light though the mountains surrounding them are already ablaze with glowing splendor on the side opposite the sun. And just as the shadows in the hollows on earth diminish in size as the sun rises higher, so these spots on the moon lose their blackness as the illuminated region grows larger and larger.

Again, not only are the boundaries of shadow and light in the moon seen to be uneven and wavy, but still more astonishingly many bright points appear within the darkened portion of the moon, completely divided and separated from the illuminated part and at a considerable distance from it. After a time these gradually increase in size and brightness, and an hour or two later they become joined with the rest of the lighted part which has now increased in size. Meanwhile more and more peaks shoot up as if sprouting now here, now there, lighting up within the shadowed portion; these become larger, and finally they too are united with that same luminous surface which extends ever further. An illustration of this is to be seen in the figure above. And on the earth, before the rising of the sun, are not the

*Galileo Galilei*

FIGURE 1.12: The moon, on the fourth or fifth day after new moon.

highest peaks of the mountains illuminated by the sun's rays while the plains remain in shadow? Does not the light go on spreading while the larger central parts of those mountains are becoming illuminated? And when the sun has finally risen, does not the illumination of plains and hills finally become one? But on the moon the variety of elevations and depressions appears to surpass in every way the roughness of the terrestrial surface, as we shall demonstrate further on.

At present I cannot pass over in silence something worthy of consideration which I observed when the moon was approaching first quarter, as shown in the previous figure. Into the luminous part there extended a great dark gulf in the neighborhood of the lower cusp. When I had observed it for a long time and had seen it completely dark, a bright peak began to emerge, a little below its center, after about two hours. Gradually growing, this presented itself in a triangular shape, remaining completely detached and separated from the lighted surface. Around it three other small points soon began to shine, and finally, when the moon was about to set, this triangular shape (which had meanwhile become more widely extended) joined with the rest of the illuminated region and suddenly burst into the gulf of shadow like a vast promontory of light, surrounded still by the three bright peaks already mentioned. Beyond the ends of the cusps, both above and below, certain bright points emerged which were quite detached from the remaining lighted part, as may be seen depicted in the same figure. There were also a great number of dark spots in both the horns, especially in the lower one; those nearest the boundary of light and shadow appeared larger and darker, while those more distant from the boundary were not so dark and distinct. But in all cases, as we have mentioned earlier, the blackish portion of each spot is turned toward the source of the sun's radiance, while a bright rim surrounds the spot on the side away from the sun in the direction of the shadowy region of the moon. This part of the moon's surface, where it is spotted as the tail of a peacock is sprinkled with azure eyes, resembles those glass vases which have been plunged while still hot into cold water and have thus acquired a crackled and wavy surface from which they receive their common name of "ice-cups."

As to the large lunar spots, these are not seen to be broken in the above manner and full of cavities and prominences; rather, they are even and uniform, and brighter patches crop up only here and there. Hence if anyone wished to revive the old Pythagorean opinion that the moon is like another earth, its brighter part might very fitly represent the surface of the land and its darker region that of the water. I have never doubted that if our globe were seen from afar when flooded with sunlight, the land regions would appear brighter and the watery regions darker. The large spots in the moon are also seen to be less elevated than the brighter tracts, for whether the moon is waxing or waning there are always seen, here and there along its boundary of light and shadow, certain ridges of brighter hue around the large spots (and we have attended to this in preparing the diagrams); the edges of these spots are not only lower, but also more uniform, being uninterrupted by peaks or ruggedness.

Near the large spots the brighter part stands out particularly in such a way that before first quarter and toward last quarter, in the vicinity of a certain spot in the upper (or northern) region of the moon, some vast prominences arise both above and below as shown in the figures reproduced below. Before last quarter this same spot is seen to be walled about with certain blacker contours which, like the loftiest mountaintops, appear darker on the side away from the sun and brighter on that which faces the sun. (This is the opposite of what happens in the cavities, for there the part away from the sun appears brilliant, while that which is turned toward the sun is

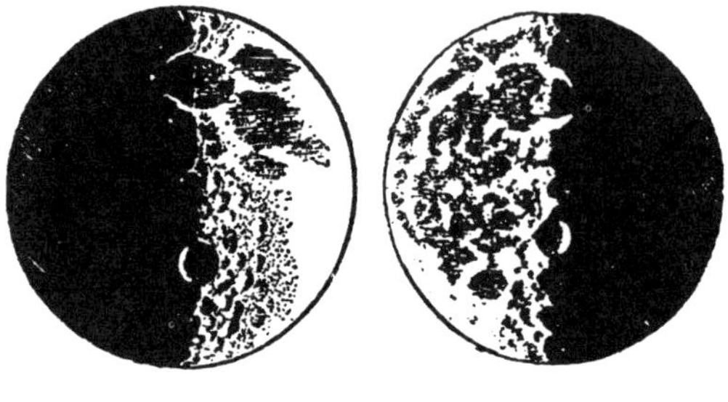

FIGURE 1.13: The moon, near its 1st and 4th quarters, showing bright ridges around large darker spots.

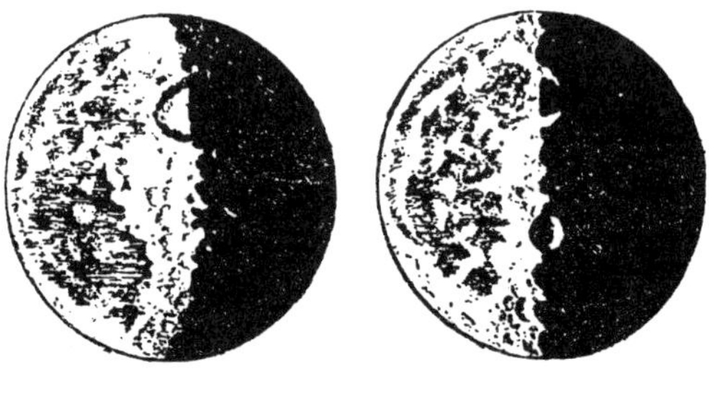

FIGURE 1.14: The moon, just before the last quarter, showing in the northern region a large spot "walled in" with black contours (left picture) and, after a time, the "wall" in the dark region emerges as bright ridges (right picture). Also shown is the large cavity in the southern region with the appearance of "Bohemia" viewed from the sky.

dark and in shadow.) After a time, when the lighted portion of the moon's surface has diminished in size and when all (or nearly all) the said spot is covered with shadow, the brighter ridges of the mountains gradually emerge form the shade. This double aspect of the spot is illustrated in the ensuing figures.

There is another thing which I must not omit, for I beheld it not without a certain wonder; this is that almost in the center of the moon there is a cavity larger than all the rest, and perfectly round in shape. I have observed it near both first and last quarters, and have tried to represent it as correctly as possible in the second of the above figures. As to light and shade, it offers the same appearance as would a region like Bohemia[1] if that were enclosed on all sides by very lofty mountains arranged exactly in a circle. Indeed, this area on the moon is surrounded by such enormous peaks that the bounding edge adjacent to the dark portion of the moon is seen to be bathed in sunlight before the boundary of light and shadow reaches halfway across the same space. As in other spots, its shaded portion faces the sun while its lighted part is toward the dark side of the moon; and for a third time I draw attention to this as a very cogent proof of the ruggedness and unevenness that pervades all the bright region of the moon. Of these spots, moreover, those are always darkest which touch the boundary line between light and shadow, while those farther off appear both smaller and less dark, so that when the moon ultimately becomes full (at opposition[2] to the sun), the shade of the cavities is distinguished from the light of the places in relief by a subdued and very tenuous separation.[3]

The things we have reviewed are to be seen in the brighter region of the moon. In the large spots, no such contrast of depressions and pominences is

[1] This casual comparison between a part of the moon and a specific region on earth was later the basis of much trouble for Galileo.... Even in antiquity the idea that the moon (or any other heavenly body) was of the same nature as the earth had been dangerous to hold. The Athenians banished the philosopher Anaxagoras for teaching such notions, and charged Socrates with blasphemy for repeating them.

[2] Opposition of the sun and moon occurs when they are in line with the earth between them (full moon, or lunar eclipse); conjunction, when they are in line on the same side of the earth (new moon, or eclipse of the sun).

[3] Galileo was on to more than he realized. The "cavity" he described was a giant crater formed by the impact of a huge meteorite. His comparison of its appearance to Bohemia (which he had never seen from the ground, let alone from the air) was to establish that the moon was rocky and uneven, like parts of the earth. Bohemia was ringed by mountains and would, Galileo surmised, look much the same through a telescope from a comparable distance as the "cavity" on the moon did. Indeed it does, as we now know. But more than that, some scientists are now of the opinion that the "Bohemian Plateau" in western Czechoslovakia is itself a crater formed by the impact of a huge meteorite. See Robert Kunzig, "Back to Bohemia," *Discover* 10 (June 1989): 22–23 *[Editor's note.]*

perceived as that which we are compelled to recognize in the brighter parts by the changes of aspect that occur under varying illumination by the sun's rays throughout the multiplicity of positions from which the latter reach the moon. In the large spots there exist some holes rather darker than the rest, as we have shown in the illustrations. Yet these present always the same appearance, and their darkness is neither intensified nor diminished, although with some minute difference they appear sometimes a little more shaded and sometimes a little lighter according as the rays of the sun fall on them more or less obliquely. Moreover, they join with the neighboring regions of the spots in a gentle linkage, the boundaries mixing and mingling. It is quite different with the spots which occupy the brighter surface of the moon; these, like precipitous crags having rough and jagged peaks, stand out starkly in sharp contrasts of light and shade. And inside the large spots there are observed certain other zones that are brighter, some of them very bright indeed. Still, both these and the darker parts present always the same appearance; there is no change either of shape or of light and shadow; hence one may affirm beyond any doubt that they owe their appearance to some real dissimilarity of parts. They cannot be attributed merely to irregularity of shapes, wherein shadows move in consequence of varied illuminations from the sun, as indeed is the case with the other, smaller, spots which occupy the brighter part of the moon and which change, grow, shrink, or disappear from one day to the next, as owing their origin only to shadows of prominences.

From *Discoveries and Opinions of Galileo*, trans. and ed. by Stillman Drake, 27–38 (New York: Doubleday Anchor, 1957).

## Questions and Topics for Discussion and Writing

1. Why was it so important for Galileo to demonstrate that the moon was not a perfect sphere?
2. Referring to Galileo's drawings of the appearance of the moon at different times, explain how these drawings indicate aspects of the surface of the moon. What, in particular, does Galileo show by comparing views of the moon in its first and last quarters?
3. Galileo repeatedly compared his observations of shadows and reflections on the moon to similar observations on earth. Why was this dangerous and provocative for him to do?

# * The Sentence of Galileo by the Inquisition, 1633

## The Inquisition

In 1632, Galileo published his *Dialogue Concerning the Two Chief World Systems* in which his three interlocutors discuss the merits of the Copernican system versus the Aristotelian-Ptolemaic system and in which clearly the Aristotelian-Ptolemaic earth-centered system gets much the worst of the argument. Within five months the book was banned. Galileo, in poor health, was called before the Inquisition, tried and convicted of suspected heresy. The possible punishments were severe including torture and burning at the stake. Under the circumstances Galileo's sentence was gentle, ordering him to recant and placing him under house arrest for the rest of his life. This is the decree of the Inquisition.

Whereas you, Galileo, son of the late Vincenzo Galilei, Florentine, aged seventy years, were in the year 1615 denounced to this Holy Office for holding as true the false doctrine taught by some that the Sun is the centre of the world and immovable and that the Earth moves, and also with a diurnal motion; for having disciples to whom you taught the same doctrine; for holding correspondence with certain mathematicians of Germany concerning the same: for having printed certain letters, entitled "On the Sunspots," wherein you developed the same doctrine as true; and for replying to the objections from the Holy Scriptures, which from time to time were urged against it, by glossing the said Scriptures according to your own meaning: and whereas there was thereupon produced the copy of a document in the form of a letter, purporting to be written by you to one formerly your disciple, and in this divers propositions are set forth, following the position of Copernicus, which are contrary to the true sense and authority of Holy Scripture:

This Holy Tribunal being therefore of intention to proceed against the disorder and mischief thence resulting, which went on increasing to the prejudice of the Holy Faith, by command of His Holiness and of the Most Eminent Lords Cardinals of this supreme and universal Inquisition, the two propositions of the stability of the Sun and the motion of the Earth were by the theological Qualifiers qualified as follows:

The proposition that the Sun is the centre of the world and does not move from its place is absurd and false philosophically and formally heretical, because it is expressly contrary to the Holy Scripture.

The proposition that the Earth is not the centre of the world and immovable but that it moves, and also with a diurnal motion, is equally absurd and false philosophically and theologically considered at least erroneous in faith.

But whereas it was desired at that time to deal leniently with you, it was decreed at the Holy Congregation held before his Holiness on the twenty-fifth of February, 1616, that his Eminence the Lord Cardinal Bellarmine should order you to abandon altogether the said false doctrine and, in the event of your refusal, that an injunction should be imposed upon you by the Commissary of the Holy Office to give up the said doctrine and not to teach it to others, not to defend it, nor even discuss it; and failing acquiescence in this injunction, that you should be imprisoned. And in execution of this decree, on the following day, at the Palace, and in the presence of his Eminence, the said Lord Cardinal Bellarmine, after being gently admonished by the said Lord Cardinal, the command was enjoined upon you by the Father Commissary of the Holy Office of that time, before a notary and witnesses, that you were altogether to abandon the said false opinion and not in future to hold or defend or teach it in any way whatsoever, neither verbally nor in writing; and upon your promising to obey, you were dismissed.

And, in order that a doctrine so pernicious might be wholly rooted out and not insinuate itself further to the grave prejudice of Catholic truth, a decree was issued by the Holy Congregation of the Index prohibiting the books which treat of this doctrine and declaring the doctrine itself to be false and wholly contrary to the sacred and divine Scripture.

And whereas a book appeared here recently, printed last year at Florence, the title of which shows that you were the author, this title being: "Dialogue of Galileo Galilei on the Great World Systems"; and whereas the Holy Congregation was afterward informed that through the publication of the said book the false opinion of the motion of the Earth and the stability of the Sun was daily gaining ground, the said book was taken into careful consideration, and in it there was discovered a patent violation of the aforesaid injunction that had been imposed upon you, for in this book you have defended the said opinion previously condemned and to your face declared to be so, although in the said book you strive by various devices to produce the impression that you leave it undecided, and in express terms as probable: which, however, is a most grievous error, as an opinion can in no wise be probable which has been declared and defined to be contrary to divine Scripture.

Therefore by our order you were cited before this Holy Office, where, being examined upon your oath, you acknowledged the book to be written and published by you. You confessed that you began to write the said book about ten or twelve years ago, after the command had been imposed upon you as above; that you requested license to print it without, however, intimating to those who granted you this license that you had been commanded not to hold, defend, or teach the doctrine in question in any way whatever.

You likewise confessed that the writing of the said book is in many places drawn up in such a form that the reader might fancy that the arguments brought forward on the false side are calculated by their cogency to compel conviction rather than to be easy of refutation, excusing yourself for having fallen into an error, as you alleged, so foreign to your intention, by the fact that you had written in dialogue and by the natural complacency that every man feels in regard to his own subtleties and in showing himself more clever than the generality of men in devising, even on behalf of false propositions, ingenious and plausible arguments.

And, a suitable term having been assigned to you to prepare your defense, you produced a certificate in the handwriting of his Eminence the Lord Cardinal Bellarmine, procured by you, as you asserted, in order to defend yourself against the calumnies of your enemies, who charged that you had abjured and had been punished by the Holy Office, in which certificate it is declared that you had not abjured and had not been punished but only that the declaration made by His Holiness and published by the Holy Congregation of the Index had been announced to you, wherein it is declared that the doctrine of the motion of the Earth and the stability of the Sun is contrary to the Holy Scriptures and therefore cannot be defended or held. And, as in this certificate there is no mention of the two articles of the injunction, namely, the order not "to teach" and "in any way," you represented that we ought to believe that in the course of fourteen or sixteen years you had lost all memory of them and that this was why you said nothing of the injunction when you requested permission to print your book. And all this you urged not by way of excuse for your error but that it might be set down to a vainglorious ambition rather than to malice. But this certificate produced by you in your defense has only aggravated your delinquency, since, although it is there stated that said opinion is contrary to Holy Scripture, you have nevertheless dared to discuss and defend it and to argue its probability; nor does the license artfully and cunningly extorted by you avail you anything, since you did not notify the command imposed upon you.

And whereas it appeared to us that you had not stated the full truth with regard to your intention, we thought it necessary to subject you to a rigorous examination at which (without prejudice, however, to the matters confessed by you and set forth as above with regard to your said intention) you answered like a good Catholic. Therefore, having seen and maturely considered the merits of this your cause, together with your confessions and excuses above-mentioned, and all that ought justly to be seen and considered, we have arrived at the under-written final sentence against you:

Invoking, therefore, the most holy name of our Lord Jesus Christ and of His most glorious Mother, ever Virgin Mary, by this our final sentence, which sitting in judgment, with the counsel and advice of the Reverend Masters of sacred theology and Doctors of both Laws, our assessors, we deliver in these writings, in the cause and causes at present before us between the Magnificent Carlo Sinceri, Doctor of both Laws, Proctor Fiscal of this Holy Office, of the one part, and you Galileo Galilei, the defendant, here present, examined, tried, and confessed as shown above, of the other part—

We say, pronounce, sentence, and declare that you, the said Galileo, by reason of the matters adduced in trial, and by you confessed as above, have rendered yourself in the judgment of this Holy Office vehemently suspected of heresy, namely, of having believed and held the doctrine — which is false and contrary to the sacred and divine Scriptures — that the Sun is the centre of the world and does not move from east to west and that the Earth moves and is not the centre of the world; and that an opinion may be held and defended as probable after it has been declared and defined to be contrary to the Holy Scripture; and that consequently you have incurred all the censures and penalties imposed and promulgated in the sacred canons and other constitutions, general and particular, against such delinquents. From which we are content that you be absolved, provided that, first, with a sincere heart and unfeigned faith, you abjure, curse, and detest before us the aforesaid errors and heresies and every other error and heresy contrary to the Catholic and Apostolic Roman Church in the form to be prescribed by us for you.

And, in order that this your grave and pernicious error and transgression may not remain altogether unpunished and that you may be more cautious in the future and an example to others that they may abstain from similar delinquencies, we ordain that the book of the "Dialogue of Galileo Galilei" be prohibited by public edict.

We condemn you to the formal prison of this Holy Office during our pleasure, and by way of salutary penance we enjoin that for three years to come you repeat once a week the seven penitential Psalms. Reserving to ourselves liberty to moderate, commute, or take off, in whole or in part, the aforesaid penalties and penance.

And so we say, pronounce, sentence, declare, ordain, and reserve in this and in any other better way and form which we can and may rightfully employ.

Quoted in Giorgio de Santillana, *The Crime of Galileo, 306-310 (Chicago: The University of Chicago Press, 1955).*

## Questions and Topics for Discussion and Writing

1. Why was the Inquisition so upset about Galileo's views? What difference did it make to them what people thought about the motions of the planets and the earth?
2. Was this sentence the inevitable result of Galileo's activities? Could there have been a different resolution of Galileo's disagreement with the Church? If so, what could that have been and why? If not, why not?

# Newton: The Man


## Thomas H. Leith



It was Isaac Newton who brought together the pieces of the new scientific theories and discoveries about the physical world made by Copernicus, Kepler, Galileo, Descartes, and others. His great synthesis, which we still call Newtonian physics, completely replaced Aristotle's physics as fundamental scientific laws. The key points: the laws of inertia, acceleration, action and reaction, and the principle of universal gravitation are the foundation of physical science. Also Newton invented the calculus and was the first to analyze the composition of white light. These are the main achievements for which he is remembered today. But this is but a tiny fraction of Newton's work, the vast bulk of which was not about science at all and most of which was never published.

Amazing as Newton's scientific achievements were, his personal life is equally amazing. He certainly ranks as one of the great eccentrics in history. Here, Professor Leith sketches some of the main events in Newton's life.


Nature, and Nature's laws lay hid in night:
God said, *Let Newton be!* and all was light.
*Alexander Pope,*

Newton was born at Woolsthorpe within a year of Galileo's death—1642—in Lincolnshire, England. His father was a farmer, who died before Newton's birth. At 12 he moved away to school in Grantham where he lived with a druggist who taught him skills in the manual arts. Thus he built windmills to scale, water clocks, sundials, kites, and lanterns. These were largely copied from books which we still have, just as we have his notes from them. He did not show his genius early (nothing like Pascal who at 16 had made major discoveries in geometry; or Hamilton who at 14 spoke 8 languages, at 16 wrote a famous paper in mathematics, around 20 was Royal Astronomer of Ireland and was the leading mathematician of his age; or Maxwell whose first paper in mathematics was published at 15), though we know he stood first in his tiny school. At 18 he went to Trinity College at Cambridge where he worked his way doing odd jobs. We know little of him until his third year when he met the holder of a mathematics chair at the University and who turned his interests further to mathematics and to optics.

In 1665 he took his B.A. without any great distinction. The University closed because of the plague and he returned home and spent much of his 23rd and 24th years there. He tells us little of this period but indications are that the period was of great import in his life. At home he seems to have had much leisure time and spent it in intense concentrated thought for hours on end. Here he developed his law of gravitational attraction and likely his famous 3 laws of motion, not published until over 20 years later. Here too he made major discoveries in the differential and integral calculus and he reports that he solved areas between curves and lines to many decimals *in his head.* (It was 4 years later in 1669 at 26 years of age that he sent his first result to his mathematics professor friend, but they weren't published for 30 more years. This reticence to publish led to disputes whether he or Leibniz had discovered the calculus first and as to possible plagiarism). Here, too, he bought his prism to study light and colours.

In 1667 he returned to Cambridge as a minor fellow, becoming a major fellow the next year. Two other major fellows were injured falling drunk downstairs and another went insane. When his mathematics friend resigned the next year, Newton became the Lucasian Professor of Mathematics (the position held recently by the great Dirac and now held by Stephen W. Hawking). One of his friends at this time said of him that he was "a man of quite exceptional ability and singular skill." At this time he built a number of lenses and discovered chromatic aberration. To get around it he built a reflecting telescope (an early form of those in many observatories today) which aroused great interest in Cambridge and at the Royal Society in London. Indeed, he gave it to the latter and they still have it. Feeling that was too easy a task to fit the honour, he wrote his first published paper in 1672 (at 30) on light and prisms. Robert Hooke reviewed it and suggested some criticisms (it opposed certain ingenious ideas of his own) which Newton listened to politely. Later these two great experimenters and theoreticians were prodded into a nasty dispute by an enemy of Hooke's. Huygens criticized Newton also, partly from lack of understanding of Newton's point that he wanted to measure the behaviour of colors and not argue about their *nature*. Several others were so obnoxious that Newton swore in 1676 to thenceforth leave science only for his private satisfaction. Fortunately he relented.

Before this, in 1674, Hooke wrote him about planetary motion and renewed his interest in it. In 1684 Christopher Wren asked Hooke to prove a certain theory Hooke had proposed that the forces pulling the planets out of straight paths into curves around the sun obeyed an inverse square law. While Hooke was correct in his guess he couldn't prove it. Later Wren and Halley (of comet fame) reversed the question and asked Newton what kind of curve such a law would give. He replied at once—an ellipse. They asked in amazement how he knew. He said he had calculated it but lost his papers. He did the work again, however and in two different ways! Halley saw a major discovery in it and asked Newton for a thorough treatise on his work.

As a result, in 1686 Newton sent the first book, of what was to be three to the Royal Society. The minutes of the meeting record its presentation as *Philosophiae Naturales Principia Mathematica*—among other things a demonstration of the Copernican hypothesis and of Kepler's laws for elliptical orbits, using the inverse square law of attraction toward the sun. Within 18 months all three books were finished and publishing was begun. Just earlier, Hooke had noticed that his suggestions of 6 years previous was used by Newton without acknowledgement. Of course he didn't know Newton had had the ideas as long as 15–20 years before, though it is true that his letter in 1674 had renewed Newton's interest. Newton took offense when he was told of Hooke's chagrin, painted worse than it was, likely because of his prior experience with Hooke and threatened not to finish the third book just as he had threatened to leave science earlier. Eventually Newton saw Hooke had been misrepresented and was asking only for justice from the Society, and Newton mentioned Hooke in the *Principia* when it was published.

We are told that at this time Newton was very absent minded—indeed accounts suggest he always was. He'd forget to eat or he'd go out for wine for a party and forget to return. He'd dash up to his rooms to write an idea while fresh in his mind—standing beside a chair without noticing it. The *Principia* came out in 1687, financed by Halley (the Royal Society was too poor), and it sold for six shillings. It is much like Euclid's "Elements"—the laws are derived from geometry. He could have done it far more easily using the calculus but he didn't. Just why isn't clear. He does tell us that it was made abstruse purposely so that only mathematicians could read it and he wouldn't have to deal with questions from lesser mortals. He suggests these lesser mortals read Book Three which is a more popular presentation.

Scientists immediately saw the great value of the work and we have extant glowing reviews of the work calling Newton the greatest living mathematician. However, it wasn't used at once. Six years later Cambridge was still using the old Cartesian arguments. Part of the reason lay in its difficulty and one famous mathematician reported that he had to tear it up and carry a few pages at a time so he could digest it.

In 1689 he went to Parliament as the representative of Cambridge University. In London he met Pepys, John Locke, and many other great figures of the time. Later, Newton sought one administrative post after another. Failing, he neared a nervous breakdown and accused his friends of various evils toward him. His interest in science at this time also lagged or resulted in quarrels. However, in 1696 he was made Warden of the Mint and 3 years later the Master (or Head) with a good salary and considerable leisure time. (There are stories of a love affair but all evidence points to it being untrue. Indeed, we have no record of Newton's ever having been in love.) He turned out to be a good administrator but did little science, though we are told that Jean Bernoulli (the great Swiss mathematician) set a problem about this time as a challenge to all European mathematicians and Newton did it in one evening.

In 1703 he became president of the Royal Society and so remained till his death in 1727. Usually the president was not a scientist and this was a great honour. In 1704 he published his *Opticks*—his great work on light—based largely upon studies long before but delayed to avoid controversy. It is coincidence that Hooke died just before it appeared. The second edition, however, has various new sections of great thought and worth—evidence he could still contribute to science when he desired. One is on polarized light and another on electrical attraction—both major ideas.

In 1705 Queen Anne knighted him *and went to Cambridge to do it*. He was now 62. Most of his time was spent in the Mint or dealing with the problems of the Royal Society (mostly quarrels). The weekly meetings today are on Thursday—because they were changed from Wednesday for Newton's convenience at the Mint. Part of his time he spent fighting on his own account, especially with an astronomer called Flamsteed. He treated him rather highhandedly to say the least and in later editions left out portions of the first edition of the *Principia* where he was mentioned. Newton had a proud side! Leibniz too had caused trouble before Newton left for London and after the publication of the *Opticks*. Most of the inciting was on Newton's part however: he accused Leibniz of plagiarism and sponsored a book against him at the Royal Society. The trouble died only with Leibniz's death in 1716.

The second edition of the *Principia* appeared in 1713 with revisions by a brilliant young friend of Newton's. The first edition was already scarce. The new preface points out that many still didn't believe it, being unconvinced by experimental and observational data. A third edition appeared in 1726 and now it was written in other than Latin for the first time. In 1711 Newton published his calculus treatise written many years before and just before this a work on algebra. Another book on his optical lectures came out after his death (he never actually did the work of preparing any of his books for publication anyway). Most of his interest at this time lay in theology and history, though he had corresponded with Locke as early as 1691 on the prophecies in Daniel. His notes contain a million and a half words on religion and theology (some of it has been only recently published) and another half million words on Alchemy. Two books on this part of his work were published, one before his death. They deal with Biblical, Greek, Egyptian, and Assyrian chronologies and with the Biblical Books of Daniel and Revelation. The latter was a concerted attempt at a correlation of prophecy with history, evidencing deep learning in church history. He said of it "it is the noblest use to which the human intellect can be applied."

In 1727 he went to London to preside at the Royal Society. The trip was too much for him and he died a month later in March—aged 85. His body lay in state and was buried in Westminster Abbey. The casket was carried by 2 dukes and 3 earls, and the Lord High Chancellor. Present at his funeral also was Voltaire, who did much to popularize his views on the Continent.

In closing let us briefly review his character.

1. He could almost intuit the solution to a mathematical problem, often unproven for many years.
2. He anticipated much later theory in varied areas.
3. His scientific contributions came in rather brief periods in a long life. One wonders what might have resulted had he sustained his attack. Indeed 75% of his time he spent with the Mint, religious matters, and alchemy.
4. He was deeply religious but was not capable of full agreement with the 39 articles of the Church of England. His works on prophecy were meant to show the stupidity of trying to foreknow but the great use of historical data in interpreting prophecy *post facto*. Allied with this were his mystical and alchemical interests. His library had many works by the chemist Boyle and by the Rosicrucians among others. Even while writing his scientific works he spent much time in the alchemical laboratory. "What his aim might be I was not able to penetrate into," a friend tells us, and his writings never make it clear. Certainly it was a great driving passion in his life but, typically, he kept the public and the printer

out of his interests. Remember, others always had to drag him into publications and apparently had not sufficient interest or understanding to lead him to publish his chemical work.

5. He could be very spiteful and ungenerous when his authority was questioned but, to his students of capability, he was most helpful. Toward truth itself, rather than toward the portion he felt he had, he was most humble. Before he died he made his famed remark "I do not know what I may appear to the world but to myself I seem to have been only like a boy, playing on the seashore, and diverting myself in now and then finding a smoother pebble or a prettier shell than ordinary, while the great ocean of truth lay all undiscovered before me." Likely his religious and alchemical interests are part of this broader view.

Thomas H. Leith, *The Nature and Growth of Science*, 113-117 (Toronto: Wall & Thompson, 1989).

### Questions and Topics for Discussion and Writing

1. Comment on the different personalities of Copernicus, Galileo, and Newton, based on what you have read about each of them. Were their personalities suited to the scientific work they did?
2. Newton spent much more of his time on theology and alchemy than on the mathematics and physics for which we remember him. Do you think he would have accomplished more if he had spent more of his time on mathematics and physics? Why or why not?

# * Einstein

## James R. Newman

Newton's physics provided a secure foundation for the development of modern physical science. The Newtonian model of the universe is a good enough description of the world that an enormous amount of physical theory and application has been built upon it, including most of the practical applications known to us today. But the Newtonian world view is not quite satisfactory at the extremes. Newton's laws do not describe physical phenomena well at a very small scale, such as at a subatomic level, nor at very large scales, such as at astronomical distances or at speeds close to the speed of light.

Near the end of the nineteenth century, Newtonian physics had been pursued so far that some of the anomalies of large and small scale phenomena began to be noticed. Leading scientists turned their attention to these problems and began to record and analyze the unusual observations. In the first half of the twentieth century a whole new physics arose, not so much replacing Newtonian physics, but extending and enlarging it. At the subatomic level, quantum mechanics took over where Newton's laws no longer were adequate. At the level of the very large scale, relativity theory explained events that Newtonian physics could not. Both quantum mechanics and relativity theory posed new questions, not just about physics, but about the nature of knowledge and of existence itself. Thus the new physics of this century has opened a vast set of problems and issues for scientists, philosophers, and laymen to contemplate.

Though quantum mechanics is associated with several different important scientists, relativity theory is primarily the creation of one man, Albert Einstein. (Quantum mechanics is based upon a number of fundamental implications of atomic physics, many of which were inferred by Einstein. But Einstein himself could not accept some of the implications of quantum mechanics.) The following review of Einstein's work by James R. Newman gives an excellent overview of the scope, development, and importance of relativity, and some insight into the mind and person of this century's most renowned scientific genius.

To the eyes of the Man of Imagination, Nature is Imagination Itself.

*William Blake*

Einstein died in 1955. Some fifty years earlier, when he was twenty-six, he put forward an idea that changed the world. His idea revolutionized our conception of the physical universe; its consequences have shaken human society. Since the rise of science in the seventeenth century, only two other men, Newton and Darwin, have produced a comparable upheaval in thought.

Everybody knows that Einstein did something remarkable, but what exactly did he do? Even among educated men and women, few can answer. We are resigned to the importance of his theory, but we do not comprehend it. It is this circumstance that is largely responsible for the isolation of modern science. This is bad for us and bad for science; therefore more than curiosity is at stake in the desire to understand Einstein.

Relativity is a hard concept, prickly with mathematics. there are many popular accounts of it, a small number of which are good, but it is a mistake to expect they will carry the reader along—like a prince stretched on his palanquin. One must tramp one's own road. Nevertheless, relativity is in some respects simpler than the theory it supplanted. It makes the model of the physical world more susceptible to proof by experiment; it replaces a grandiose scheme of space and time with a more practical scheme. Newton's majestic system was worthy of the gods; Einstein's system is better suited to creatures like ourselves, with limited intelligence and weak eyes.

But relativity is radically new. It forces us to

change deeply rooted habits of thought. It requires that we free ourselves from a provincial perspective. It demands that we relinquish convictions so long held that they are synonymous with common sense, that we abandon a picture of the world that seems as natural and as obvious as that the stars are overhead. it may be that Einstein's ideas will seem easy to a generation that has grown up with them; but for our generation, as Bertrand Russell has said, a complete "imaginative reconstruction is unavoidable." Anyone who seeks to understand the world of the twentieth century must make this effort.

* * *

In 1905, while working as an examiner in the Swiss Patent Office, Einstein published in the *Annalen der Physik* a thirty-page paper with the title "On the Electrodynamics of Moving Bodies." The paper embodied a vision. Poets and prophets are not alone in their visions; a young scientist—it happens mostly to the young—may in a flash glimpse a distant peak that no one else has seen. He may never see it again, but the landscape is forever changed. The single flash suffices; he will spend his life describing what he saw, interpreting and elaborating his vision, giving new directions to other explorers. Einstein had not been a wonder child. He had not astonished his teachers or his parents. His father showed him a compass at the age of five; the event, he tells us, made a profound and lasting impression upon him. For the first time he realized that "something deeply hidden had to be behind things." When he was twelve, a book on Euclidean geometry came into his hands; this too was an unforgettable event. His "holy geometry booklet," as he called it, opened a new delight: it was possible to prove things, to attain certainty by pure thinking. He continued his studies in mathematics and he also read with "breathless attention" a six-volume popular work that comprised the entire field of the natural sciences. At the age of seventeen, having already dipped into theoretical physics, he entered the Polytechnic Institute of Zurich. By then the problem that was to culminate in his vision had begun to stir in him.

At the heart of the theory of relativity are questions connected with the velocity of light. The young Einstein began to brood about these while still a high-school student. Suppose, he asked himself, a person could run as fast as a beam of light, how would things look to him? Imagine that he could ride astride the beam, holding a mirror just in front of him. Then, like a fictional vampire, he would cast no image; for since the light and the mirror are traveling in the same direction at the same velocity, and the mirror is a little ahead, the light can never catch up to the mirror and there can be no reflection.

But this applies only to *his* mirror. Imagine a stationary observer, also equipped with a mirror, who watches the rider flashing by. Obviously the observer's mirror will catch the rider's image. In other words the optical phenomena surrounding this event are purely relative. They exist for the observer; they do not exist for the rider. This was a troublesome paradox, which flatly contradicted the accepted views of optical phenomena. We shall have to see why.

The velocity of light had long engaged the attention of physicists and astronomers. In the seventeenth century the Danish astronomer Römer discovered that light needed time for its propagation. Thereafter, increasingly accurate measurements of its velocity were made, and by the end of the nineteenth century the established opinion was that light in a vacuum always travels at a certain constant rate, about 186,000 miles a second.

But now a new problem arose. In the mechanics of Galileo and Newton, rest and uniform motion (i.e., constant velocity) are regarded as indistinguishable. Of two bodies, A and B, it can only be said that one is in motion *relative* to the other. The train glides by the platform; or the platform glides by the train. the earth approaches the fixed stars; or the stars approach the earth. There is no way of deciding which of these alternatives is true. And in the science of mechanics it makes no difference.

One of the questions, therefore, was whether, in respect of motion, light itself was like a physical body; that is, whether its motion was relativistic in the Newtonian sense, or absolute.

The wave theory of light appeared to answer this question. A wave is a progressive motion in some kind of medium; a sound wave, for example, is a movement of air particles. Light waves, it was supposed, move in an all-pervasive medium called the ether. The ether was assumed to be a subtle jelly with marvellous properties. It was colorless, odorless, without detectable features of any kind. It could penetrate all matter. It quivered in transmitting light. Also, the body of the ether as a whole was held to be stationary. To the physicist this was its most important property, for, being absolutely at rest, the ether offered a unique frame of reference for determining the velocity of light. Thus while it was hopeless to attempt to determine the absolute motion of a physical body, because one could find no absolutely stationary frame of reference against which to measure it, the attempt was not hopeless for light: the ether, it was thought, met the need.

The ether, however, did not meet the need. Its marvellous properties made it a terror for experimentalists. How could motion be measured against an ectoplasm, a substance with no more substantiality than an idea? Finally, in 1887, two American physicists, A.A. Michelson (the first American, by the way, to win the Nobel prize in physics) and E.W. Morley, rigged up a beautifully precise instrument, called an interferometer, with which they hoped to discover some evidence of the relationship between light and the hypothetical ether. If the earth moves through the ether, a beam of light traveling in the direction of the earth's motion should move faster through the ether than a beam traveling in the opposite direction. Moreover, just as one can swim across a river and back more quickly than one can swim the same distance up and down stream, it might be expected that a beam of light taking analogous paths through the ether would complete the to-and-fro leg of the journey more quickly than the up-and-down leg.

This reasoning was the basis of the Michelson-Morley experiment. With great care and ingenuity they carried out a number of trials in which they compared the velocity of a beam of light moving through the ether in the direction of the earth's motion, and another beam traveling at right angles to this motion. There was, as I have indicated, every reason to believe that these velocities would be different. Yet no difference was observed. The results were entirely negative. The light beam seemed to move at the same velocity in either direction. The possibility that the earth dragged the ether with it having been ruled out, the inquiry had come to a dead end. Perhaps there was no difference; perhaps there was no ether. The Michelson-Morley findings were a major paradox.

Various ideas were advanced to resolve it. The most imaginative of these, and also the most fantastic, was put forward by the Irish physicist G.F. FitzGerald. He suggested that since matter is electrical in essence and held together by electrical forces, it may contract in the direction of its motion as it moves through the ether. the contraction would be very small; nevertheless in the direction of motion the unit of length would be shorter. This hypothesis would explain the Michelson-Morley result. The arms of their interferometer might contract as the earth rotated; this would shorten the unit of length and cancel out the added velocity imparted to the light by the rotation of the earth. The velocities of the two beams—in the direction of the earth's motion and at right angles to it—would appear equal. FitzGerald's idea was elaborated by the famous Dutch physicist H. A. Lorentz. He put it in mathematical form and connected the contraction caused by motion with the velocity of light. According to his arithmetic the contraction was just enough to account for the negative results of the Michelson-Morley experiment. There the subject rested until Einstein took it up anew.

He knew of the Michelson-Morley findings (though there is evidence to show that they played a surprisingly small part in leading him to his own theory). He knew also of other inconsistencies in the contemporary model of the physical world. One was the slight but persistent misbehavior (by classical standards) of the planet Mercury as it moved in its orbit; it was losing time (at a trifling rate, to be sure—43 seconds of arc per century), but Newton's theory of its motion was exact and there was no way of accounting for the discrepancy. Another was the bizarre antics of electrons, which, as W. Kaufmann and J.J. Thomson discovered, increased in mass as they went faster. The question was, could these inconsistencies be overcome by patching and mending classical theories, by improvising in the vein of the astronomers who for centuries tacked on epicycles to Ptolemy's system to keep it alive? Or had the time come for a Copernican renovation?

Making his own way, Einstein turned to another aspect of the velocity problem. Velocity measurements involve time measurements, and time measurements, as he perceived, involve the concept of simultaneity. Is this concept simple and intuitively clear? No one doubted that it was, but Einstein demanded proof.

I enter my study in the morning as the clock on the wall begins to strike. Obviously these events are simultaneous. Assume, however, that on entering the study I hear the first stroke of the town-hall clock, half a mile away. It took time for the sound to reach me; therefore while the sound wave fell on my ears at the moment I entered the study, the event that produced the wave was not simultaneous with my entry.

Consider another kind of signal. I see the light from a distant star. An astronomer tells me that the image I see is not of the star as it is today, but of the star as it was the year Brutus killed Caesar. What does simultaneity mean in this case? Is my *here-now* simultaneous with the star's *there-then*? Can I speak meaningfully of the star as it was the day Joan of Arc was burned, even though ten generations will have to pass before the light emitted by the star on that day reaches the earth? How can I be sure it will ever get here? In short, is the concept of simultaneity for different places exactly equivalent to the concept for one and the same place?

Einstein soon convinced himself that the answer is no. Simultaneity, as he realized, depends on signals; the speed of light (or other signal) must therefore enter into the meaning of the concept. Not only does the separation of events in *space* becloud the issue of simultaneity in *time*, but relative motion may do so. A pair of events, which one observer pronounces simultaneous, may appear to another observer, in motion with respect to the first, to have happened at different times. In his own popular account of relativity (see Figure 1.15), Einstein gave a convincing and easy example, which showed that *any* measurement of time is a measurement with respect to a given observer. A measurement valid for one observer may not be valid for another. Indeed, the measurement is certain not to be valid if one attempts to extend it from the system where the measurement was made to a system in motion relative to the first.

We approach the consummation of Einstein's revolutionary theory. He was now aware of these facts. Measuring the velocity of light requires a time measurement. This involves a judgment of simultaneity. Simultaneity is not an absolute fact, the same for all observers. The individual observer's judgement depends on relative motion.

Einstein went further. To measure a time interval of any kind requires the use of a clock (or its equivalent). The method consists of matching a signal, marking the beginning of the interval, against a certain position of the hands of the clock, and matching another signal, marking the end of the interval, against another position of the hands. The distance between the two points of coincidence is a measure of the duration of the interval. Simultaneity is clearly involved.

The sequence does not end there. Simultaneity may also be involved in measuring distances. A passenger on a moving train who wants to measure the length of his car has no difficulty. With a yardstick he can do the job as easily as if he were measuring his room at home. Not so for a stationary observer watching the train go by. The car is moving and he cannot measure it simply by laying yardstick end on end. He must use light signals, which will tell him when the ends of the car coincide with certain arbitrary points. Therefore, problems of time arise. Suppose the thing to be measured is an electron, which is in continual motion at high speed. Light signals will enter the experiment, judgements of simultaneity will have to be made, and once again it is obvious that observers of the electron, who are in motion relative to each other, will get different results. The whole comfortable picture of reality begins to disintegrate: neither space nor time is what it seems.

The clarification of the concept of simultaneity thrust upon Einstein the task of challenging two assumptions, assumptions hedged with the divinity of Isaac Newton. "Absolute, true, and mathematical time, of itself and from is own nature, flows equably without relation to anything external...." This was Newton's sonorous definition in his great book, *Principia Mathematica*. To this definition he added the equally majestic "Absolute space, in its own nature, without relation to anything external, remains always similar and immovable." These assumptions, as Einstein saw, were magnificent but untenable. they were at the bottom of the paradoxes of contemporary physics. They had to be discarded. Absolute time and absolute space were concepts that belonged to an outworn metaphysic. They went beyond observation and experiment; indeed, they were refuted by the nasty facts. Physicists had to

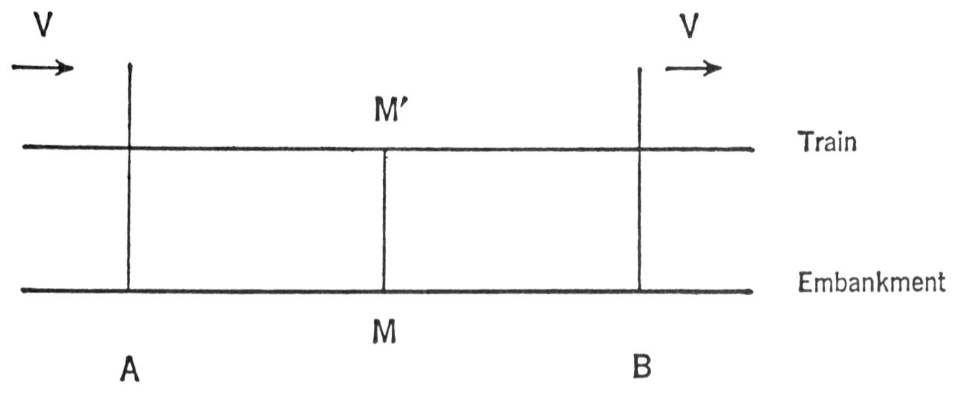


The diagram shows a long railroad train traveling along the rails with velocity V, in the direction toward the right of the page. The bottom line denotes the embankment running parallel to the rails. The letters A and B mark two places on the rails, and the letter M marks a point on the embankment directly midway between A and B. At M stands an observer, equipped with a pair of mirrors that are joined in a V and inclined at 90°. By means of this device he can observe both places A and B at the same time. We imagine two events at A and B, say two flashes of lightning, which the observer perceives in his mirror device at the same time. These he pronounces to be simultaneous, by which he means that the rays of light emitted at A and B by the lightning bolts meet at the midpoint M of the length A → B along the embankment. Now consider the moving train, and imagine a passenger seated in it. As the train proceeds along the rails, the passenger will arrive at a point *M′*, which is directly opposite M, and therefore exactly midway between the length A → B along the rails. Assume further that the passenger arrives at *M′* just when the flashes of lightning occur. We have seen that the observer at M correctly pronounces the lightning bolts to be simultaneous; the question is, will the train passenger at *M′* make the same pronouncement? It is easily shown that he will not. Obviously if the point *M′* were stationary with respect to M, the passenger would have the same impression of simultaneity of the lightning flashes as the observer on the embankment. But *M′* is not stationary; it is moving toward the right with the velocity V of the train. Therefore (considered with reference to the embankment) the passenger is moving toward the beam of light coming from B, and away from the beam coming from A. It seems clear then that he will see the beam emitted by the flash at B sooner than the beam emitted by the flash at A. Accordingly he will pronounce the flash at B as earlier in time than the flash at A.

Which of the two pronouncements is correct, the observer's or the passenger's? The answer is that each is right in his own system. The observer is right with respect to the embankment, the passenger with respect to the train. The observer may say that he alone is right because he is at rest, while the passenger is moving and his impressions are therefore distorted. To this the passenger can reply that motion does not distort the signals, and that, in any case, there is no more reason to believe that he was moving and the observer at rest than that the passenger was at rest and the observer moving.

There is nothing to choose between these views, and they can be logically reconciled only by accepting the principle that simultaneity of events is meaningful only with respect to a particular reference system; moreover, that every such system has its own particular time, and unless, as Einstein says, we are told the reference system to which the statement of time refers, a bare statement of the time of an event is meaningless.

*James R. Newman*

live with these facts.

In taking this bold step Einstein was animated by the spirit of certain philosophers of science. He acknowledged his debt to the Austrian positivist Ernst Mach, to the brilliant French mathematician Henri Poincaré, and of course to the great Scottish skeptic of the eighteenth century, David Hume. Hume took the cheerful view that we know nothing and that there is no such thing as a rational belief. But, he said, in our pathetic groping—and to ease our discomfort—we ought at least to pretend that experience and observation are useful guides. Einstein was not primarily a philosopher, but he saw that there were problems of physics that could not be solved without taking seriously the basic criticisms of scientific procedure that these philosophers had uttered. The central point of his approach was that *nothing* had meaning that could not be verified. A new philosophy of physics had to precede a reform of physics. Without ever explicitly formulating such a philosophy, Einstein put it into his work.

* * *

The clear way out of the Michelson-Morley paradox was to accept it. From the point of view of common sense the results were extraordinary, yet they had been verified. It was not the first time that science had had to overrule common sense. The evidence showed that the velocity of light measured by *any* observer, whether at rest or in motion relative to the light source, is the same. Einstein embodied this fact in a principle from which a satisfactory theory of the interaction between the motion of bodies and the propagation of light could be derived. This principle, or first postulate, of his Special Theory of Relativity states that *the velocity of light in space is a constant of nature, unaffected by the motion of the observer or of the source of the light.*

The hypothesis of the ether thus became unnecessary. One did not have to try to measure the velocity of light against an imaginary frame of reference, for the plain reason that whenever light is measured against *any* frame of reference its velocity is the same. Why then conjure up ethereal jellies? The ether simply lost its reason for being.

A second postulate was needed. Newtonian relativity applied to the motion of material bodies; but light waves, as I mentioned earlier, were thought not to be governed by this principle. Einstein pierced the dilemma in a stroke. He simply extended Newtonian relativity to include optical phenomena. The second postulate says: *In any experiment involving mechanical or optical phenomena it makes no difference whether the laboratory where the experiment is being performed is at rest or in uniform motion; the results of the experiments will be the same.* More generally, one cannot by any method distinguish between rest and uniform motion, except in relation to each other.

* * *

Is that all there is to the Special Theory of Relativity? The postulates are deceptively simple. Moreover, to the sharp-eyed reader they may appear to contradict each other. The contradictions, however, are illusory, and the consequences are revolutionary.

Consider the first point. From the postulates one may infer that on the one hand light has the velocity $c$, and, on the other hand, even when, according to our traditional way of calculating, it should have the velocity $c + q$ (where $q$ is the velocity of the source), its velocity is still $c$. Concretely, light from a source in motion with respect to a given frame of reference has the same velocity as light from a source at rest with respect to the same frame. (As one physicist suggested, this is as if we were to say that a man walking up an escalator does not get to the top any sooner than a man standing still on the escalator.) This seems absurd. But the reason it seems absurd is that we take it for grated that the velocity of the moving source must be added to the normal velocity of light to give the correct velocity of the beam emitted by the source. Suppose we abandon this assumption. We have already seen, after all, that motion has a queer effect on space and time measurements. It follow that the established notions of velocity must be reconsidered. The postulates were not inherently contradictory; the trouble lay with the classical laws of physics. They had to be changed. Einstein did not hesitate. To preserve his postulates he consigned the old system to the flames. In them were consumed the most cherished notions of space, time and matter.

One of the clichés about Einstein's theory is that it shows that everything is relative. The statement that every thing is relative is as meaningful as the statement that everything is bigger. As Russell pointed out, if everything were relative, there would be nothing for it to be relative to. The name "relativity" is misleading. Einstein was in fact concerned with finding something that is *not* relative, something that mathematicians call an invariant. With this as a fixed point, it might be possible to formulate physical laws that would incorporate the "objective residue" of an observer's experience; that is, that part of the space and time characteristics of a physical event which, though perceived by him, are independent of the observer and might therefore be expected to appear the same to all observers. The constancy principle of the velocity of light provided Einstein with the invariant he needed. it could be maintained, however, only at the expense of the traditional notion of time. And even this offering was not enough. Space and time are intertwined. They are part of the same reality. Tinkering with the measure of time unavoidably affects the measure of space.

Einstein, you will notice, arrived at the same conclusion as FitzGerald and Lorentz without adopting their electrical hypotheses. It was a consequence of his postulates that clocks and yardsticks yield different measurements in relative motion than at rest. Is this due to an actual physical change in the instruments? The question may be regarded as irrelevant. the physicist is concerned only with the difference in measurements. If clock springs and yardsticks contract, why is it not possible to detect the change? Because any scales used to measure it would suffer the same contraction. It has been said that this argument resembles the plan of the White Knight in *Through the Looking-Glass* "to dye one's whiskers green, and always use so large a fan that they could not be seen." But what is at issue is nothing less than the foundations of rational belief.

Science uses assumptions and theories to explain the physical world, and tests them by observation and experiment. When the evidence contradicts a theory, it is modified or discarded. For a long time it was useful and even comfortable to believe that length and time are absolute quantities. This was an assumption subject to verification. When the evidence failed to support the assumption, it had to be changed. Shrinking rulers and lagging clocks are a sad business; it affected men's minds like the Lisbon earthquake. The preacher had said that the earth would abide, but now no one could be sure. The subversion of space and time was no less shocking. but there was no alternative if physics was to remain a fruitful activity.

* * *

Earlier I mentioned Kaufmann's and Thomson's discovery that an electron increases in mass as it goes faster. Relativity explains this astonishing fact. The first postulate sets an upper limit to velocity. Specifically only light is mentioned, but one can deduce from the postulate that no material body can move faster than light. In Newton's system there were no such limits. He defined mass as the "quantity of matter," and gave the formula that acceleration (which means change—or, more accurately, rate of change—of velocity) is the quotient of the force acting on a body divided by its mass. In his view the mass of a body is the same at all speeds. But just as his laws of motion have been shown not to be universally true, his concept of the permanence of mass turns out to be flawed. According to Einstein's Special Theory, the resistance of a body to changes in velocity increases with velocity. Thus, for example, more force is required to increase a body's velocity from 50,000 to 50,001 miles per hour than from 100 to 101 miles per hour. The resistance to change is called *inertia*, and its measure is mass. (This jibes with the intuitive notion that the amount of force needed to accelerate a body depends on its "quantity of matter"—Newton's definition of mass.) The ideas fall neatly into place: with increased speed, inertia increases; increased inertia evinces itself as increased mass. The increase in inertia or mass is, to be sure, very small at ordinary speeds, and therefore undetectable, which explains why Newton and his successors, though a brilliant company, did not discover it. This circumstance also explains why Newton's laws are perfectly valid for all ordinary instances of matter in motion: even a rocket moving at 10,000 miles an hour is a tortoise compared to a beam of light at 186,000 miles a second. but the increase in mass becomes a major factor where high-speed nuclear particles are concerned; for example, the electrons in a hospital X-ray tube are speeded up to a point where their normal mass is doubled, and in an ordinary TV picture tube the electrons have 5 per cent extra mass due to their energy of motion. And at the speed of light the push of even an unlimited accelerating force against a body is completely frustrated, because its mass, in effect, becomes infinite.

It is only a step now to Einstein's fateful equation.

Energy is converted into mass by motion. The bookkeeping is quite exact. The quantity of additional mass, multiplied by an enormous number, the square of the speed of light, is equal to the energy that was turned into mass. In a moving body, therefore, mass and energy are equivalent. But is this equivalence of mass and energy a special circumstance attendant upon motion? What about a body at rest? does its mass also represent energy? Einstein concluded that it does. The Special Theory does not furnish a complete answer, but Einstein went beyond his theory. "The mass of a body is a measure of its energy content," he wrote in 1905, and gave his now famous formula, ($E = mc^2$, where $E$ is energy content, $m$ is mass (which varies according to speed) and $c$ is the velocity of light.

"It is not impossible," Einstein said in this same paper, "that with bodies whose energy content is variable to a high degree (*e.g.*, with radium salts) the theory may be successfully put to the test." In the 1930s many physicists were making this test, measuring atomic masses and the energy of products of many nuclear reactions. All the results verified his idea. A distinguished physicist, Dr. E. U. Condon, tells a charming story of Einstein's reaction to this triumph: "One of my most vivid memories is of a seminar at Princton (1934) when a graduate student was reporting on researches of this kind and Einstein was in the audience. Einstein had been so preoccupied with other studies that he had not realized that such confirmation of his early theories had become an everyday affair in the physical laboratory. He grinned like a small boy and kept saying over and over, "*Ist dass wirklich so?*" Is it really true?—as more and more specific evidence of his $E = mc^2$ relation was being presented."

* * *

The year 1905 was for Einstein a marvellous year, like Newton's *annus mirabilis* (1665–6), when he discovered the law of gravitation, invented the calculus and laid the foundations of his great optical discoveries. Besides the paper on relativity, Einstein in 1905 wrote two other major papers: on the photoelectric effect, and on the dance of tiny particles in liquids and gases, a phenomenon known as Brownian motion. The paper on Brownians motion greatly illuminated the problem of molecular motion and provided a formula with which one could estimate the number of molecules in a given volume, though they themselves are invisible, by measuring the journeys of the tiny visible particles of liquid or gas. The other paper, for which he won the Nobel prize, brilliantly extended Max Planck's quantum theory by showing that light itself, like other forms of energy, consists of tiny packets (which later got the name photons).

For ten years after he formulated the Special Theory, Einstein grappled with the task of generalizing relativity to include accelerated motion. This essay cannot carry the weight of the details, but I shall describe the matter briefly.

While it is impossible to distinguish between rest and uniform motion by observations made within a system, it seems quite possible, under the same circumstances, to determine *changes* in velocity, i.e., acceleration. In a train moving smoothly in a straight line, at constant velocity, one feels no motion. But if the train speeds up, slows down or takes a curve, the change is felt immediately. One has to make an effort to keep from falling, to prevent the soup from sloshing out of the plate and so on. These effects are ascribed to what are called *inertial forces*, producing acceleration (the name is intended to convey the fact that the forces arise from the inertia of a mass, i.e., its resistance to changes in its state). It would seem, then, that any one of several simple experiments should furnish evidence of such acceleration, and distinguish it from uniform motion or rest. Moreover, it should even be possible to determine the effect of acceleration on a beam, of light. For example, if a beam were set parallel to the floor of a laboratory at rest or in uniform motion, and the laboratory were accelerated upward or downward, the light would no longer be parallel to the floor, and by measuring the deflection one could compute the acceleration.

When Einstein turned these points over in his mind, he perceived a loose end in the reasoning, which others had not noticed. How is it possible in either a mechanical or an optical experiment to distinguish between the effects of gravity, and of acceleration produced by inertial forces? Take the light-beam experiment. At one point the beam is parallel to the floor of the laboratory; then suddenly it is deflected. The observer ascribes the deflection to acceleration caused by inertial forces, but how can he be sure? He must make his determination entirely on the basis of what he sees *within* the laboratory, and he is therefore unable to tell whether inertial forces are at work—as in the moving train—or whether the observed effects are produced by a large (though unseen) gravitating mass.

Here then, Einstein realized, was the clue to the problem of generalizing relativity. As rest and uniform motion are indistinguishable, so are acceleration and the effects of gravitation. Neither mechanical nor optical experiments conducted within a laboratory can decide whether the system is accelerated, or in uniform motion and subjected to a gravitational field. (The poor wretch in tomorrow's space ship, suddenly thrown to the ground, will be unable to tell whether his vehicle is starting its rocket motor for the home journey or falling into the gravitational clutches of Arcturus.) Einstein formulated his conclusion in 1911 in his "principle of equivalence of gravitational forces and inertial forces."

His ideas invariably had startling consequences. From the principle of equivalence he deduced, among other things, that gravity must affect the path of a ray of light. This follows from the fact that acceleration would affect the ray, and gravity is indistinguishable from acceleration. Einstein predicted that this gravity effect would be noticeable in the deflection of the light from the fixed stars whose rays pass close to the huge mass of the sun. He realized, of course, that it would not be easy to observe the bending, because under ordinary conditions the sun's brilliant light washes out the light of the stars. But during a total eclipse the stars near the sun would be visible, and circumstances would be favorable to checking his prediction. "It would be extremely desirable," Einstein wrote in his paper enunciating the equivalence principle, "if astronomers would look into the problem presented here, even though the consideration developed above may appear insufficiently founded or even bizarre."

Eight years later, in 1919, a British eclipse expedition, headed by the famous astronomer Arthur Eddington, confirmed Einstein's astounding prediction. The weather on the small island of Principe was stormy, and the crucial photographs had to be taken through breaks in the clouds. For six nights Eddington and his companion developed the pictures, the days being spent in measurement. At last came a moment that Eddington later said was the greatest in his life: "The one plate I measured gave a result agreeing with Einstein."

Alfred North Whitehead described the occasion when the expedition results were announced to the Royal Society of London. "The whole atmosphere of tense interest was exactly that of the Greek drama: we were the chorus commenting on the decree of destiny as disclosed in the development of a supreme incident. There was dramatic quality in the very staging:—the traditional ceremonial, and in the background the picture of Newton to remind us that the greatest of scientific generalisations was now, after more than two centuries, to receive its first modification. Nor was the personal interest wanting: a great adventure in thought had at length come safe to shore."

* * *

Einstein saw with the eye of reason. He was so sure of himself that he did not feel he had to wait for the physical confirmation of his equivalence principle. In 1916 he announced his General Theory of Relativity, a higher synthesis incorporating both the Special Theory and the principle of equivalence. Two profound constructs are developed in the General Theory: the union of time and space into a four-dimensional continuum (a consequence of the Special Theory), and the curvature of space.

It was to one of his former professors at Zurich, the Russian-born mathematician Hermann Minkowski, that Einstein owed the idea of the union of space and time. "From henceforth," Minkowski had said in 1908, "space in itself and time in itself sink to mere shadows, and only a kind of union of the two preserves an independent existence." To the three familiar dimensions of space, a fourth, of time, had to be added, and thus a single new medium, space-time, replaced the orthodox frame of absolute space and absolute time. An event within this medium—one may, for example, think of a moving object as an "event"—is identified not only by three spatial co-ordinates denoting *where* it is, but by a time coordinate denoting *when* the event is there. *Where* and *when* are, as we have seen, judgments made by an observer, depending on certain interchanges of light signals. It is for this reason that the time co-ordinate includes as one of its elements the number for the velocity of light.

With the absolute space and time discarded, the old picture of the universe proceeding moment by moment from the past through the present into the future must also be discarded. In the new world of Minkowski and Einstein, there is neither absolute past nor absolute future; nor is there an absolute present dividing past from future and "stretching everywhere at the same moment through space." The motion of an object is represented by a line in space-time, called a "world line." The event makes its own history. The signals it emits take time to reach the observer; since he can record only what he sees, an event present for one observer may be past for another, future for a third. In Eddington's words, the absolute "here-now" of former beliefs has become a merely relative "seen-now."

But this must not be taken to mean that every observer can portray only his own world, and that in place of Newtonian order we have Einsteinian anarchy. Just as it was possible in the older sense to fix precisely the distance between two points in three-dimensional space, so it is possible in the four-dimensional continuum of space-time to define and measure distance between events. This distance is called an "interval" and has a "true absolute value," the same for all who measure it. Thus after all "we have found something firm in a shifting world."

How is the concept of curved space related to this picture? The concept itself sticks in the craw. A vase, a pretzel, a line can be curved. but how can empty space be curved? Once again we must think not in terms of metaphysical abstractions but of testable concepts.

Light rays in empty space move in straight lines. Yet in some circumstances (e.g., where the ray is close to the sun) the path of motion is seen to be curved. A choice of explanations offers itself. We may, for example, say that a gravitational mass in the neighborhood of the ray has bent it; or we may say that this gravitational mass has curved the space through which the ray is traveling. There is no logical reason to prefer one explanation to the other. Gravitational fields are no less an imaginative construct than space-time. The only relevant evidence comes from measuring the path of the light itself—not the field or space-time. It turns out to be more fruitful to explain the curved path of the light ray as an effect of curved space-time, rather than as an effect of the direct action of gravity on light.

Let me suggest an analogy. A thin sheet of rubber is stretched over a large drum kettle. I take a very light marble and permit it to roll over the sheet. I observe that the path of its motion is a straight line. I now take several lead weights and place them at different points on the rubber sheet. Their weight dimples it, forming small slopes and hollows. Suppose I release the marble on this surface. The path of motion will no longer be straight but will curve toward the slopes and eventually fall into one of the hollows. Now think of space-time as corresponding to the sheet of rubber, and large gravitational masses to the lead weights; think also of any "event"—a moving particle, a beam of light, a planet—as the counterpart of the marble rolling on the membrane. Where there are no masses, space-time is "flat" and paths of motion are straight lines. But in the neighborhood of large masses space-time is distorted into "slopes" and "hollows," which affect the path of any object entering upon them. This is what used to be called the attraction of gravitation. But gravitation in Einstein's theory is merely an aspect of space-time. The starlight bent toward the sun "dips" into the "slope" around it but has enough energy not to be trapped in the "hollow"; the earth circling the sun is riding on the "rim" of its "hollow" like a cyclist racing round a velodrome; a planet that gets too deep into the "hollow" may fall to the bottom. (This is one of the hypotheses astronomers make about collisions that may have formed new planets in our universe.) There are slopes and hollows wherever there is matter; and since astronomical evidence seems to favor the hypothesis that matter is on average uniformly distributed throughout the universe, and finite—though not necessarily constant—Einstein suggested the possibility that the whole of space-time is gently curved, finite, but unbounded. It is not inconsistent with this hypothesis that the universe is expanding, in which case the density of matter would decrease. A finite but unbounded universe is roughly analogous—though it is of higher dimension—to the two-dimensional curved surface of the earth. The area is finite without boundaries, and if one travels in a straight line in a given direction one must, after a time, return to the original point of departure.

* * *

Einstein spent his life searching for what is changeless in an incessantly changing world. He searched for unity in multiplicity. In his model of physical reality, space, time, energy, matter are bound together in a single continuum. The crown of his efforts—to find a set of field equations that would unite gravitational and electromagnetic phenomena—may have eluded him. But his achievement is beyond measure or praise. Two points about his work are worth making. The first is that his model of the world was not a machine with man outside it as observer and interpreter. The observer is part of the reality he observes; by observation he shapes it. "Event, signal and observation," as Bronowski has written, "is the relationship which Einstein saw as the fundamental unit in physics. Relativity is the understanding of the world not as events but as relations."

The second point is that his theory did much more than answer questions. As a living theory it forced new questions upon us. Einstein challenged unchallengeable writs; he would have been the last to claim that his own writs were beyond challenge. He broadened the human mind.

* * *

I want to say a few words about the man.

He was a thoroughly good man. He was kindly and tolerant. He was gentle. He felt for humanity and deeply sensed himself a part of it, sharing its hopes and fears. He hated war—the atavist instincts that produce it, the degradation it causes, the suffering it inflicts. By tragic chance one of his ideas paved the way for the most terrible weapon of all times. His innocent vision became a nightmare. For the last ten years of his life he abjured men to reflect upon their peril, to avert what he sadly suspected their folly would encompass—universal death.

He was a skeptical man, contemptuous of dogma, properly disrespectful of authority. His passion for understanding drove his researches. It was a passion not only to decode the patterns of the physical world, but also to grasp the deeper meaning of man's relation to nature and of man's dependence on man. He regarded science as a discoverer and a liberator. It would open the world and teach men how to live in it and to be happy.

He was a courageous man. He esteemed only truth and would not bow to tyrants, to mob opinion, to the clamor of irresponsible journalists, to the threats of squalid politicians. He respected the conscience of free-thinking men, whatever their beliefs, and he did not hesitate to speak out on behalf of heretics and others who in our unhappy time have been persecuted for their independence.

"It is the moral qualities of its leading personalities that are perhaps of even greater significance for a generation and for the course of history than purely intellectual accomplishments. Even these latter are, to a far greater degree than is commonly credited, dependent on the stature of character." Einstein's words, spoken on the occasion of a memorial celebration for Madame Curie, are appropriate to an estimate of his place in our civilization. The values he cherished are the values of science no less than of theirs. I do not mean one must be noble to be creative. I do mean that science as one of the glories of man is animated by the values of a decent society: the love of truth for its own sake, respect for dissent, independence, free expression. Einstein was an incorruptible human being. Scientific thought was for him the guide of action. He knew that, in Clifford's noble phrase, "the truth which it arrives at is not that which we can ideally contemplate without error, but that which we may act upon without fear." If the values he lived by were better served, mankind could look forward to a brighter future.

*James R. Newman*

From James R. Newman, *Science and Sensibility* (New York: Simon & Schuster, 1961).

### Questions and Topics for Discussion and Writing

1. What does Newman mean when he says that young scientists may have a flash of insight when they are young and spend the rest of their lives interpreting and elaborating their vision? Give some examples of this phenomena.

2. Why did scientists postulate the existence of the ether? What was it supposed to do?
3. What did the Michelson-Morley experiment show?
4. Newman says that Einstein took the philosophical position that nothing had meaning that could not be verified. Explain what this meant for Einstein's work.
5. Why is "relativity" a misleading name for relativity theory? What are the invariants in relativity theory?
6. How is the space-time continuum of relativity different from space and time in the Newtonian model of the world?
7. Newman says that Einstein was "properly disrespectful of authority." In what way was this disrespect essential to his work?

Charles Darwin

*Issue*

# The Development of the Life Sciences

The life sciences have a long history, beginning with Aristotle's *History of Animals*, if not before. Parallel to the early development of biology is the development of medicine, which also has roots in the ancient Greeks and, of course, in many other cultures also. However, none of the early history of the life sciences will be considered in this book. Instead, we will jump right to the mid-nineteenth century and examine Darwin and the theory of evolution.

A strong argument can be made that biology really began to be a "scientific" discipline with Darwin, that the previous work in the biological subjects was interesting study but should not be called a science. That point need not be resolved here, but it certainly is true that the theory of evolution is one of the most interesting issues in the history of the life sciences from the point of view of its interaction with society.

In the *physical* sciences, the Copernican Revolution caused a considerable stir outside of scientific and scholarly circles because it called into question a matter of everyday life that had been taken for granted, that the earth was stationary and at the center of the universe. Most of the fuss about the Copernican viewpoint was not over its scientific credibility, but over the displacement of humanity from center stage. Likewise, in the *life* sciences, Darwin's theory took away yet another special place for humans and made us just another animal. In both cases objections to the theory were often made on religious grounds because the Bible, read literally, says that the sun moves (not the earth) and that God specifically created man and woman. The heliocentric theory of Copernicus has triumphed completely. Its present detractors are few and far between. Not so with evolution, which still has lively opponents who claim that evolution is not proved and that creationism is a viable alternate view.

Most of the articles in this section are concerned with the theory of evolution, its proponents and detractors. The final selection is an article on the beginnings of genetics. But these are far from being all of the articles in this book on the life sciences, as much of Unit III on the Environment and parts of Unit IV on Current Issues are about current problems in society that concern the sciences of living things.

In this section, we begin with "What Darwin Explained," by Byron Wall, which examines Darwin's work, what he tried to do and what he succeeded in doing. This is followed by one of the original and clearest statements of the theory of evolution by Darwin's rival and colleague, Alfred Russel Wallace. (Darwin himself is too long-winded.) Next is one of the most famous challenges to the teaching of evolution, the "Scopes Trial" in the State of Tennessee in

1925; the selection is an excerpt of the trial transcript. The theory of evolution and the evidence of evolution have both come a long way since the work of Darwin himself. The article "Darwinism Defined: The Difference between Fact and Theory" by prominent evolutionist Stephen Jay Gould brings the scientific issues up to date. The final essay, Ruth Moore's "The Discovery of Units of Heredity," reviews the work of Mendel and the early steps of a science of genetics. Without a viable theory of inheritance, Darwin was unable to explain the actual mechanism of evolution. Mendel's work, though done and published in Darwin's own time, was unknown to Darwin and his followers.

# What Darwin Explained

## Byron E. Wall

Darwin was a remarkable character. In his youth he travelled all around the world to exotic places, but then returned to England and lived for the rest of his long life as a recluse on his country estate. Early in his life he conceived the essential structure of the theory for which he is famous, but refrained from publishing it for years until his hand was forced by circumstance. Once his theory was published it caused an enormous stir among scientists and laymen, and in a short period of time came to dominate biological thought.

Given Darwin's eccentricities, the success of his approach warrants some examination. Just what did Darwin accomplish? What did the theory of evolution as he presented it explain?

To clarify these questions, first consider claims which might be made for Darwin but which are false.

### Evolution Before Darwin

First, consider whether Darwin was an originator of ideas. Darwin was certainly not the first to consider that species had evolved. That idea had been around for ages. The most popular evolutionary theory current in Darwin's youth was that of Jean Lamarck, which Darwin had studied in detail. But there were many minor evolutionists around too, including Darwin's grandfather, Erasmus Darwin, all of whose ideas were known to Darwin.

But of course, Darwin's theory of evolution differed from these rivals by attributing the direction of evolutionary change to natural selection. Where others had claimed an inner perfectability or striving within living things to be the cause of modifications of structure, or attributed them to direct climactic or other environmental influences, Darwin saw the primary factor to be the weeding out of the unfit from the fit in populations of species that naturally are born with a wide variety of characteristics and capabilities. The unfit would simply not survive and propagate. But Darwin was hardly the first to think of nature operating this way.

Here for example is a well-known passage from Aristotle's *Physics* in which Aristotle discusses the distinction between necessity and accident in his analysis of purpose in nature. He sketches out the principle of natural selection, which he held to be erroneous, in order to show how some possible objections might be disposed of. The discussion of natural selection goes as follows:

> Why should nature work, not for the sake of something, nor because it is better so, but just as the sky rains, not in order to make the corn grow, but of necessity? What is drawn up must cool, and what has been cooled must become water and descend, the result of this being that the corn grows. Similarly if a man's crop is spoiled on the threshing-floor, the rain did not fall for the sake of this—in order that the crop might be spoiled—but that result just followed. Why then should it not be the same with the parts in nature, e.g. that our teeth should come up *of necessity*—the front teeth sharp, fitted for tearing, the molars broad and useful for grinding down food—since they did not arise for this end, but it was merely a coincident result; and so with all other parts in which we suppose that there is purpose? Wherever then all the parts came about just what they would have been if they had come to be for an end, such things survived, being organized spontaneously in a fitting way; whereas those which grew otherwise perished and continued to perish, as Empedocles says his "man-faced ox-progeny" did.[1]

[1] Aristotle, *Physics* Bk II Ch 8, 198b 17-33, in *The Basic Works of Aristotle,* ed. Richard McKeon, 249 (New York: Random House, 1941).

As a principle in and of itself, natural selection is clearly stated in this passage from Aristotle. It even appears to have been part of a full theory of evolution in the works of the Greek philosopher Empedocles, who lived about 450 B.C. Empedocles' own works have not survived intact. Scholars have pieced together fragments of his writings from various sources; they believe they have about one fifth of the original long poem that was the source of Aristotle's remarks. From those fragments it is obvious that the rest of Empedocles' evolutionary theory was very different from Darwin's. Empedocles speaks of disjointed limbs wandering around seeking other limbs to unite with and form whole bodies. Natural selection operated on the monsters that were formed from these chance unions of straying body parts. Small wonder that Aristotle had no trouble dismissing the idea.

More recent statements of natural selection also can be found. Darwin claimed to have hit upon the concept after having read Thomas Malthus' *Essay on Population*, but the concept is clearly expressed in Lyell's *Principles of Geology* and attributed to the French botanist Augustin de Candolle. In fact, in the *Origin of Species*, Darwin himself refers the idea to both Lyell and de Candolle.

## Darwin on Heredity

What about heredity? If Darwin's theory of evolution explains how species change little by little over millions of years, it must also explain how they in general remain the same from generation to generation. Well, Darwin did have something to say on the subject of heredity, but it was not very profound and it was soon rejected. He called his idea the "hypothesis of pangenesis." In Darwin's own words, the basic idea was as follows:

> According to this hypothesis, every unit or cell of the body throws off gemmules or undeveloped atoms, which are transmitted to the offspring of both sexes, and are multiplied by self-division. They may remain undeveloped during the early years of life or during successive generations; and their development into units or cells, like those from which they were derived, depends on their affinity for, and union with other units or cells previously developed in the due order of growth.[1]

Actually even this poor hypothesis was more than Darwin said on the subject in the *Origin of Species*. In the first editions of the *Origin*, Darwin implicitly adopted the then current notions of blending inheritance, wherein the characteristics of both parents were somehow mixed together like different colours of ink in the offspring. A few years later it was pointed out that on this way of looking at things it was virtually impossible for natural selection to have any measurable effect on the characteristics of species. This so troubled Darwin that in later editions of the *Origin*, he softened his insistence on the importance of natural selection.

Apart from blending inheritance and pangenesis, Darwin had no way to explain why acorns don't grow into monkeys. Genetics as we know it is essentially a twentieth century science. With modern genetics, we can explain Darwin's theory of evolution, but the reverse in not true.

What about the origin of life itself? Despite the title, *On the Origin of Species*, Darwin had little to say on the subject. Darwin's choice of title refers to his account of how new species arise from earlier species. But for the original species, the first forms of life itself, Darwin believed that one or a few life forms were created directly by God. The concept of life arising directly from inorganic matter by natural causes is an extension of Darwin's theory beyond the limits he put on it himself.

The foregoing remarks on what Darwin did not accomplish were designed to clear the stage for the analysis of what he did accomplish and what his theory did explain. Before turning directly to that topic, we must digress once more to discuss the scientific and intellectual background of Darwin's work.

## Darwin's Scientific Milieu

Darwin lived after the Copernican revolution. That fact has some consequences bearing on scientific explanations. In the article in this volume entitled "What the Copernican Revolution is All About," we looked at the dominant scientific world

[1] Charles Darwin, *The Descent of Man and Selection in Relation to Sex*, Ch VIII "Principles of Sexual Selection," Section: "The Laws of Inheritance." (London, 1871).

view before the Copernican revolution, in particular about Aristotle's separation of the universe into two parts, that below the moon and that beyond the moon. The sublunar world was characterized by change, generation and corruption, potentiality and actuality. All things had a beginning and an end, were born and died, and did so in fulfillment of some purpose. In the superlunar world, everything was eternal. The stars and planets moved endlessly in repeating cycles. Whatever purpose they had was fulfilled by maintaining their motion in a changeless pattern without determinate goals.

After the Copernican revolution, the distinction between above and below the moon disappeared. Galileo showed that the superlunar world displayed characteristics of the sublunar world, and Newton developed a unified physics that applied both to things on earth and to celestial bodies. But did this mean that the sublunar categories of fulfillment of purposes were then applied to the rest of the universe? In fact it was just the reverse; Aristotle's characterization of the celestial motions as continuing without final goals became the model for explanations both in the heavens and on earth.

### Where Newton Saw Purpose at Work

Newton's system was based on a few simple axioms which he claimed to be universally true for all time and places and from which he could deduce logically necessary conclusions. For example, the principle of inertia stated that if a body was moving at a constant speed it would retain that speed in the same direction until some external force stopped it or altered its motion. Thus for both Newton and Aristotle, the planets moved continuously and eternally because it was logically necessary that they should do so, and not because they were seeking some goal. Of course, for Newton, their motion due to inertia would be in a straight line which had to be corrected by gravitational attraction to other bodies, such as the sun. But unlike Aristotle's conception, Newton's system applied with the same changeless, purposeless necessity to motions on earth.

There were certain aspects of events in the world that did not appear to Newton to follow necessarily from his premises and which therefore suggested that purposeful action was taking place. One was the principle of universal gravitation itself. Newton had no way of adequately explaining it in mechanical terms. He left open the question whether it was due to a material or an immaterial cause. That question was not given an answer until Einstein formulated general relativity theory in the twentieth century. Meanwhile it was largely forgotten that this *was* a problem. Newton cited two other indications of purpose afoot and claimed they showed evidence of Divine intervention. These were much noted and discussed after Newton's death. One concerned the solar system. Newton saw no necessity according to his physics that all the planets went around the sun in the same direction and all in approximately the same plane. Moreover, by his calculations, the planets ought not to be moving precisely as they had been observed over centuries to move, so something must be correcting their motions to keep the system going as it does. The other evidence of purpose was the living world. These two problem areas are discussed by Newton in this passage from the famous Query 31 of the *Opticks*:

> All material Things seem to have been composed of the hard and solid Particles...variously associated in the first Creation by the Counsel of an intelligent Agent. For it became him who created them to set them in order. And if he did so, it's unphilosophical to seek for any other Origin of the World, or to pretend that it might arise out of a Chaos by the mere Laws of Nature; though being once form'd it may continue by those Laws for many Ages. For while Comets move in very excentric Orbs in all manner of Positions, blind Fate could not make all the Planets move one and the same way in Orbs concentrick, some inconsiderable Irregularities excepted, which may have risen from the mutual Actions of Comets and Planets upon one another, and which will be apt to increase, till this System wants a Reformation. Such a wonderful Uniformity in the Planetary System must be allowed the Effect of Choice. And so must the Uniformity in the Bodies of Animals, they having generally a right and a left side shaped alike, and on either side of their Bodies two Legs behind, and upon their Shoulders, and between their Shoulders a Neck running down into a Back-Bone, and a Head upon it; and in the Head two Ears, two Eyes, a Nose, a Mouth, and a Tongue, alike situated. Also the first Contrivance of those very artificial Parts of Animals, the Eyes, Ears, Brain, Muscles, Heart, Lungs, Midriff, Glands, Larynx, Hands, Wings, swimming Bladders, natural Spectacles, and other Organs of Sense and Motion; and the Instinct of

> Brutes and Insects, can be the effect of nothing else than the Wisdom and Skill of a powerful everliving Agent, who being in all Places, is more able by his Will to move the Bodies within his boundless uniform Sensorium, and thereby to form and reform the Parts of the Universe, than we are by our Will to move the Parts of our own Bodies.[1]

Newton's argument is that since for both the planetary orbits and animal adaptation, events could have, under his principles of motion, turned out any number of ways, and since in fact they have turned out to be so well ordered, the particular way things are *must* have been chosen by some supreme intelligence. There is a curious leap in logic here. On the one hand, implicit in this argument is the assumption that the axioms of Newton's physics were merely *discovered* by Newton. The were *formulated* by God; or another way of putting it is that they are absolute principles of nature, not just the best way of thinking about it that a frail human mind could devise. That assumption is not unexpected. What may be more difficult for the modern reader is why Newton concludes that just because the way things are cannot be deduced as necessary consequences of nature's laws it must be that the way things are was chosen for some end by an agent able to do the choosing.

The reasoning seems to lie in Aristotle's account of causation. The passage quoted earlier from Aristotle having to do with natural selection was, as mentioned, part of a discussion of necessity and accident. He put forth the natural selection argument in order to dispose of it, but how did he dispose of it? His refutation is that the bodies and parts of animals are formed normally in a given regular fashion, but no things that happen spontaneously or by coincidence happen necessarily, and are therefore the necessary result of some cause. Aristotle recognized four categories of causes. By showing that animal adaptation did not follow necessarily from material, formal, and efficient causes, and also could not happen by chance, it must be that it follows necessarily from the remaining cause, the final cause or purpose. Newton's reasoning seems to be along the same lines.

### The Design Argument

Darwin's age was completely dominated by the Newtonian viewpoint in science, especially in Great Britain. Newtonian physics was taken as revealed truth, at least in its general formulation, and along with it, Newton's analysis of what he called the effect of choice. Any regularity that could not be accounted for by necessary implication of physical laws was evidence of the hand of God.

This is the Design argument. The design in nature implied the existence of a designer. The study of nature, therefore, was analogous to the study of Scripture. Thus the study of nature could be, and was, called Natural Theology.

In 1829, the eighth Earl of Bridgewater died, leaving in his will a bequest for works "on the power, wisdom and goodness of God as manifested in the Creation." In other words, for works of Natural Theology that developed the Design argument. Eight "Bridgewater Treatises" were written by notable authorities in the 1830s. One of the most famous was by Sir Charles Bell, a distinguished anatomist at the University of Edinburgh. It bore the title, *The Hand: Its Mechanism and Vital Endowments as Evincing Design.* Bell's approach was to consider not just the human hand, but to compare the structure and function of many animals that had hand-like appendages, and to show that every difference in visible structure was accompanied by a multitude of internal relations that made the whole animal adapted to its function in nature. Hence only God could have designed it so. As Bell expressed it,

> It must now be apparent that nothing less than the Power which originally created is equal to effect those changes on animals, which adapt them to their conditions: and that their organization is predetermined; not consequent on the condition of the earth or of the surrounding elements...
>
> It has been shown, that...there is nothing like chance or irregularity in the composition of the system. In proportion indeed as we comprehend the principles of mechanics, or of hydrolics, as applicable to the animal machinery, we shall be satisfied of the perfection of the design. If anything appear disjointed or thrown in by chance, let the student

---

[1] Isaac Newton, *Opticks,* 4th ed., Bk III, Pt. I, Query 31 (London, 1704).

> mark that for contemplation and experiment, and most certainly, when it comes to be understood, other parts will receive the illumination, and the whole design stand more fully disclosed.[1]

The force of the design argument according to its own criteria depended upon the impossibility of accounting for the evident design with uniform physical laws. But by Darwin's age, many of the mysteries of nature *had* been deduced from specifiable physical laws, even some of those mysteries mentioned by Newton.

### The Nebular Hypothesis and Uniformitarianism

Newton's conclusion that the planets would not stay precisely in their orbits due to gravitational effects alone was shown to be false by the French mathematical astronomer, Pierre Laplace. Laplace carried the calculations further and showed that while the planets do deviate from their mean orbits, the combined gravitational effect of the different bodies in the system bring them back to their regular paths over time. In answer to Newton's other planetary mystery, Laplace offered a solution which came to be called the nebular hypothesis. The reason that all the planets move in the same direction, lie in about the same plane, and indeed rotate in the same direction on their individual axes, Laplace suggested, was because the solar system was originally a mass of hot gas which rotated about some central axis. As the gas cooled and condensed, it formed itself into the sun, planets and their satellites, which, due to the conservation of angular momentum, continued to rotate in the same direction. As far as Laplace was concerned, he no longer needed the hypothesis of God.

Closer to home, another realm in which design was being accounted for by physical law was geology. The Uniformitarian account of the shape of the earth, set forth in magnificent detail in Charles Lyell's *Principles of Geology*, which Darwin read aboard H.M.S. Beagle, argued that mountains, valleys, rivers, seas, deserts, forests, and so on, were not created as they are now, but instead formed naturally over eons of time from such causes as wind, rain, erosion, earthquakes, droughts, etc.—in short, by the slow accumulation of small changes, all wrought by motions that do follow uniform universal laws.

Uniformitarianism was but one of the new geological theories that began to appear in the late eighteenth century, partly as the result of new knowledge of the earth's shape and structure that followed from voyages of discovery to distant lands and from the churning up of land at home in the transformation of the environment during the industrial revolution. The new information did not confine itself to rock formations. It also provided a rich source for a new problem of interpretation through the discovery of fossil remains of presumably extinct species. The kinds of fossils found differed from layer to layer in the earth's strata. In the deeper strata were found fossils of animals no longer known. In the less deep, the fossils resembled living species more, and in the shallowest strata they resembled living species most. This pattern was especially marked in the remains of the higher vertebrates—those closest to the human anatomy.

A uniformitarian way of coping with the fossil evidence was to deny it. Lyell himself argued, unconvincingly, that what appeared to be new species were really the result of migrations of species that previously existed in other parts of the world. All the species were created at the beginning. The theistic aspect of this interpretation was that the Creator worked his wonders through natural laws, and did not have to intervene in the system. But for those not wedded to uniformitarianism, the evidence of new species was further evidence of God at work. Species were created, not all at once, but in successive stages as conditions became fitting for them to exist. This view could even be accommodated to the creation story in the Book of Genesis by taking the seven days of creation to be metaphors for different epochs of the earth's history.

### Darwin's Dilemma

Now, taking the historical sketch as a general context, let us turn to Darwin himself. The events of Darwin's early career posed a dilemma to him. Consider the following hypothetical question.

---

[1] Charles Bell, *The Hand: Its Mechanism and Vital Endowments as Evincing Design* (London, 1833).

Suppose yourself to be a young English chap named Charles Darwin. You go to Cambridge in the early nineteenth century and while you are there you learn a lot about animal and plant adaptation, about the design argument and the evidence for it, under the benevolent tutelage of a clergyman-naturalist. You learn above all else to examine the evidence of nature slowly and carefully before leaping to conclusions. You take a position as a naturalist on a five-year voyage around the world, during which you see with your own eyes more and more of the diversity of nature and the amazing balance of flora and fauna in each region you visit.

You especially notice the propensity of species to exhibit variations in their external characteristics, and you also notice that those variations differ from region to region. You know that all this can be accounted for by supposing special creations of species in different times and places. You also know of theories that account for species differences due to the effects of the surrounding environment. But on some islands, miles apart, you see the differences in the animal and plant life occurring between islands that have much the same environmental conditions.

You have taken with you Lyell's *Principles of Geology*, and part of your duties are to do geological surveys. Your work makes a competent geologist of you, and you see a great deal of evidence to support the uniformitarian view that you read expounded in Lyell's book. But you also learn a lot about fossils of extinct species and note their relationship to living species in the same areas. You suspect that the fossils represent an earlier form that has become modified over time into the present living species.

Eventually you return home, write up your observations as required, and begin to ponder what you have seen and read. The inorganic world *does* seem to be governed by uniform laws. The living world *does* exhibit a truly amazing ordered design. If species have been modified over time so as to be wonderfully adapted to their environments, perhaps the cause of their modifications is the same set of physical laws that govern geological changes. But what about those islands? Uniform laws are supposed to work uniform effects.

Suppose after mulling this over for a couple of years, you happen to read a certain essay on population by Thomas Malthus and that gives you an idea, or reminds you of one you read somewhere else and have forgotten. You conceive the principle of natural selection. You see that if each generation of living things varies naturally, some of the variations will be more favourably suited to surviving in the *physical* environment than others, and also, some variations will be more favourably suited for surviving in the competitive *living* environment. The cumulative effect over several generations can explain why species appear to be adapted to their physical surroundings, and it can also explain the differences on those isolated islands having the same physical conditions, because the living environment was different on each island—different predators, different sources of food, different proportions of each species. The uniform cause of the evolution of species is neither the physical nor the living environment, but the variability of the species themselves. Natural selection is merely a result of conditions, and it merely happens to direct the course of evolution the way it does. There's nothing supernatural about it.

Now, suppose all this is clear to you in 1838. Would you do as Alfred Russell Wallace did when he came to the same idea twenty years later and rush to announce it to the world? Or would you do as Charles Darwin did when he reached this position and sit on it for twenty years? Perhaps if you had been through what Darwin had been through and you knew what Darwin knew, you would not rush your idea into print.

Many reasons have been offered to explain why Darwin kept his idea to himself, reasons having to do with Darwin's personality, his childhood, his overbearing father, his timidity about open controversy, his wish not to offend his wife by the theological implications that could be drawn from his theory, and so on. These are valuable insights, but put them aside and consider the case purely on its own merits.

### Darwin's Failure to Explain Evolution

Natural selection *could* explain the species phenomena that Darwin had observed. But it *did not* explain them. Darwin knew enough to know that it did not. There were too many other possible explanations. Consider the evidence itself. Darwin had seen a lot and read a lot, but of course he had not encountered the whole living world. The fossil record was woefully incomplete. Maybe all species were created at once, and just appeared in different places at different times. Even if they had originally appeared at different times and places, they could have been specially created then and there. Even if natural selection could explain the preservation of some favourable characteristics, it did not necessarily explain all of the amazing adaptive design exhibited by living forms.

Then there is variation in species. Darwin had seen that the offspring of species do exhibit varying characteristics, but he did not know what the extent of this variation was, nor did he have any idea of the cause of variation. Could it continue indefinitely or was it restricted to a limited range?

Laplace had been able to show that the apparent design in the solar system was the necessary result of uniform laws because he could *calculate* the results from those laws. (Actually Laplace did not quite succeed to the extent he thought he did.) But Darwin neither knew the relevant laws nor had any means to calculate their effects. The process of induction is quite different in the living as opposed to the inorganic world. In the latter, it is not unreasonable to assume that the properties exhibited by a few samples of a substance will also be exhibited by all of that substance. But in the living world, every instance is somewhat different, even within the same species. A pound of lead is a pound of lead, but every rose is unique—except in poetry.

What, then, would be accomplished by tossing out one more speculative theory of evolution to the world? It proved nothing. To prove the theory Darwin would have to accomplish two tasks: On the one hand, he would have to be able to state the detailed physical and biological laws that are the cause of variation. He would need a detailed knowledge of reproduction and heredity. He would need to be able to show exactly how natural selection works, to specify how long is required for successive stages of evolution and the factors which determine the direction of evolution. On the other hand, he would have to show that the results he could deduce did indeed happen. He would have to reconstruct the natural history of the world; show what species existed when and where, when they became extinct, when new species arose, what their different characteristics were. For this he would have to rely on a multitude of indirect evidence to interpret the fossil record and fill in the gaps. He would need many subsidiary theories to establish the validity of the historical record that was put together on this indirect evidence.

Darwin completed neither of these tasks. They were far beyond the powers of any one man. That is what was required to prove the theory of evolution, and Darwin failed to do it. But that is precisely what he was trying to do!

With immense patience, Darwin sat down to his tasks and worked away at them piecemeal, as though he would live forever. He spent years dissecting and classifying barnacles. The patient matter-of-fact way he went about his study of barnacles is revealed in the anecdote told about one of his young children during these years. The child took it for granted that dissecting barnacles was a perfectly ordinary and normal thing for a father to do. Upon visiting the home of a playmate and being shown around the living quarters, the child detected that something was missing and exclaimed, "But where does your father do *his* barnacles? " But barnacles were only one of many studies undertaken by Darwin. He did extensive experiments cultivating orchids and breeding pigeons to study the variation process. He had long discussions with animal breeders to learn more about the rules that govern the preservation of desirable characteristics. He read widely in geology, paleontology, zoology and botany—and not just what were considered scientific treatments of these subjects; he also consulted such publications as gardening magazines to gain practical knowledge that others might not think important. He kept up a correspondence with leading scientists to pick their brains.

He published some of his subsidiary studies and gained a modest reputation, but except for a few trusted colleagues he kept his main theoretical conception to himself. Twenty years had elapsed since Darwin read Malthus and got his flash of insight, and he was still at work collecting evidence and working out the details of his theory. Twice he had written up a tentative sketch of his theory of evolution, but he kept these short essays to himself. At the urging of the colleagues who knew of his theory, he had reluctantly begun to write out the details when disaster struck.

## Wallace Rushes In

The young upstart naturalist Alfred Russel Wallace, whom Darwin had previously corresponded with, sent Darwin a short paper that Wallace had hurriedly composed on an idea that occurred to him after reading Malthus' *Essay on Population*. The essay was titled "On the Tendency of Varieties to Depart Indefinitely from the Original Type." As Darwin remarked at the time, it was an excellent short abstract of one of Darwin's brief sketches of his own theory, which of course Wallace knew nothing about.

With the threat of Wallace completely upstaging him and getting all the credit, Darwin finally took the leap and published his own major theory and research. At first, an excerpt from the second of his sketches was published along with Wallace's paper by the Linnean Society. Then, the next year, 1859, Darwin published a longer, more detailed version of his theory. It was not the huge, multivolume treatise that Darwin had begun to write, to be entitled "Natural Selection." Instead, because of that blasted Wallace, he had to make do with a mere 490 pages, which hc wished to give the cautious title of "An Abstract of an Essay on the Origin of Species." Pushed to a little more boldness, the full title of the book as it first appeared was *On the Origin of Species by Means of Natural Selection; or, the Preservation of Favoured Races in the Struggle for Life*. In the sixth edition of the book, the word "On" was dropped from the main title, which gives the impression that it is intended as a definitive treatment of the subject. The book was a great success and made Darwin famous. By the time of the sixth edition, Darwin probably did feel it was definitive.

It is important to remember, however, that Darwin never did complete what he set out to do. He did publish other books that would have been individual volumes of his planned great work and he revised the *Origin* five times, but with all of this he fell short of his original aim—a logically flawless, deductive theory of evolution by natural selection.

If Darwin failed to achieve his great goal, it is a fair question to ask just what he did achieve. The theory of evolution is now regarded as correct and plays a central role in modern biology, though many of the details of Darwin's own account and even some of his major conceptions have been discarded, particularly Darwin's view of inheritance.

## What Darwin Did Explain

Darwin certainly deserves the credit for stating his theory in enough detail to make people take him seriously, and to prompt the further research that has led to the establishment of evolutionary theory. Darwin had to be taken seriously because of the sheer amount of evidence he amassed that favoured his interpretation. Philosophers have frequently remarked that the theory of evolution has become established not by infallible deductive logic, but because the evidence for it far outweighs the evidence for any other interpretation. Historians have interpreted Darwin's long delay in publishing the theory to a desire to overwhelm the opposition. Certainly Darwin showed himself to be thinking along these lines. In the introduction to the *Origin of Species*, Darwin, apologizing to the reader for the omission of many relevant details, expressed his hope of someday publishing all the facts and references on which he based his conclusions. "For," he stated, "I am well aware that scarcely a single point is discussed in this volume on which facts cannot be adduced, often apparently leading to conclusions directly opposite to those at which I have arrived. A fair result can be obtained only by fully stating and balancing the facts and arguments on both sides of each question..."[1]

So, on this way of looking at it, Darwin's accomplishment was to tip the scales of evidence in favour of evolution, and in so doing spawn so much research that confirmed his theory that other views were driven to extinction. This might be called natural selection among scientific theories, or the survival of the fittest *theory*. But here is another accomplishment of Darwin, one which he himself may have been unaware of. It concerns a point of logic.

## Darwin's Contribution to Logic

The opposition to Darwin's theory can be divided into three main types: emotional, scientific and logical. The emotional opposition stemmed from

[1] Charles Darwin, *On the Origin of Species By Means of Natural Selection*, 1st ed., 1-2 (London, 1859).

repugnance to the idea that man may have descended from some lower animal form. On the subject of *human* evolution, Darwin wrote to Wallace that though it was "the highest and most interesting problem for the naturalist," it was "so surrounded with prejudices" that he thought he would avoid it altogether. He finally did go ahead and write *The Descent of Man*. Nevertheless, Darwin let others fight the emotional furor that followed. He had no interest in any other than scientific disputes. The scientific opposition stemmed from other interpretations of the phenomena of nature. This is the arena in which Darwin's weight of evidence counted most, where he produced so many facts that could be interpreted by *his* theory, but not by other evolutionary or non-evolutionary theories.

The third type of opposition, the logical, lurked behind the other two. It has to do with the design argument. The design argument gave solace to those who would oppose Darwin on human evolution because it implied that God had specially intervened in his system to create all species, including man. The same argument propped up rival scientific theories because it held that the complexities of animal and plant adaptation could not have arisen from natural causes.

Consider once again of the logic of the design argument. It does *not* assert that it is highly improbable that complex living things were formed solely according to simple physical laws. It asserts that it is *impossible* for them to have done so. It is impossible because it would require many coincidences, and things that happen as a result of coincidence or chance do not happen with regularity. Since living things exhibit marvelous regularity in their structures and fit so well together in the order of nature, they could not have resulted from coincidental events that took place for no purpose and were merely the result of random physical motion.

No amount of evidence can assail this argument until the fallacy in it is seen. For me, Darwin's greatest achievement was to reveal that fallacy. The fallacy is that the design argument fails to take account of time. Random causes, such as whatever caused the variations in plants and animals that Darwin observed, produce random effects—in the *first* instance. But if one of those random variations, however infrequent, has a greater tendency to be preserved than other variations, then in successive repetitions of the generating process that variation becomes more and more frequent until its occurrence becomes the regular, expected effect.

In Sir Charles Bell's "Bridgewater Treatise" on *The Hand*, the staggering complexity and intricacies of structure of hand-like appendages in different animals that make them so perfectly adapted to their functions provide all the evidence of design necessary for Bell to conclude that only God could have made them so; they could certainly not have arisen as chance artifacts of natural processes. We can compare that argument with one given by Darwin in *The Origin of Species*, where the instrument under discussion is not the hand, but one of even greater complexity and perfection, the eye. In the passage below, notice that the thrust of Darwin's argument is not to prove that the eye evolved from natural causes, but only to defeat the logic of the design argument by showing that it is possible that it did evolve naturally.

Darwin begins as follows:

> To suppose that the eye, with all its inimitable contrivances for adjusting the focus to different distances, for admitting different amounts of light, and for the correction of spherical and chromatic aberration, could have been formed by natural selection, seems, I freely confess, absurd in the highest possible degree. Yet reason tells me, that if numerous gradations from a perfect and complex eye to one very imperfect and simple, each grade being useful to its possessor can be shown to exist; if further, they eye does vary ever so slightly, and the variations be inherited, which is certainly the case; and if any variation of modification in the organ be ever so useful to an animal under changing conditions of life, then the difficulty of believing that a perfect and complex eye could be formed by natural selection, though insuperable by our imagination, can hardly be considered real. How a nerve comes to be sensitive to light, hardly concerns us more than how life itself first originated; but I may remark that several facts makes me suspect that any sensitive nerve may be rendered sensitive to light...[1]

Darwin goes on to give some examples of light

[1] *Ibid.*, 186-87.

sensitive nerves in various crustaceans, showing that eye-like contrivances exist in many degrees of complexity and relative perfection. This is to establish the plausibility of the idea that something like an eye evolved from a mere light-sensitive nerve. Then Darwin continues:

> With these facts, here far too briefly and imperfectly given, which show that there is much graduated diversity in the eyes of living crustaceans, and bearing in mind how small the number of living animals is in proportion to those which have become extinct, I can see no very great difficulty...in believing that natural selection has converted the simple apparatus of an optic nerve merely coated with pigment and invested by transparent membrane, into an optical instrument as perfect as is possessed by any member of the great Articulate class.
>
> He who will go thus far, if he find on finishing this treatise that large bodies of facts, otherwise inexplicable, can be explained by the theory of descent, ought not to hesitate to go further, and to admit that a structure even as perfect as the eye of an eagle might be formed by natural selection, although in this case he does not know any of the transitional grades. His reason ought to conquer his imagination; though I have felt the difficulty far too keenly to be surprised at any degree of hesitation in extending the principle of natural selection to such startling lengths.[1]

In both Darwin's analysis of the eye and Bell's analysis of the hand, the chief source of evidence is comparative anatomy of eyes and hands of living creatures. Bell's argument is that this great complex diversity proves the existence of a purposeful agent of design. Darwin's argument is that it doesn't prove that, because it is *possible* that these marvelous effects are the result of gradual evolution by natural selection over enormous periods of time.

Time as a factor in a scientific theory was not new with Darwin. The uniformitarian geology that guided Darwin's thoughts depended on the accumulation of small causes to produce large effects. But while the arrangement of geological things in the world was noticeably different from ages past, the uniformitarians did not claim that the level of order in the present was different from that in the past. The nebular hypothesis required a vast amount of time in its explanation of how the apparent order of the solar system arose from a chaos of gas. But the order that was to be explained was the direction of orbit and rotation of the planets, and that was already present in the rotating nebula that was its former state. Similarly, the chief rival theory of evolution in Darwin's day held that species slowly changed over long periods of time, but the changes were effected by the individuals in each generation adapting themselves to the environment and then passing on that adaptation to the next generation. Hence the chief phenomenon to be explained, from the point of view of the design argument, the *orderedness* of living things, was given to them as a capacity from the start.

In Darwin's use of time, the resulting order is far greater than that of its causes, and the random causes are unrelated to the ordered effects. Darwin of course did not point out the fallacy in the design argument in this abstract fashion; he presented his case in terms of concrete biological phenomena. Because he argued in concrete terms, the truth or falsity of his conclusions depended on the facts. Because he did not have all the facts, he could not prove he was right, as he readily admitted. But his logic stands, independent of the theory of evolution.

The importance of the logical argument is that it not only cleared away the obstacle of the design argument from the advance of biological theory, it also opened new vistas for many subjects in which long sequences of events might produce unexpected results. The great popularity of Darwin's theory in the late nineteenth century assured that both his conclusions and his arguments would get careful scrutiny. From direct analysis of Darwin's works and from that indirect process through which bits and pieces of the thought of an influential person get assimilated into the common notions of an era, an evolutionary approach has become part of the analysis of many problems that have nothing to do with biological species.

An example is the work of Darwin's son, Sir George Darwin on lunar evolution. The younger Darwin argued that the reason that the moon is always turned the same way toward the earth, instead of rotating on its axis, is due to the long-term

[1] *Ibid.*, 188.

effect of the tides on earth. It is the moon which causes the tides in the first place, but since the earth rotates, the gravitational pull of the earth produces a slight drag on the moon. That drag, argued George Darwin, could suffice to slow down and eventually stop a rotation of the moon on its own axis. The subject matter was not biological, but the reasoning followed the pattern of his father's logic, namely to show that the known effect *can* follow the proposed cause, given sufficient time. Then anyone who wishes to oppose the suggested theory cannot do so by claiming it is impossible.

## Conclusion

This chapter began by robbing Darwin of credit for either originating or proving the theory of evolution. Now it seems on the verge of giving him credit for half the important ideas of the last one hundred years. Let's stop short of that. Evolutionary thinking was commonplace throughout the nineteenth century. It is absurd to suggest that ever since 1859, everyone who thought of change over time got the idea from Darwin. What Darwin does deserve credit for is laying a solid groundwork for the later establishment of the theory of evolution by natural selection, causing the demise of the design argument, and turning the attention of scientists, social scientists, philosophers, and the general public to a closer look at the effect of time. To understand that, think about the overriding difference between the results predicted by Darwin's theory and by other evolutionary concepts. The other evolutionary theories, prominent in the nineteenth century, including not only biological theories but also social theories, had a common view that evolution meant change in a given direction, in short, progress. Their conceptions of the cause of evolution imply a certain direction due to an inner striving and the like.

What Darwin explained was that systematic change over time did not imply progress. The process Darwin saw was a blind production of diversities pruned down by a ruthless competition. The process may give the appearance of progress because it can result in increasing order within the system. But there is no inherent direction to the process and no inherent reason to suppose that natural changes are changes for the better. According to Darwin, when the environment changes, those who happen to have what it takes to survive do, and all the others just become extinct. Design and order in the world today is an accident of circumstance, not a product of forethought, and there is no guarantee that the order of today will be of much use in the world of tomorrow. That is the most important thought that Darwin explained to us.

*Byron E. Wall*

## Questions and Topics for Discussion and Writing

1. What are some of the accomplishments and explanations often attributed to Darwin, but for which he was not responsible?
2. Explain the principle of natural selection.
3. What did Laplace mean when he said he no longer needed the hypothesis of God?
4. Define uniformitarianism. Why was the evolution of species a problem for uniformitarianism?
5. Discuss the design argument. How did it arise; what did it explain; what is wrong with it?
6. Discuss and compare the concepts of evolution and progress.
7. Why is time such a crucial component of the theory of evolution?
8. The author asserts that Darwin's greatest achievement was to establish a point in logic, not to establish the theory of evolution. Is he being serious? How could the point in logic possibly outweigh the enormous work that Darwin devoted his entire life to? Make a case for or against the author's viewpoint.

# * On the Tendency of Varieties to Depart Indefinitely from the Original Type

## Alfred Russel Wallace

In 1858, Charles Darwin had been patiently working up his theory of evolution by natural selection for twenty years without publishing anything on the subject—despite pleas by several of his scientific colleagues who knew of his work that he should commit some of it to print. Then a young naturalist acquaintance, Alfred Russel Wallace, who was then working in Indonesia, sent him a paper he had written, asking for Darwin's comments and asking Darwin to forward it on if he thought it worthy of publication. The paper, "On the Tendency of Varieties to Depart Indefinitely from the Original Type," contained to Darwin's amazement, a more or less complete statement of the theory of evolution by natural selection that Darwin had been labouring over all these years and was not yet ready to publish.

Faced with the possibility of being completely upstaged by his younger colleague, Darwin was finally persuaded to publish *something.* A few months later, on July 1, 1858, Wallace's paper was read to the Linnean Society (and then published in its Journal) and along with it, two pieces by Darwin, one an extract from his notebook of 1839, the other an abstract of a letter Darwin wrote to the American biologist, Professor Asa Gray, in 1857. The main purpose of publishing Darwin's pieces was to show that he had been thinking along the same lines as Wallace long before Wallace had. They certainly would not have been satisfactory to Darwin as statements of his theory.

The next year, 1859, Darwin, compelled now to commit himself, published what *he* viewed as merely an abstract of the work he was writing. The abstract, running 490 pages, is the work we know as *On the Origin of Species,* and is the work which marks the beginning of biology as a modern science. Clearly Darwin is not a man of few words. It is difficult to do justice to Darwin in a collection such as this by excerpting a few pages from one of his works. Those readers with more than a passing interest in the issues should certainly consult *The Origin of Species* itself, and one of Darwin's several other subsequent works, to see the power and depth of his arguments and explanations. Here, to serve the purpose of giving the reader a sense of the theory of evolution as it was understood by Darwin is the paper by Wallace as it appeared in the *Journal of the Linnean Society* in 1858. It was, after all, even in Darwin's view, an excellent précis of Darwin's own work.

One of the strongest arguments which have been adduced to prove the original and permanent distinctness of species is, that *varieties* produced in a state of domesticity are more or less unstable, and often have a tendency, if left to themselves, to return to the normal form of the parent species; and this instability is considered to be a distinctive peculiarity of all varieties, even of those occurring among wild animals in a state of nature, and to constitute a provision for preserving unchanged the originally created distinct species.

In the absence or scarcity of facts and observations as to *varieties* occurring among wild animals, this argument has had great weight with naturalists, and has led to a very general and somewhat prejudiced belief in the stability of species. Equally general, however, is the belief in what are called "permanent or true varieties,"—races of animals which continually propagate their like, but which differ so slightly (although constantly) from some other race, that the one is considered to be a *variety* of the other. Which is the *variety* and which the original *species*, there is generally no means of determining, except in those rare cases in which the one race has been known to produce an offspring unlike itself and resembling the other. This, however, would seem quite incompatible with the "permanent invariability of species," but the difficulty is overcome by assuming that such varieties have strict limits, and can never again vary further from the original type, although they may return to it, which, from the analogy of the domesticated animals, is considered to be highly probable, if not certainly proved.

It will be observed that this argument rests entirely on the assumption, that *varieties* occurring in a state of nature are in all respects analogous to or even identical with those of domestic animals, and are governed by the same laws as regards their permanence or further variation. But it is the object of the present paper to show that this assumption is altogether false, that there is a general principle in nature which will cause many *varieties* to survive the parent species, and to give rise to successive variations departing further and further from the original type, and which also produces, in domesticated animals, the tendency of varieties to return to the parent form.

The life of wild animals is a struggle for existence. The full exertion of all their faculties and all their energies is required to preserve their own existence and provide for that of their infant offspring. The possibility of procuring food during the least favourable seasons, and of escaping the attacks of their most dangerous enemies, are the primary conditions which determine the existence both of individuals and of entire species. These conditions will also determine the population of a species; and by a careful consideration of all the circumstances we may be enabled to comprehend, and in some degree to explain, what at first sight appears so inexplicable—the excessive abundance of some species, while others closely allied to them are very rare.

The general proportion that must obtain between certain groups of animals is readily seen. Large animals cannot be so abundant as small ones; the carnivora must be less numerous than the herbivora; eagles and lions can never be so plentiful as pigeons and antelopes; the wild asses of the Tartarion deserts cannot equal in numbers the horses of the more luxuriant prairies and pampas of America. The greater or less fecundity of an animal is often considered to be one of the chief causes of its abundance or scarcity; but a consideration of the facts will show us that it really has little or nothing to do with the matter. Even the least prolific of animals would increase rapidly if unchecked, whereas it is evident that the animal population of the globe must be stationary, or perhaps, through the influence of man, decreasing. Fluctuations there may be; but permanent increase, except in restricted localities, is almost impossible. For example, our own observation must convince us that birds do not go on increasing every year in a geometrical ratio, as they would do, were there not some powerful check to their natural increase. Very few birds produce less than two young ones each year, while many have six, eight, or ten; four will certainly be below the average; and if we suppose that each pair produce young only four times in their life, that will also be below the average, supposing them not to die either by violence or want of food. Yet at this rate how tremendous would be the increase in a few years from a single pair! A simple calculation will show that in fifteen years each pair of birds would have increased to nearly ten millions! whereas we have no reason to believe that the number of the birds of any country increases at all in fifteen or in one hundred and fifty years. With such powers of increase the population must have reached its limits, and have become stationary, in a very few years after the origin of each species. It is evident, therefore, that each year an immense number of birds must perish—as many in fact as are born; and as on the lowest calculation the progeny are each year twice as numerous as their parents, it follows that, whatever be the average number of individuals existing in any given country, *twice that number must perish annually,*—a striking result, but one which seems at least highly probable, and is perhaps under rather than over the truth. It would therefore appear that, as far as the continuance of the species and the keeping up the average number of individuals are concerned, large broods are superfluous. On the average all above *one* become food for hawks and kites, wild cats and weasels, or perish of cold and hunger as winter comes on. This is strikingly proved by the case of particular species; for we find that their abundance in individuals bears no relation whatever to their fertility in producing offspring. Perhaps the most remarkable instance of an immense bird population is that of the passenger pigeon of the United States, which lays only one, or at most two eggs, and is said to rear generally but one young one. Why is this bird so extraordinarily abundant,[1] while others producing two or three times as many young are much less plentiful? The explanation is not difficult. The food most congenial to this species, and on which it thrives best, is abundantly distributed over a very extensive region, offering such difference of soil and climate, that in one part or another of the area the supply never fails. The bird is capable of a very rapid and long-continued flight, so that it can pass without fatigue over the whole of the district it inhabits, and as soon as the supply of food begins to fail in one place is able to discover a fresh feeding-ground. This example strikingly shows us that the procuring a constant supply of wholesome food is almost the sole condition requisite for ensuring the rapid increase of a given species, since neither the limited fecundity, nor the unrestrained attacks of birds of prey and of man are here sufficient to check it. In no other birds are these peculiar circumstances so strikingly combined. Either their food is more liable to failure, or they have not sufficient power of wing to search for it over an extensive area, or during some season of the year it becomes very scarce, and less wholesome substitutes have to be found; and thus, though more fertile in offspring, they can never increase beyond the supply of food in the least favourable seasons. Many birds can only exist by migrating, when their food becomes scarce, to regions possessing a milder, or at least a different climate, though, as these migrating birds are seldom excessively abundant, it is evident that the countries they visit are still deficient in a constant and abundant supply of wholesome food. Those whose organization does not permit them to migrate when their food becomes periodically scarce, can never attain a large population. This is probably the reason why woodpeckers are scarce with us, while in the tropics they are among the most abundant of solitary birds. Thus the house sparrow is more abundant that the redbreast, because its food is more constant and plentiful,—seeds of grasses being preserved during the winter, and our farm-yards and stubble-fields furnishing an almost inexhaustible supply. Why, as a general rule, are aquatic, and especially sea birds, very numerous in individuals? Not because they are more prolific than others, generally the contrary; but because their food never fails, the sea-shores and river-banks daily swarming with a fresh supply of small mollusca and crustacea. Exactly the same laws will apply to mammals. Wild cats are prolific and have few enemies; why then are they never as abundant as rabbits? The only intelligible answer is, that their supply of food is more precarious. It appears evident, therefore, that so long as a country remains physically unchanged, the numbers of its animal population cannot materially increase. If one species does so, some others requiring the same kind of food must diminish in proportion. The numbers that die annually must be immense; and as the individual existence of each animal depends upon itself, those that die must be the weakest—the very young, the aged, and the diseased,—while those that prolong their existence can only be the most perfect in health and vigour—those who are best able to obtain food regularly, and avoid their numerous enemies. It is, as we commenced by remarking, "a struggle for existence," in which the weakest and least perfectly organized must always succumb.

Now it is clear that what takes place among the individuals of a species must also occur among the several allied species of a group—viz. that those which are best adapted to obtain a regular supply of food, and to defend themselves against the attacks of their enemies and the vicissitudes of the seasons, must necessarily obtain and preserve a superiority in population; while those species which from some defect of power or organization are the least capable of counteracting the vicissitudes of food, supply, etc., must diminish in numbers, and, in extreme cases, become altogether extinct. Between these extremes the species will present various degrees

[1] It is ironic and saddening that Wallace chose to illustrate this point about an abundant species, possessing strong survival characteristics, with the passenger pigeon, which is now extinct. *[The Editor.]*

of capacity for ensuring the means of preserving life; and it is thus we account for the abundance or rarity of species. Our ignorance will generally prevent us from accurately tracing the effects to their causes; but could we become perfectly acquainted with the organization and habits of the various species of animals, and could we measure the capacity of each for performing the different acts necessary to its safety and existence under all the varying circumstances by which it is surrounded, we might be able even to calculate the proportionate abundance of individuals which is the necessary result.

If now we have succeeded in establishing these two points—1st, *that the animal population of a country is generally stationary, being kept down by a periodical deficiency of food, and other checks*; and, 2nd, *that the comparative abundance or scarcity of the individuals of the several species is entirely due to their organization and resulting habits, which, rendering it more difficult to procure a regular supply of food and to provide for their personal safety in some cases than in others, can only be balanced by a difference in the population which have to exist in a given area*—we shall be in a condition to proceed to the consideration of *varieties*, to which the preceding remarks have a direct and very important application.

Most or perhaps all the variations from the typical form of a species must have some definite effect, however, slight, on the habits or capacities of the individuals. Even a change of colour might, by rendering them more or less distinguishable, affect their safety; a greater or less development of hair might modify their habits. More important changes, such as an increase in the power or dimensions of the limbs or any of the external organs, would more or less affect their mode of procuring food or the range of country which they inhabit. It is also evident that most changes would affect, either favourably or adversely, the powers of prolonging existence. An antelope with shorter or weaker legs must necessarily suffer more from the attacks of the feline carnivora; the passenger pigeon with less powerful wings would sooner or later be affected in its powers of procuring a regular supply of food; and in both cases the result must necessarily be a diminution of the population of the modified species. If, on the other hand, any species should produce a variety having slightly increased powers of preserving existence, that variety must inevitably in time acquire a superiority in numbers. These results must follow as surely as old age, intemperance, or scarcity of food produce an increased mortality. In both cases there may be many individual exceptions; but on the average the rule will invariably be found to hold good. All varieties will therefore fall into two classes—those which under the same conditions would never reach the population of the parent species, and those which would in time obtain and keep a numerical superiority. Now, let some alteration of physical conditions occur in the district—a long period of drought, a destruction of vegetation by locusts, the irruption of some new carnivorous animal seeking "pastures new"—any change in fact tending to render existence more difficult to the species in question, and tasking its utmost powers to avoid complete extermination; it is evident that, of all the individuals composing the species, those forming the least numerous and most feebly organized variety would suffer first, and, were the pressure severe, must soon become extinct. The same causes continuing in action, the parent species would next suffer, would gradually diminish in numbers, and with a recurrence of similar unfavourable conditions might also become extinct. The superior variety would then alone remain, and on a return to favourable circumstances would rapidly increase in numbers and occupy the place of the extinct species and variety.

The *variety* would now have replaced the *species*, of which it would be a more perfectly developed and more highly organized form. It would be in all respects better adapted to secure its safety, and to prolong its individual existence and that of the race. Such a variety *could not* return to the original form; for that form is an inferior one, and could never compete with it for existence. Granted, therefore, a "tendency" to reproduce the original type of the species, still the variety must ever remain preponderant in numbers, and under adverse physical conditions *again alone survive*. But this new, improved, and populous race might itself, in course of time, give rise to new varieties, exhibiting several diverging modifications of form, any of which, tending to increase the facilities for preserving existence, must by the same general law, in their turn become predominant. Here, then, we have *progres-*

*sion and continued divergence* deduced from the general laws which regulate the existence of animals in a state of nature, and from the undisputed fact that varieties do frequently occur. It is not, however, contended that this result would be invariable; a change of physical conditions in the district might at times materially modify it, rendering the race which had been the most capable of supporting existence under the former conditions now the least so, and even causing the extinction of the newer and, for a time, superior race, while the old or parent species and its first inferior varieties continued to flourish. Variations in unimportant parts might also occur, having no perceptible effect on the life-preserving powers; and the varieties so furnished might run a course parallel with the parent species, either giving rise to further variations or returning to the former type. All we argue for is, that certain varieties have a tendency to maintain their existence longer than the original species, and this tendency must make itself felt; for though the doctrine of chances or averages can never be trusted to on a limited scale, yet, if applied to high numbers, the results come nearer to what theory demands, and, as we approach to an infinity of examples, become strictly accurate. Now the scale on which nature works is so vast—the numbers of individuals and periods of time with which she deals approach so near to infinity, that any cause, however slight, and however liable to be veiled and counteracted by accidental circumstances, must in the end produce its full legitimate results.

Let us now turn to domesticated animals, and inquire how varieties produced among them are affected by the principles here enunciated. The essential difference in the condition of wild and domestic animals is this,—that among the former, their well-being and very existence depend upon the full exercise and healthy condition of all their senses and physical powers, whereas, among the latter, these are only partially exercised, and in some cases are absolutely unused. A wild animal has to search, and often to labour, for every mouthful of food—to exercise sight, hearing, and smell in seeking it, and in avoiding dangers, in procuring shelter from the inclemency of the seasons, and in providing for the subsistence and safety of its offspring. There is no muscle of its body that is not called into daily and hourly activity; there is no sense or faculty that is not strengthened by continual exercise. The domestic animal, on the other hand, has food provided for it, is sheltered, and often confined, to guard it against the vicissitudes of the seasons, is carefully secured from the attacks of its natural enemies, and seldom even rears its young without human assistance. Half of its senses and faculties are quite useless; and the other half are but occasionally called into feeble exercise, while even its muscular system is only irregularly called into action.

Now when a variety of such an animal occurs, having increased power or capacity in any organ or sense, such increase is totally useless, is never called into action, and may even exist without the animal ever becoming aware of it. In the wild animal, on the contrary, all its faculties and powers being brought into full action for the necessities of existence, any increase becomes immediately available, is strengthened by exercise, and must even slightly modify the food, the habits, and the whole economy of the race. It creates as it were a new animal, one of superior powers, and which will necessarily increase in numbers and outlive those inferior to it.

Again, in the domesticated animal all variations have an equal chance of continuance; and those which would decidedly render a wild animal unable to compete with its fellows and continue its existence are no disadvantage whatever in a state of domesticity. Our quickly fattening pigs, short-legged sheep, pouter pigeons, and poodle dogs could never have come into existence in a state of nature, because the very first step towards such inferior forms would have led to the rapid extinction of the race; still less could they now exist in competition with their wild allies. The great speed but slight endurance of the race horse, the unwielding strength of the ploughman's team, would both be useless in a state of nature. If turned wild on the pampas, such animals would probably soon become extinct, or under favorable circumstances might each lose those extreme qualities which would never be called into action, and in a few generations would revert to a common type, which must be that in which the various powers and faculties are so proportioned to each other as to be best adapted to procure food and secure safety,—that in which by the full exercise of every part of his organization the animal can alone continue to live. Domestic varieties, when turned wild, *must* return to something near the type of the original wild stock, or *become altogether extinct.*

We see, then, that no inferences as to varieties in a state of nature can be deduced from the observation of those occurring among domestic animals. The two are so much opposed to each other in every circumstance of their existence, that what applies to the one is almost sure not to apply to the other. Domestic animals are abnormal, irregular, artificial; they are subject to varieties which never occur and never can occur in a state of nature; their very existence depends altogether on human care: so far are many of them removed from that just proportion of faculties, that true balance of organization, by means of which alone an animal left to its own resources can preserve its existence and continue its race.

The hypothesis of Lamarck—that progressive changes in species have been produced by the attempts of animals to increase the development of their own organs, and thus modify their structure and habits—has been repeatedly and easily refuted by all writers on the subject of varieties and species, and its seems to have been considered that when this was done the whole question has been finally settled; but the view here developed renders such an hypothesis quite unnecessary, by showing that similar results must be produced by the action of principles constantly at work in nature. The powerful retractile talons of the falcon and the cat-tribes have not been produced or increased by the volition of those animals; but among the different varieties which occurred in the earlier and less highly organized forms of these groups, *those always survived longest which had the greatest facilities for seizing their prey*. Neither did the giraffe acquire its long neck by desiring to reach the foliage of the more lofty shrubs, and constantly stretching its neck for the purpose, but because any varieties which occurred among its antitypes with a longer neck than usual *at once secured a fresh range of pasture over the same ground as their shorter-necked companions, and on the first scarcity of food were thereby enabled to outlive them.* Even the peculiar colours of many animals, especially insects, so closely resembling the soil or the leaves or the trunks on which they habitually reside, are explained on the same principle; for though in the course of ages varieties of many tints may have occurred, *yet those races having colours best adapted to concealment from their enemies would inevitably survive the longest*. We have also here an acting cause to account for that balance so often observed in nature,—a deficiency in one set of organs always being compensated by an increased development of some others—powerful wings accompanying weak feet, or great velocity making up for the absence of defensive weapons; for it has been shown that all varieties in which an unbalanced deficiency occurred could not long continue their existence. The action of this principle is exactly like that of the centrifugal governor of the steam engine, which checks and corrects any irregularities almost before they become evident; and in like manner no unbalanced deficiency in the animal kingdom can ever reach any conspicuous magnitude, because it would make itself felt at the very first step, by rendering existence difficult and extinction almost sure to follow. An origin such as is here advocated will also agree with the peculiar character of the modifications of form and structure which obtain in organized beings—the many lines of divergence from a central type, the increasing efficiency and power of a particular organ through a succession of allied species, and the remarkable persistence of unimportant parts such as colour, texture of plumage and hair, form of horns or crests, through a series of species differing considerably in more essential characters. It also furnishes us with a reason for that "more specialized structure" which Professor Owen states to be a characteristic of recent compared with extinct forms, and which would evidently be the result of the progressive modification of any organ applied to a special purpose in the animal economy.

We believe we have now shown that there is a tendency in nature to the continued progression of certain classes of *varieties* further and further from the original type—a progression to which there appears no reason to assign any definite limits—and that the same principle which produces this result in a state of nature will also explain why domestic varieties have a tendency to revert to the original type. This progression, by minute steps, in various directions, but always checked and balanced by the necessary conditions, subject to which alone existence can be preserved, may, it is believed, be followed out so as to agree with all the phenomena presented by organized beings, their extinction and succession in past ages, and all the extraordinary modifications of form, instinct, and habits which they exhibit.

Ternate, February, 1858.

*Journal of the Linnean Society, Zoology* (1858) III, 45-62.

## Questions and Topics for Discussion and Writing

1. What is the importance, for evolutionary theory, of the existence of "varieties" within a species (i.e. animals or plants belonging to a single species, but having certain different characteristics such as size, shape, colour, etc., of various parts of the body)?
2. What point is Wallace making when he discusses the reproductive rates of animals?
3. Explain the "struggle for existence."
4. How is it possible that Wallace, who knew nothing of Darwin's evolutionary ideas, came to virtually the identical theory of evolution?

# The Scopes Trial

## State of Tennessee

In 1925, the "Butler Act" was passed in the state of Tennessee, prohibiting the teaching of evolution in the public schools. To test the law before the courts, a high school science teacher named John Thomas Scopes agreed to be the defendant. He got himself arrested and indicted and stood trial in July of that year. The American Civil Liberties Union provided for his defence, led by the famous trial lawyer, Clarence Darrow. The prosecution lawyers included former presidential candidate William Jennings Bryan.

The trial riveted the attention of America on Tennessee. After Scopes' conviction Mississippi and Arkansas followed suit passing laws similar to the Butler Act. The Butler Act was finally repealed in 1967 and in 1968 the United States Supreme Court declared the Arkansas law unconstitutional.

The text which follows is excerpted from the official transcript of *State of Tennessee* vs. *John Thomas Scopes*. The trial went on for eleven days, so the transcript is quite long. These excerpts give some of the more memorable arguments on both sides, at the expense of some abrupt jumps to different parts of the transcript. The attorneys for the prosecution quoted here are William Jennings Bryan and General A. T. Stewart, the Attorney General of Tennessee. Counsel for the defence included Clarence Darrow, John R. Neal, Dudley Field Malone, and Arthur Garfield Hays.

GEN. STEWART—[*Reading*: ] State of Tennessee, County of Rhea. Circuit Court. July Special Term, 1925.

The grand jurors for the state aforesaid, being duly summoned, elected, empanelled, sworn, and charged to inquire for the body of the county aforesaid, upon their oaths present:

That John Thomas Scopes, heretofore on the 24th day of April, 1925, in the county aforesaid, then and there, unlawfully did willfully teach in the public schools of Rhea County, Tennessee, which said public schools are supported in part and in whole by the public school fund of the state, a certain theory and theories that deny the story of the divine creation of man as taught in the Bible, and did teach instead thereof that man has descended from a lower order of animals, he, the said John Thomas Scopes, being at the time, and prior thereto, a teacher in the public schools of Rhea County, Tennessee, aforesaid, against the peace and dignity of the State.

THE COURT—What is your plea, gentlemen?

* * *

MR. NEAL—[*Reading*: ] The defendant moves the court to quash the indictment in this case for the following reasons:

First—(A) Because the act which is the basis of the indictment, and which the defendant is charged with violating, is unconstitutional and void in that it violates...Section 3, Article I of the constitution of Tennessee:

> Section 3. Right of Worship Free—That all men have a natural and indefeasible right to worship Almighty God according to the dictates of his own conscience; that no man can of right, be compelled to attend, erect or support any place of worship, or to maintain any minister against his consent; that no human authority can, in any case whatever, control or interfere with the rights of conscience; and that no preference shall ever be given, by law, to any religious establishment or mode of worship.

* * *

GEN. STEWART—The next one, and the one which Dr. Neal referred to as one of the most important ones, Section 3, Article I, still of the constitution, the right of free worship...

If your Honor please, this law is as far removed from that interference with the provision in the constitution as it is from any other that is not even cited. This does not interfere with the religious worship—it does not even approach interference with religious worship...

MR. DARROW—...The part we claim is that last clause, "no preference shall ever be given, by law, to any religious establishment or mode of worship."

GEN. STEWART—Yes, that "no preference shall ever be given, by law, to any religious establishment or mode of worship." Then, how could that interfere, Mr. Darrow?

MR. DARROW—That is the part we claim is affected.

GEN. STEWART—In what wise?

MR. DARROW—Giving preference to the Bible.

* * *

GEN. STEWART—...There is as little in that as in any of the rest. If your Honor please, the St. James Version of the Bible is the recognized one in this section of the country. The laws of the land recognize the Bible; the laws of the land recognize the law of God and Christianity as a part of the common law.

MR. MALONE—Mr. Attorney General, may I ask a question?

GEN. STEWART—Certainly.

MR. MALONE—Does the law of the land or the law of the state of Tennessee recognize the Bible as part of a course in biology or science?

GEN. STEWART—I do not think the law of the land recognizes them as confusing one another in any particular.

* * *

GEN. STEWART—...The question involved here is, to my mind, the question of the exercise of the police power.

MR. NEAL—It does not mention the Bible?

GEN. STEWART—Yes, it mentions the Bible. The legislature, according to our laws, in my opinion, would have the right to preclude the teaching of geography. That is—

MR. NEAL—Does not it prefer the Bible to the Koran?

GEN. STEWART—It does not mention the Koran.

MR. MALONE—Does not it prefer the Bible to the Koran?

GEN. STEWART—We are not living in a heathen country.

MR. MALONE—Will you answer my question? Does not it prefer the Bible to the Koran?

GEN. STEWART—We are not living in a heathen country, so how could it prefer the Bible to the Koran?...

MR. MALONE—...I would say to base a theory set forth in any version of the Bible to be taught in the public school is an invasion of the rights of the citizen, whether exercised by the police power or by the legislature.

GEN. STEWART—Because it imposes a religious opinion?

MR. MALONE—Because it imposes a religious opinion, yes. What I mean is this: If there be in the state of Tennessee a single child or young man or young woman in your school who is a Jew, to impose upon any course of science a particular view of creation from the Bible is interfering, from our point of view, with his civil rights under our theory of the case. That is our contention.

* * *

MR. DARROW—...This case we have to argue is a case at law, and hard as it is for me to bring my mind to conceive it, almost impossible as it is to put my mind back into the sixteenth century, I am going to argue it as if it was serious, and as if it was a death struggle between two civilizations.

...We have been informed that the legislature has the right to prescribe the course of study in the public schools. Within reason, they no doubt have, no doubt. They could not prescribe it, I am inclined to think, under your constitution, if it omitted arithmetic and geography and writing, neither under the rest of the constitution, if it shall remain in force in the State, could they prescribe it if the course of study was only to teach religion, because several hundred years ago, when our people believed in freedom, and when no man felt so sure of their own sophistry that they were willing to send a man to jail who did not believe them, the people of Tennessee adopted a constitution, and they made it broad and plain, and said that the people of Tennessee would always enjoy religious freedom in its broadest terms; so, I assume, that no legislature could fix a course of study which violated that...

I remember, long ago, Mr. Bancroft wrote this sentence, which is true: "That it is all right to preserve freedom in constitutions, but when the spirit of freedom has fled from the hearts of the people, then its matter is easily sacrificed under law." And so it is, unless there is left enough of the spirit of freedom in the state of Tennessee, and in the United States, there is not a single line of any constitution that can withstand bigotry and ignorance when it seeks to destroy the rights of the individual; and bigotry and ignorance are ever active. Here, we find today as brazen and as bold an attempt to destroy learning as was ever made in the Middle Ages, and the only difference is we have not provided that they shall be burned at the stake, but there is time for that, your Honor; we have to approach these things gradually.

Now, let us see what we claim with reference to this law. If this proceeding, both in form and substance, can prevail in this court, then your Honor, no law—no matter haw foolish, wicked, ambiguous, or ancient, but can come back to Tennessee. All the guarantees go for nothing. All of the past has gone, will be forgotten, if this can succeed...

The statue should be comprehensible. It should not be written in Chinese anyway. It should be in passing English, as you say, so that common, human beings would understand what it meant, and so a man would know whether he is liable to go to jail when he is teaching, not so ambiguous as to be a snare or a trap to get someone who does not agree with you. If should be plain, simple and easy. Does this statute state what you shall teach and what you shall not? Oh, no! Not at all. Does it say you cannot teach the earth is round, because Genesis says it is flat? No. Does it say you cannot teach that the earth is millions of ages old, because the account in Genesis makes it less than six thousand years old? Oh, no. It doesn't state that. If it did you could understand it. It says you shan't teach any theory of the origin of man that is contrary to the divine theory contained in the Bible.

Now let us pass up the word "divine"! No legislature is strong enough in any state in the Union to characterize and pick any book as being divine. Let us take it as it is. What is the Bible?...The Bible is not one book. The Bible is made up of sixty-six books written over a period of about one thousand years, some of them very early and some of them comparatively late. It is a book primarily of religion and morals. It is not a book of science. Never was and was never meant to be. Under it there is nothing prescribed that would tell you how to build a railroad or a steamboat or to make anything that would advance civilization. It is not a textbook or a text on chemistry. It is not big enough to be. It is not a book on geology; they knew nothing about it. It is not a work on evolution; that is a mystery. It is not a work on astronomy. The man who looked out at the universe and studied the heavens had no thought but that the earth was the center of the universe. But we know better than that. We know that the sun is the center of the solar system. And that there are an infinity of other systems around about us. They thought the sun went around the earth and gave us night. We know better. We know the earth turns on its axis to produce days and nights. They thought the earth was [created] 4,004 years before the Christian Era. We know better. They told it the best they knew. And while suns may change all you may learn of chemistry, geometry and mathematics, there are no doubt certain primitive, elemental instincts in the organs of man that remain the same; he finds out what he can and yearns to know more and supplements his knowledge with hope and faith.

That is the province of religion and I haven't the slightest fault to find with it. Not the slightest in the world. One has one thought and one another, and instead of fighting each other as in the past, they should support and help each other. Let's see now. Can your Honor tell what is given as the origin of man as shown in the Bible? Is there any human being who can tell us? There are two conflicting accounts in the first two chapters. There are scattered all through it various acts and ideas, but to pass that up for the sake of argument, no teacher in any school in the state of Tennessee can know that he is violating a law, but must test every one of its doctrines by the Bible, must he not? You cannot say two times two equals four or make a man an educated man if evolution is forbidden. It does not specify what you cannot teach, but says you cannot teach anything that conflicts with the Bible. Then just imagine making it a criminal code that is so uncertain and impossible that every man must be sure that he has read everything in the Bible and not only read it but understands it, or he might violate the criminal code. Who is the chief mogul that can tell us what the Bible means? He or they should write a book and make it plain and distinct, so we would know. Let us look at it. There are in America at least five hundred different sects or churches, all of which quarrel with each other on the importance and nonimportance of certain things or the construction of certain passages. All along the line they do not agree among themselves and cannot agree among themselves. They never have and probably never will. There is a great division between the Catholics and the Protestants. There is such a disagreement that my client, who is a school-teacher, not only must know the subject he is teaching, but he must know everything about the Bible in reference to evolution. And he must be sure that he expresses it right or else some fellow will come along here, more ignorant perhaps than he, and say, "You made a bad guess and I think you have committed a crime." No criminal statute can rest that way...

It cannot stand a minute in this court on any theory than that it is a criminal act, simply because they say it contravenes the teaching of Moses, without telling us what those teachings are. Now, if this is the subject of a criminal act, then it cannot make a criminal out of a teacher in the public schools and leave a man free to teach it in a private school. It cannot make it criminal for a teacher in the public schools to teach evolution, and for the same man to stand among the hustings and teach it. It cannot make it a criminal act for this teacher to teach evolution and permit books upon evolution to be sold in every store in the state of Tennessee and to permit the newspapers from foreign cities to bring into your peaceful community the horrible utterances of evolution. Oh, no, nothing like that. If the state of Tennessee has any force in this day of Fundamentalism, in this day when religious bigotry and hatred is being kindled all over our land, see what can be done...

...Ignorance and fanaticism is ever busy and needs feeding. Always it is feeding and gloating for more. Today it is the public school teachers, tomorrow the private. The next day the preachers and the lecturers, the magazines, the books, the newspapers. After a while, your Honor, it is the setting of man against man and creed against creed, until with flying banners and beating drums we are marching backward to the glorious ages of the sixteenth century, when bigots lighted fagots to burn the men who dared to bring any intelligence and enlightenment and culture to the human mind...

MR. BRYAN—Little Howard Morgan—and, your Honor, that boy is going to make a great lawyer someday. I didn't realize it until I saw how a 14–year–old boy understood the subject so much better than a distinguished lawyer who attempted to quiz him. The little boy understood what he was talking about and, to my surprise, the attorneys didn't seem to catch the significance of the theory of evolution and the thought—and I'm sure he wouldn't have said it if he hadn't had thought it—he thought that little boy was talking about the individuals coming up from one cell. That wouldn't be evolution—that is growth, and one trouble about evolution is that it has been used in so many different ways that people are confused about it...

Tell me that the parents of this day have not any right to declare that children are not to be taught this doctrine? Shall not be taken down from the high plane upon which God put man? Shall be detached from the throne of God and be compelled to link their ancestors with the jungle; tell that to these children? Why, my friends, if they believe it, they go back to scoff at the religion of their parents! And the parents have a right to say that no teacher paid by their money shall rob their children of faith in God and send them back to their homes, skeptical, infidels, or agnostics, or atheists!

This doctrine that they want taught, this doctrine that they would force upon the schools, where they will not let the Bible be read—why, up in the state of New York they are now trying to keep the schools from adjourning for one hour in the afternoon, not that any teacher shall teach them the Bible, but that the children may go to the churches to which they belong and there have instruction in the Word. And they are refusing to let the school do that. These lawyers who are trying to force Darwinism and evolution on your children do not go back to protect the children of New York in their right to even have religion taught to them outside of the schoolroom, and they want to bring their experts in here...

Now, my friends, I want you to know that they not only have no proof, but they cannot find the beginning. I suppose this distinguished scholar who came here shamed them all by his number of degrees. He did not shame me, for I have more than he has, but I can understand how my friends felt when he unrolled degree after degree. Did he tell you where life began? Did he tell you that back of all these that there was a God? Not a word about it. Did he tell you how life began? Not a word, and not one of them can tell you how life began. The atheists say it came some way without a God; the agnostics say it came in some way, they know not whether with a God or not. And the Christian evolutionists say we came away back there somewhere, but they do not know how far back—they do not give you the beginning—not that gentleman that tried to qualify as an expert; he did not tell you whether it began with God or how. No, they take up life as a mystery that nobody can explain, and they want you to let them commence there and ask no questions. They want to come in with their little padded up evolution that commences with nothing and ends nowhere. They do not dare to tell you that it began with God and...ended with God. They come here with this bunch of stuff that they call evolution, that they tell you that everybody believes in, but do not know that everybody knows as a fact, and nobody can tell how it came, and they do not explain the great riddle of the universe—they do not deal with the problems of life—they do not teach the great science of how to live—and yet they would undermine the faith of these little children in that God who stands back of everything and whose promise we have that we shall live with Him forever bye and bye. They shut God out of the world. They do not talk about God. Darwin says the beginning of all things is a mystery unsolvable by us. He does not pretend to say how these things started...

And your Honor asked me whether it has anything to do with the principle of the virgin birth. Yes, because this principle of evolution disputes the miracle; there is no place for the miracle in this train of evolution, and the Old Testament and the New are filled with miracles, and if this doctrine is true, this logic eliminates every mystery in the Old Testament and the New, and eliminates everything supernatural; and that means they eliminate the virgin birth—that means that they eliminate the resurrection of the body—that means that they eliminate the doctrine of atonement. And they believe man has been rising all the time, that man never fell; that when the Savior came there was not any reason for His coming; there was no reason why He should not go as soon as He could, that He was born of Joseph or some other correspondent, and that He lies in his grave. And when the Christians of this state have tied their hands and said, "We will not take advantage of our power to teach religion to our children, by teachers paid by us," these people come in from the outside of the state and force upon the people of this state and upon the children of the taxpayers of this state a doctrine that refutes not only their belief in God, but their belief in a Savior and belief in heaven, and takes from them every moral standard that the Bible gives us...

Your Honor, we first pointed out that we do not need any experts in science....And, when it comes to Bible experts, every member of the jury is as good an expert on the Bible as any man that they could bring, or that we could bring. The one beauty about the Word of God is, it does not take an expert to understand it. They have translated that Bible into five hundred languages; they have carried it into nations where but few can read a word, or write, to people who never saw a book, who never read, and yet can understand that Bible, and they can accept the salvation that the Bible offers, and they can know more about that book by accepting Jesus and feeling in their hearts the sense of their sins forgiven than all of the skeptical outside Bible experts that could come in here to talk to the people of Tennessee about the construction that they place upon the Bible, that is foreign to the construction that the people here place upon it.

MR. HAYS—The Defense desires to call Mr. Bryan as a witness, and, of course, since the only question here is whether Mr. Scopes taught what these children said he taught, we recognize what Mr. Bryan says as a witness would not be very valuable. We think there are other questions involved, and we should want to take Mr. Bryan's testimony for the purposes of our record, even if your Honor thinks it is not admissible in general, so we wish to call him now.

THE COURT—Mr. Bryan, you are not objecting to going on the stand?

MR. BRYAN—Not at all.

THE COURT—Do you want Mr. Bryan sworn?

MR. DARROW—no.

MR. BRYAN—I can make affirmation; I can say, "So help me God, I will tell the truth."

MR. DARROW—No, I take it you will tell the truth, Mr. Bryan.

Q [MR. DARROW]—You have given considerable

study to the Bible, haven't you, Mr. Bryan?

A [MR. BRYAN]—Yes, sir, I have tried to.

* * *

Q—Do you claim that everything in the Bible should be literally interpreted?

A—I believe everything in the Bible should be accepted as it is given there; some of the Bible is given illustratively. For instance: "Ye are the salt of the earth." I would not insist that man was actually salt, or that he had flesh of salt, but it is used in the sense of salt as saving God's people.

Q—But when you read that Jonah swallowed the whale—or that the whale swallowed Jonah—excuse me please—how do you literally interpret that?

A—When I read that a big fish swallowed Jonah—it does not say whale.

Q—Doesn't it? Are you sure?

A—That is my recollection of it. A big fish, and I believe it; and I believe in a God who can make a whale and can make a man and make both do what He pleases.

Q—Mr. Bryan, doesn't the New Testament say a whale?

A—I am not sure. My impression is that it says fish; but it does not make so much difference; I merely called your attention to where it says fish—it does not say whale.

Q—But in the New Testament it says whale, doesn't it?

A—That may be true; I cannot remember in my own mind what I read about it.

Q—Now, you say, the big fish swallowed Jonah, and he there remained how long? three days? and then he spewed him upon the land. You believe that the big fish was made to swallow Jonah?

A—I am not prepared to say that; the Bible merely says it was done.

Q—You don't know whether it was the ordinary run of fish, or made for that purpose?

A—You may guess; you evolutionists guess.

Q—But when we do guess, we have a sense to guess right.

A—But do not do it often.

Q—You are not prepared to say whether that fish was made especially to swallow a man or not?

A—The Bible doesn't say, so I am not prepared to say.

Q—You don't know whether that was fixed up specially for the purpose?

A—No, the Bible doesn't say.

Q—But you do believe He made them—that He made such a fish and that it was big enough to swallow Jonah?

A—Yes, sir. Let me add: one miracle is just as easy to believe as another.

Q—It is for me.

A—It is for me.

Q—Just as hard?

A—It is hard to believe for you, but easy for me. A miracle is a thing performed beyond what man can perform. What you get beyond what man can do, you get with the realm of miracles; and it is just as easy to believe the miracle of Jonah as any other miracle in the Bible.

Q—Perfectly easy to believe that Jonah swallowed the whale?

A—If the Bible said so; the Bible doesn't make as extreme statements as evolutionists do.

MR. DARROW—That may be a question, Mr. Bryan, about some of those you have known.

A—The only thing is, you have a definition of fact that includes imagination.

Q—And you have a definition that excludes everything but imagination.

GEN. STEWART—I object to that as argumentative.

THE WITNESS—You—

MR. DARROW—The witness must not argue with me, either.

Q—Do you consider the story of Jonah and the whale a miracle?

A—I think it is.

Q—Do you believe Joshua made the sun stand still?

A—I believe what the Bible says. I suppose you mean that the earth stood still?

Q—I don't know. I am talking about the Bible now.

A—I accept the Bible absolutely.

Q—The Bible says Joshua commanded the sun to stand still for the purpose of lengthening the day, doesn't it? and you believe it?

A—I do.

Q—Do you believe at that time the entire sun went around the earth?

A—No, I believe that the earth goes around the sun.

Q—Do you believe that men who wrote it thought that the day could be lengthened or that the sun could be stopped?

A—I don't know what they thought.

Q—You don't know?

A—I think they wrote the fact without expressing their own thoughts.

* * *

MR. DARROW—Have you an opinion as to whether—whoever wrote the book, I believe it is, Joshua, the Book of Joshua, thought the sun went around the earth or not?

A—I believe that he was inspired.

MR. DARROW—Can you answer my question?

A—When you let me finish the statement.

Q—It is a simple question, but finish it.

THE WITNESS—You cannot measure the length of my answer by the length of your question.

[*Laughter in the courtyard.*]

MR. DARROW—No, except that the answer be longer.

[*Laughter in the courtyard.*]

A—I believe that the Bible is inspired, an inspired author, whether one who wrote as he was directed to write understood the things he was writing about, I don't know.

Q—Whoever inspired it? Do you think whoever inspired it believed that the sun went around the earth?

A—I believe it was inspired by the Almighty, and He may have used language that could be understood at that time.

Q—Was—

THE WITNESS—Instead of using language that could not be understood until Mr. Darrow was born.

[*Laughter and applause in the courtyard.*]

* * *

Q—You believe the story of the flood to be a literal interpretation?

A—Yes, sir.

Q—When was that flood?

A—I would not attempt to fix the date. The date is fixed, as suggested this morning.

Q—About 4004 B.C.?

A—That has been the estimate of a man that is accepted today. I would not say it is accurate.

* * *

MR. DARROW—How long ago was the flood, Mr. Bryan?

* * *

THE WITNESS—Oh, I would put the estimate where it is, because I have no reason to vary it. But I would have to look at it to give you the exact date.

Q—I would, too. Do you remember what book the account is in?

A—Genesis.

MR. HAYS—Is that the one in evidence?

MR. NEAL—That will have it; that is the King James Version.

MR. DARROW—The one in evidence has it.

THE WITNESS—It is given here, as 2348 years B.C.

Q—Well, 2348 years B.C. You believe that all the living things that were not contained in the ark were destroyed.

A—I think the fish may have lived.

Q—Outside of the fish?

A—I cannot say.

Q—You cannot say?

A—No, I accept that just as it is; I have no proof to the contrary.

Q—I am asking you whether you believe?

A—I do.

Q—That all living things outside of the fish were destroyed?

A—What I say about the fish is merely a matter of humor.

Q—I understand.

* * *

Q—Don't you know that the ancient civilizations of China are 6,000 or 7,000 years old, at the very least?

A—No; but they would not run back beyond the creation, according to the Bible, 6,000 years.

Q—You don't know how old they are, is that right?

A—I don't know how old they are, but probably you do. [*Laughter in the courtyard.*] I think you would give preference to anybody who opposed the Bible, and I give the preference to the Bible.

Q—I see. Well, you are welcome to your opinion. Have you any idea how old the Egyptian civilization is?

A—No.

Q—Do you know of any record in the world, outside of the story of the Bible, which conforms to any statement that it is 4,200 years ago or thereabouts that all life was wiped off the face of the earth?

A—I think they have found records.

Q—Do you know of any?

A—Records reciting the flood, but I am not an authority on the subject.

Q—Now, Mr. Bryan, will you say if you know of any record, or have ever heard of any records, that describe that a flood existed 4,200 years ago, or about that time, which wiped all life off the earth?

A—The recollection of what I have read on that subject is not distinct enough to say whether the records attempted to fix a time, but I have seen in the discoveries of archaeologists where they have found records that described the flood.

Q—Mr. Bryan, don't you know that there are many old religions that describe the flood?

A—No, I don't know.

Q—You know there are others besides the Jewish?

A—I don't know whether these are the record of any other religion or refer to this flood.

Q—Don't you ever examine religion so far to know that?

A—Outside of the Bible?

Q—Yes.

A—No; I have not examined to know that, generally.

Q—You have never examined any other religions?

A—Yes, sir.

Q—Have you ever read anything about the origins of religions?

A—Not a great deal.

Q—You have never examined any other religion?

A—Yes, sir.

Q—And you don't know whether any other religion ever gave a similar account of the destruction of the earth by the flood?

A—The Christian religion has satisfied me, and I have never felt it necessary to look up some competing religions.

---

*State of Tennessee* vs. *John Thoman Scopes,* Nos. 5231 and 5232, in the Circuit Court of Rhea County, Tennessee.

---

## Questions and Topics for Discussion and Writing

1. Why does Clarence Darrow compare the Scopes trial to the Inquisition of the sixteenth century?
2. What is Darrow trying to achieve by putting William Jennings Bryan—the *counsel* for the prosecution—on the witness stand and questioning him?
3. What is actually at issue in this trial (judging from the small excerpt at hand)? Is it the fact that Scopes taught evolution in the schools? Is it the truth or falsity of evolution? Is it the issue of whether evolution is in conflict with the Bible? Is it some other issue?

# Darwinism Defined: The Difference Between Fact and Theory

## Stephen Jay Gould

In this essay, Stephen Jay Gould, professor of biology, geology, and the history of science at Harvard University and a very popular writer on evolution, takes to task those commentators who confuse the fact of evolution with theories that explain it.

Charles Darwin, who was, perhaps, the most incisive thinker among the great minds of history, clearly divided his life's work into two claims of different character: establishing the fact of evolution, and proposing a theory (natural selection) for the mechanism of evolutionary change. He also expressed, and with equal clarity, his judgment about their different status: confidence in the facts of transmutation and genealogical connection among all organisms, and appropriate caution about his unproved theory of natural selection. He stated in the *Descent of Man*: "I had two distinct objects in view; firstly, to show that species had not been separately created, and secondly, that natural selection had been the chief agent of change...If I have erred in...having exaggerated its [natural selection's] power...I have at least, as I hope, done good service in aiding to overthrow the dogma of separate creations."

Darwin wrote those words more than a century ago. Evolutionary biologists have honored his fundamental distinction between fact and theory ever since. Facts are the world's data; theories are explanations proposed to interpret and coordinate facts. The fact of evolution is as well established as anything in science (as secure as the revolution of the earth about the sun), though absolute certainty has no place in our lexicon. Theories, or statements about the causes of documented evolutionary change, are now in a period of intense debate—a good mark of science in its healthiest state. Facts don't disappear while scientists debate theories. As I wrote in an early issue of this magazine (*Discover*, May 1981), "Einstein's theory of gravitation replaced Newton's, but apples did not suspend themselves in mid-air pending the outcome."

Since facts and theories are so different, it isn't surprising that these two components of science have had separate histories ever since Darwin. Between 1859 (the year of publication for the *Origin of Species*) and 1882 (the year of Darwin's death), nearly all thinking people came to accept the fact of evolution. Darwin lies beside Newton in Westminster Abbey for this great contribution. His theory of natural selection has experienced a much different, and checkered, history. It attracted some notable followers during his lifetime (Wallace in England, Weismann in Germany), but never enjoyed majority support. It became an orthodoxy among English-speaking evolutionists (but never, to this day, in France or Germany) during the 1930s, and received little cogent criticism until the 1970s. The past fifteen years have witnessed a revival of intense and, this time, highly fruitful debate as scientists discover and consider the implications of phenomena that expand the potential causes of evolution well beyond the unitary focus of strict Darwinism (the struggle for reproductive success among organisms within populations). Darwinian selection will not be overthrown; it will remain a central focus of more inclusive evolutionary theories. But new findings and interpretations at all levels, from molecular change in genes to patterns of overall diversity in geological time, have greatly expanded the scope of important causes—from random, selectively neutral change at the genetic level, to punctuated equilibria and catastrophic mass extinction in geological time.

In this period of vigorous pluralism and intense debate among evolutionary biologists, I am greatly saddened to note that some distinguished commentators among non-scientists, in particular Irving Kristol in a *New York Times* Op Ed piece[1] of Sept. 30, 1986 ("Room for Darwin and the Bible"), so egregiously misunderstand the character of our discipline and continue to confuse this central distinction between secure fact and healthy debate about theory.

I don't speak of the militant fundamentalists who label themselves with the oxymoron "scientific creationists," and try to sneak their Genesis literalism into high school classrooms under the guise of scientific dissent. I'm used to their rhetoric, their dishonest mis- and half-quotations, their constant repetition of "useful" arguments that even they must recognize as nonsense (disproved human footprints on dinosaur trackways in Texas, visible misinterpretation of thermodynamics to argue that life's complexity couldn't increase without a divine boost). Our struggle with these ideologues is political, not intellectual. I speak instead of our allies among people committed to reason and honorable argument.

Kristol, who is no fundamentalist, accuses evolutionary biologists of bringing their troubles with creationists upon themselves by too zealous an insistence upon the truths of Darwin's world. He writes: "...the debate has become a dogmatic crusade on both sides, and our educators, school administrators, and textbook publishers find themselves trapped in the middle." He places the primary blame upon a supposedly anti-religious stance in biological textbooks: "There is no doubt that most of our textbooks are still written as participants in the 'warfare' between science and religion that is our heritage from the 19th century. And there is also little doubt that it is this pseudoscientific dogmatism that has provoked the current religious reaction."

Kristol needs a history lesson if he thinks that current creationism is a product of scientific intransigence. Creationism, as a political movement against evolution, has been a continually powerful force since the days of the Scopes trial. Rather than using evolution to crusade against religion in their texts, scientists have been lucky to get anything at all about evolution into books for high school students ever since Scopes's trial in 1925. My own high school biology text, used in the liberal constituency of New York City in 1956, didn't even mention the word evolution. The laws that were used against Scopes and cowed textbook publishers into submission weren't overturned by the Supreme Court until 1968 (*Epperson v. Arkansas*).

But what about Kristol's major charge—anti-religious prejudice and one-dimensional dogmatism about evolution in modern textbooks? Now we come to the heart of what makes me so sad about Kristol's charges and others in a similar vein. I don't deny that some texts have simplified, even distorted, in failing to cover the spectrum of modern debates; this, I fear, is a limitation of the genre itself (and the reason why I, though more of a writer than most scientists, have never chosen to compose a text). But what evidence can Kristol or anyone else provide to demonstrate that evolutionists have been worse than scientists from other fields in glossing over legitimate debate within their textbooks?

Consider the evidence. Two textbooks of evolution now dominate the field. One has as its senior author Theodosius Dobzhansky, the greatest evolutionist of our century, and a lifelong Russian Orthodox; nothing anti-religious could slip past his watchful eye. The second, by Douglas Futuyma, is a fine book by a kind and generous man who could never be dogmatic about anything except intolerance. (His book gives a fair hearing to my own heterodoxies, while dissenting from them.)

When we come to popular writing about evolution, I suppose that my own essays are as well read as any. I don't think that Kristol could include me among Darwinian dogmatists, for most of my essays focus upon my disagreements with the strict version of natural selection. I also doubt that Kristol

[1] An invited editorial article published on the page opposite to the newspaper's own editorials—hence an "Op Ed" piece. *[The Editor.]*

would judge me anti-religious, since I have campaigned long and hard against the same silly dichotomy of science versus religion that he so rightly ridicules. I have written laudatory essays about several scientists (Burnet, Duvier, Buckland, and Gosse, among others) branded as theological dogmatists during the nineteenth-century reaction; and, while I'm not a conventional believer, I don't consider myself irreligious.

Kristol's major error lies in his persistent confusion of fact with theory. He accuses us—without giving a single concrete example, by the way—of dogmatism about *theory* and sustains his charge by citing our confidence in the *fact* of transmutation. "It is reasonable to suppose that if evolution were taught more cautiously, as a conglomerate idea consisting of conflicting hypotheses rather than as an unchallengeable certainty, it would be far less controversial."

Well, Mr. Kristol, evolution (as theory) is indeed "a conglomerate idea consisting of conflicting hypotheses," and I and my colleagues teach it as such. But evolution is also a fact of nature, and so do we teach it as well, just as our geological colleagues describe the structure of silicate minerals, and astronomers the elliptical orbits of planets.

Rather than castigate Mr. Kristol any further, I want to discuss the larger issue that underlies both this incident and the popular perception of evolution in general. If you will accept my premise that evolution is as well established as any scientific fact (I shall give the reasons in a moment), then why are we uniquely called upon to justify our chosen profession; and why are we alone subjected to such unwarranted infamy? To this central question of this essay, I suggest the following answer. We haven't received our due for two reasons: (1) a general misunderstanding of the different methods used by all historical sciences (including evolution), for our modes of inference don't match stereotypes of "*the* scientific method"; and (2) a continuing but unjustified fear about the implication both of evolution itself and of Darwin's theory for its mechanism. With these two issues resolved, we can understand both the richness of science (in its pluralistic methods of inquiry) and the absence of any conflict, through lack of common content, between proper science and true religion.

Our confidence in the fact of evolution rests upon copious data that fall, roughly, into three great classes. First, we have the direct evidence of small-scale changes in controlled laboratory experiments of the past hundred years (on bacteria, on almost every measurable property of the fruit fly *Drosophila*), or observed in nature (color changes in moth wings, development of metal tolerance in plants growing near industrial waste heaps), or produced during a few thousand years of human breeding and agriculture. Creationists can scarcely ignore this evidence, so they respond by arguing that God permits limited modification within created types, but that you can never change a cat into a dog (who ever said that you could, or that nature did?).

Second, we have direct evidence for large-scale changes, based upon sequences in the fossil record. The nature of this evidence is often misunderstood by non-professionals who view evolution as a simple ladder of progress, and therefore expect a linear array of "missing links." But evolution is a copiously branching bush, not a ladder. since our fossil record is so imperfect, we can't hope to find evidence for every tiny twiglet. (Sometimes, in rapidly evolving lineages of abundant organisms restricted to a small area and entombed in sediments with an excellent fossil record, we do discover an entire little bush—but such examples are as rare as they are precious.) In the usual case, we may recover the remains of side branch number 5 from the bush's early history, then bough number 40 a bit later, then the full series of branches 156–161 in a well preserved sequence of younger rocks, and finally surviving twigs 250 and 287.

In other words, we usually find sequences of structural intermediates, not linear arrays of ancestors and descendants. Such sequences provide superb examples of temporally ordered evolutionary trends. Consider the evidence for human evolution in Africa. What more could you ask from a record of rare creatures living in terrestrial environments that provide poor opportunity for fossilization? We have a temporal sequence displaying clear trends in a suite of features, including threefold increase of brain size and corresponding decrease of jaws and teeth. (We are missing direct evidence for an earlier transition to upright posture, but wide-ranging and unstudied sediments of the right age have been found in East Africa, and we have an excellent chance to fill in this part of our story.) What alternative can we suggest to evolution? Would God—for some inscrutable reason, or merely to test our faith—create five species, one after the other (*Australopithecus afarensis, A. Africanus, Homo habilis, H. erectus, and H. sapiens*), to mimic a continuous trend of evolutionary change?

Or, consider another example with evidence of structurally intermediate stages—the transition from reptiles to mammals. The lower jaw of mammals contains but a single bone, the dentary. Reptiles build their lower jaws of several bones. In perhaps the most fascinating of those quirky changes in function that mark pathways of evolution, the two bones articulating the upper and lower jaws of reptiles migrate to the middle ear and become the malleus and incus (hammer and anvil) of mammals.

Creationists, ignorant of hard evidence in the fossil record, scoff at this tale. How could jaw bones become ear bones, they ask. What happened in between? An animal can't work with a jaw half disarticulated during the stressful time of transition.

The fossil record provides a direct answer. In an excellent series of temporally ordered structural intermediates, the reptilian dentary gets larger and larger, pushing back as the other bones of a reptile's lower jaw decrease in size. We've even found a transitional form with an elegant solution to the problem of remaking jaw bones into ear bones. This creature has a double articulation—one between the two bones that become the mammalian hammer and anvil (the old reptilian joint), and a second between the squamosal and dentary bones (the modern mammalian condition). With this built-in redundancy, the emerging mammals could abandon one connection by moving two bones into the ear, while retaining the second linkage, which becomes the sole articulation of modern mammals.

Third, and most persuasive in its ubiquity, we have the signs of history preserved within every organism, every ecosystem, and every pattern of biogeographic distribution, by those pervasive quirks, oddities, and imperfections that record pathways of historical descent. These evidences are indirect, since we are viewing modern results, not the processes that caused them, but what else can we make of the pervasive pattern? Why does our body, from the bones of our back to the musculature of our belly, display the vestiges of an arrangement better suited for quadrupedal life if we aren't the descendants of four-footed creatures? Why do the plants and animals of the Galapagos so closely resemble, but differ slightly from the creatures of Equador, the nearest bit of land 600 miles to the east, especially when cool oceanic currents and volcanic substrate make the Galapagos such a different environment from Ecuador (thus removing the potential argument that God makes the best creatures for each place, and small differences only reflect a minimal disparity of environments)? The similarities can only mean that Ecudorian creatures colonized the Galapagos and then diverged by a natural process of evolution.

This method of searching for oddities as vestiges of the past isn't peculiar to evolution, but a common procedure of all historical science. How, for example, do we know that words have histories, and haven't been decreed by some all-knowing committee in Mr. Orwell's bureau of Newspeak? Doesn't the bucolic etymology of so many words testify to a different life style among our ancestors? In this article, I try to "broadcast" some ideas (a mode of sowing seed) in order to counter the most "egregious" of creationist sophistries (the animal *ex grege*, or outside the flock), for which, given the *quid pro quo* of business, this fine magazine pays me an "emolument" (the fee that millers once received to grind corn).

I don't want to sound like a shrill dogmatist shouting"rally round the flag boys," but biologists have reached a consensus, based on these kinds of data, about the fact of evolution. When honest critics like Irving Kristol misinterpret this agreement, they're either confusing our fruitful consonance about mechanisms of change, or they've misinterpreted part of our admittedly arcane technical literature.

Once such misinterpretation has gained sufficient notoriety in the last year that we crave resolution both for its own sake and as an illustration of the frustrating confusion that can arise when scientists aren't clear and when commentators, as a result of hidden agendas, don't listen. Tom Bethell argued in *Harper's* (February 1985) that a group of young taxonomists called pattern cladists have begun to doubt the existence of evolution itself.

This would be truly astounding news, since cladistics is a powerful method dedicated to reforming classification by using only the branching order of lineages on evolutionary trees ("propinquity of descent" in Darwin's lovely phrase), rather than vague notions of overall similarity in form or function. (For example, in the cladistic system, a lungfish is more closely related to a horse than to a salmon because the common ancestor of lungfish and horse is more recent in time than the link point of the lungfish-horse lineage with the branch leading to modern bony fishes (including salmon).

Cladists use only the order of branching to construct their schemes of relationships; it bothers them not a whit that lungfish and salmon look and work so much alike. Cladism, in other words, is the purest of all genealogical systems for classification, since it works only with closeness of common ancestry in time. How preciously ironic then, that this most rigidly evolutionary of all taxonomic systems should become the subject of such extraordinary misunderstanding—as devised by Bethell, and perpetuated by Kristol when he writes: "...many younger biologists (the so-called 'cladists') are persuaded that the differences among species—including those that seem to be closely related—are such as to make the very concept of evolution questionable."

This error arose for the following reason. A small splinter group of cladists (not all of them, as Kristol claims)—"transformed" or "pattern" cladists by their own designation—have adopted what is to me an ill-conceived definition of scientific procedure. They've decided, by misreading Karl Popper's philosophy, that patterns of branching can be established unambiguously as a fact of nature, but that processes causing events of branching, since they can't be observed directly, can't be known with certainty. Therefore, they say, we must talk only of pattern and rigidly exclude all discussion of process (hence "pattern cladistics").

This is where Bethell got everything arse-backwards and began the whole confusion. A philosophical choice to abjure all talk about process isn't the same thing as declaring that no reason for patterns of branching exists. Pattern cladists don't doubt that evolution is the cause behind branching; rather, they've decided that our science shouldn't be discussing causes at all.

Now I happen to think that this philosophy is misguided; in unguarded moments I would even deem it absurd. Science, after all, is fundamentally about process; learning why and how things happen is the soul of our discipline. You can't abandon the search for cause in favor of a dry documentation of pattern. You must take risks of uncertainty in order to probe the deeper questions, rather than stopping with sterile security. You see, now I've blown our cover. We scientists do have our passionate debates—and I've just poured forth an example. But as I wrote earlier, this is a debate about the proper approach to causes, not an argument about whether causes exist, or even whether the cause of branching is evolution or something else. No cladist denies that branching patterns arise by evolution.

This incident also raises the troubling issue of how myths become beliefs through adulterated repetition without proper documentation. Bethell began by misunderstanding pattern cladistics, but at least he reports the movement as a small splinter, and tries to reproduce their arguments. then Kristol picks up the ball and recasts it as a single sentence of supposed fact—and all cladists have now become doubters of evolution by proclamation. Thus a movement, by fiat, is turned into its opposite—as the purest of all methods for establishing genealogical connections becomes a weapon for denying the mechanism that all biologists accept as the cause of branching on life's tree: evolution itself. Our genealogy hasn't been threatened, but my geniality has almost succumbed.

When I ask myself why the evidence for evolution, so clear to all historical scientists, fails to impress intelligent nonscientists, I must believe that more than simple misinformation lies at the root of our difficulty with a man like Irving Kristol. I believe that the main problem centers upon a restrictive stereotype of scientific method accepted by most non-practitioners as the essential definition of all scientific work.

We learn in high school about *the* scientific method—a cut-and-dried procedure of simplication to essential components, experiment in the controlled situation of a laboratory, prediction and replication. But the sciences of history—not just evolution but a suite of fundamental disciplines ranging from geology, to cosmology, to linguistics—can't operate by this stereotype. We are charged with explaining events of extraordinary complexity that occur but once in all their details.

We try to understand the past, but don't pretend to predict the future. We can't see past processes directly, but learn to infer their operation from preserved results.

Science is a pluralistic enterprise with a rich panoply of methods appropriate for different kinds of problems. Past events of long duration don't lie outside the realm of science because we cannot make them happen in a month within our laboratory. Direct vision isn't the only, or even the usual, method of inference in science. We don't see electrons, or quarks, or chemical bonds, any more than we see small dinosaurs evolve into birds, or India crash into Asia to raise the Himalayas.

William Whewell, the great English philosopher of science during the early nineteenth century, argued that historical science can reach conclusions as well confirmed as any derived from experiment and replication in laboratories, by a method he called "consilience" (literally "jumping together") of inductions. Since we can't see the past directly or manipulate its events, we must use the different tactic of meeting history's richness head on. We must gather its wondrously varied results and search for a coordinating cause that can make sense of disparate data otherwise isolated and uncoordinated. We must see if a set of results so diverse that no one had ever considered their potential coordination might jump together as the varied products of a single process. Thus plate tectonics can explain magnetic stripes on the sea floor, the rise and later erosion of the Appalachians, the earthquakes of Lisbon and San Francisco, the eruption of Mount St. Helens, the presence of large flightless ground birds only on continents once united as Gondwanaland, and the discovery of fossil coal in Antarctica.

Darwin, who understood the different rigor of historical science so well, complained bitterly about those critics who denied scientific status to evolution because they couldn't see it directly or reproduce its historical results in a laboratory. He wrote to Hooker in 1861: "Change of species cannot be directly proved...The doctrine must sink or swim according as it groups and explains phenomena. It is really curious how few judge it in this way, which is clearly the right way." And later, in 1868: "This hypothesis may be tested...by trying whether it explains several large and independent classes of facts; such as the geological succession of organic beings, their distribution in past and present times, and their mutual affinities and homologies."

If a misunderstanding of the different methods of historical inquiry has impeded the recognition of evolution as a product of science at its best, then a residual fear for our own estate has continued to foster resentment of the fact that our physical bodies have ancient roots in ape-like primates, waddling reptiles, jawless fishes, worm-like invertebrates, and other creatures deemed even lower or more ignoble. Our ancient hopes for human transcendence have yet to make their peace with Darwin's world.

But what challenge can the facts of nature pose to our own decisions about the moral value of our lives? We are what we are, but we interpret the meaning of our heritage as we choose. Science can no more answer the questions of how we ought to live than religion can decree the age of the earth. Honorable and discerning scientists (most of us, I trust) have always understood that the limits to what science can answer also describe the power of its methods in their proper domain. Darwin himself exclaimed that science couldn't touch the problem of evil and similar moral conundrums: "A dog might as well speculate on the mind of Newton. Let each man hope and believe what he can."

There is no warfare between science and religion, never was except as a historical vestige of shifting taxonomic boundaries among disciplines. Theologians haven't been troubled by the fact of evolution, unless they try to extend their own domain beyond its proper border (hubris and territorial expansionism aren't the sins of scientists alone, despite Mr. Kristol's fears). The Reverend Henry Ward Beecher, our greatest orator during Darwin's century, evoked the most quintessential of American metaphors in dismissing the entire subject of conflict between science and religion with a single epithet: "Design by wholesale is grander than design by retail"—or, general laws rather than creation of each item by fiat will satisfy our notion of divinity.

Similarly, most scientists show no hostility to religion. Why should we, since our subject doesn't intersect the concerns of theology? I strongly dispute Kristol's claim that "the current teaching of evolution in our public schools does indeed have an ideological bias against religious belief." Unless at least half my colleagues are inconsistent dunces, there can be—on the most raw and direct empirical grounds—no conflict between science and religion. I know hundreds of scientists who share a conviction about the fact of evolution, and teach it in much the same way. Among these people I note an entire spectrum of religious attitudes—from devout daily prayer and worship to resolute atheism. Either there's no correlation between religious belief and confidence in evolution—or else half these people are fools.

The common goal of science and religion is our shared struggle for wisdom in all its various guises. I know no better illustration of this great unity than a final story about Charles Darwin. This scourge of fundamentalism had a conventional church burial—in Westminster Abbey no less. J. Frederick Bridge, Abbey organist and Oxford don, composed a funeral anthem especially for the occasion. It may not rank high in the history of music, but it is, as my chorus director opined, a "sweet piece." (I've made what may be the only extant recording of this work, marred only by the voice of yours truly within the bass section.) Bridge selected for this text the finest biblical description of the common aim that will forever motivate both the directors of his building and the inhabitants of the temple of science—wisdom. "Her ways are ways of pleasantness and all her paths are peace" (Proverbs 3: 17).

I am only sorry that Dr. Bridge didn't set the very next metaphor about wisdom (Proverbs 3: 18), for it describes, with the proper topology of evolution itself, the greatest dream of those who followed the God of Abraham, Isaac, and Jacob: "She is a tree of life to them that lay hold upon her."

From *Discover* (January, 1987), 64–70.

## Questions and Topics for Discussion and Writing

1. Explain the distinction that Gould draws between evolution as a historical fact and the theory of evolution.
2. Why is it so incredible that cladists would be reported to have doubts about evolution?
3. Does Gould believe that science and religion are mutually opposed to each other? What does Gould think of scientific creationism?

# The Discovery of Units of Heredity


## Ruth Moore


The theory of evolution by natural selection as expounded by Darwin or Wallace had one very serious flaw: there was no adequate explanation of the mechanism of inheritance. Not only could Darwin not explain how species could slowly change characteristics, he could not really explain how species in general stayed the same from generation to generation. It was a problem that troubled all the theoretically minded evolutionists in Darwin's day, and drove some of them back toward accepting Lamarck's view that animals acquired characteristics during their own lifetimes and then passed on those new characteristics to their offspring. The irony of all this was that all the while the key insight and its experimental confirmation had been worked out by a quiet monk and published in a scientific journal where anyone might have read it.

In the selection that follows, Ruth Moore tells the story of that monk, Gregor Mendel, and his work. In 1900, 34 years after Mendel's research was published, its importance was finally realized and follow-up work was begun. This was the beginning of the science of genetics.

Mendel's work indicated that inheritable traits were transmitted in discrete units which could only combine in certain ways. Because the units (he called them "factors," we would call them "genes") were discrete entities, they persisted in the population until the last living being containing them died without passing them on. Darwin had thought that inherited traits would blend together in the offspring and not remain discrete. Were that the case, it is difficult to see how the population could ever *evolve* because it would always be tending back to the average traits. But if inherited traits persisted, then those with survival value could eventually become a significant percentage of the "gene pool" of the population.

Mendel never knew exactly what a "factor" or "gene" was, only how it behaved from generation to generation. For that nearly a hundred years had to pass from when Mendel began his work—until 1953 when James Watson and Francis Crick discovered the structure of the DNA molecule. Then it was possible to see that the "factors" passed on were really codes embodied in the protein structure of DNA.

Gregor Johann Mendel was born on July 22, 1822, into a long-settled peasant family living in the tiny village of Heinzendorf in what was then Austrian Silesia and is now Czechoslovakia. As a boy he aided his father in the grafting of trees and in cultivating the garden that was a part of "peasant holding no. fifty-eight." He received additional training in horticulture and beekeeping at the village school. Johann did so well in this work and in his academic studies that his teachers urged that he be sent on to the high school at Troppau. Mendel never lost his love of growing things as he struggled to make his way at Troppau and later at the Philosophical Institute at Olmütz. Gradually this early interest turned into a wider interest in science, although little science was included in the institute's course....

Mendel was so excellent a teacher that the head of the school confidently urged him to try for a teacher's certificate. Mendel wrote the necessary essay and presented himself in Vienna for the oral examinations in the natural history and physics that he wanted to teach. He failed to pass the examination, partly, it seemed, because he lacked certain formal knowledge of his subjects. In part his failure may have been due to the conviction of the friendly chief examiner that the unusual young teacher should be given an opportunity for "higher scientific training" at the University of Vienna.

The abbot of the monastery agreed, and Mendel went to the university for four terms. At the end, he returned to teach in the Brünn "Modern School."

Soon after entering the monastery Mendel began trying to develop new colors in flowers. He also took up the breeding of gray and white mice. Although the latter experiments were dropped after a while, probably at a hint of ecclesiastical disapproval, they were a certain indication that the work with flowers was not a fancy but the first approach of a serious scientist to a fundamental problem.

Mendel's own account of how he happened to begin his studies is a simple one. As he tried to produce new colors in his flowers, he acquired considerable experience in artificial fertilization. His interest was aroused by the surprising and unaccountable results that he sometimes obtained. Whenever he crossed certain species, the same hybrid forms cropped up with striking regularity. But when he crossed one of his hybrids with another, some very different characters sometimes appeared among their progeny.

These phenomena sent Mendel to the literature. He searched all the books available in the monastery and elsewhere in Brünn. A number of scientists were working with the problem of hybrids, but no one had discovered any law governing their formation—if, indeed, anyone could conceive of a law capable of explaining the infinitude of forms in the multitude of plants, Inheritance in all its profusion seemed beyond any nailing down.

Mendel nevertheless had seen certain forms appearing with regularity. Others too had noted this phenomenon. They had not, as far as Mendel could find, counted the forms and classified them. Nor had anyone arranged hybrids according to their generations or worked out their statistical relationships. Mendel, in his remote Austrian town, did not realize that even his concept of counting and figuring the mathematical relationships of hybrid plants was original and entirely his own.

To study heredity properly, Mendel saw, a large number of generations would have to be bred. He also recognized that many plants would have to be included in each generation. A few might produce a misleading picture. "It indeed required some courage to undertake such far-reaching labors," Mendel wrote later in the introduction to one of his monographs. "It appears, however, to be the only way in which we can finally reach the solution of a problem which is of great importance in the evolution of organic forms."

The year was 1854. Many people were asking how the great number of species could have arisen, and yet Charles Darwin would not publish *The Origin of Species* for another five years. It seems unlikely that the Austrian monk was planning a direct study of the great problem of evolution. His interest was concentrated on heredity and how traits could be handed on from parent to offspring. Nevertheless, as his words show, Mendel was well aware of the evolutionary significance of the work he was about to launch.

Mendel planned the big experiment with care and foresight. His first requirement was a plant with varied characters, each of which bred true to form. If a plant produced green peas, he had to be sure that it would go on producing green peas if crossed with another green-pea plant. Mendel also had to have a plant that could be protected from foreign pollen during its flowering period. If even one bee carried in some pollen other than that he would dust on, the whole experiment would be upset. Equally important, the characters had to be easily observable and countable. It was a huge task Mendel was laying out for himself....

...From the thirty-four varieties [of seeds] Mendel selected twenty-two, and from these, seeds with seven sharply contrasting paris of characters. As he wanted to make no fine-line decisions about whether or not a part of a plant had changed, he selected the following clear-cut differences with which to work:

1. The form of the ripe seeds—round or wrinkled;
2. The color of the peas (the seed)—yellow or intense green;
3. The color of the transparent seed cover or skin—white or grayish;
4. The form of the ripe pods—inflated or constructed between the peas;
5. The color of the unripe pods—green or yellow;
6. The position of the flowers—distributed along the stem or bunched at the top; in other words, axial or terminal;

7. The length of the stem—tall (six or seven feet) or dwarf (three-fourths to one and a half feet).

* * *

Each kind of seed went into its own jar—to await the spring of 1856.

Mendel was at last ready to launch his experiment proper. In one section of his long narrow garden he planted round seeds, and in the next section, wrinkled ones. In other sections went the yellow and the green, and, in their own places, all of the others. Keeping the characters that he wanted to compare next to one another would simplify the work that lay ahead...

As soon as the pollen had ripened in the...round peas, Mendel collected a bit on a fine camel's-hair brush and, removing the bags on the wrinkled peas, dusted it on their stigmas. The little bags then were tied on again to keep away the bees and other pollen-carrying insects. To make certain that his experiments were not affected by which plant served as the seed parent, Mendel also reversed his fertilization procedure. Some of the pollen from the wrinkled peas was deposited on the prepared stigmas of the round...

His first indication of what was happening would come when the pods formed in the "unripe pod color" section of his garden. As the pods appeared and filled out and grew longer, Mendel saw with elation that all of them were green. Whether they grew in the "yellow" half of the plot and came from parents that had produced yellow pods or in the "green" half and sprang from parents with green pods, all of them hung green on the vines. Yellow and green parents alike had produced green offspring. Mendel searched his plants. He could not find a single yellow among the unripe pods...

Confirmation of this striking result came as the summer drew to a close and the pods dried on the vines...

During the winter, as he studied his jars of peas and the way all had completely taken on the character of one parent, it was evident that one characteristic in each pair had been entirely "dominant" over the other. Mendel decided to designate the hereditary trait that prevailed as a "dominant." Later he named the other factor, the one lost to sight, a "recessive."

Only the first step had been completed in the big experiment. The next, the crossing of hybrid with hybrid, had to await another spring. As the sun grew warm and the spring rains fell, Mendel planted his hybrid seeds. Each carefully marked group went into its own plot, but this time his procedure would be different. He would not operate on the buds, He would permit the peas to fertilize themselves in the natural way, and thus would obtain a cross of two identical hybrids.

The plantation was again filled with pea vines, and Mendel waited as patiently as he could for the pods and peas to form and dry.

As the young pods grew, Mendel saw a few yellows appearing among the greens. He counted both colors and entered the figures in his notebooks. When the pods dried in the "pod form" plot, some were crinkled around the peas they contained; they were, in the terms he was using, constricted. The others were inflated. Again Mendel counted both types, and again he reserved judgement...

Mendel continued his plantings through six or seven generations in all cases. As he did so, he learned that the 3-to-1 ratio applied to the appearance of the peas only. When Mendel planted the hybrid parents, they produced both round and wrinkled seeds. One fourth were pure rounds and, as long as they were planted, would yield only round peas. Two fourths were round in appearance, but were actually hybrids and in the next generation would produce round and wrinkled peas in a steady 3-to-1 ratio. The wrinkled peas from the union were again pure wrinkled and never would yield anything but their own kind. The wrinkled peas from the first hybrids yielded only wrinkled peas and continued to produce only wrinkled as long as he planted them and permitted them to fertilize themselves. They were pure recessives also...

"It can be seen," said Mendel, "how rash it may be to draw from the external resemblances conclusions as to their internal nature."

Appearance meant little or nothing. It was not surprising that human beings had eternally been bewildered when they tried to understand heredity. The surface differences of offspring were legion in themselves, and yet below the surface lay still other differences and possibilities for variation.

With the clarity and simplicity of genius, Mendel labeled the "dominant" in the union with a capital

*Ruth Moore*

*A*, and the recessive with a small *a*. A constant dominant thus would be formed by the coming together of two *A*'s, and it would be described as *AA*; a hybrid by either *Aa* or *aA*; and recessive by *aa*.

With this understanding, it was possible for him to chart this coming together and separation and the clear-cut results that it produced. Mendel drew, and the world has since followed, his chart of heredity...

Charles Darwin also had experimented with peas and had noticed that the hybrid divided 3 to 1, but Darwin was no mathematician and did not pursue this revealing indication of order in heredity. Mendel, on the other hand, was an excellent mathematician. The new experiment confirmed what he had already glimpsed in his first experiments. He was simply obtaining every combination that could be formed by the separate factors present. If A and *a* were combined, only one combination could be formed: *Aa*. If *Aa* and *Aa* came together, four combinations could be made: *AA, Aa, aA, aa*. And this was exactly what had happened...

Mendel had no way to look into the egg and pollen cells to search for the hereditary factors whose existence he inferred from his experiments. His results, however, were explainable in no other way. Just as Mendel had grasped the secrets of the combinations of these factors, so did he reason out the biological basis that had to underlie them. He formulated three laws of inheritance:

1. All living things are a complex of a large number of independent heritable units.
2. When each parent contributes the same kind of factor and the two come together in the offspring, a constant character is produced. But if one parent contributes one kind of factor, say *A*, and the other another, say *a*, a hybrid results. When the hybrid forms reproductive cells, the two differentiating elements "liberate themselves" again and thus are free to form new combinations in the next union.
3. The factors are unaffected by their long association in the individual. They emerge from the union as distinct as when they entered it.

In a letter to a fellow scientist Mendel enlarged upon this radical idea: "The course of development consists simply in this: that in each successive generation the two primal characters issue distinct and unadulterated out of the hybridized pair, there being nothing whatever to show that either of them had inherited or taken over anything from the others."

This was Mendel's ultimately famous law of the purity of the gametes, or reproductive cells. Mendel regarded it only as a hypothesis, and felt that he must subject it to additional tests...

At last the time had come to report on his eight years of unremitting work. during the fall and early winter of 1864, Mendel checked and rechecked his findings. In his fine copperplate script he wrote the paper that eventually would explain to the world the all-important phenomenon of heredity.

The Brünn Society for the Study of Natural Science was holding its February 1865 meeting at the Modern School, the school in which Mendel taught. The night was cold and snowy, but most of the forty members came to hear Mendel report on the work he had so long been carrying on at the monastery. Curiosity about it was keen.

As Mendel read, this curiosity gave way to incomprehension. The several botanists in the society were as much confused by the report on invariable hereditary ratios in peas as were the other members—a chemist, an astronomer, a geologist, and an authority on cryptograms. Mendel spoke for the hour allotted to him, and then announced that at the next meeting he would explain why the peculiar and regular segregation of characteristics occurred.

The next month, as Mendel presented his algebraic equations, attention wavered. The combination of mathematics and botany which Mendel was expounding was unheard of. And the idea that lay behind it, that heredity was a giant shuffling and reshuffling of separate and invisible hereditary factors, stood in such diametrical contrast to all that had been taught that it probably could not be grasped. No one rose to ask a question. The minutes recorded no discussion.

A number of the members spoke to Mendel afterward about his experimental work. "I encountered various views," Mendel told his friends. "No one undertook a repetition of the experiment."

The editor of the *Proceedings* of the society extended the usual invitation to Mendel to prepare his paper for publication in the society's journal. "I agreed to do so," said Mendel, "after I had once more looked through my notes relating to the various years of the experiment without being able to discover any sort of mistake."

Mendel's monograph—*Versuche über Pflanzenhybriden*—appeared in 1866. According to the society's custom, copies of the *Proceedings* in which it was included were sent to Vienna, Berlin, Rome, St. Petersburg, and Uppsala. But the brief paper that could have altered all ideas of heredity, and at a time when Darwin was still at work on the role of heredity in evolution, attracted no attention. It sat all but unread on library shelves.

From Ruth Moore, *The Coil of Life (New York: Alfred A. Knopf, 1961).*

## Questions and Topics for Discussion and Writing

1. Why was it so important that Mendel thought to count and record the characteristics of each generation of plants?
2. Explain the concept of dominant and recessive genes. Explain how the 3-to-l ratio in visible characteristics comes about.
3. Why, in your opinion, was no notice taken of Mendel's work for 34 years, despite its great importance for evolutionary theory (which *was* being given lots of notice)?

*Unit 2*

# The Nature of Technology

- The Role of Technology in Civilization
- Case Studies of Technologies
- Technological Innovation and Failure
- The Dependence of Society on Technological Solutions

The Astronomical Clock of Prague.

*Issue*

# The Role of Technology in Civilization

Though there is much said about the role of science in civilization, as well as the roles of the arts, politics, and other realms of the world of ideas, less is said about the role of technology—perhaps because technology is in its nature a thing or things, and not an idea.

The two articles in this section make the case for technology as a major influence in our culture. Lewis Mumford's "Technics and Civilization," is an excerpt from his classic book of the same title that showed the tremendous importance of technology in shaping all of society's institutions. Much the same sentiment is expressed by Peter F. Drucker in "The Technological Revolution: Notes on the Relationship of Technology, Science, and Culture." Drucker is at pains to show that it is not science that has led to the advance in the human condition, but technology.

# The Monastery and the Clock

## Lewis Mumford

Science is hard to define adequately and, in consequence, a never ending debate is always in progress on what constitutes the boundaries between science and non-science. With technology there is probably even less agreement among those who try seriously to discuss and define its nature. But in fact not nearly so many do try to define technology as try to define science—and that is not because technology has less influence or importance in our lives, but because we assimilate technology so readily that it takes an effort to think of technology as something apart from the general stream of civilization. Of those who *have* thought deeply about technology, Lewis Mumford is one of the most interesting and provocative. In this selection from his book *Technics and Civilization,* first published in 1934, Mumford describes what, to him, is the key machine of industrial civilization—one that it is too easy for us to take for granted and forget—the clock.

Where did the machine first take form in modern civilization? There was plainly more than one point of origin. Our mechanical civilization represents the convergence of numerous habits, ideas, and modes of living, as well as technical instruments; and some of these were, in the beginning, directly opposed to the civilization they helped to create. But the first manifestation of the new order took place in the general picture of the world: during the first seven centuries of the machine's existence the categories of time and space underwent an extraordinary change, and no aspect of life was left untouched by this transformation. The application of quantitative methods of thought to the study of nature had its first manifestation in the regular measurement of time; and the new mechanical conception of time arose in part out of the routine of the monastery. Alfred Whitehead has emphasized the importance of the scholastic belief in a universe ordered by God as one of the foundations of modern physics: but behind that belief was the presence of order in the institutions of the church itself.

The technics of the ancient world were still carried on from Constantinople and Baghdad to Sicily and Cordova: hence the early lead taken by Salerno in the scientific and medical advances of the Middle Age. It was, however, in the monasteries of the West that the desire for order and power, other than that expressed in the military domination of weaker men, first manifested itself after the long uncertainty and bloody confusion that attended the breakdown of the Roman Empire. Within the walls of the monastery was sanctuary: under the rule of the order surprise and doubt and caprice and irregularity were put at bay. Opposed to the erratic fluctuations and pulsations of the worldly life was the iron discipline of the rule. Benedict added a seventh period to the devotions of the day, and in the seventh century, by a bull of Pope Sabinianus, it was decreed that the bells of the monastery be rung seven times in the twenty-four hours. These punctuation marks in the day were known as the canonical hours, and some means of keeping count of them and ensuring their regular repetition became necessary.

According to a now discredited legend, the first modern mechanical clock, worked by falling weights, was invented by the monk named Gerbert who afterwards became Pope Sylvester II near the close of the tenth century. This clock was probably only a water clock, one of those bequests of the ancient world either left over directly from the days of the Romans, like the water-wheel itself, or coming back again into the west through the Arabs. But the legend, as so often happens, is accurate in its implications if not in its facts. the monastery was the seat of a regular life, and an instrument for striking the hours at intervals or for reminding the bell-ringer that it was time to strike the bells, was an almost inevitable product of this life. If the mechanical clock did not appear until the cities of the thirteenth century demanded an orderly routine, the habit of order itself and the earnest regulation of time-sequences had become almost second nature in the monastery. Coulton agrees with Sombart in looking upon the Benedictines, the great working order, as perhaps the original founders of modern capitalism: their rule certainly took the curse off work and their vigorous engineering enterprises may even have robbed warfare of some of its glamor. So one is not straining the facts when one suggests that the monasteries—at one time there were 40,000 under the Benedictine rule—helped to give human enterprise the regular collective beat and rhythm of the machine; for the clock is not merely a means of keeping track of the hours, but of synchronizing the actions of men.

Was it by reason of the collective Christian desire to provide for the welfare of souls in eternity by regular prayers and devotions that time-keeping and the habits of temporal order took hold of men's minds: habits that capitalist civilization presently turned to good account? One must perhaps accept the irony of this paradox. At all events, by the thirteenth century there are definite records of mechanical clocks, and by 1370 a well-designed "modern" clock had been built by Heinrich von Wyck at Paris. Meanwhile, bell towers had come into existence, and the new clocks, if they did not have, till the fourteenth century, a dial and a hand that translated the movement of time into a movement through space, at all events struck the hours. The clouds that could paralyze the sundial, the freezing that could stop the water clock on a winter night, were no longer obstacles to time-keeping: summer or winter, day or night, one was aware of the measured clank of the clock. The instrument presently spread outside the monastery; and the regular striking of the bells brought a new regularity into the life of the workman and the merchant. The bells of the clock tower almost defined urban existence. Time-keeping passed into time-serving and time-accounting and time-rationing. As this took place, Eternity ceased gradually to serve as the measure and focus of human actions.

The clock, not the steam-engine, is the key-machine of the modern industrial age. For every phase of its development the clock is both the outstanding fact and the typical symbol of the machine: even today no other machine is so ubiquitous. Here, at the very beginning of modern technics, appeared prophetically the accurate automatic machine which, only after centuries of further effort, was also to prove the final consummation of this technic in every department of industrial activity. There had been power-machines, such as the water-mill, before the clock; and there had also been various kinds of automata, to awaken the wonder of the populace in the temple, or to please the idle fancy of some Moslem caliph: machines one finds illustrated in Hero and Al-Jazari. But here was a new kind of power-machine, in which the source of power and the transmission were of such a nature as to ensure the even flow of energy throughout the works and to make possible regular production and a standardized product. In its relationship to determinable quantities of energy, to standardization, to automatic action, and finally to its own special product, accurate timing, the clock has been the foremost machine in modern technics: and at each period it has remained in the lead: it marks a perfection toward which other machines aspire. The clock, moreover, served as a model for many other kinds of mechanical works, and the analysis of motion that accompanied the perfection of the clock, with the various types of gearing and transmission that were elaborated, contributed to the success of quite different kinds of machine. Smiths could have hammered thousands of suits of armor or thousands of iron cannon, wheelwrights could have shaped thousands of great water-wheels or crude gears, without inventing any of the special types of movement developed in clockwork, and without any of the accuracy of measurement and fineness of articulation that finally produced the accurate eighteenth century chronometer.

The clock, moreover, is a piece of power-machinery whose "product" is seconds and minutes: by its essential nature it dissociated time from human events and helped create the belief in an independent world of mathematically measurable sequences: the special world of science. There is relatively little foundation for this belief in common human experience: throughout the year the days are of uneven duration, and not merely does the relation between day and night steadily change, but a slight journey from East to West alters astronomical time by a certain number of minutes. In terms of the human organism itself, mechanical time is even more foreign: while human life has regularities of its own, the beat of the pulse, the breathing of the lungs, these change from hour to hour with mood and action, and in the longer span of days, time is measured not by the calendar but by the events that occupy it. The shepherd measures from the time the ewes lambed; the farmer measures back to the day of sowing or forward to the harvest: if growth has its own duration and regularities, behind it are not simply matter and motion but the facts of development: in short, history. And while mechanical time is strung out in a succession of mathematically isolated instants, organic time—what Bergson calls duration—is cumulative in its effects. Though mechanical time can, in a sense, be speeded up or run backward, like the hands of a clock or the images of a moving picture, organic time moves in only one direction—through the cycle of birth, growth, development, decay, and death—and the past that is already dead remains present in the future that has still to be born.

Around 1345, according to Thorndike, the division of hours into sixty minutes and of minutes into sixty seconds became common: it was this abstract framework of divided time that became more and more the point of reference for both action and thought, and in the effort to arrive at accuracy in this department, the astronomical exploration of the sky focused attention further upon the regular, implacable movements of the heavenly bodies through space. Early in the sixteenth century a young Nuremberg mechanic, Peter Henlein, is supposed to have created "many-wheeled watches out of small bits of iron" and by the end of the century the small domestic clock had been introduced in England and Holland. As with the motor car and the airplane, the richer classes first took over the new mechanism and popularized it: partly because they alone could afford it, partly because the new bourgeoisie were the first to discover that, as Franklin later put it, "time is money." To become "as regular as clockwork" was the bourgeois ideal, and to own a watch was for long a definite symbol of success. The increasing tempo of civilization led to a demand for greater power: and in turn power quickened the tempo.

Now, the orderly punctual life that first took shape in the monasteries is not native to mankind, although by now Eastern peoples are so thoroughly regimented by the clock that it is "second nature" and they look upon its observance as a fact of nature. Many Eastern civilizations have flourished on a loose basis in time: the Hindus have in fact been so indifferent to time that they lack even an authentic chronology of the years. Only yesterday, in the midst of the industrializations of Soviet Russia, did a society come into existence to further the carrying of watches there and to propagandize the benefits of punctuality. The popularization of timekeeping, which followed the production of the cheap standardized watch, first in Geneva, then in America around the middle of the last century, was essential to a well-articulated system of transportation and production.

To keep time was once a peculiar attribute of music: it gave industrial value to the workshop song or the tattoo or the chantey of the sailors tugging at a rope. But the effect of the mechanical clock is more pervasive and strict: it presides over the day from the hour of rising to the hour of rest. When one thinks of the day as an abstract span of time, one does not go to bed with the chickens on a winter's night: one invents wicks, chimneys, lamps, gaslights, electric lamps, so as to use all the hours belonging to the day. When one thinks of time, not as a sequence of experiences, but as a collection of hours, minutes, and seconds, the habits of adding time and saving time come into existence. Time took on the character of an enclosed space: it could be divided, it could be filled up, it could even be expanded by the invention of labor-saving instruments.

Abstract time became the new medium of existence. Organic functions themselves were regulated by it: one ate, not upon feeling hungry, but when prompted by the clock: one slept, not when one was tired, but when the clock sanctioned it. A generalized time-consciousness accompanied the wider use of clocks: dissociating time from organic sequences, it became easier for the men of the Renascence to indulge the fantasy of reviving the classic past or of reliving the splendors of antique Roman civilization: the cult of history, appearing first in daily ritual, finally abstracted itself as a special discipline. In the seventeenth century journalism and periodic literature made their appearance: even in dress, following the lead of Venice as fashion-center, people altered styles every year rather than every generation.

The gain in mechanical efficiency through co-ordination and through the closer articulation of the day's events cannot be overestimated: while this increase cannot be measured in mere horse-power, one has only to imagine its absence today to foresee the speedy disruption and eventual collapse of our entire society. The modern industrial régime could do without coal and iron and steam easier than it could do without the clock.

From Lewis Mumford, *Technics and Civilization*, 9–18 (New York: Harcourt, Brace & World, 1963).

## Questions and Topics for Discussion and Writing

1. Why were monasteries the logical place for clocks to have developed and become an indispensable part of human life?
2. Why does Mumford call the clock the key machine of the modern industrial age?
3. Discuss the conception of abstract time and the significance that Mumford attributes to it.

# * The Technological Revolution

## Peter F. Drucker

It is often said that modern civilization, with all its creature comforts and amenities, is the result of science. This is said without much attention to the distinction between science and technology. In the late twentieth century it may be difficult to decide where science ends and technology begins. If science is a theoretical understanding of the nature of the world and technology is an organization of the form and materials of nature to accomplish some predefined goal, then clearly the two get very closely intertwined in very complex applications such as space exploration and biotechnology. But usually the distinction is easier to make.

In this article, Peter F. Drucker, guru of modern corporate management, argues forcefully that the progress of society is thanks to technology, not science. To help the reader see his point he traces the development of technology from antiquity, showing how technological developments, not scientific ones, have improved the human condition.

*The standard answer to the question: What brought about the explosive change in the human condition these last two hundred years, is "The progress of science."*

This paper enters a demurrer. It argues that the right answer is more likely: "A fundamental change in the concept of technology." Central to this was the re-ordering of old technologies into systematic public disciplines with their own conceptual equipment, e.g., the "differential diagnosis" of nineteenth century medicine. In the century between 1750 and 1857 the three main technologies of man—agriculture, the mechanical arts (today's engineering), and medicine—went in rapid succession through this process, which resulted almost immediately in an agricultural, an industrial, and a medical "revolution," respectively.

This process owed little or nothing to the new knowledge of contemporary science. In fact, in every technology the practice with its rules of thumb, was far ahead of science. Technology therefore became the spur to science; it took, for instance, 75 years until Clausius and Kelvin could give a scientific formulation to the thermodynamic behavior of Watt's steam engine. Science could indeed have had no impact on the Technological Revolution until the transformation from craft to technological discipline had first been completed.

But technology had an immediate impact on science which was transformed by the emergence of systematic technology. The change was the most fundamental one—a change in science's own definition and image of itself. From being "natural philosophy" science became a social institution. The words in which science defined itself remained unchanged: "the systematic search for rational knowledge." But "knowledge" changed its meaning from being "understanding," i.e., focused on man's mind, to being "control," i.e., focused on application in and through technology. Instead of raising, as science had always done, fundamental problems of metaphysics it came to raise, as it rarely did before, fundamental social and political problems.

It would be claiming too much to say that technology established itself as the paramount power over science. But it was technology that built the future home, took out the marriage license, and hurried a rather reluctant science through the ceremony. And it is technology that gives the union of the two its character; it is a coupling of science *to* technology, rather than a coupling of science and technology.

The evidence indicates that the key to this change lies in new basic concepts regarding technology, that is, in a genuine Technological Revolution with its own causes and its own dynamics.

## I

Of all major technologies medicine alone has been taught systematically for any length of time. An unbroken line leads back for 1000 years, from the medical school of today to the medical schools of the Arab Caliphates. The trail, though partly overgrown, goes back, another 1400 years, through the School of Alexandria to Hippocrates. From the beginning, medical schools taught both theoretical knowledge and clinical practice, engaged simultaneously therefore in "science" and "technology." Unlike any other technologist in the West, the medical practitioner has continuously enjoyed social esteem and position.

Yet, until very late—1850 or thereabouts—there was no organized or predictable relationship between scientific knowledge and medical practice. The one major contribution to health care which the West made in the Middle Ages was the invention of spectacles. The generally accepted date is 1286; by 1290 the use of eyeglasses is fully documented. This invention was, almost certainly, based directly upon brand-new scientific knowledge, most probably on Roger Bacon's optical experiments. Yet Bacon was still alive when spectacles came in—he died in 1294. Until the nineteenth century there is no other example of such all but instantaneous translation of new scientific knowledge into technology—least of all in medicine. Yet Galen's theory of vision which ruled out any mechanical correction was taught in the medical schools until 1700.

Four hundred years later, in the Age of Galileo, medicine took its next big step—Harvey's discovery of the circulation of the blood, the first major new knowledge since the ancients. Another hundred years, and Jenner's smallpox vaccination brought both the first specific treatment and the first prevention of a major disease.

Harvey's findings disproved every single one of the theoretical assumptions that underlay the old clinical practice of bleeding. By 1700 Harvey's findings were taught in every medical school and repeated in every medical text. Yet bleeding remained the core of medical practice and a universal panacea for another hundred years, and was still applied liberally around 1850. What killed it finally was not scientific knowledge—available and accepted for 200 years—but clinical observation.

In contrast to Harvey, Jenner's achievement was essentially technological and without any basis in theory. It is perhaps the greatest feat of clinical observation. Smallpox vaccination had hard sledding—it was, after all, a foolhardy thing deliberately to give oneself the dreaded pox. But what no one seemed to pay any attention to was the complete incompatibility of Jenner's treatment with any biological or medical theory of the time, or of any time thereafter until Pasteur, 100 years later. That no one, apparently, saw fit to try explaining vaccination or to study the phenomenon of immunity appears to us strange enough. But how can one explain that the same doctors who practiced vaccination, for a century continued to teach theories which vaccination had rendered absurd?

The only explanation is that science and technology were not seen as having anything to do with one another. To us it is commonplace that scientific knowledge is being translated into technology, and vice versa. This assumption explains the violence of the arguments regarding the historical relationship between science and the "useful arts." But the assumptions of the debate are invalid: the presence of a tie proves as little as its absence—it is our age, not the past, which presumes consistency between theory and practice.

The basic difference was not in the content but in the focus of the two areas. "Science" was a branch of philosophy, concerned with understanding. Its object was to elevate the human mind. It was misuse and degradation of science to use it—Plato's famous argument. Technology, on the other hand, was focused on use. Its object was increase of the human capacity to do. Science dealt with the most general, technologies with the most concrete. Any resemblances between the two was "purely coincidental."

## II

There are no hard and fast dates for a major change in an attitude, a world view. And the Technological Revolution was nothing less. We do know, however, that it occurred within the half century 1720 to 1770—the half century that separates Newton from Benjamin Franklin.

Few people today realize that Swift's famous encomium on the man who makes two blades grow where one grew before, was not in praise of the scientists. On the contrary, it was the final, crushing argument in a biting attack on them, and especially against the august Royal Society. It was meant to extol the sanity and benefits of nonscientific technology against the arrogant sterility of idle enquiry into nature concerned with understanding; this is against Newtonian Science, for Swift was, as always, on the unpopular side. But his basic assumption—that science and application were radically different and worlds apart—was clearly the prevailing one in the opening decades of the eighteenth century. Not one scientist spoke out against the weirdest technological "projects" of the South Sea Bubble of 1720, even though their theoretical infeasibility must have been obvious to them. And while Newton, as Master of the Royal Mint, reformed its business practices, he did not much bother with its technology.

Fifty years later, around 1770, Dr. Franklin is the "philosopher" par excellence and the West's scientific lion. Franklin though a first-rate scientist, owed his fame to his achievements as a technologist—"artisan" in eighteenth century parlance. He was a brilliant gadgeteer, as witness the Franklin stove and bifocals. Of his major scientific exploits, one—the investigation of atmospheric electricity—was immediately turned into useful application: the lightning rod. Another, his pioneering work in oceanography with its discovery of the Gulf stream, was undertaken for the express purpose of application, viz., to speed up the transatlantic mail service. Yet the scientists hailed Franklin as enthusiastically as did the general public.

In the fifty years between 1720 and 1770—not a particularly distinguished period in the history of science, by the way—a fundamental change in the attitude towards technology, both of laity and of scientists, must have taken place. One indication is the change in English attitude towards patents. During the South Sea Bubble they were still unpopular and attacked as "monopolies." They were still given to political favorites rather than to an inventor. By 1775 when Watt obtained his patent, they had become the accepted means of encouraging and rewarding technological progress.

We know in detail what happened to technology in the period which includes both the Agricultural Revolution and the opening of the Industrial Revolution. Technology as we know it today, that is, systematic, organized work on the material tools of man, was born then. It was produced by collecting and organizing existing knowledge, by applying it systematically, and by publishing it. Of these steps the last one was both the most novel—craft skill was not for nothing called a "mystery"—and the most important.

The immediate effect of the emergence of technology was not only rapid technological progress: it was the establishment of technologies as systematic disciplines to be taught and learned and, finally, the re-orientation of science towards feeding these new disciplines of technological application.

Agriculture and the mechanical arts changed at the same time, though independently.

Beginning with such men as Jethro Tull and his systematic work on horse-drawn cultivating machines in the early years of the seventeenth century and culminating towards its end in Coke of Holkham's work on balanced large-scale farming and selective live-stock breeding, agriculture changed from a "way of life" into an "industry." Yet this work would have had little impact but for the systematic publication of the new approach, especially by Arthur Young. This assured both rapid adoption and continuing further work. As a result, yields doubled while manpower needs were cut in half—which alone made possible that large-scale shift of labor from the land into the city and from producing food to consuming food on which the Industrial Revolution depended.

Around 1780, Albrecht Thaer in Germany, an enthusiastic follower of the English, founded the first agricultural college—a college not of "farming" but of "agriculture." This in turn, still in Thaer's lifetime, produced the first, specifically application-focused new knowledge, namely, Liebig's work on the nutrition of plants, and the first science-based industry, fertilizer.

The conversion of the mechanical arts into a technology followed the same sequence and a similar time table. The hundred years between the 1714 offer of the famous £20,000 prize for a reliable chronometer and Eli Whitney's standardization of parts was, of course, the great age of mechanical invention—of the machine tools, of the prime movers, and of industrial organization. Technical training, though not yet in systematic form, began with the founding of the *École des Ponts et Chaussées* in 1747. Codification and publication in organized form goes back to Diderot's *Encyclopédie,* the first volume of which appeared in 1750. In 1776—that miracle year that brought the Declaration of Independence, *The Wealth of Nations,* Blackstone's *Commentaries,* and Watt's first practical steam engine—the first modern technical university opened: the *Bergakademie* (Mining Academy) in Freiberg, Saxony. Significantly enough, one of the reasons for its establishment was the need for technically trained managers created by the increasing use of the Newcomen steam engine, especially in deep-level coal mining.

In 1794, with the establishment of the *École Polytechnique* in Paris, the profession of engineer was established. And again, within a generation, we see a re-orientation of the physical sciences—organic chemistry and electricity begin their scientific career, being simultaneously "sciences" and "technologies." Liebig, Woehler, Faraday, Henry, Maxwell were great scientists whose work was quickly applied by great inventors, designers, and industrial developers.

Only medicine, of the major technologies, did not make the transition in the eighteenth century. The attempt was made—by the Dutchman Gerhard Van Swieten, not only a great physician but politically powerful as advisor to the Hapsburg Court. Van Swieten attempted to marry the clinical practice which his teacher Boerhaave had started at Leyden around 1700, with the new scientific methods of such men as the Paduan Morgagni whose *Pathological Anatomy* (1761) first treated diseases as afflictions of an organ rather than as "humours." But—a lesson one should not forget—the very fact that medicine (or rather, something by that name) was already respectable and organized as an academic faculty defeated the attempt. Vienna relapsed into medical scholasticism as soon as Van Swieten and his backer, the Emperor Joseph II, died.

It was only after the French Revolution had abolished all medical schools and medical societies that a real change could be effected. Then another court physician, Corvisart, Napoleon's doctor, accomplished, in Paris around 1820, what Van Swieten had failed in. Even then opposition to the scientific approach remained powerful enough to drive Semmelweis out of Vienna and into exile when he found, around 1840, that traditional medical practices were responsible for lying-in fever with its ghastly death toll. Not until 1850, with the emergence of the "modern" medical school in Paris, Vienna, and Wuerzburg, did medicine become a genuine technology and an organized discipline.

This too happened, however, without benefit of science. What was codified and organized was primarily old knowledge, acquired in practice. Immediately *after* the re-orientation of the practice of medicine, the great medical scientists appeared—Claude Bernard, Pasteur, Lister, Koch. And they were all application-focused, all driven by a desire to do, rather than by a desire to know.

We know the results of the Technological Revolution, and its impacts. We know that, contrary to Malthus, food supply in the last two hundred years has risen a good deal more than an exploding human population. We know that the average lifespan of man a hundred fifty years ago, was still close to the "natural life span": the 25 years or so needed for the physical reproduction of the species. In the most highly developed and prosperous areas, it has almost tripled. And we know the transformation of our lives through the mechanical technologies, their potential, and their dangers.

Most of us also know that the Technological Revolution has resulted in something even more unprecedented: a common world civilization. It is corroding and dissolving history, tradition, culture, and values throughout the world, no matter how old, how highly developed, how deeply cherished and loved.

And underlying this is a change in the meaning and nature of "knowledge" and of our attitude to it. Perhaps one way of saying this is that the non-Western world does not want Western science primarily because it wants better understanding. It wants Western science because it wants technology and its fruits. It wants control, not understanding. The story of Japan's "Westernization" between 1867 and her emergence as a "modern nation" in the Chinese War of 1894, is the classical, as it is the earliest, example.

But this means that the Technological Revolution endowed technology with a power which none of the "useful arts"—whether agricultural, mechanical, or medical—had ever had before: impact on man's mind. Previously, the "useful arts" had to do only with how man lives and dies, how he works plays, eats, and fights. How and what he thinks, how he sees the world and himself in it, his beliefs and values, lay elsewhere—in religion, in philosophy, in the arts, in science. To use technological means to affect these areas was traditionally "magic"—considered at least evil, if not asinine to boot.

With the Technological Revolution, however, application and cognition, matter and the mind, tool and purpose, knowledge and control have come together for better or worse.

## III

There is only one thing we do not know about the Technological Revolution—but it is essential: What happened to bring about the basic change in attitudes, beliefs, and values which released it? "Scientific progress," I have tried to show, had little to do with it. But how responsible was the great change in world outlook which, a century earlier, had brought about the great Scientific Revolution? What part did the rising capitalism play? And what was the part of the new, centralized national state with its mercantilistic policies on trade and industry and its bureaucratic obsession with written, systematic, rational procedures everywhere? (After all, the eighteenth century codified the laws as it codified the useful or applied arts.) Or do we have to do here with a process, the dynamics of which lie in technology? Is it the "progress of technology" which piled up to the point when it suddenly turned things upside down, so that the "control" which nature had always exercised over man now became, at least potentially, "control" which man exercises over nature?

This should be, I submit, a central question both for the general historian and for the historian of technology.

For the first, the Technological Revolution marks one of the great turning points—whether intellectually, politically, culturally, or economically. In all four areas the traditional—and always unsuccessful—drives of systems, powers, and religions for world domination is replaced by a new and highly successful world-imperialism, that of technology. Within a hundred years, it penetrates everywhere and puts, by 1900, the symbol of its sovereignty, the steam engine, even into the Dalai Lama's palace in Lhasa.

For the historian of technology, the Technological Revolution is not only the cataclysmic event within his chosen field; it is the point at which such a field as "technology" emerges. Up to that point there is, of course, a long and exciting history of crafts and tools, artifacts and mechanical ingenuity, slow, painful advances and sudden, rapid diffusion. But only the historian, endowed with hindsight, sees this as "technology," and as belonging together. To contemporaries, these were separate things, each belonging to its own sphere, application, and way of life.

Neither the general historian nor the historian of technology has yet, however, concerned himself much with the Technological Revolution. The first—if he sees it at all—dismisses technology as the bastard child of science. The only general historian of the first rank (excepting only that keen connoisseur of techniques and tools, Herodotus) who devotes time and attention to technology, its role and impact is, to my knowledge, Franz Schnabel. That Schnabel taught history at a technical university (Karlsruhe) may explain his interest. The historians of technology, for their part, tend to be historians of materials, tools, and techniques rather than historians of technology. The rare exceptions tend to be non-technologists such as Lewis Mumford or Roger Burlingame who, understandably, are concerned more with the impact of technology on society and culture than with the development and dynamics of technology itself.

Yet technology is important today precisely because it unites both the universe of doing and that of knowing, connects both the intellectual and the natural histories of man. How it came thus to be in the center—when it always before had been scattered around the periphery—has yet to be probed, thought through, and reported.

From Peter F. Drucker, *Technology, Management, and Society* (New York: Harper & Row, 1970).

*Peter F. Drucker*

## Questions and Topics for Discussion and Writing

1. Give some examples in which relevant scientific knowledge had no effect on changing a technological process which was misguided.
2. Give some examples where a useful technology existed long before the scientific theory which explained it was known.
3. Describe in some detail the "Technological Revolution" named by Drucker. Discuss the difference between the Technological Revolution and the Scientific Revolution.

*Issue*

# Case Studies of Technologies

One way to get a good feel for the place of technology in society is to look very closely at one technology to see how it worked, how it affected people's lives, and how it changed over time and in relation to other technologies.

There is no limit to the number of technologies that could be profitably and interestingly examined. Major inventions are frequently written about: the steam engine, the railroad, the computer, the wheel. The two chosen for study here are less obvious, but no less interesting: the lighthouse and bread making. "Nineteenth Century Lighthouse Technology" by Judith Tulloch discusses the lighthouse in its most romantic period, when lighthouse keeping was a way of life. "Mechanization and Organic Substance: Bread" by Siegfried Giedion examines the history of mechanized bread making and what it has done to public tastes and preferences.

# Nineteenth Century Lighthouse Technology

## Judith Tulloch

The lighthouse is an interesting example of a complex technology with a long history, and an obvious value to society. Though lighthouses have existed since at least 300 B.C. and continue in use today, their internal workings have changed a great deal. Probably the most interesting period in lighthouse history is the nineteenth century, when tending a lighthouse was a way of life for whole families and when the techniques used were ingenious enough to produce sufficient light and the correct signal *without the benefit of electricity* for either producing light or for running a motor to revolve the light.

The safety of its mariners has long been a matter of concern to seagoing societies. The identification of specific hazards to navigation and the provision of marks to indicate entrances to harbours were as much a concern to the Egyptians when they built the great pharos at Alexandria *ca.* 300 BC as they were to British lighthouse authorities when they ringed the coast of the British Isles with lights in the 19th century. Indeed, though equipment and techniques have changed, these two problems remain the essential concern of modern lighthouse services throughout the world.

Early lights were usually crude: most consisted of open flames, candles, or as at Louisbourg in 1737, a pan of oil with a number of wicks set in it. Some of the early towers were open to the elements but more often the light source was protected by a glass enclosed room at the top of the light tower. This room, known as the "lantern," housed the light source itself and, usually, the clockwork machinery which controlled its rotation.

The increasing use of oil lamps late in the 18th century gave the lightkeeper greater control over the light source. Lamps could easily be lit and extinguished and were both more manageable and economical than open fires. Many early lamps used a flat wick, like the oil lamp in domestic use. These wicks often smoked badly and deposited a layer of soot on the lantern windows, thus obscuring the light. In 1782, however, the Swiss inventor Aimé Argand developed the cylindrical wick burner which still bears his name. This lamp was adopted rapidly for use in lighthouses, and with only minor modifications, it remained in use throughout the 19th century and into the 20th century. The hollow wick used in these lamps burned more efficiently since a current of air rose through the centre of the wick to improve combustion and decrease smoking. A specially designed cylindrical chimney which directed air inward to the flame further enhanced the effect of the hollow wick to produce a brighter, more intense flame that was possible with flat wick burners.

The development of a more effective light source was only part of the answer to the problem of increasing the amount of light seen by sailors as well as the distance at which it could be sen. Light rays given off by the flame had to be collected and redirected to form a horizontal beam of light which would be visible at a distance. The principles of the science of optics were to be applied to meet the needs of the lighthouse. Development of the parabolic reflector was the first major advance in the field of lighthouse optics. By placing the light source at the focal point of the parabola formed by the metal reflector, the majority of light rays from the flame, which had been scattered formerly in all directions, were reflected as a horizontal beam. Some light was still lost but approximately $^{12}/_{17}{}^{ths}$ of the light was reflected. This was a vast improvement over earlier equipment. Crude parabolic reflectors had been used by the Swedes in the mid 18th century, and in England, Wiliam Hutchinson installed wooden reflectors lined with segments of mirror at a lighthouse in Liverpool in the 1760s. Improvements were soon made to increase the efficiency of the reflectors and by 1788, the type of reflector which remained in common use throughout the world in the 19th century had been adopted by Trinity House, the agency in charge of English lighthouses and the trendsetter for lighthouse development throughout the century.

These reflectors, many of which were still to be found in Canadian lighthouses into this century, formed what is known as the catoptric system (from the Greek "mirror" or "reflector"). Typically, 19th century catoptric equipment consisted of a number of silver-lined copper reflectors (usually 21 inches or 25 inches in diameter) mounted on a metal framework which was rotated through the action of large clockwork machinery, usually made of cast brass. At the focal point of the reflectors was set a brass argand burner with a flame of about 1 inch in diameter. The wide-spread acceptance of this arrangement, developed by Robert Stevenson around 1810, was a result of the preeminence of the Stevenson family in lighthouse engineering. Throughout the 19th century, members of this Scottish family served as the chief engineers for the Northern Lighthouse Board (responsible for the Scottish lights). In addition to building lighthouses, such as the famous Bell Rock and Skerryvore lights, they made improvements in both the catoptric and dioptric (refracting lens) systems and served as lighthouse consultants not only in the British Isles but in North America. For example, in 1835, Robert Stevenson prepared an estimate for the Newfoundland government of the type of light required for Cape Spear, then under construction near St. John's.

Catoptric equipment was adopted for use in most lighthouses in British North America and remained in use here long after advances in technology had rendered it obsolete in Great Britain. Its continued use is explained by a number of factors, among them the fact that the colonial lighthouse services functioned with less money than their British counterparts and catoptric lights were usually cheaper to buy than lens lights. Indeed, some of the catoptric lights in use in the British North American provinces had been bought second hand from British lighthouse authorities after their replacement by the more efficient lens light. For example, three of Newfoundland's earliest lighthouses—were equipped with Stevenson catoptric machinery, respectively from Inchkeith (1816), the Isle of May (1816) and Bell Rock (1811). In these three cases, the equipment remained in use until the end of the 19th century. Another reason for the continued use of reflector lights in the colonies was the fact that it was a simpler system to manage. colonial lightkeepers were less well-trained than their British counterparts and in some cases were only part-time workers. Hence, equipment requiring less technical skill was more suitable. Moreover, the colonial light services adopted kerosene as a fuel soon after it became commercially available in the 1860s and this inexpensive fuel proved to be extremely effective when used with catoptric equipment. In 1872, a group of visitors from the English Trinity House commented favourably on the combination of economy and efficiency found in the kerosene-fueled catoptric lights of the colonies and concluded that the system was "admirably adapted for a young country."[1]

Long before this time, however, advances in lighthouse technology had resulted in the adoption in Britain, France and the United States of a new lighting system based on the principles of the refraction of light. This system, known as the dioptric or lens system (from the Greek "to see through"), consisted of a series of glass lenses and prisms which refracted the light to the desired focal plane. The lenses and prisms were joined together to form a ring within which the light source was placed. The early dioptric lights were further improved by the addition first of mirrors and then of prisms above and below the light source, forming what was in essence a bee-hive of glass enclosing the lamp. With this development, virtually all light produced was utilized and the equipment was known as a "holophotal"light (from the Greek holo—"all"). These great lenses are as much a work of art as a technical artifact. Alan Stevenson of the Stevenson

[1] Canada, Parliament, *Sessional Papers*, 1873, No. 8, App. 37, 361.

clan described them from that viewpoint,

> Nothing can be more beautiful than an entire apparatus for a fixed light of the first order. It consists of a central belt of refractors, forming a hollow cylinder six feet in diameter and thirty inches high; below it are six triangular rings of glass, ranged in a cylindrical form and above a crown of thirteen rings of polished glass ten feet high and six feet in diameter. I know of no work of art more beautifully creditable to the boldness, ardor, intelligence and zeal of the artist.[1]

In 1823, the first effective lens equipment was installed at the Tour de Corduouan lighthouse at the mouth of the Gironde River in France by one of the pioneers in lighthouse optics, Augustin Fresnel. Recognizing the value of the equipment, English and Scottish lighthouse authorities immediately undertook studies of how the refractive machinery could be adapted best for use in the lighthouses under their jurisdiction. Not until 1835, however, was the first dioptric equipment installed at a Scottish lighthouse—at Inchkeith on the Firth of Forth on the east coast of Scotland. This installation in fact enabled the catoptric equipment in use there since 1816 to be sold to the Newfoundland government for installation at Cape Spear.

After this first venture, dioptric machinery was adopted quickly in lighthouses around the coasts of Britain and, after mid-century, in the United States. Because of the great technical skill required in cutting and grinding the glass, manufacture of the equipment remained concentrated in France until the mid 19th century when French craftsmen were employed by the Birmingham firm of Chance Brothers, a major supplier of lighthouse equipment to countries around the world.

Dioptric lights are classified according to their size, from the 1st Order with an interior diameter of sic feet which was considered a major sea light, down to the tiny 6th Order harbour light with an interior diameter of twelve inches. In Canada, there were only a few 1st Order lights: among them Sambro at the entrance to Halifax harbour and Cap des Rosiers at the tip of the Gaspé Peninsula on the St. Lawrence. The ultimate development in dioptric lights was the hyper-radial light. Similar in design to conventional dioptric equipment, the hyper-radial lights were much larger than 1st Order lenses. One of these massive lights was installed in 1907 at Cape Race on the southeast tip of Newfoundland. Established in 1855, this station traditionally was the most important landfall light in Canada since it is the first light met by shipping westbound into the gulf of St. Lawrence and conversely the last light seen by east-bound vessels as they head across the Atlantic. The hyper-radial light here had four lenses each twelve feet high with a total weight of more than three tons of glass.

The advent of electricity made the great lenses obsolete. By increasing the candlepower of the electric bulb, simple plastic "bulldog" lenses could be as effective as the glass prisms and were substantially earlier to maintain. The first Canadian light to be lit electrically was at Reed Point, near Saint John, New Brunswick. Electricity was installed there in 1895, and during the next half century, almost all Canadian lights were converted. In many cases, the old lens has remained in use with an electric bulb providing the light source, but this equipment is gradually being replaced by a light similar to an airport beacon. Since the 1950s, mercury vapour bulbs have generally replaced incandescent ones since they produce higher intensity light and have a much longer life, an important characteristic in theses days when most lights are unwatched.

The catoptric (reflector) and dioptric (lens) lights in use in the 19th century were capable of producing a variety of signals. The light might be "fixed," i.e. it might be an unchanging beam of light visible through a defined arc of the horizon. Such lights often were used to mark harbour entrances where two of them could be lined up as leadlights to indicate the safe channel. More common was the revolving light in which the reflectors or lens was placed on a metal framework which rotated at a constant speed to produce a recognizable signal. At Cape Spear, for example, the light's signal was a 17 second flash followed by 43 seconds of darkness.

[1] Quoted in Dudley Witney, *The Lighthouse* (Toronto, 1975), 42. The superiority of the dioptric light over the catoptric was illustrated by the fact that even with the best reflector only 17% of the light produced reached a ship's bridge while the equivalent figure with dioptric equipment was 83%.

In this way a light could be distinguished from its neighbour whose signal might be a steady beam or a 40 second flash followed by 20 seconds of darkness. This latter example is known as an occulting light in which the period of darkness is less than the period of light. Mariners therefore could use the lights to verify their position and it was a matter of great concern to lighthouse authorities to ensure that a light could not reasonably be mistaken for any other. Indeed one of the most serious complaints against a keeper was that he had allowed his light to display an incorrect signal.

For most of the 19th century, the rotatory machinery consisted of a series of geared brass wheels powered by heavy weights, essentially a large version of the clockwork used in a grandfather clock. The weights usually fell perpendicularly from the lightroom through the tower, and in most cases had to be rewound several times during the night. Later in the century, heavy springs were used to power the light and in some of the smaller lights this type of equipment was in use until after the Second World War. With the coming of electricity, the rotatory machinery has been replaced gradually by electric motors.

Like the equipment itself, the type of fuel used at lighthouse establishments changed over the course of the 19th century. Early in the century, sperm whale oil was the accepted fuel for use with both argand and flat wick burners. Indeed, when the catoptric equipment from Inchkeith was sent to Newfoundland in 1835 for use at Cape Spear, a sample of sperm oil was sent as representing the only type of suitable fuel. By the 1840s, the cost of sperm oil had increased substantially as the supply of sperm whales diminished through over-hunting. Hence, lighthouse authorities began to search for alternatives. In Britain and Europe, a variety of vegetable oils were adopted, principally rapeseed oil and colza oil, which was made from a type of wild cabbage specifically grown for its oil. In North America, both colza and sperm oil were used until their replacement by kerosene in the 1860s. Along the east coast, lighthouse authorities had a range of fish and mammal oils for consideration. In Newfoundland, seal oil and porpoise oil were available. Later in the century, most Atlantic lighthouses were converted to use kerosene and this remained the principal fuel until the 20th century.

Cape Spear light near St. John's, Newfoundland, has served as an example to illustrate some of the general trends discussed above. Constructed in 1835, this light is Newfoundland's oldest extant lighthouse and it is now the principal resource of a national historic park. Cape Spear serves as an excellent model since it has experienced a full sequence of development in lighthouse technology. The first light installed here in 1836 was a second hand catoptric light designed by Robert Stevenson which had been in service at Inchkeith on the east coast of Scotland since 1816. This light was purchased by the Newfoundland government after a new dioptric light was installed at Inchkeith in 1835. No detailed description or pictures of the original equipment survive, but it appears to have been the conventional Stevenson design as set forth above, utilizing a 25 inch silvered-copper reflector with an argand burner. One point of rarity was the fact that it consisted of seven reflectors set on a seven-sided metal framework: the more usual arrangement was an even number of reflectors, for example, two on each side of a triangular or square frame. When first installed, the lamps burned sperm oil, but this was changed to seal oil in 1840 as a result of the difficulty and cost of acquiring sperm oil. Seal oil was used as the fuel until 1874 when kerosene was introduced. At this time, a refinement of the argand burner, the doty burner (consisting of a number of wicks arranged in concentric circles) was adopted since it had been found to be particularly effective with kerosene.

This combination of kerosene lamps and parabolic reflectors remained unchanged for thirty-five years. In 1912, the Newfoundland lighthouse authorities installed a small 4th Order dioptric light (interior diameter of about 220 inches). This equipment continued to be rotated by clockwork machinery similar to that in use since 1835 until the light was electrified in 1930. A 500 W bulb supplied the light source while the clockwork was replaced by a small electric motor. This system is still in use today in the concrete tower built nearby in 1955.

## Life in the Lighthouse Family

One of the aspects of lighthouse history which is of particular interest is the daily life of the lighthouse family. Traditionally lightkeeping has been considered a way of life rather than a job. This certainly was the case in the 19th century and to a large degree still seems to be true today. Recently a magazine article on west coast lighthouses quoted the keeper of Estevan Point on Vancouver Island:

> I don't see this as a job so much as a way of life. I don't even really think about it. There's that light to be kept going and the horns to be kept blowing—that's what I said I would do when I signed the contract, and that's what I'm going to do.[1]

Particularly in the days before automation of the lights, the demands of lighthouse technology had a direct impact on day to day life. Most North American lightstations served not only as beacons to mariners but also as homes for the keepers and their families. Because the home was also the workplace, there appears to have been a closer connection between work and family life than was common in other occupations, even in an era which saw a greater integration between the two than is experienced generally today. With this close connection it is not surprising that lightkeeping became, in many instances a family occupation. Children, especially sons, helped their father from an early age and frequently succeeded to the position on his death or retirement. At Cape Spear, for example, members of the Cantwell family have tended the light from 1846 to the present and this type of record is not unique.

Most of the lightkeeper's activities centred on the accurate and reliable operation of the light. Duties and responsibilities were set forth in detailed instructions which were to be read to staff by the principal keeper at least semi-annually. Lighthouses in British North America were managed by Boards of Lighthouse commissioners in the Maritimes and Newfoundland and by Trinity Houses at Quebec and Montreal. These authorities adopted regulations based on those established by the Scottish and English lighthouse boards. By 1855, the English Board of Trade and Trinity House had prepared a booklet on lighthouse management with instructions to keepers specifically for use in "the colonies." Similar regulations were in force long before this date. In 1823, Robert Stevenson had set forth detailed rules for the management of his recently completed tower light at Bell Rock. Even after Confederation, the new Department of Marine and Fisheries continued to base its regulations on British models.

The regulations and instructions varied very little over the years. This is not surprising in view of the continued use of the same type of equipment. How closely the rules were followed cannot be determined. In may of the smaller "family stations" which characterized the North American lighthouse service, they were probably not enforced as rigidly as in the major tower lights of Great Britain. Yet the fact that most of the regulations had to be followed in order for the light to function properly appears to have resulted in a relatively close observance. There are few documents which view 19th century lighthouse life from the inside, and to analyze day to day routine, it is necessary to turn to instructions to keepers and to later memoirs of lighthouse families.

Although lightkeeping was a twenty-four hour a day occupation, it was clearly during the night that closest attention to the light was required. Preparations for lighting the lamps began shortly before sunset so that by nightfall the apparatus was functioning correctly. The evening duties consisted of ensuring that all the equipment was in order, removing the curtains which covered the lantern windows, lighting the oil lamps and putting the rotatory machinery into operation. The lightkeeper then remained in the lantern for approximately 30 minutes to make sure that the light was giving its correct signal and that all parts of the apparatus were running smoothly.

Even after the light was in operation, the keeper could not leave the equipment unattended. At many 19th century lightstations, including Cape Spear, the clockwork machinery had to be wound up every three hours. Moreover, frequent checks were required to make sure that the wicks were not smoking, that the lamps had sufficient oil and that the speed of rotation had not changed. This of course would alter the signal given out and could be disastrous if one light was mistaken for another. If, for example, part of the rotatory equipment broke during the night and could not be repaired easily, the keeper and his helpers would have to turn the light

[1] Andreas Schroeder, "On the Rocks" *Today Magazine* (11 October 1980), 9.

by hand, a task which demanded concentration to maintain the correct speed. As late as the 1940s at a light on the south coast of Nova Scotia, Evelyn Richardson described how she and her young son worked through a winter night in the unheated lantern turning the light while her lightkeeper husband repaired a broken spring on the revolving mechanism.[1]

The oil lamps were extinguished at daybreak and, according to British lighthouse procedure, the equipment and lightroom were then cleaned and prepared for relighting in the evening. In view of the general adoption of British lightkeeping customs and regulations in the colonies, it is likely that the keepers here maintained essentially the same routine. The argand burners were withdrawn from the focal point of the reflectors so that the silvered interior could be thoroughly cleaned. Mild abrasives, such as rouge or rottenstone, were applied with chamois skins and the interior was then carefully polished, usually with linen rubbing cloths. Every three weeks, the reflectors were rubbed with turpentine and olive oil. Extreme care was necessary in cleaning the interior of the reflectors since the silver could easily be scratched and the reflecting power diminished. The 1904 instructions to Canadian lightkeepers noted that "the great art of keeping the reflectors in order consists in the daily, patient, and skilful application of manual labour in rubbing their surfaces, beginning at the centre and gradually working outwards with a circular motion of the hand," and concluded, "Any scratches in the silvering must be due to dust or careless work, and the keeper will be held accountable for them."[2]

The copper exterior of the reflectors was cleaned carefully as well and the oil lamps were emptied, washed thoroughly, polished and refilled. Their wicks were trimmed or replaced and the lamps were again ready for use. The same care was required in a station with a dioptric light. The fine glass prisms had to be wiped with spirits of wine and polished with a dry chamois cloth. In this case as well, close attention to cleanliness was required to prevent scratches to the glass during the cleaning process.

After preparing the light for re-lighting, the keeper then went on to clean the rest of the lantern. The keeper was required to wash and polish the windows to remove any soot which might have accumulated from the burning oil and frequently required to crawl out onto the exterior catwalk around the tower to wash the outside of the windows to clear away salt spray, or snow and ice, a constant winter menace. In some lighthouses calico curtains then were hung to cover the windows of the lantern both to prevent the sun reflecting off the metal of the reflectors and to make the lighthouse more visible as a daymark. The rotatory mechanism was cleaned and wound up again and the entire lightroom was swept and dusted. A Scottish lightkeeper estimated in 1900 that these duties took both the keeper and assistant from one to three hours a day.[3]

With this work finished, the lightkeeper was able to devote time to any repairs or maintenance needed in the rest of the building. Since most lighthouses are in exposed locations, maintenance to the structure was a constant concern. Repairs from wind damage were required regularly and in most cases the exterior of the lighthouse was painted every second year. In England and Scotland, lightkeepers were required to take meteorological readings several times during the day and night. Colonial lightkeepers seem to have kept similar entries for temperature, weather conditions, tides and the time of sunset and sunrise. A circular sent to Newfoundland lightkeepers in 1861, which appears to have reflected established practice, indicated that they were to keep a log of daily events by recording the weather, wind and any other items of importance. Maintaining this journal formed part of the keeper's daily routine. Lightkeepers also were required to keep a close inventory of lighthouse stores. Forms were provided to record the amount of oil used each

1 Evelyn Richardson, *We Keep a Light* (Toronto, 1961), 39.

2 Canada, Department of Marine and Fisheries, *Rules and Instructions for the Guidance of Light Keepers and of Engineers in Charge of Steam Fog Alarms in the Dominion of Canada 1904* (Ottawa, 1904), 9–10.

3 Patrick Beaver, *A History of Lighthouses* (Secaucus, New Jersy, 1973), Appendix A, "Life as a Lighthouse Keeper, written by a light-keeper under the Northern Lights commissioners, *circa* 1900," 144.

night, number of lamp chimneys broken and number of wicks used. Similarly, an inventory of the cleaning supplies was taken on a quarterly basis. The need to keep such records meant that at least one member of the lighthouse establishment had to be literate, a condition which could set them apart from many of their neighbours, particularly in the early 19th century and even later in more isolated settlements.

Since a lightkeeper's workplace was also the family home, the duties of the occupation would not but affect domestic life. For example, in stations where access was difficult the building could not be left unattended even during the day since a sudden storm might prevent the lightkeeper's return before nightfall. Hence in these situations, either the assistant keeper or a responsible family member remained at home ready to light the lamps and maintain them through the night if required. Similarly in stations with a fog horn, the keeper's wife might have to operate the alarm in her husband's absence if a ship was lost in the fog nearby. At larger stations where there was an assistant keeper, the family involvement would be less since the keeper and assistant would seldom be absent at the same time. Even in these cases, however, the demands of the light were never forgotten and were clearly part of everyday life. As an example, the wife of a former keeper of Point Riche light on the northwest coast of Newfoundland mentioned that she was never able to hang curtains in the windows of her home because her husband insisted on being able to see the beam of light from the adjacent tower at all times to make sure it did not change speed or go out.[1]

The daily cleaning and maintenance chores and the nightly routine of watches to ensure that the signal was bright and regular formed the constant background for day to day life in lighthouses around the world in the 19th century. This need to provide accurate navigational aids and the similarity of the equipment in use resulted in the establishment of a way of life which was common throughout the lighthouse service. It has been altered significantly only with increasing automation of lights and the gradual disappearance of resident lightkeepers.

From Richard A. Jarrell and Arnold E. Roos, eds., *Critical Issues in the History of Canadian Science, Technology and Medicine,* 47–58 (Thornhill, Ontario: HSTC Publications, 1983).

### Questions and Topics for Discussion and Writing

1. Describe and give approximate dates for the major improvements in lighthouse technology described in this article.
2. Lighthouse keeping is largely an occupation of the past, due to electricity and automation. Comment on what you can infer from the article and any other information available to you on what was the life of a lighthouse keeper's family: its advantages and disadvantages, its adventure and/or boredom, its challenge, dangers, responsibilities, and so on.

[1] Interview with Margaret Breton, Port aux Choix, 24 July 1979.

*Judith Tulloch*

# * Mechanization and Bread

## Siegfried Giedion

One way to get a glimpse of the way that technology has touched our lives is to consider a very basic and ordinary function which has been part of human life since the beginnings of civilization and trace the development and changes in how that function has been performed over the course of civilization, in particular noting the effects of technological changes. In his book, *Mechanization Takes Command: A Contribution to Anonymous History,* published in 1948, Siegfried Giedion picks a variety of commonplace products and processes, all easily taken for granted, and examines their extraordinary historical development. This article, a selection from that book, is about the making of bread and what has happened to it as the process became mechanized. The article, though long, is exceptionally rewarding and thought provoking. Giedion shows how mechanization of the bread-making process has led to totally new taste preferences for bread that suited the machine-made product at the expense of natural tastes and nutritional value.

Food is one of man's most immediate points of contact with nature. Man's sustenance must be suited to the laws that govern his body. Our ability to adapt cannot stretch beyond narrow limits. When these limits are exceeded, our organs rebel—depending on the dosage—either immediately, or later in life, or it may be, imperceptibly and in the course of generations.

No field within the entire scope of mechanization is so sensitive to mishandling as that of nutrition. Here mechanization encounters the human organism (whose laws of health and disease are still incompletely known). The step from the sound to the unsound is nowhere so short as in the matter of diet.

This is not always immediately perceptible. The ultimate effects cannot usually be foreseen. If man deviates too long from the constant of nature, his taste becomes slowly vitiated and his whole organism threatened. Unwittingly, he impairs judgment and instinct, without which balance is so easily lost.

### THE MECHANIZATION OF KNEADING

Toward the end of the handicraft era, the pharmacologist and agronomist Antoine Augustin Parmentier (1737–1813), who in the universal manner of his time combined scientific precision with a knowledge of the crafts, came to this definition: "Kneading is an operation by which the mixing together of yeast, flour, water, and air yields a new substance with special properties, soft, flexible and homogeneous."[1]

This kneading of dough is a strenuous activity, consisting at once of pulling, pushing, and beating. It was performed by hand and, for large quantities of dough, with the feet as well. With the advent of industrialization, the end of the guilds, and the expansion of cities, there arose a demand for machine kneading. The mechanical kneader can produce more quickly and more hygienically.

The beginnings reach a long way back. The Romans used rotating kneaders, and we have a number of experiments from the late Renaissance, with its penchant for applied mechanics. All these are primitive implements, which, among the various phases of kneading, are restricted almost entirely to striking. Boards or wooden bars attached to a crank shaft strike the dough. Elementary kneading devices are found even in pre-industrial times. There is, for example, the Castilian *braga*, a large roller sus-

[1] Parmentier, *L'Art du boulanger*, Paris, 1778, p. 361

pended from the ceiling, which swings back and forth over enormous bread boards. This Renaissance idea was used in the Vatican kitchens, although here the dough was worked by boards instead of rollers. Certain characteristics, which became more and more pronounced in the latest stages of mechanized bread making, can already be observed in the product of this simple Spanish machine; the bread was "whiter than that kneaded by hand; its crust is not crisp but extremely delicate and not too elastic."

Late in the eighteenth century, Italy, the land of floury foods, began to employ kneading machines on a larger scale. By 1789 kneading machines were operating in the municipal bakeries of Genoa. Here already a rotary mechanism was used. A tread wheel, such as we shall find in the first mechanical bakery in Paris toward the middle of the nineteenth century, drove vertical bars placed at intervals in the tub. This procedure, it was stressed, could provide a "pain léger et délicat." The bread of the municipal bakeries of Genoa was the only one sold on the market. And even today large-scale bread manufacture in Europe has been adopted mainly by co-operatives...

The inventions, despite their astonishing number, were extremely slow to be adopted. The reason given for this is that "the ordinary needs of the baker were not large enough for the advantageous and profitable use of machines; for even with machines, many stages of dough-making must be performed by hand." This observation, referring to conditions in Germany in the 'forties, does not essentially differ from a report, in the *Scientific American* for 1885, on a French mixer then to be seen in operation at a fashionable bakery on the Avenue de l'Opéra in Paris: "kneading by hand usually employed in all bakeries, is here replaced by machinery."[1] Indeed, the first rotary mixer was not patented in America until the late 'sixties, at the time when attempts were being made to carry mechanization into every sphere of daily life.

More than half a century was needed before the decisive step could be taken toward complete mechanization. This came with the introduction of the high-speed mixer, which did not enter general use until surprisingly late: after 1925. The high-speed mixer no longer attempts to imitate the various movements of the human hand. The agitator usually consists of two arms attached to simple steel bars, which perform sixty to eighty revolutions a minute. "High speed" refers to the greater number of revolutions, but even more to the incredible speed and agitation that the dough receives with each revolution, that is, the accelerated mixing speed. The shocks administered in this process are so great that the more delicate European wheat cannot withstand them. High-speed mixers have not proved successful in Europe.

But in America they have become one of the mainstays of full mechanization. This is perhaps not only because they allow accelerated production but also because they achieve a more thorough mixing and interpenetration of the ingredients, which are thrown back and forth with incredible speed between steel arms, producing a dough of more uniform structure. If "in 1925 the whole industry had sold itself on the desirability of the high speed mixer," as a contemporary has stated, the main reason seems to have been that the energetic mixing made possible the manufacture of a bread even whiter than before.

## THE MECHANIZATION OF BAKING

### The Baker's Oven in the Handicraft Age

The form of the baking oven has passed down almost unchanged through the centuries. Like the axe and the knife, it is a basic tool of the human inventory. The oven developed into an egg-shaped chamber, which is excellently adapted to retaining and distributing heat...

Thus the baking oven was an oval chamber encased in a thick, fireproof vault of clay, brick, or stone. A fire of logs and faggots was built within. Once the stones had stored enough heat, the ashes were raked out; then the dough was put in and slowly baked by the heat which the stones retained. The highest temperature struck the dough when it was first put into the oven, and the heat gradually abated in the course of baking: a natural process,

[1] *Scientific American*, 17 Oct. 1885.

coinciding with the needs of bread baking. Every detail of this simple device—the vaulting, the inclined hearth, the position of the flue—was the outcome of unfathomably ancient experience.

## The Influence of Technology: The Indirectly Heated Oven

The method in which the chamber was heated, the embers swept out, and the bread inserted came to be regarded as too slow for the large quantities of bread required by the nineteenth century. No continuous flow of production was possible. The bake oven could not be mechanized in its traditional form. The first step, then, was to make the chamber in which the bread was baked separate from the fire. The baking chamber and the furnace chamber became independent of one another. The hot gases were made to pass, often in ingenious ways, around and over the baking chamber...

The nineteenth century developed the technique of heating air in special chambers before conducting it into the baking chamber. This type was called the *aerotherme*. In principle, this is nothing other than the hot-air heating favored in dwellings throughout the second half of the nineteenth century, with the difference that the aerotherme ovens were a closed heat circuit. The air, heated in special channels and reservoirs, did not come into contact with the fire gases. Not only was the method more efficient, but the air in the baking chamber remained entirely pure. This oven was used with success in French hospitals in the 'forties.

The aerotherme oven (1832) of the Frenchman Aribert retains some characteristics of the old baking oven. The bread trays were slowly carried on rails through the straight or circular hot-air tunnels from a hotter to a cooler zone. This aerotherm oven, which shows beginnings of distinctly mechanical operation, was to be found in several French cities around 1840...

## The Oven and the Endless Belt

Aribert, whose aerotherme oven (1832) with its plates running on rails pointed to a clearly mechanizing trend, called his invention a *four continu*, continuously working oven. Around 1850 efforts in this field focused on bringing about continuous operation by means of a moving mechanism within the baking chamber. The ordinary hot-air ovens could be heated continuously, but changing the loaves took too much time. Trained workers were needed to manipulate the peel, the flat wooden shovel with the long handle used to introduce and remove the loaves.

Attempts were made to increase production in two ways.

One was the use of draw plates, which could be thrust into the oven or taken out by unskilled workers with a single motion. This type has been preserved and developed in the most varied forms...

But far more persistent were the efforts to obtain a continuous flow by building a chain conveyor through the baking chamber. Between 1850 and 1860, the most diverse solutions were attempted: chains running horizontally, chains running vertically, singly, or in series, until, by the early 'sixties, the bake oven of 1800 had become a highly complicated mechanism...

At the time these automatic ovens came into being, Boston, Chicago, New York, and especially Philadelphia had mechanical bakeries using ovens, sometimes several storeys high, fed by an endless belt. Some of these were successful, others ill fated, either burning down or, when they failed to pay their way, passing from hand to hand. But successful or otherwise, they were on the whole show pieces. Small bakers were the rule, and, much longer than the European, American housewives continued to make their bread at home.

There was another factor, too: the automatic manufacture of bread is an unusually exacting task. In the second half of the nineteenth century only the first and last phases of the process were at hand: kneading machines and conveyor ovens. The middle links were lacking: machines to weigh the dough automatically, to divide it, to make it into balls, roll it, and carry it on endless belts through glass-paneled passages exactly controlled for warmth and humidity. And modern yeast, which cuts fermentation time almost by half, was not yet available.

After 1900, as mechanical baking approached readiness for final mechanization, all experiments had to go back to the beginning. The astonishingly diverse baking ovens of the 'fifties and 'sixties, which clearly developed the principles underlying all types, had been forgotten. They belonged to the great realm of shelved inventions. This broken continuity impresses the historian much as a mine shaft left unworked and unexploited.

Around 1907, before a Canadian bakers' convention, a Mr. Roberts, representative of a London baking-oven firm, envisioned what he called an "oven of the future" in which the dough would be put in at one end and the finished bread come out of the other. These are the very words we read in the specifications of 1850. In his audience was an enterprising baker, Dent Harrison of Westmount, Quebec, who became the initiator of the present-day development. At that time traveling ovens existed, but they were used only in biscuit-making factories. In one of these, at Montreal, Harrison and Roberts baked experimental batches on the traveling oven. The loaves coming out of the other end were burned black. Dent Harrison nevertheless continued to think the "new" idea practicable, and gave Roberts an order "for the first travelling oven in the world for baking bread."

In 1913 this oven, constructed in England, was set up in Montreal. Its baking plate was 50 feet long and 6 feet wide. Fired with coal, it functioned from the very start...

Soon gas took the place of coal in heating the ovens; in 1917 a series of gas burners was placed at fixed intervals above and below the hearth plate throughout the length of the oven.

Finally, in the 'twenties, high pressure steam was used, which circulated through a series of tubes around the hearth...

## THE MECHANIZATION OF BREAD MAKING

Thus far we have been dealing with individual phases in the process of bread manufacture. So long as an uninterrupted production line was lacking, mass production remained impossible.

The first assembly line came into being...in an English "victualling office," when the various machines for making ships' biscuits were inter-related. This was in 1833, just a half century after Oliver Evans had devised his mechanical mill. In this royal victualling office, the amounts of flour and water flowing into the mixer, as well was the rolling of the dough, were automatically regulated. "Heavy cast iron rollers...run alternately with great rapidity from one end of a table to the other by means of a beam connected with the steam-engine below." These heavy rollers were inspired by the Castilian *braga*, French writers say. "When this operation is finished, the dough is conveyed on rollers to a second table"—where it is divided into pieces—"and finally, still running on friction tables, to the machine which at the same moment cuts and stamps the biscuits."

Thus the three main phases of modern mass production—mixing, rolling, and molding—were at once mechanized, and connected in a single production line.

The unfermented dough of flat ships' biscuit is of far simpler structure than the sensitive dough of bread. But it was not long before bread too was being turned out in continuous manufacture and even in mass production. The French had such establishments successfully operating as early as the 1840s. The machine bakery of the Mouchot brothers in Paris won continental fame. It used the latest hot-air ovens and large efficient kneading machines. These kneaders were driven by a tread wheel, as used in the Renaissance and in the municipal bakeries of Genoa in the late eighteenth century. The wheel was suspended outside the baking room and was trodden by well-drilled dogs. When the kneader had completed the necessary revolutions and the mixing was completed, a whistle automatically blew and the animals ceased their labor. The new aerotherme ovens were heated by coke which was 50 per cent cheaper and offered greater "production capacity." Two bake ovens turned out 6240 kilograms of bread in 24 hours. The gas used to illuminate this cellar bakery was manufactured by the ingenious baker himself.

In England at this time one could have found no Renaissance tread wheels for bread making. In 1850 a mechanism was put into operation in Glasgow, which was said to manufacture one and a half tons of bread an hour, or approximately four times as much as the Paris bakery of the Mouchot brothers. This was a bread factory packed within a few square yards. Everything was automatic and concentrated in one machine. A fly-wheel regulator, as used in steam engines, controlled the flow of flour and water into the mixer; rotary or sliding knives cut the dough into pieces which were mechanically molded into loaves and propelled into the oven. This bread factory in miniature recalls those queer musical contrivances that unite a whole orchestra in one instrument. But this grotesque scheme embraced two ingenious ideas: coiled pipes were used to heat two of its four baking chambers, and soda water to saturate the dough. In France this device would have been unthinkable, for the people would never have accepted such bread. Nor did it enjoy much success in England. It is one of the starting points, however, for more refined methods.

## Bread and Gas

To make the dough tasty and porous, two kinds of ferment were generally used: leaven and yeast. In bread making with leaven, some of the dough from one baking is set aside and used for the next. The leaven acts as a sort of starter. It yields a full-bodied and tasty bread, with the slightly sour aroma that many people still prefer. Leaven was used throughout the Middle Ages, and long after. The Italians still insist on leaven for their white bread, the Germans for their mixed rye breads, the Russians for their black bread.

In many highly mechanized countries, leaven has been altogether supplanted by yeast. First brewer's yeast was used. Opinions vary about when this began. Some say that yeast did not come into general use until the middle of the nineteenth century, while others point out that brewer's yeast was used in Paris bakeries as early as the seventeenth. Yeast dough, like leaven dough, is left to rise overnight. During this time it slowly rises as a result of the gas formed by fermentation, and a natural warmth develops.

On the Continent, special high-fermenting types of yeast were later developed and sold in compressed form. These ultimately led to the modern bakery yeast, which American concerns produce in large quantities. High-fermenting yeast reduces the time of fermentation from ten hours to five.

Prior to the mechanization of baking, the bread's aroma was held in great esteem. It comes from the natural bouquet of baked grain, from the volatile esters developed in fermentation or during baking, and sometimes from added aromatic herbs (caraway seed). These volatile esters, together with carbonic acid gas, are slowly produced in the course of fermentation. Chemists have stressed that a large and rapid generation of carbonic acid results in a loss of flavor.

In the 'fifties, bread, in its very substance as a material, was attacked by mechanization. For the sake of increased production, carbonic acid, instead of the slower-working ferments, was mixed with the dough.

From mid-eighteenth century on, chemicals were resorted to in bread manufacture in order to make the bread weigh more than its flour content would allow, or to make it look whiter than was to be expected from the quality of the flour: plaster, alum, sulphate of copper—"a liqueur glass of weak copper sulphate solution for 200 loaves of bread"—or such substitutes as potato or bean flour. This tampering gave the illusion of greater quantity or of better quality. Punitive measures were taken to reduce or eliminate adulteration.

In the 1850s, attempts to increase production resorted not only to mechanical means, but also to the natural sciences. The laws of the expansion and contraction of gases, which John Dalton and Gay-Lussac discovered shortly after 1800, found their practical application in bread manufacture half a century later. John Dalton's researches into the "Absorption of Gases by Water" (1803) are especially relevant.

The solubility of gases increases in ratio with the pressure. The greater the pressure over a liquid, the more gas it will absorb. Why should this not apply to dough?

Dr. John Dauglish, a British physician (1824–66), proceeded from this idea. While studying in Edinburgh (1852–5), he experimented with driving carbonic acid into dough under a pressure of 10 to 12 atmospheres. In the specification of his *Improved Method of Making Bread*, for which he obtained a patent in 1856, he states that "water charged with carbonic acid" had been used before in the manufacture of bread, but that the substance of his invention was the mixing of charged water and flour under high pressure. High-pressure apparatus to saturate water with gas had long been used in the manufacture of soda water. There would be little difference, between manufacturing soda water or bread by this process; the apparatus would merely have to be somewhat adapted to material of a different consistency. Dauglish had no monopoly over such endeavors. In the very month of 1857 when his second patent was granted, two Americans came forward with "Improvements in the preparation of dough [mixing with gas]." A few years later they brought out a pamphlet, its title a striking slogan for the aerated product: "Bread for the Millions."

Dauglish too pointed out the advantages of his unfermented bread. In a paper read at the Society of Arts in London in 1860, he showed that it could be kept indefinitely and proved that it was more healthful. The American authors of *Bread for the Millions* go one step further. They raise the alarm against bread made with yeast. "A ferment or yeast," they write, "may be characterized as a body in a rotting condition." Pasteur, who had not long before (1857) discovered the yeast bacilli, might have shown some surprise at this definition.

John Dauglish was a fanatic who wore himself out crusading for his bread, and died at the age of forty-two. His name is still associated with the idea.

His first apparatus was simple. Soda water and flour were mingled under high atmospheric pressure in a strong-walled mixer. When the kneading was complete, the pressure was removed. This allowed the gas to expand, and within a few minutes its tiny bubbles would raise the dough. The high pressure and sudden expansion caused coldness to spread through the dough as it might in a corpse. The natural warmth developed when gas is generated by fermentation was lacking. Formerly, however, the dough would rise in six to eight hours. Now it rose instantaneously…

The gas bread manufactured in various American cities around 1860 never became truly popular. A single company, the Aerated Bread Company, instigated by Dr. Dauglish himself, still sports the letters A.B.C. over its chain restaurants in London.

Dr. Dauglish was not one of the really creative inventors. He did perfect the thought he pursued so fanatically throughout his career, an idea, however, that was not new. There is something quixotic about the goal and the fanaticism of the man who devoted his lifetime to attaining it. His apparatus cannot be dismissed as a mere mechanical plaything. It revealed at an early date how mechanization was later to change the nature of bread.

## THE HUMAN ASPECT: BREAD UNDER MECHANIZATION

Two questions will be considered here.

- How was flour, the material of bread, affected by the advance of mechanization?
- How did mechanization alter the structure of bread, for so many centuries the symbol of human sustenance?

### Flour in Mass Production

The flour used by bakers today has a whiteness far exceeding that of 1850. This whiteness is not the result of adulteration with foreign substances (alum, copper sulphate, etc.), but of a revolutionized milling process. The millstones crushed the grain in such manner as to mix inseparably at the first milling most of the components of the berry, the starchy interior, the nutritious layer of gluten cells, and the valuable oily germ or embryo. The oily substance of the germ penetrated the mixture, making it oily to the touch and frequently unpleasant in appearance. Moreover, there was the risk that the oily ingredients would turn rancid in prolonged storage. In this low or flat grinding process, the millstones with their sharp grooves were set as close together as possible.

The millstones, which crushed the grain as small as possible at the first milling, are now replaced by rollers, relatively far apart, which squeeze the grain far more gradually. This "high grinding" is performed in stages. After every passage through rollers, the grain is separated by sifting machines into its various components. The process is repeated four, six, eight, ten times.

When did this begin?

The gradual reduction of the wheat into flour by means of rollers was called the "Hungarian system." Although France and other countries worked on this process, it was systematically developed in a country having much wheat and a fondness for elaborate pastries requiring extra fine flour—Hungary. This took place between 1834 and 1873… In America the transformation of the milling process was carried out between 1870 and 1880 in the wheat-growing Middle West (Minneapolis). It had begun in the late 'sixties with attempts to separate the thin glutinous middlings surrounding the kernel.

The year 1871 brought a cleaner, such as the French had sought around 1860, by which a hitherto poor grade of wheat could be made to yield the most highly priced flour. Minnesota flour was the outcome of this "new purifying process," popularly known as the "new process." "Minnesota fancy reigned supreme in every market in which it was introduced as the sharpest, whitest, brightest flour ever produced," as an anonymous contemporary pamphlet points out. This cleansing process was devised before the rollers were tried out in America in 1873; but from then on, the tempo as well as the size of the mills increased enormously. Around 1881 all the large Minneapolis mills were adapted to the new process and extensively automatized.

Hand in hand came large-scale production and increasing concentration of ownership. It was the period in which the Chicago packing houses moved toward monopoly.

No radical change had taken place in the century from Oliver Evans' "new process of flour making by automatic machinery" to the "new process of middling flour." Likewise no basic improvement in milling machinery occurred after 1890. Instead there were technical changes in the direction of novelty. Interest centered around obtaining still finer and whiter flour. Efforts bore on elaborate, ever more complicated apparatus for the artificial bleaching of flour. The public—the millers maintained—demanded a flour of greater whiteness than could be obtained by anything but artificial bleaching. This may have been the case. But the decisive reason lay elsewhere. It had formerly been considered necessary to age the flour for a period of months. During this time it lost its natural creamy color and became pure white. But the time factor involved in this aging process was onerous to large-scale production. It meant large warehouses and capital outlays lying idle. "Millers sought for a way to escape the burden," specialists inform us. "The remedy they found was the artificial bleaching and aging of flour." This was done by high-voltage currents or infiltration with gases (chlorine). Commercial bleaching was first introduced in France around the turn of the century, then successfully carried out in England, finally finding its large-scale application in America. The cabinet in which the bleaching takes place within a matter of minutes is no bulkier than a small chest of drawers. Blown in through tubes, the chlorine gas instantaneously penetrates the whirling particles; and the flour drops another storey downwards straight into its paper sack.

Artificial bleaching was not accepted without debate. There were long and bitter controversies and investigations. The experts often took diametrically opposite stands. It is not for us to judge whether they are right who maintain that the bleaching process causes no deleterious effects, or they who claim that the present milling methods remove the most important nutritional values from the grain. We need only note that the new methods arose in the demand for greater output, the human considerations having little voice in the matter.

Mechanization of the milling process yielded a brilliant façade and a more or less artificial product. The oleaginous germ that formerly made the flour somewhat greasy to the touch, and which contains the most valuable elements, has been rigorously excluded. More recently, we have seen attempts to make good the values removed from the flour by vitamin reinforcements added to the yeast or the dough. The whiteness of the flour remains unspotted. But such measures remind one of a dentist who extracts good natural teeth, filling the void with a bright and handsome set of artificial ones. The device of adding vitamins is a simple one. It consists of a metal cabinet slotted like a letter box. The vitamin paste dropping in small clusters into the flour is thoroughly mixed with it by a screw conveyor such as Oliver Evans used.

### Full Mechanization: Bread on the Production Line

Many of the machines by which the mass manufacture of bread and the continuous production line became possible were first devised in Europe. Bread baking is a complicated craft. Although kneading machines and other labor-saving devices were used long before, nowhere was the passage to full mechanization more hesitant than in the food products.

Even in England, where this mechanization had its beginnings, "bread factories are less numerous than small bakeries. After 1900, the proportion is 80 : 1 in favor of the small bakeries." Since that time, in comparison with America, things have remained much as they were.

Such is certainly the case on the Continent, even in countries with a high living standard. Their small bakeries are largely equipped with mechanical implements and electric ovens. Everyone tastes differences in the bread of the various local bakers and selects his baker accordingly.

Only in countries where the most advanced stages of mechanization have been reached, as in the United States and, in this respect, Canada, has the small baker almost disappeared and mass production become the rule. According to the U.S. Census of 1939, corporate enterprise manufactured bread and other baked goods (biscuits, crackers, and pretzels excepted) to the value of 514 million dollars as against 20 million dollars manufactured by non-corporate, i.e. small establishments. This went hand in hand with the growing mechanization of daily life between 1914 and 1930, and especially in the period 1925–30.

In this period the bread factory was consolidated into a production unit. All operations and machines were co-ordinated and synchronized with clockwork precision. The manufacture of dough became automatized in all its stages; the loaf obtained an extremely even structure and absolute uniformity.

The dimensions of the ovens have increased over those of 1850; this is true of all types—steam ovens, draw-plate ovens, rotary ovens—above all of the endless-belt type, which has grown to the length of 100 to 130 feet. The term "tunnel oven" is no exaggeration.

Engineers use their thermodynamic experience to insulate the ovens and to regulate the heat with great precision. All dirt is kept out; no stoking or coal-shoveling pollutes the atmosphere, for gas, electricity, or oil have replaced coal. The baking chamber is now as clean as a hydroelectric power plant.

The kneaders with their shiny enamel housings often reveal a surprising beauty of form. All sharp edges and corners disappear, the forms have the fluidity of a melting ice block. We are in the period of streamline design. The assembly line, which rolls almost without human intervention, has developed to a point rarely reached in other fields.

Beyond a doubt there is something impressive in the precision, synchronization, and cleanliness of a modern bread factory with its hourly output of 30,000 loaves. In Philadelphia we visited one of the largest plants of the kind—a five-storey reinforced concrete building, its façade a vast expanse of glass, its auto park accommodating 500 delivery trucks.

The production method is the same as in almost all modern factories: that is, production starts in the upper storeys; conveyor systems link one operation to another. Oliver Evans, in his mill at Red Clay Creek (1784), was the first to couple the force of gravity with mechanical conveyor systems. In the modern bread factory, the flour is first elevated to the top storey, where batteries of high-speed mixers are aligned. Each of these mixers holds three and a half barrels (or some 600 pounds). "Mammoth" mixers have been abandoned; they overheated the dough. The plant we visited formerly had the largest mixer of this kind, capable of mixing 30 barrels, or nearly 6000 pounds, of dough at a time. But giant mixers and giant troughs were unsuited to so sensitive a material.

After a few minutes of mixing, the heavy mass of sponge—or first mix—is poured from each machine into a trough. The troughs swing on pivots much as miners' towcars, but much longer, and hang by strong rods from trollies which glide on a monorail close to the ceiling. The system is the same in principle as that used for moving carcasses in packing houses, but far more massive. The troughs are shunted into the next room, where fermentation takes place. Trough follows trough along the ceiling, train after train, monorail after monorail, as in a suspended freight yard. In four and a half hours the sponge rises from the bottom of the trough to the brim. The slowly forming gas bloats the mass like a balloon, and if the elastic, membrane-like layer spanning the vessel is broken open, the warm carbonic-acid gas rushes out and it is possible to look into countless invisibly working gas chambers. At this stage, the living mass still has the irregular hollow, bubbles, and cavities of the hand-made loaf.

After the mass stands for four and a half hours, the train is set in motion. Trough after trough is pushed to the rear of the room and tipped over; the sponge flows through large openings in the floor down to the storey below.

There it goes into a second battery of mixers, where the irregular sponge is given the ingredients it needs to become a uniform mass. Before fermentation, only about 60 per cent of the flour is put in. Now the other 40 per cent is added, along with milk, water, fats, six parts per hundred of sugar, vitamins, and what other ingredients may seem desirable. The dough now has its final composition.

After this second passage through the high-speed mixers, the dough must be allowed half an hour or an hour's rest, to appease the shaken-up molecules. Then it is ready to be divided. Again it drops one storey groundwards, where a divider severs the mass into pieces, each the weight of a loaf. A second machine rounds these into balls, as the baker formerly did by hand. This encourages a thin skin to form around the pieces of dough to prevent the gas from escaping, or, as bakers say, to prevent "bleeding." The balls of dough fall each into its hollow on the bucket proofer, an endless chain, and glide for a quarter of an hour through warm glassed-in chambers, where they are sheltered against any draft.

The final preparation of the dough takes place in a molder which, mechanically imitating and breaking down that last movements of the hand-baker, rolls the balls into cylindrical shape. Now they are ready to be laid into the pan, an operation that can be automatically performed but which even large plants prefer to do by hand. Fermentation is again stimulated. In pan-proofing cabinets with high humidity, the yeast works vigorously and in the course of an hour drives the dough over the top of its pan. Now the baking process begins. The pans disappear into the 10-foot-wide slot of the white-walled ovens, standing side by side. For half an hour they travel on their endless belt through the 130-foot tunnel. Fully baked, they are automatically pushed out at the other end. A second time the human hand intervenes, and heavily gloved workers pluck the loaves from the hot pans. A conveyor moves them to the floor below where, in about two hours, they slowly cool in overhead coolers—moving racks in chambers with exact control of temperature and humidity.

Methods appear in the 'thirties by which the cooling time is reduced by division into two stages: "The first step at atmospheric pressure, and thereafter quicker cooling of the bread by evaporation of moisture under subatmospheric pressure."

There remains only to slice the loaves, to pack and distribute them.

The automatic slicing machines, which cut sixty loaves a minute into uniform slices, are late comers in bread manufacture. Invented in 1928, hesitantly accepted in 1930, by 1940 they were regarded as indispensable. When in 1942 the United States Government, being at war, banned the use of bread-slicing machines to save steel, this so displeased the housewives that the order had to be canceled.

The idea of cutting whole loaf of bread at once, we may note, had appeared in America as early as the 'sixties, when a bread-cutting device with ten parallel sickle-shaped blades attached to a lever was patented. But this machine, like the vacuum cleaner and the dishwashing machine, joined the shelved inventions. There were reasons for this. To work satisfactorily, the modern bread-slicing machine, as a leading expert remarks, requires bread such as only present-day mechanization can produce, loaves of absolute uniformity, soft in crust and regular in crumb.

As a rule the slicers form a unit with the wrapping machines, also introduced relatively late, in 1913–14. From these the wrapped product is conveyed directly to trucks waiting in rows on the ground floor.

Viewing the entire cycle, we find that the time saved by these mechanisms is out of proportion to their complexity. In the double mixing (sponge-dough) method we have described, some eight and a half hours are required before the flour is transformed into finished bread. The main time saving over earlier methods lies in the use of quicker rising yeast, which reduces the fermentation period from nine to approximately four and a half hours. There are quicker ways—the straight dough method—in which one mixes all the ingredients at once. Fermentation time can thus be cut to three and a half hours. But general preference is for the longer, sponge-dough method.

The decisive saving rests not in acceleration of the process but in the enormous output that mechanization and the assembly line render possible.

The bread-making process refuses to be hastened beyond narrow limits; for mechanization here encounters an organic substance whose laws are inviolable. Antoine Augustin Parmentier defined dough (1778) as the soft, flexible, and uniform substance obtained by thoroughly mixing flour, water, air, and leaven (for yeast was not yet used in bread baking). Today one might add that this substance is not a static one, but an organic body ever in flux, a highly sensitive culture whose incubation must be carefully controlled in every phase of its mass manufacture with the aid of thermometers and hot-air chambers. We have had occasion to see how after each mechanical intervention a pause of varying length is required to restore the interrupted activity of the

FIGURE 2.1.: Advertisement for wrapped and sliced bread. 1944.

enzymes. In the mechanized process, more subdivisions and pauses, and far greater caution, are necessary than in hand baking.

Wherever mechanization encounters a living substance, bacterial or animal almost indifferently, it is the organic substance that determines the law.

### The Changed Nature of Mechanized Bread

The question still remains: How was bread changed under the influence of mechanization?

Dr. Dauglish's carbonated bread had an extremely uniform, finely porous structure, produced by the stream of gas injected under pressure. The crust was relatively hard, but the inside was elastic as a rubber sponge and completely tasteless. It did little good to mix the dough with wine instead of with water, as was later done in the hope that the acid ingredients would heighten its flavor.

Dr. Daughlish's actual formula did not prevail, but characteristics distinguishing his mechanical bread from that of the pre-industrial period have.

#### *Uniformity*

How did it become possible, after 1900, for the large baking corporations to eliminate the small bakers who had hitherto possessed the market? The experts always offer this explanation: the product of the small bakers, they say, was always changing; one day it was such, another day it was such. Mechanization, on the other hand, furnished bread that was entirely uniform. That the public demanded uniformity may be true, but economic factors were doubtless more decisive. After 1900 we enter the period when anonymous corporations penetrate nearly every province of living.

Bread as completely homogeneous as Dr. Daughlish obtained by his gas process remained impossible with yeast, that is, fermentation, until the high-speed mixer was generally introduced (1928). The tremendous beating power of its agitator scattered the yeast particles throughout the dough. Uniformity and the increasing stress on appearance entered hand in hand.

Today the public demands that their eggs shall have yolks of uniform color. One city prefers a light yellow, another a deep orange. Industry provides the corresponding chicken feed, which, with the help of artificial coloring, never fails to produce yolks of the same shade.

Public taste is not so particular about the color of bread crust, but here too any required tint can be precisely obtained by means of thermostats and added ingredients. The neat wrapper, often highly effective from the advertising standpoint in addition to keeping the bread clean and fresh, satisfies the demand for uniformity (See Figure 2.1).

#### *Changes in Structure*

The bread of full mechanization has the resiliency of a rubber sponge. When squeezed it returns to its former shape. The loaf becomes constantly whiter, more elastic, and frothier. This is not the result of mechanization alone. With the increasingly complex machinery and greater capital outlays, new ways of pushing consumption had to be devised.

Since mechanization, it has often been pointed out, white bread has become much richer in fats, milk, and sugar. But these are added largely to stimulate sales by heightening the loaf's eye-appeal. The shortenings used in bread, a leading authority states, are "primarily for the purpose of imparting desirable tender eating or chewing qualities to the finished product." they produce the "soft velvet crumb," a cakelike structure, so that the bread is half-masticated, as it were, before reaching the mouth.

About 6 per cent of sugar is usually added to white bread. This too makes for looseness of structure and imparts a slight sweetness. Moreover it stimulates fermentation. But above all, sugar is "the source of crust color." If the amount is reduced, the thin crust becomes "pale and unattractive" in appearance instead of being infused with that golden-yellow gloss, like those bright red apples whose appetizing exterior has almost driven out other varieties of less dazzling appearance but more delicate flavor.

Mechanization does not stand still. We have mentioned those proposals of the 1930s for cooling the bread more rapidly, under atmospheric, followed by subatmospheric, pressure. In a later patent the inventor turns to good advantage the opportunity of the cooling phase for further work upon the make-up. While the loaf is having its dampness extracted in the vacuum chamber, it is pierced by "an injector device provided with a needle valve" much as the human skin is pierced by the doctor's hypodermic. One can, the inventor assures us, "provide any desired penetration." His discourse has something of Dr. Daughlish's fantasy and seems worth recording for its symptomatic value.

> Another object of the invention is to provide for injecting substances such as lemon, orange, raisin or other flavors or flavoring extracts, colouring materials... Thus bread may be treated with a colouring material so that it will assume the desired tint. This material may have any desired flavor and hence the loaf, in addition to being coloured, will also have its taste fixed... Similarly vitamins may be injected by volatile carriers....For reinforcing the keeping qualities such gases as ozone may be introduced.

As the loaf becomes ever softer and frothier, there is developed the demand for crust of the utmost thinness and for bread that is absolutely fresh. In the 'thirties of the past century, Sylvester Graham, the great reformer..., condemned his countrymen's taste for bread as it comes steaming from the oven. Mechanization may have fostered the wish by using its skills to develop the soft airy texture that is characteristic of freshly baked bread. But it is certainly not the sole cause. Such texture concurs with the whole process of rapid production: long before mechanization set in, Sylvester Graham made it clear that thin-crusted bread, made of finely milled flour, bakes more quickly than any other. Imperceptibly, public taste adapted itself to this fact. Today's arbiter of taste in the bread industry indicates to bakers what he considers the main defects in crust: toughness, thickness, cracks. And he recommends above all tenderness and uniformity.

Extreme freshness has grown into a demand that it does not pay to disregard. "Storekeepers find stale bread, that is, bread that has stood for a day or even less, unsaleable." To preserve its fresh-baked quality the bread is wrapped in thick waxed paper bearing a carefully printed design—a usage that was respected through the paper shortage of the Second World War. The housewife, according to an American expert, feels the bread through the wrapper, and if it is not so soft and elastic that her fingers can almost meet in the middle, puts it back. Only thoroughly fresh bread is tolerated in the home—a factor conducive to waste. This schooling in waste works more in favor of expanded production than to the advantage of human digestion.

There has been no dearth of efforts to restore artificially the natural values removed for appearance's sake. In 1916 a large raisin firm began to advertise, urging bakers to use raisins in bread. Within two years the sale of raisins increased over tenfold.

Suggestions for adding vitamins to the dough date from the late 'twenties. They were not carried out on a large scale until 1940, when vitamins were raised into a public preoccupation. "The American public has been told that they can now obtain a new white bread containing vitamins and minerals natural to whole wheat."

### Mechanization Alters Public Taste

Doubtless fine bread, like many other luxuries, was so cheapened in price through mass production as to become available to all. No King of France had breads of such white and silken quality on his table. But bread's position as the staff of life has been undermined. The complicated machinery of full mechanization has altered its structure and converted it into a body that is neither bread nor cake, but something half-way between the two. Whatever new enrichments may be devised, nothing can really help as long as this sweetish softness continues to haunt its structure.

If the type of bread preferred in the United States were put to the public vote, the verdict would not be hard to foresee. The 1939 Census tells the story: the quantity of white pan bread consumed is about four times that of rye or whole wheat. And we may note that the whole-wheat bread is not unlike the white pan bread in frothiness and flavor, and that as a rule 40 to 50 per cent of highly refined white flour is mixed with rye bread; here too the crust is kept as thin as possible.

The change of taste through mechanization forms an even clearer picture when the quantity of white flour used in bread and other bakery products in 1939 is compared to that of whole wheat (including graham). The ratio is 27 : 1.

The adulteration of taste, similar to the one that accompanied the mechanizing of bread, was witnessed a century earlier in a quite different field. The ruling artistic taste in the nineteenth century was formed by exploiting certain dormant wishes of the public. The public loves what is sweet, smooth, and outwardly appealing. These desires can be strengthened, weakened, or diverted into positive channels. The painters of the ruling taste set out to satisfy this public demand more and more, thus securing a market and rewards. The outcome was a blurring of the instincts in all classes of society, a disorientation that is doing its damage even today.

How ancient instincts were warped when bread met the impact of mechanization cannot be demonstrated in detail. There was an impenetrable thicket of action and reaction, the stimulating of wishes and their satisfaction. The changed characteristics of bread always turned out to the benefit of the producer. It was as if the consumer unconsciously adapted his taste to the type of bread best suited to mass production and rapid turn-over.

From Siegfried Giedion, *Mechanization Takes Command, 169–201.*

## Questions and Topics for Discussion and Writing

1. Describe the process of baking and the qualities of the resulting bread as produced in the handicraft age.
2. What are the advantages of mechanizing bread-making?
3. Discuss the changes in public tastes for bread that accompanied the changes in the mechanization of bread-making.
4. Name some other basic products or processes that have undergone drastic changes due to mechanization over the centuries and speculate on how the mechanization process has changed the public's use of product or process, expectations of it, and preferences associated with each.

**Photograph on next page:**
Crash of the dirigible Hindenburg, 1937.

*Issue*

# Technological Innovation and Failure

All technology is a product of human ingenuity that has been put to the test of application. Wherever there is ingenuity—thinking of something that others have not thought of before—there will also be resistance from those who think otherwise. And wherever the ingenuity is put to the test, there is the possibility of failure, sometimes at the expense of people's lives.

The three articles in this section examine three sides of innovation, resistance, and failure: Simon Newcomb's "The Outlook for the Flying Machine" explains why a flying machine (airplane) is a wrongheaded idea that just won't work. Elting Morison's "Gunfire at Sea: A Case Study of Innovation" examines the difficulty of getting other people to accept an important innovation, especially those people with vested interests in the status quo. Samuel C. Florman, writing about the space shuttle Challenger disaster in "Engineering Under Pressure," blames engineers who put engineering considerations aside in favour of management concerns.

# The Outlook for the Flying Machine

## Simon Newcomb

If there is any general lesson to be learned from the history of invention, it is that predicting what will or will not be invented in the future is nearly impossible. Nonetheless, people continue to do it, even people who should know better. Looking back at past prognostications, it is hard to resist a chuckle at some of the more pompous pronouncements of what *can't* be done. This article by the very eminent American scientist, Professor Simon Newcomb, LL.D., explains with convincing scientific certainty why airplanes are a virtual impossibility—at least, an impossibility, he asserted, given the technology of the times. The article appeared on October 22, 1903—in other words, two months before the Wright brothers' first successful powered flight.

Mr. Secretary Langley's trial of his flying machine, which seems to have come to an abortive issue last week, strikes a sympathetic chord in the constitution of our race. Are we not the lords of creation? Have we not girdled the earth with wires through which we speak to our antipodes? Do we not journey from continent to continent over oceans that no animal can cross, and with a speed of which our ancestors would never have dreamed? Is not all the rest of the animal creation so far inferior to us in every point that the best thing it can do is to become completely subservient to our needs, dying, if need be, that its flesh may become a toothsome dish on our tables? And yet here is an insignificant little bird, from whose mind, if mind it has, all conceptions of natural law are excluded, applying the rules of aerodynamics in an application of mechanical force to an end we have never been able to reach, and this with entire ease and absence of consciousness that it is doing an extraordinary thing. Surely our knowledge of natural laws, and that inventive genius which has enabled us to subordinate all nature to our needs, ought also to enable us to do anything that the bird can do. Therefore we must fly. If we cannot yet do it, it is only because we have not got to the bottom of the subject. Our successors of the not distant future will surely succeed.

This is at first sight a very natural and plausible view of the case. And yet there are a number of circumstances of which we should take account before attempting a confident forecast. Our hope for the future is based on what we have done in the past. But when we draw conclusions from past successes we should not lose sight of the conditions on which success has depended. There is no advantage which has not its attendant drawbacks; no strength which has not its concomitant weakness. Wealth has its trials and health its dangers. We must expect our great superiority to the bird to be associated with conditions which would give it an advantage at some point. A little study will make these conditions clear.

We may look on the bird as a sort of flying machine complete in itself, of which a brain and nervous system are fundamentally necessary parts. No such machine can navigate the air unless guided by something having life. Apart from this, it could be of little use to us unless it carried human beings on its wings. We thus meet with a difficulty at the first step—we cannot give a brain and nervous system to our machine. These necessary adjuncts must be supplied by a man, who is no part of the machine, but something carried by it. The bird is a complete machine in itself. Our aerial machine must be ship plus man. Now, a man is, I believe, heavier than any bird that flies. The limit which the rarity of the air places upon its power of supporting wings, taken in connection with the combined weight of a man and a machine, make a drawback which we should not too hastily assume our ability to overcome. The example of the bird does not prove that man can fly. The hundred and fifty pounds of dead weight which the manager of the machine must add to it over and above that necessary in the bird may well prove an insurmountable obstacle to success.

I need hardly remark that the advantage possessed by the bird has its attendant drawbacks when we consider other movements than flying. Its wings are simply one pair of its legs, and the human race could not afford to abandon its arms for the most effective wings that nature or art could supply.

Another point to be considered is that the bird operates by the application of a kind of force which is peculiar to the animal creation, and no approach to which has ever been made in any mechanism. This force is that which gives rise to muscular action, of which the necessary condition is the direct action of a nervous system. We cannot have muscles or nerves for our flying machine. We have to replace them by such crude and clumsy adjuncts as steam engines and electric batteries. It may certainly seem singular if man is never to discover any combination of substances which, under the influence of some such agency as an electric current, shall expand and contract like a muscle. But, if he is ever to do so, the time is till in the future. We do not see the dawn of the age in which such a result will be brought forth.

Another consideration of a general character may be introduced. As a rule it is the unexpected that happens in invention as well as discovery. There are many problems which have fascinated mankind ever since civilization began which we have made little or no advance in solving. The only satisfaction we can feel in our treatment of the great geometrical problems of antiquity is that we have shown their solution to be impossible. The mathematician of to-day admits that he can neither square the circle, duplicate the cube, nor trisect the angle. May not our mechanicians, in like manner, be ultimately forced to admit that aerial flight is one of that great class of problems with which man can never cope, and give up all attempts to grapple with it?

I do not claim that this is a necessary conclusion from any past experience. But I do think that success must await progress of a different kind from that of invention. It is an unfortunate fact that we do not always appreciate the distinction between progress in scientific discovery and ingenious application of discovery to the wants of civilization. The name of Marconi is familiar to every ear; the names of Maxwell and Herz, who made the discoveries which rendered wireless telegraphy possible, are rarely recalled. Modern progress is the result of two factors: discoveries of the laws of nature and of actions or possibilities in nature, and the application of such discoveries to practical purposes. The first is the work of the scientific investigator, the second that of the inventor.

In view of the scientific discoveries of the past ten years, which, after bringing about results that would have been chimerical if predicted, leading on to the extraction of a substance which seems to set the laws and limits of nature at defiance by radiating a flood of heat, even when cooled to the lowest point that science can reach—a substance, a few specks of which contain power enough to start a railway train, and which seems to embody perpetual motion itself, he would be a bold prophet who would set any limit to possible discoveries in the realm of nature. We are binding the universe together by agencies which pass from sun to planet and from star to star. We are peering into the law of gravitation itself with the full hope of discovering something in its origin which may enable us to evade its action. We are determined to find out all we can about the mysterious ethereal medium which seems to fill all space, and which conveys light and heat from one heavenly body to another, but which yet evades all direct investigation. Quite likely the twentieth century is destined to see the natural forces which will enable us to fly from continent to continent with a speed far exceeding that of the bird.

But when we inquire whether aerial flight is possible in the present state of our knowledge; whether, with such materials as we possess, a combination of steel, cloth and wire can be made which, moved by the power of electricity or steam, shall form a successful flying machine, the outlook may be altogether different. To judge it sanely, let us bear in mind the difficulties which are encountered in any flying machine. The basic principle on which any such machine must be constructed is that of the aeroplane. This, by itself, would be the simplest of all flyers, and therefore, the best if it could be put into operation. The principle involved may be readily comprehended by the accompanying figure. (Figure 2.2) *AM* is the section of a flat plane surface, say, a thin sheet of metal or a cloth supported by wires. It moves through the air, the latter being represented by the horizontal rows of dots. The direction of the motion is that of the horizontal line, *AP*. The aeroplane has a slight inclination measured by the proportion between the perpendicular *MP* and the length, *AP*.

FIGURE 2.2: An inclined flat plane moving horizontally through the air.

We may raise the edge *M* up or lower it at pleasure. Now the interesting point, and that on which the hopes of inventors are based, is that if we give the plane any given inclination, even one so small that the perpendicular *MP* is only two or three per cent of the length *AM*, we can also calculate a certain speed of motion through the air which, if given to the plane, will enable it to bear any required weight. A plane ten feet square, for example, would not need any great inclination, nor would it require a speed higher than a few hundred feet a second to bear a man. What is of yet more importance, the higher the speed the less the inclination required, and, if we leave out of consideration the friction of the air and the resistance arising from any object which the machine may carry, the less the horse-power expended in driving the plane.

Maxim exemplified this by experiment several years ago. He found that, with a small inclination, he could readily give his aeroplane, when it slid forward upon ways, such a speed that it would rise from the ways of itself. The whole problem of the successful flying machine is, therefore, that of arranging an aeroplane that shall move through the air with the requisite speed.

The practical difficulties in the way of realizing the movement of such an object are obvious. The aeroplane must have its propellers. These must be driven by an engine with a source of power. Weight is an essential quality of every engine. The propellers must be made of metal, which has its weakness, and which is liable to give way when its speed attains a certain limit. And, granting complete success, imagine the proud possessor of the aeroplane darting through the air at a speed of several hundred feet per second! It is the speed alone that sustains him. How is he ever going to stop? Once he slackens his speed, down he begins to fall. He may, indeed, increase the inclination of his aeroplane. Then he increases the resistance necessary to move it. Once he stops he falls a dead mass. How shall he reach the ground without destroying his delicate machinery? I do not think the most imaginative inventor has yet even put upon paper a demonstrative, successful way of meeting this difficulty. The only ray of hope is afforded by the bird. The latter does succeed in stopping and reaching the ground safely after its flight. But we have already mentioned the great advantages which the bird possesses in the power of applying force to its wings, which, in its case, form the aeroplanes. But we have already seen that there is no mechanical combination, and no way of applying force, which will give to the aeroplanes the flexibility and rapidity of movement belonging to the wings of a bird. That this difficulty is insurmountable would seem to be a very fair deduction, not only from the failure of all attempts to surmount, but from the fact that Maxim has never, so far as we are aware, followed up his seemingly successful experiment.

It may be surmounted in a way which may, at first sight, seem plausible. In order that the aeroplane may have its full sustaining power, there is not need that this motion be continuously forward. A nearly horizontal surface, swinging around in a circle, on a vertical axis, like the wings of a windmill moving horizontally, will fulfill all the conditions. In fact, we have a machine on this simple principle in the familiar toy which, set rapidly whirling, rises in the air. Why more attempts have not been made to apply this system I do not know. Were there any hopeful possibility of making any flying machine whatever, it would seem that we should look in this direction.

*Simon Newcomb*

The difficulties which I have pointed out are only preliminary ones, patent on the surface. A more fundamental one still, which the writer feels may approve insurmountable, is based on a law of nature which we are bound to accept. It is that when we increase the size of such a machine without changing its model we increase the weight in proportion to the cube of the linear dimensions, while the effectiveness of the supporting power of the air increases only as the square of those dimensions. For example, suppose that an inventor succeeds, as well he may, in making a machine which would go into a match case, yet complete in all its parts, able to fly around the room. It may carry a button, but nothing heavier. Elated by his success, he makes one on the same model twice as large in every dimension. The parts of the first, which are one inch in length, he increases to two inches. Every part is twice as long, twice as broad, and twice as thick. The result is that his machine is eight times as heavy as before. But the sustaining surface is only four times as great. As compared with the smaller machine, its ratio of effectiveness is reduced to one-half. It may carry two or three buttons, but will not carry over four, because the total weight, machine plus buttons, can only be quadrupled, and if he more than quadruples the weight of the machine, he must less than quadruple that of the load. How many such enlargements must he make before his machine will cease to sustain itself, before it will fall as an inert mass when we seek to make it fly through the air? Is there any size at which it will be able to support a human being? We may well hesitate before we answer this question in the affirmative.

Dr. Graham Bell, with a cheery optimism very pleasant to contemplate, has pointed out that the law I have just cited may be evaded by not making a larger machine on the same model, but changing the latter in a way tantamount to increasing the number of small machines. This is quite true, and I wish it understood that, in laying down the law I have cited, I limit it to two machines of different sizes on the same model throughout. Quite likely the most effective flying machine would be one carried by a vast number of little birds. The veracious chronicler who escaped from a cloud of mosquitoes by crawling into an immense metal pot and then amused himself by clinching the antennae of the insects which bored through the pot until to his horror they became so numerous as to fly off with the covering, was more scientific than he supposed. Yes, a sufficient number of humming birds, if we could combine their forces, would carry an aerial excursion party of human beings through the air. If the watchmaker can make a machine which will fly through the room with a button, then, by combining ten thousand such machines he may be able to carry a man. But how shall the combined forces be supplied?

It is of interest to notice that the law is reversed in the case of a body which is not supported by the resistance of a fluid in which it is immersed, but floats in it, the ship or balloon, for example. When we double the linear dimensions of a steamship in all its parts, we increase not only her weight, but her floating power, her carrying capacity and her engine capacity eight-fold. But the resistance which she meets with when passing through the water at a given speed is only multiplied four times. Hence, the larger we build the steamship the more economical the application of the power necessary to drive it. The proportionately diminishing resistance which, in the flying machine, represents the floating power is, in the ship, something to be overcome. Thus there is a complete reversal of the law in its practical application to the two cases.

The balloon is in the same class with the ship. Practical difficulties aside, the larger it is built the more effective it will be, and the more advantageous will be the ratio of the power which is necessary to drive it and the resistance to be overcome.

If, therefore, we are ever to have aerial navigation with our present knowledge of natural capabilities, it is to the airship floating in the air, rather than the flying machine resting on the air, to which we are to look. In the light of the law which I have laid down, the subject, while not at all promising, seems worthy of more attention than it has received. It is not at all unlikely that if a skilful and experienced naval constructor, aided by an able corps of assistants, should design an airship of a diameter of not less than two hundred feet, and a length at least four or five times as great, constructed, possibly, of a textile substance impervious to gas and borne by a light framework, but, more likely, of exceedingly thin plates of steel carried by a frame fitted to secure the greatest combination of strength and lightness, he might find the result to be, ideally at least, a ship which would be driven through the air by a steam engine with a velocity far exceeding that of the fleetest Atlantic liner. Then would come the practical problem of realizing the ship by overcoming the mechanical difficulties involved in the construction of such a huge and light framework. I would not be at all surprised if the result of the exact calculation necessary to determine the question should lead to an affirmative conclusion, but I am quite unable to judge whether steel could be rolled into parts of the size and form which would be required in the mechanism.

I may, in conclusion, caution the reader on one point. I should be very sorry if this suggestion leads to the subject being taken-up by any other than skilful engineers or constructors, able to grapple with all problems relating to the strength and resistance of materials. As a single example of what is to be avoided I may mention the project which sometimes has been mooted, of making a balloon by pumping the air from a very thin, hollow receptacle. Such a project is as futile as can well be imagined; no known substance would begin to resist the necessary pressure. Our aerial ship must be filled with some substance lighter than air. Whether heated air could be made to answer the purpose, or whether we should have to use a gas, is the question for a designer.

From *The Independent* 55 (1903): Part 2; Oct. 22, 1903.

## Questions and Topics for Discussion and Writing

1. Why is Professor Newcomb so convinced that powered flight is impossible in his day? Sketch the main points of his argument.
2. Given that we know Professor Newcomb was wrong, what is wrong with his reasoning? What was he not aware of, or what did he incorrectly assess?
3. Put yourself in the position of a Professor Newcomb living today. Think of a device or invention which would be very useful if it existed, but which seems impossible or at least highly improbable. Make the strongest case you can as to why it will not and can not be invented.

# * Gunfire at Sea: A Case Study of Innovation

## Elting E. Morison

On the one hand, this is an article about the history of the technique of continuous-aim cannon firing at sea—which may seem a bit arcane to be included in this book. On the other hand, this is a very penetrating and enlightening (as well as entertaining) analysis by Elting Morison of the sort of person who is likely to propose innovations and the way human societies adapt to change.

In the early days of the last war when armaments of all kinds were in short supply, the British, I am told, made use of a venerable field piece that had come down to them from previous generations. The honorable past of this light artillery stretched back, in fact, to the Boer War. In the days of uncertainty after the fall of France, these guns, hitched to trucks, served as useful mobile units in the coast defense. But it was felt that the rapidity of fire could be increased. A time-motion expert was, therefore, called in to suggest ways to simplify the firing procedures. He watched one of the gun crews of five men at practice in the field for some time. Puzzled by certain aspects of the procedures, he took some slow-motion pictures of the soldiers performing the loading, aiming, and firing routines.

When he ran these pictures over once or twice, he noticed something that appeared odd to him. A moment before the firing, two members of the gun crew ceased all activity and came to attention for a three-second interval extending throughout the discharge of the gun. He summoned an old colonel of artillery, showed him the pictures, and pointed out this strange behavior. What, he asked the colonel, did it mean. The colonel, too, was puzzled. He asked to see the pictures again. "Ah," he said when the performance was over, "I have it. They are holding the horses."

This story, true or not, and I am told it is true, suggests nicely the pain with which the human being accommodates himself to changing conditions. The tendency is apparently involuntary and immediate to protect oneself against the shock of change by continuing in the presence of altered situations the familiar habits, however incongruous, of the past.

Yet, if human beings are attached to the known, to the realm of things as they are, they also, regrettably for their peace of mind, are incessantly attracted to the unknown and things as they might be. As Ecclesiastes glumly pointed out, men persist in disordering their settled ways and beliefs by seeking out many inventions.

The point is obvious. Change has always been a constant in human affairs; today, indeed, it is one of the determining characteristics of our civilization. In our relatively shapeless social organization, the shifts from station to station are fast and easy. More important for our immediate purpose, America is fundamentally an industrial society in a time of tremendous technological development. We are thus constantly presented with new devices or new forms of power that in their refinement and extension continually bombard the fixed structure of our habits of mind and behavior. Under such conditions, our salvation, or at least our peace of mind, appears to depend upon how successfully we can in the future become what has been called in an excellent phrase a completely "adaptive society."

It is interesting, in view of all this, that so little investigation, relatively, has been made of the process of change and human responses to it. Recently, psychologists, sociologists, cultural anthropologists, and economists have addressed themselves to the subject with suggestive results. But we are still far from a full understanding of the process and still further from knowing how we can set about simplifying and assisting an individual's or a group's accommodation to new machines or new ideas.

With these things in mind, I thought it might be interesting and perhaps useful to examine historically a changing situation within a society; to see if from this examination we can discover how the new machines or ideas that introduced the changing situation developed; to see who introduces them, who resists them, what points of friction or tension in the social structure are produced by the innovation, and perhaps why they are produced and what, if anything, may be done about it. For this case study the introduction of continuous-aim firing in the United States Navy has been selected. The system, first devised by an English officer in 1898, was introduced in our Navy in the years 1900 to 1902.

I have chosen to study this episode for two reasons. First, a navy is not unlike a society that has been placed under laboratory conditions. Its dimensions are severely limited; it is beautifully ordered and articulated; it is relatively isolated from random influences. For these reasons the impact of change can be clearly discerned, the resulting dislocations in the structure easily discovered and marked out. In the second place, the development of continuous-aim firing rests upon mechanical devices. It therefore presents for study a concrete, durable situation. It is not like many other innovating reagents—a Manichean heresy, or Marxism, or the views of Sigmund Freud—that can be shoved and hauled out of shape by contending forces or conflicting prejudices. At all times we know exactly what continuous-aim firing really is. It will be well now to describe, as briefly as possible, what it really is. This will involve a short investigation of certain technical matters. I will not apologize, as I have been told I ought to do, for this preoccupation with how a naval gun is fired. For one thing, all that follows is understandable only if one understands how the gun goes off. For another thing, a knowledge of the underlying physical considerations may give a kind of elegance to the succeeding investigation of social implications. And now to the gun and the gunfire.

The governing fact in gunfire at sea is that the gun is mounted on an unstable platform, a rolling ship. This constant motion obviously complicates the problem of holding a steady aim. Before 1898 this problem was solved in the following elementary fashion. A gun pointer estimated the range of the target, ordinarily in the nineties about 1600 yards. He then raised the gun barrel to give the gun the elevation to carry the shell to the target at the estimated range. This elevating process was accomplished by turning a small wheel on the gun mount that operated the elevating gears. With the gun thus fixed for range, the gun pointer peered through open sights, not unlike those on a small rifle, and waited until the roll of the ship brought the sights on the target. He then pressed the firing button that discharged the gun. There were by 1898, on some naval guns, telescope sights which naturally greatly enlarged the image of the target for the gun pointer. But these sights were rarely used by gun pointers. They were lashed securely to the gun barrel, and, recoiling with the barrel, jammed back against the unwary pointer's eye. Therefore, when used at all, they were used only to take an initial sight for purposes of estimating the range before the gun was fired.

Notice now two things about the process. First of all, the rapidity of fire was controlled by the rolling period of the ship. Pointers had to wait for the one moment in the roll when the sights were brought on the target. Notice also this: There is in every pointer what is called a "firing interval"—that is, the time lag between his impulse to fire the gun and the translation of this impulse into the act of pressing the firing button. A pointer, because of this reaction time, could not wait to fire the gun until the exact moment when the roll of the ship brought the sights onto the target; he had to will to fire a little before, while the sights were off the target. Since the firing interval was an individual matter, varying obviously from man to man, each pointer had to estimate from long practice his own interval and compensate for it accordingly.

These things, together with others we need not here investigate, conspired to make gunfire at sea relatively uncertain and ineffective. The pointer, on a moving platform, estimating range and firing interval, shooting while his sight was off the target, became in a sense an individual artist.

In 1898, many of the uncertainties were removed from the process and the position of the gun pointer radically altered by the introduction of continuous-aim firing. The major change was that which enabled the gun pointer to keep his sight and gun barrel on the target throughout the roll of the ship. This was accomplished by altering the gear ratio in the elevating gear to permit a pointer to compensate for the roll of the vessel by rapidly elevating and depressing the gun. From this change another followed. With the possibility of maintaining the gun always on the target, the desirability of improved sights became immediately apparent. The advantages of the telescope sight as opposed to the open sight were for the first time fully realized. But the existing telescope sight, it will be recalled, moved with the recoil of the gun and jammed back against the eye of the gunner. To correct this, the sight was mounted on a sleeve that permitted the gun barrel to recoil through it without moving the telescope.

These two improvements in elevating gear and sighting eliminated the major uncertainties in gunfire at sea and greatly increased the possibilities of both accurate and rapid fire.

You must take my word for it, since the time allowed is small, that this changed naval gunnery from an art to a science, and that gunnery accuracy in the British and our Navy increased, as one student said, 3000% in six years. This does not mean much except to suggest a great increase in accuracy. The following comparative figures may mean a little more. In 1899 five ships of the North Atlantic Squadron fired five minutes each at a lightship hulk at the conventional range of 1600 yards. After twenty-five minutes of banging away, two hits had been made on the sails of the elderly vessel. Six years later one naval gunner made fifteen hits in one minute at a target 75 by 25 feet at the same range—1600 yards; half of them hit in a bull's-eye 50 inches square.

Now with the instruments (the gun, elevating gear, and telescope), the method, and the results of continuous-aim firing in mind, let us turn to the subject of major interest: how was the idea, obviously so simple an idea, of continuous-aim firing developed, who introduced it into the United States Navy, and what was its reception?

The idea was the product of the fertile mind of the English officer Admiral Sir Percy Scott. He arrived at it in this way while, in 1898, he was the captain of H.M.S. *Scylla*. For the previous two or three years he had given much thought independently and almost alone in the British Navy to means of improving gunnery. One rough day, when the ship, at target practice, was pitching and rolling violently, he walked up and down the gun deck watching his gun crews. Because of the heavy weather, they were making very bad scores. Scott noticed, however, that one pointer was appreciably more accurate than the rest. He watched this man with care, and saw, after a time, that he was unconsciously working his elevating gear back and forth in a partially successful effort to compensate for the roll of the vessel. It flashed through Scott's mind at that moment that here was the sovereign remedy for the problem of inaccurate fire. What one man could do partially and unconsciously perhaps all men could be trained to do consciously and completely.

Acting on this assumption, he did three things. First, in all the guns of the *Scylla*, he changed the gear ratio in the elevating gear, previously used only to set the gun in fixed position for range, so that a gunner could easily elevate and depress the gun to follow a target throughout the roll. Second, he rerigged his telescopes so that they would not be influenced by the recoil of the gun. Third, he rigged a small target at the mouth of the gun, which was moved up and down by a crank to simulate a moving target. By following this target as it moved and firing at it with a subcaliber rifle rigged in the breech of the gun, the pointer could practice every day. Thus equipped, the ship became a training ground for gunners. Where before the good pointer was an individual artist, pointers now became trained technicians, fairly uniform in their capacity to shoot. The effect was immediately felt. Within a year the *Scylla* established records that were remarkable.

At this point I should like to stop a minute to notice several things directly related to, and involved in, the process of innovation. To begin with, the personality of the innovator. I wish there were time to say a good deal about Admiral Sir Percy Scott. He was a wonderful man. Three small bits of evidence must here suffice, however. First, he had a certain mechanical ingenuity. Second, his personal life was shot through with frustration and bitterness. There was a divorce and a quarrel with that ambitious officer Lord Charles Beresford, the sounds of which, Scott liked to recall, penetrated to the last outposts of empire. Finally, he possessed, like Swift, a savage indignation directed ordinarily at the inelastic intelligence of all constituted authority, especially the British Admiralty.

There are other points worth mention here. Notice first that Scott was not responsible for the invention of the basic instruments that made the reform in gunnery possible. This reform rested upon the gun itself, which as a rifle had been in existence on ships for at least forty years; the elevating gear, which had been, in the form Scott found it, a part of the rifled gun from the beginning; and the telescope sight, which had been on shipboard at least eight years. Scott's contribution was to bring these three elements appropriately modified into a combination that made continuous-aim firing possible for the first time. Notice also that he was allowed to bring these elements into combination by accident, by watching the unconscious action of a gun pointer endeavoring through the operation of this elevating gear to correct partially for the roll of his vessel. Scott, as we have seen, had been interested in gunnery; he had thought about ways to increase accuracy by practice and improvement of existing machinery; but able as he was, he had not been able to produce on his own initiative and by his own thinking the essential idea and modify instruments to fit his purpose. Notice here, finally, the intricate interaction of chance, the intellectual climate, and Scott's mind. Fortune (in this case, the unaware gun pointer) indeed favors the prepared mind, but even fortune and the prepared mind need a favorable environment before they can conspire to produce sudden change. No intelligence can proceed very far above the threshold of existing data or the binding combinations of existing data.

All these elements that enter into what may be called "original thinking" interest me as a teacher. Deeply rooted in the pedagogical mind often enough is a sterile infatuation with "inert ideas"; there is thus always present in the profession the tendency to be diverted from the *process* by which these ideas, or indeed any ideas, are really produced. I well remember with what contempt a class of mine which was reading Leonardo da Vinci's *Notebooks* dismissed the author because he appeared to know no more mechanics than, as one wit in the class observed, a Vermont Republican farmer of the present day. This is perhaps the expected result produced by a method of instruction that too frequently implies that the great generalizations were the result, on the one hand, of chance—an apple falling in an orchard or a teapot boiling on the hearth—or, on the other hand, of some towering intelligence proceeding in isolation inexorably toward some prefigured idea, like evolution, for example.

This process by which new concepts appear, the interaction of fortune, intellectual climate, and the prepared imaginative mind, is an interesting subject for examination offered by any case study of innovation. It was a subject as Dr. Walter Cannon pointed out, that momentarily engaged the attention of Horace Walpole, whose lissome intelligence glided over the surface of so many ideas. In reflecting upon the part played by chance in the development of new concepts, he recalled the story of the three princes of Serendipity who set out to find some interesting object on a journey through their realm. They did not find the particular object of their search, but along the way they discovered many new things simply because they were looking for *something*. Walpole believed this intellectual method ought to be given a name, in honor of the founders, serendipity; and serendipity certainly exerts a considerable influence in what we call original thinking. There is an element of serendipity, for example, in Scott's chance discovery of continuous-aim firing in that he was, and had been, looking for some means to improve his target practice and stumbled upon a solution by observation that had never entered his head.

Serendipity, while recognizing the prepared mind, does tend to emphasize the role of chance in intellectual discovery. Its effect may be balanced by an anecdote that suggests the contribution of the adequately prepared mind. There has recently been much posthaste and romage in the land over the question of whether there really was a Renaissance. A scholar has recently argued in print that since the Middle Ages actually possessed many of the instruments and pieces of equipment associated with the Renaissance, the Renaissance could be said to exist as a defined period only in the mind of the historians such as Burckhardt. This view was entertainingly rebutted by the historian of art Panofsky, who pointed out that although Robert Grosseteste indeed did have a very rudimentary telescope, he used it to examine stalks of grain in a field down the street, Galileo, a Renaissance intelligence, pointed his telescope at the sky.

Here Panofsky is only saying in a provocative way that change and intellectual advance are the products of well-trained and well-stored inquisitive minds, minds that relieve us of "the terrible burden of inert ideas by throwing them into a new combination." Educators, nimble in the task of pouring the old wine of our heritage into the empty vessels that appear before them, might give thought to how to develop such independent, inquisitive minds.

But I have been off on a private venture of my own. Now to return to the story, the introduction of continuous-aim firing. In 1900 Percy Scott went out to the China Station as commanding officer of H.M.S. *Terrible*. In that ship he continued his training methods and his spectacular successes in naval gunnery. On the China Station he met up with an American junior officer, William S. Sims. Sims had little of the mechanical ingenuity of Percy Scott, but the two were drawn together by temperamental similarities that are worth noticing here. Sims had the same intolerance for what is called spit and polish and the same contempt for bureaucratic inertia as his British brother officer. He had for some years been concerned, as had Scott, with what he took to be the inefficiency of his own Navy. Just before he met Scott, for example, he had shipped out to China in the brand new pride of the fleet, the battleship *Kentucky*. After careful investigation and reflection he had informed his superiors in Washington that she was "not a battleship at all—but a crime against the white race." The spirit with which he pushed forward his efforts to reform the naval service can best be stated in his own words to a brother officer: "I am perfectly willing that those holding views differing from mine should continue to live, but with every fibre of my being I loathe indirection and shiftiness, and where it occurs in high place, and is used to save face at the expense of the vital interests of our great service (in which silly people place such a child-like trust), I want that man's blood and I will have it no matter what it costs me personally."

From Scott in 1900 Sims learned all there was to know about continuous-aim firing. He modified, with the Englishman's active assistance, the gear on his own ship and tried out the new system. After a few months' training, his experimental batteries began making remarkable records at target practice. Sure of the usefulness of his gunnery methods, Sims then turned to the task of educating the Navy at large. In thirteen great official reports he documented the case for continuous-aim firing, supporting his arguments at every turn with a mass of factual data. Over a period of two years, he reiterated three principal points: first, he continually cited the records established by Scott's ships, the *Scylla* and the *Terrible*, and supported these with the accumulating data from his own tests on an American ship; second, he described the mechanisms used and the training procedures instituted by Scott and himself to obtain these records; third, he explained that our own mechanisms were not generally adequate without modification to meet the demands placed on them by continuous-aim firing. Our elevating gear, useful to raise or lower a gun slowly to fix it in position for the proper range, did not always work easily and rapidly enough to enable a gunner to follow a target with his gun throughout the roll of the ship. Sims also explained that such few telescope sights as there were on board our ships were useless. Their cross wires were so thick or coarse they obscured the target, and the sights had been attached to the gun in such a way that the recoil system of the gun plunged the eyepiece against the eye of the gun pointer.

This was the substance not only of the first but of all the succeeding reports written on the subject of gunnery from the China Station. It will be interesting to see what response these met with in Washington. The response falls roughly into three easily identifiable stages.

First stage: At first, there was no response. Sims had directed his comments to the Bureau of Ordnance and the Bureau of Navigation; in both bureaus there was dead silence. The thing—claims and records of continuous-aim firing—was not credible. The reports were simply filed away and forgotten. Some indeed, it was later discovered to Sims's delight, were half-eaten-away by cockroaches.

Second stage: It is never pleasant for any man's best work to be left unnoticed by superiors, and it was an unpleasantness that Sims suffered extremely ill. In his later reports, beside the accumulating data he used to clinch his argument, he changed his tone.

He used deliberately shocking language because, as he said, "They were furious at my first papers and stowed them away. I therefore made up my mind I would give these later papers such a form that they would be dangerous documents to leave neglected in the files." To another friend he added, "I want scalps or nothing and if I can't have 'em I won't play."

Besides altering his tone, he took another step to be sure his views would receive attention. He sent copies of his reports to other officers in the fleet. Aware as a result that Sims's gunnery claims were being circulated and talked about, the men in Washington were then stirred to action. They responded, notably through the Chief of the Bureau of Ordnance, who had general charge of the equipment used in gunnery practice, as follows: (1) our equipment was in general as good as the British; (2) since our equipment was as good, the trouble must be with the men, but the gun pointer and the training of gun pointers were the responsibility of the officers on the ships; and most significant (3) continuous-aim firing was impossible. Experiments had revealed that five men at work on the elevating gear of a six-inch gun could not produce the power necessary to compensate for a roll of five degrees in ten seconds. These experiments and calculations demonstrated beyond peradventure or doubt that Scott's system of gunfire was not possible.

This was the second stage—the attempt to meet Sims's claims by logical, rational rebuttal. Only one difficulty is discoverable in these arguments; they were wrong at important points. To begin with, while there was little difference between the standard British equipment and the standard American equipment, the instruments on Scott's two ships, the *Scylla* and the *Terrible*, were far better than the standard equipment on our ships. Second, all the men could not be trained in continuous-aim firing until equipment was improved throughout the fleet. Third, the experiments with the elevating gear had been ingeniously contrived at the Washington Navy Yard—on solid ground. It had, therefore, been possible to dispense in the Bureau of Ordnance calculation with Newton's first law of motion, which naturally operated at sea to assist the gunner in elevating or depressing a gun mounted on a moving ship. Another difficulty was of course that continuous-aim firing was in use on Scott's and some of our own ships at the time the chief of the Bureau of Ordnance was writing that it was a mathematical impossibility. In every way I find this second stage, the apparent resort to reason, the most entertaining and instructive in our investigation of the responses to innovation.

Third stage: The rational period in the counterpoint between Sims and the Washington men was soon passed. It was followed by the third stage, that of name-calling—the *argumentum ad hominem.* Sims, of course, by the high temperature he was running and by his calculated overstatement, invited this. He was told in official endorsements on his reports that there were others quite as sincere and loyal as he and far less difficult; he was dismissed as a crackbrained egotist; he was called a deliberate falsifier of evidence.

The rising opposition and the character of the opposition were not calculated to discourage further efforts by Sims. It convinced him that he was being attacked by shifty, dishonest men who were the victims, as he said, of insufferable conceit and ignorance. He made up his mind, therefore, that he was prepared to go to any extent to obtain the "scalps" and the "blood" he was after. Accordingly, he, a lieutenant, took the extraordinary step of writing the President of the United States, Theodore Roosevelt, to inform him of the remarkable records of Scott's ships, of the inadequacy of our own gunnery routines and records, and of the refusal of the Navy Department to act. Roosevelt, who always liked to respond to such appeals when he conveniently could, brought Sims back from China late in 1902 and installed him as Inspector of Target Practice, a post the naval officer held through the remaining six years of the Administration. And when he left, after many spirited encounters we cannot here investigate, he was universally acclaimed as "the man who taught us how to shoot."

With this sequence of events (the chronological account of the innovation of continuous-aim firing) in mind, it is possible now to examine the evidence to see what light it may throw on our present interest: the origins of and responses to change in a society.

First, the origins. We have already analyzed briefly the origins of the idea. We have seen how Scott arrived at his notion. We must now ask ourselves, I think, why Sims so actively sought, almost alone among his brother officers, to introduce the idea into his service. It is particularly interesting here to notice again that neither Scott nor Sims invented the instruments on which the innovation rested. They did not urge their proposal, as might be expected, because of pride in the instruments of their own design. The telescope sight had first been placed on shipboard in 1892 by Bradley Fiske, an officer of great inventive capacity. In that year Fiske had even sketched out on paper the vague possibility of continuous-aim firing, but his sight was condemned by his commanding officer, Robely D. Evans, as of no use. In 1892 no one but Fiske in the Navy knew what to do with a telescope sight any more than Grosseteste had known in his time what to do with a telescope. And Fiske, instead of fighting for his telescope, turned his attention to a range finder. But six years later Sims, following the tracks of his brother officer, took over and became the engineer of the revolution. I would suggest, with some reservations, this explanation: Fiske, as an inventor, took his pleasure in great part from the design of the device. He lacked not so much the energy as the overriding sense of social necessity that would have enabled him to *force* revolutionary ideas on the service. Sims possessed this sense. In Fiske, who showed rare courage and integrity in other professional matters not intimately connected with the introduction of new weapons of his own design, we may here find the familiar plight of the engineer who often enough must watch the products of his ingenuity organized and promoted by other men. These other promotional men when they appear in the world of commerce are called entrepreneurs. In the world of ideas they are still entrepreneurs. Sims was one, a middle-aged man caught in the periphery (as a lieutenant) of the intricate webbing of a precisely organized society. Rank, the exact definition and limitation of a man's capacity at any given moment in his career, prevented Sims from discharging all his exploding energies into the purely routine channels of the peacetime Navy. At the height of his powers he was a junior officer standing watches on a ship cruising aimlessly in friendly foreign waters. The remarkable changes in systems of gunfire to which Scott introduced him gave him the opportunity to expend his energies quite legitimately against the encrusted hierarchy of his society. He was moved, it seems to me, in part by his genuine desire to improve his own profession but also in part by rebellion against tedium, against inefficiency from on high, and against the artificial limitations placed on his actions by the social structure, in his case, junior rank.

Now having briefly investigated the origins of the change, let us examine the reasons for what must be considered the weird response we have observed to this proposed change. Why this deeply rooted, aggressive, persistent hostility from Washington that was only broken up by the interference of Theodore Roosevelt? Here was a reform that greatly and demonstrably increased the fighting effectiveness of a service that maintains itself almost exclusively to fight. Why then this refusal to accept so carefully documented a case, a case proved incontestably by records and experience? Why should virtually all the rulers of a society so resolutely seek to reject a change that so markedly improved its chances for survival in any contest with competing societies? There are the obvious reasons that will occur to all of you—the source of the proposed reform was an obscure, junior officer 8000 miles away; he was, and this is a significant factor, criticizing gear and machinery designed by the very men in the bureaus to whom he was sending his criticisms. And furthermore, Sims was seeking to introduce what he claimed were improvements in a field where improvements appeared unnecessary. Superiority in war, as in other things, is a relative matter, and the Spanish-American War had been won by the old system of gunnery. Therefore, it was superior even though of the 9500 shots fired at various but close ranges, only 121 had found their mark.

These are the more obvious, and I think secondary or supporting, sources of opposition to Sims's proposed reforms. A less obvious cause appears by far the most important one. It has to do with the fact that the Navy is not only an armed force; it is a society. Men spend their whole lives in it and tend to find the definition of their whole being within it. In the forty years following the Civil War, this society had been forced to accommodate itself to a series of technological changes—the steam turbine, the electric motor, the rifled shell of great explosive power, case-hardened steel armor, and all the rest of it. These changes wrought extraordinary changes in ship design, and, therefore, in the concepts of how ships were to be used; that is, in fleet tactics, and even in naval strategy. The Navy of this period is a paradise for the historian or sociologist in search of evidence bearing on a society's responses to change.

To these numerous innovations, producing as they did a spreading disorder throughout a service with heavy commitments to formal organization, the Navy responded with grudging pain. For example, sails were continued on our first-line ships long after they ceased to serve a useful purpose mechanically, but like the holding of the horses that no longer hauled the British field pieces, they assisted officers over the imposing hurdles of change. To a man raised in sail, a sail on an armored cruiser propelled through the water at 14 knots by a steam turbine was a cheering sight to see.

This reluctance to change with changing conditions was not limited to the blunter minds and less resilient imaginations in the service. As clear and untrammeled an intelligence as Alfred Thayer Mahan, a prophetic spirit in the realm of strategy unfettered by personal attachments of any kind, was occasionally at the mercy of the past. In 1906 he opposed the construction of battleships with single-caliber main batteries—that is, the modern battleship—because, he argued, such vessels would fight only at great ranges. These ranges would create in the sailor what Mahan felicitously called "the indisposition to close." They would thus undermine the physical and moral courage of a commander. They would, in other words, destroy the doctrine and the spirit, formulated by Nelson a century before, that no captain could go very far wrong who laid his ship alongside an enemy. The fourteen-inch rifle, which could place a shell upon a possible target six miles away, had long ago annihilated the Nelsonian doctrine. Mahan, of course, knew and recognized this fact; he was, as a man raised in sail, reluctant only to accept its full meaning, which was not that men were no longer brave, but that 100 years after the battle of the Nile they had to reveal their bravery in a different way.

Now the question still is, why this blind reaction to technological change, observed in the continuation of sail or in Mahan's contentions or in the opposition to continuous-aim firing? It is wrong to assume, as it is frequently assumed by civilians, that it springs exclusively from some causeless Bourbon distemper that invades the military mind. There is a sounder and more attractive base. The opposition, where it occurs, of the soldier and the sailor to such change springs from the normal human instinct to protect oneself, and more especially, one's way of life. Military organizations are societies built around and upon the prevailing weapons systems. Intuitively and quite correctly the military man feels that a change in weapon portends a change in the arrangements of his society. Think of it this way. Since the time that the memory of man runneth not to the contrary, the naval society has been built upon the surface vessel. Daily routines, habits of mind, social organization, physical accommodations, conventions, rituals, spiritual allegiances have been conditioned by the essential fact of the ship. What then happens to your society if the ship is displaced as the principal element by such a radically different weapon as the plane? The mores and structure of the society are immediately placed in jeopardy. They may, in fact, be wholly destroyed. It was the silly cliché of the twenties that those naval officers who persisted in defending the battleship against the apparently superior claims of the carrier did so because the battleship was a more comfortable home. What, from one point of view, is a better argument? There is, as everyone knows, no place like home. Who has ever wanted to see the old place brought under the hammer by hostile forces whether they hold a mortgage or inhabit a flying machine?

This sentiment would appear to account in large part for the opposition to Sims; it was the product of an instinctive protective feeling, even if the reasons for this feeling were not overt or recognized. The years after 1902 proved how right, in their terms, the opposition was. From changes in gunnery flowed an extraordinary complex of changes: in shipboard routines, ship design, and fleet tactics. There was, too, a social change. In the days when gunnery was taken lightly, the gunnery officer was taken lightly. After 1903, he became one of the most significant and powerful members of a ship's company, and this shift of emphasis naturally was shortly reflected in promotion lists. Each one of these changes provoked a dislocation in the naval society, and with man's troubled foresight and natural indisposition to break up classic forms, the men in Washington withstood the Sims onslaught as long as they could. It is very significant that they withstood it until an agent from outside, outside and above, who was not clearly identified with the naval society, entered to force change.

This agent, the President of the United States, might reasonably and legitimately claim the credit for restoring our gunnery efficiency. But this restoration by *force majeure* was brought about at great cost to the service and men involved. Bitternesses, suspicions, wounds were made that it was impossible to conceal and were, in fact, never healed.

Now this entire episode may be summed up in five separate points:

1. The essential idea for change occurred in part by chance but in an environment that contained all the essential elements for change and to a mind prepared to recognize the possibility of change.
2. The basic elements, the gun, gear, and sight, were put in the environment by other men, men interested in designing machinery to serve different purposes or simply interested in the instruments themselves.
3. These elements were brought into successful combination by minds not interested in the instruments for themselves but in what they could do with them. These minds were, to be sure, interested in good gunnery, overtly and consciously. They may also, not so consciously, have been interested in the implied revolt that is present in the support of all change. Their temperaments and careers indeed support this view. From gunnery, Sims went on to attack ship designs, existing fleet tactics, and methods of promotion. He lived and died, as the service said, a stormy petrel, a man always on the attack against higher authority, a rebellious spirit; a rebel, fighting in excellent causes, but a rebel still who seems increasingly to have identified himself with the act of revolt against constituted authority.
4. He and his colleagues were opposed on this occasion by men who were apparently moved by three considerations: honest disbelief in the dramatic but substantiated claims of the new process, protection of the existing devices and instruments with which they identified themselves, and maintenance of the existing society with which they were identified.
5. The deadlock between those who sought change and those who sought to retain things as they were was broken only by an appeal to superior force, a force removed from and unidentified with the mores, conventions, devices of the society. This seems to me a very important point. The naval society in 1900 broke down in its effort to accommodate itself to a new situation. The appeal to Roosevelt is documentation for Mahan's great generalization that no military service should or can undertake to reform itself. It must seek assistance from outside.

Now with these five summary points in mind, it may be possible to seek, as suggested at the outset, a few larger implications from this story. What, if anything, may it suggest about the general process by which any society attempts to meet changing conditions?

There is, to begin with, a disturbing inference half-concealed in Mahan's statement that no military organization can reform itself. Certainly civilians would agree with this. We all know now that war and the preparation for war are too important, as Clemenceau said, to be left to the generals. But as I have said before, military organizations are really societies, more rigidly structured, more highly integrated, than most communities, but still societies. What then if we make this phrase to read, "No society can reform itself"? Is the process of adaptation to change, for example, too important to be left to human beings? This is a discouraging thought, and historically there is some cause to be discouraged. Societies have not been very successful in reforming themselves, accommodating to change, without pain and conflict.

This is a subject to which we may well address ourselves. Our society especially is built, as I have said, just as surely upon a changing technology as the Navy of the nineties was built upon changing weapon systems. How then can we find the means to accept with less pain to ourselves and less damage to our social organization the dislocations in our society that are produced by innovation? I cannot, of course, give any satisfying answer to these difficult questions. But in thinking about the case study before us, an idea occurred to me that at least might warrant further investigation by men far more qualified than I.

A primary source of conflict and tension in our case study appears to lie in this great word I have used so often in the summary, the word "identification." It cannot have escaped notice that some men identified themselves with their creations—sights, gun, gear, and so forth—and thus obtained a presumed satisfaction from the thing itself, a satisfaction that prevented them from thinking too closely on either the use or the defects of the thing; that others identified themselves with a settled way of life they had inherited or accepted with minor modification and thus found their satisfaction in attempting to maintain that way of life unchanged; and that still others identified themselves as rebellious spirits, men of the insurgent cast of mind, and thus obtained a satisfaction from the act of revolt itself.

This purely personal identification with a concept, a convention, or an attitude would appear to be a powerful barrier in the way of easily acceptable change. Here is an interesting primitive example. In the years from 1864 to 1871 ten steel companies in this country began making steel by the new Bessemer process. All but one of them at the outset imported from Great Britain English workmen familiar with the process. One, the Cambria Company, did not. In the first few years those companies with British labor established an initial superiority. But by the end of the seventies, Cambria had obtained a commanding lead over all competitors. The President of Cambria, R.W. Hunt, in seeking a cause for his company's success, assigned it almost exclusively to the labor policy. "We started the converter plant without a single man who had ever seen even the outside of a Bessemer plant. We thus had willing pupils with no prejudices and no reminiscences of what they had done in the old country." The Bessemer process, like any new technique, had been constantly improved and refined in this period from 1864 to 1871. The British laborers of Cambria's competitors, secure in the performance of their own original techniques, resisted and resented all change. The Pennsylvania farm boys, untrammeled by the rituals and traditions of their craft, happily and rapidly adapted themselves to the constantly changing process. They ended by creating an unassailable competitive position for their company.

How then can we modify the dangerous effects of this word, "identification"? And how much can we tamper with this identifying process? Our security—much of it, after all—comes from giving our allegiance to something greater than ourselves. These are difficult questions to which only the most tentative and provisional answers may here be proposed for consideration.

If one looks closely at this little case history, one discovers that the men involved were the victims of *severely limited* identifications. They were presumably all part of a society dedicated to the process of national defense, yet they persisted in aligning themselves with separate parts of that process—with the existing instruments of defense, with the existing customs of the society, or with the act of rebellion against the customs of the society. Of them all the insurgents had the best of it. They could, and did, say that the process of defense was improved by a gun that shot straighter and faster, and since they wanted such guns, they were unique among their fellows, patriots who sought only the larger object of improved defense. But this beguiling statement, even when coupled with the recognition that these men were right and extremely valuable and deserving of respect and admiration—this statement cannot conceal the fact that they were interested too in scalps and blood, so interested that they made their case a militant one and thus created an atmosphere in which self-respecting men could not capitulate without appearing either weak or wrong or both. So these limited identifications brought men into conflict with each other, and the conflict prevented them from arriving at a common acceptance of a change that presumably, as men interested in our total national defense, they would all find desirable.

It appears, therefore, if I am correct in my assessment, that we might spend some time and thought on the possibility of enlarging the sphere of our identifications from the part to the whole. For example, those Pennsylvania farm boys at the Cambria Steel company were, apparently, much more interested in the manufacture of steel than in the preservation of any particular way of making steel. So I would suggest that in studying innovation, we look further into this possibility: the possibility that any group that exists for any purpose—the family, the factory, the educational institution—might begin by defining for itself its grand object and see to it that that grand object is communicated to every member of the group. Thus defined and communicated, it might serve as a unifying agent against the disruptive local allegiances of the inevitable smaller elements that compose any group. It may also serve as a means to increase the acceptability of any change that would assist in the more efficient achievement of the grand object.

There appears also a second possible way to combat the untoward influence of limited identifications. We are, I may repeat, a society based on technology in a time of prodigious technological advance, and a civilization committed irrevocably to the theory of evolution. These things mean that we believe in change; they suggest that if we are to survive in good health we must, in the phrase that I have used before, become an "adaptive society." But the word "adaptive" is meant the ability to extract the fullest possible returns from the opportunities at hand: the ability of Sir Percy Scott to select judiciously from the ideas and material presented both by the past and present and to throw them into a new combination. "Adaptive," as here used, also means the kind of resilience that will enable us to accept fully and easily the best promises of changing circumstances without losing our sense of continuity or our essential integrity.

We are not yet emotionally an adaptive society, though we try systematically to develop forces that tend to make us one. We encourage the search for new inventions; we keep the mind stimulated, bright, and free to seek out fresh means of transport, communication, and energy; yet we remain, in part, appalled by the consequences of our ingenuity, and, too frequently, try to find security through the shoring up of ancient and irrelevant conventions, the extension of purely physical safeguards, or the delivery of decisions we ourselves should make into the keeping of superior authority like the state. These solutions are not necessarily unnatural or wrong, but they historically have not been enough, and I suspect they never will be enough to give us the serenity and competence we seek.

If the preceding statements are correct, they suggest that we might give some attention to the construction of a new view of ourselves as a society which in time of great change identified with and obtained security and satisfaction from the wise and creative accommodation to change itself. Such a view rests, I think, upon a relatively greater reverence for the mere *process* of living in a society than we possess today, and a relatively smaller respect for and attachment to any special *product* of a society, a product either as finite as a bathroom fixture or as conceptual as a fixed and final definition of our Constitution or our democracy.

Historically such an identification with *process* as opposed to *product*, with adventurous selection and adaptation as opposed to simple retention and possessiveness, has been difficult to achieve collectively. The Roman of the early republic, the Italian of the late fifteenth and early sixteenth century, or the Englishman of Elizabeth's time appears to have been most successful in seizing the new opportunities while conserving as much of the heritage of the past as he found relevant and useful to this purpose.

We seem to have fallen on times similar to theirs, when many of the existing forms and schemes have lost meaning in the face of dramatically altering circumstances. Like them we may find a least part of our salvation in identifying ourselves with the adaptive process and thus share with them some of the joy, exuberance, satisfaction, and security with which they went out to meet their changing times.

I am painfully aware that in setting up my historical situation for examination I have, in a sense, artificially contrived it. I have been forced to cut away much, if not all, of the connecting tissue of historical evidence and to present you only with the bare bones and even with only a few of the bones. Thus, I am also aware, the episode has lost much of the subtlety, vitality, and attractive uncertainty of the real situation. There has, too, in the process, been inevitable distortion, but I hope the essential if exaggerated truth remains. I am also aware that I have erected elaborate hypotheses on the slender evidence provided by the single episode. My defense here is only that I have hoped to suggest possible approaches and methods of study and also possible fruitful areas of investigation in a subject that seems to me of critical importance in the life and welfare of our changing society.

Elting E. Morison

From Elting E. Morison, *Men, Machines, and Modern Times*, 17–44 (Cambridge: M.I.T. Press, 1966).

## Questions and Topics for Discussion and Writing

1. What is continuous-aim firing? Why was it not used before Scott introduced it on his ships?
2. Newton's first law of motion states that any body tends to continue in its state of rest or motion in the same direction and at the same speed unless acted upon by a force. This also known as the principle of inertia. In what way was this a factor in continuous-aim firing and why was it overlooked by the Bureau of Ordnance?
3. Discuss the role of William Sims's personality in the acceptance of continuous-aim firing by the U.S. Navy. Was it necessary that the initiative be taken by someone like him, or would continuous-aim firing have been adopted anyway?
4. Discuss Morison's prescription that we should become an "adaptive society."

# Engineering Under Pressure

## Samuel C. Florman

Any costly technological failure needs to be investigated so that we can learn from it. The disastrous explosion of the space shuttle Challenger was investigated and the cause, failure of an O-ring gasket, was identified. But beyond the technological failure was a human failure of judgment in the decision process that could have been prevented. The well-known engineer and writer Samuel C. Florman takes his profession to task in this commentary on the *Challenger* accident.

In the heat of the moment, *Challenger*'s managers forgot that they were engineers, too. It may have been a crucial error. The *Challenger* disaster was a classic example of engineering failure, but it was also a unique event. There was the drama of the decision-making process and the ensuing debate about it in a national forum. But the real revelation to me was that the men who made the fateful decision to launch somehow assumed that they could not at the same time be both engineers and executives.

The mission carried echoes of previous engineering tragedies. Once again a minute detail undid a grandly conceived technological project. Once again it was demonstrated that engineers must deal with real materials—rubber, putty, steel—rather than idealized ones that behave as scientific formulas say they should. Real materials must be tested, and there wasn't enough testing.

We discovered anew that myriad multiplicative coincidences haunt complex engineering systems. Just as *Challenger*'s primary O-ring gasket failed, a second gasket, intended as a backup, came unseated from its groove because the rocket casing moved under pressure. Small wonder that engineers speak of Murphy's Law. As President (and engineer) Herbert Hoover wrote in his memoirs, failure for the engineer "is the phantasmagoria that haunts his nights and dogs his days."

Of course, failures have always been part of technological advance. When the first jet airliner, the British *Comet*, was put into service in 1952, it had been test flown for almost three years. After two crashes—one in Calcutta in 1953 and a second in the Mediterranean the following year—all of the *Comets* were grounded. After undergoing extensive modification, the fleet went back into service, and a third *Comet* promptly disintegrated in midair. Eventually the fault was found to lie in fatigue cracks in the corners of windows, a possibility that at the time seemed unthinkable in such relatively new machines. (Rounding the corners solved the problem.) The space shuttle is more than just a new machine; it is practically a new technology, and viewed in the light of history, some sort of accident was almost inevitable.

But the loss of *Challenger* differed from the *Comet* crashes, the failure of the Tacoma Narrows bridge, the collapse of the walkway at the Hyatt Regency in Kansas City, and other disasters in that a number of engineers had predicted that it might happen and strenuously objected to the launch. Morton Thiokol engineers knew that cold affected the gaskets. The evening before the flight, as temperatures dropped into the low 20s [Fahrenheit], 14 Thiokol engineers unanimously concluded that the flight should be postponed. A conference call, set up with NASA officials to discuss the problem, quickly turned confrontational.

A five-minute recess stretched to half an hour. The 14 engineers remained unshakable. "We have to make a management decision," senior vice-president Jerald Mason declared. He asked Robert Lund, Thiokol's vice-president for engineering, to "take off his engineering hat and put on his management hat." Subsequently Mason and Lund, along with Joseph Kilminster and Calvin Wiggins, two other Thiokol vice-presidents, approved the launch.

Much of the discussion since the event has stressed that the findings of "engineers" were overruled by the judgment of "managers." When Mason was asked what he had in mind when he told Lund to replace his engineering hat with a management hat, he said that he meant "it was going to take judgment rather than a precise engineering calculation."

That view of engineering rings false to me. The implication that engineering consists of precise calculation, while "judgment" is left to another class of person, demeans a profession that has always stressed art, imagination, and wisdom at least as much as exactitude. There are many instances in which engineering facts can and should be isolated from business or political choices, but the launch of *Challenger* was not one of these.

The four so-called Thiokol "managers" are, in fact, engineers. Mason has a degree in aeronautical engineering, Lund and Kilminster have degrees in mechanical engineering, and Wiggins has a degree in chemistry. George Hardy and Lawrence Mulloy, the two NASA officials who urged Thiokol to approve the launch, are engineers. So are the key NASA people above them. When they moved into management, they did not thereupon cease being engineers any more than a doctor who becomes director of a hospital stops being a doctor.

Were these engineer-executives under pressure to meet a launch schedule? Of course. But pressure goes with the job. It has often been said that engineering is doing for one dollar what any darn fool can do for two—combining electronic circuits so that radios can be made cheaper and smaller, say. It might with equal truth be said that engineering is doing in one year what any darn fool can do in two. Economy and schedule are inherent to all engineering projects. Given unlimited funds and time, the shuttle's design could surely be improved, but NASA and Thiokol engineers had no reason to expect either a blank check or freedom from a sense of urgency. Pressure can help explain—but can never excuse—an engineering mistake.

It is also unreasonable to cry "safety first" after the fact while ignoring the imperatives that drive the space program—national defense, scientific exploration, technological progress, even such intangibles as community morale, all to be enhanced—quickly—for an affordable price. If safety were to be the overwhelmingly dominant consideration, we would never in the first place blast human beings into space on top of rockets.

All we can fairly say is that safety was and is one very important aspect of the space program. NASA long ago established numerous safety measures; the required sign-off before launch by each major shuttle subcontractor is one of the most important. Shortly before the accident, the annual report of the Aerospace Safety Advisory Panel expressed general satisfaction with NASA's attention to safety. The system is sound. It should have worked. Fourteen engineers at Thiokol—not just one or two who may have been unduly apprehensive—recommended that the flight be postponed. It strains belief that four top executive-engineers at Thiokol, pressured by administrator-engineers at NASA, ignored the warnings and agreed to proceed. It was bad management. It was also bad engineering.

Should the junior engineers at Thiokol have blown the whistle that night—gone outside the chain of command or otherwise publicly aired their concerns? I cannot fault them for only arguing vigorously. They had no clear evidence, only clues and uneasy feelings. And they were pleading their case not to politicians or nontechnical administrators, but rather to sophisticated engineers like themselves.

There is bitter irony here. If the Thiokol managers and NASA administrators had *not* been technical experts, I submit that they never would have dared approve the launch. For administrators or politicians to go against the consensus of a scientific discipline or a profession "is extremely hazardous," says Don K. Price, former dean of Harvard's J.F. Kennedy School of Government. In his book *The Scientific Estate*, Price speaks of a social spectrum along which are situated four "estates": scientists, professionals (including engineers), administrators, and politicians. Each has a certain power and an ability to influence the others. Politicians usually have ample room to maneuver because technical experts are so often divided in their opinions. But in the case of a questionable shuttle launch, no sane politician or administrator would go ahead without solid support from professionals.

Whatever changes occur at NASA, there will always be risk decisions to be made in the midst of uncertainty. Doubtless we will hear about new techniques of risk assessment. It is tempting particularly after a dramatic failure, to pin one's hopes to a new methodology. A few years ago a bill was introduced in Congress that would have required all federal regulatory agencies to apply comparative risk analysis in reading their decisions. The regulatory agencies expressed strong and unanimous opposition. Peter W. Preuss, then the top health scientist for the Consumer Product Safety Commission, stated that statistical methods clearly are useful, but "the Commission's overall approach for setting priorities is to rely on its collegial judgment rather than on any set formula."

Collegial judgment. That, to my mind, is the heart of sound engineering decision making in the area of risk. And that is what was present in the unanimity of the 14 Thiokol engineers. Perhaps the *Challenger* tragedy will help restore respect for this time-honored feature of professional practice.

From *Science 86* (July/August, 1986): 16–17.

## Questions and Topics for Discussion and Writing

1. Why does Florman not accept the notion that the engineer-managers had to take of their "engineering hats" and put on their "management hats"?
2. Should the junior engineers who disapproved of the launch have gone outside the chain of command to try to stop the launch? If so, who would they appeal to?
3. What is Florman advocating that engineers do in such situations?

Experimenting with different processes to produce nuclear energy.

*Issue*

# The Dependence of Society on Technological Solutions

While there is no doubt that human society is greatly dependent upon technology both for its historical development and for the maintenance of its current form, there is considerable difference of opinion on whether technology will be able to solve the problems of society that are with us today and that loom in the future. Is the answer to plow on with more technology, or to back away and take another approach? Isaac Asimov's "There's No Way to Go But Ahead" claims that we must press on. E. F. Schumacher in "Technology With a Human Face" argues that a new, more human kind of technology is what is required. Samuel Florman in "Technology and the Tragic View" advocates a middle road that recognizes the good and the bad in technology.

# There's No Way to Go But Ahead

## Isaac Asimov

Scientist/science writer Isaac Asimov is adamant in this article that scientific and technological advances are the only possible solutions to world problems.

We are all now aware that some new scientific or technological advance, though useful, may have unpleasant side effects. More and more, the tendency is to exert caution before committing the world to something that may not be reversible.

The trouble is, it's not always easy to tell what the side effects will be. In 1846, Ascanio Sobrero produced the first nitroglycerine. Heated, a drop of it exploded shatteringly. The Italian chemist realized in horror its possible application to warfare and stopped his research at once. It didn't help, of course. Others followed up, and it and other high explosives were indeed being used in warfare by the close of the 19th century.

Did that make high explosives entirely bad? In 1867, Alfred Nobel learned how to mix nitroglycerine with diatomaceous earth to produce a safer-to-handle mixture he called "dynamite." With dynamite, earth could be moved at a rate far beyond that of pick and shovel and without brutalizing men by hard labour. It was dynamite that helped forge the way for railroads, that helped build dams, subways, foundations, bridges, and a thousand other grand-scale constructions of the industrial age.

A double-edged sword of good and evil has hung over human technology from the beginning. The invention of knives and spears increased man's food supply—and improved the art of murder. The discovery of nuclear energy now places all the earth under threat of destruction—yet it also offers the possibility of fusion power as an ultimate solution to man's energy problems.

Or think back to the first successful vaccination in 1796 and the germ theory of disease in the 1860s. Do we view medical advance as dangerous to humanity, or refuse to take advantage of vaccines and antitoxins, of anesthesia and asepsis, of chemical specifics and antibiotics? And yet the side effects of the last century's medical discoveries have done more to assure civilization's destruction than anything nuclear physicists have done. For the population explosion today is caused not by any rise in average birth rate but by the precipitous drop—thanks to medicine—in the death rate. Does that mean science should have avoided improving man's lot through medicine and kept mankind a short-lived race? Or does it mean we should use science to correct the possibly deleterious side effect, devise methods that would make it simpler to reduce the birth rate and keep it matching the falling death rate? The latter, obviously!

Science and technology are getting a bad press these days. Increasingly scornful of the materialism of our culture, young people speak about returning to a simpler, pre-industrial, pre-scientific day. They fail to realize that the "good old days" were really the horrible bad old days of ignorance, disease, slavery and death. They fancy themselves in Athens, talking to Socrates, listening to the latest play by Sophocles—never as a slave brutalized in the Athenian silver mines. They imagine themselves as medieval knights on armored chargers—never as starving peasants.

Yet, right down to modern times, the wealth and prosperity of a relative few have been built on the animal-like labour and wretched existence of many—peasants, serfs and slaves. What's more, nothing could be done about it. Slavery and peonage were taken for granted. Not until science became prominent did slavery come to be recognized as a dreadful wrong, to be abolished. It was the scientist, supposedly cold and concerned with things rather than ideals, who brought this about. His investigations made possible the harnessing of the energy of the inanimate world. With steam, electricity and radio beams to do our work for us, there was less need for the comparatively weak and fumbling human muscle—and slavery began to vanish.

It is also a fact that, before modern technology, the full flower of art and human intellect was reserved for the few. It was the technical advances of printing that scattered books widely, made universal literacy practical. It was the movies, radio, the record player and television that brought many of the marvels of mankind (along with much of the refuse) to even the poorest.

Yes, science has helped create problems, too—serious ones. And we must labour to solve them—in the only way history tells us problems have been solved: by science. If we were to turn away now, if a noble young generation abandoned the materialism of an industry, what would happen? Without the machinery of that industry, we would inevitably drift back to slavery.

In these days of urban decay and energy crisis, there is a constant longing to return to the land and flee back to a simpler way of life. But it can't be done. We have a tiger by the tail and we can't go home again. We never could.

When mankind learned how to make use of fire some 50,000 years ago, it meant protection against predators, and more and better food. It also meant that man could venture out of the tropics into colder climates. Do you suppose this didn't bring problems? When the fire went out in the cave on a winter night and could not be relighted, there was the danger of freezing. Or the smoke would ruin one's lungs.

Why not give up fire, then, go back to the tropics and the simpler, carefree ways? Ah, one could not. Extending his range, man had increased his numbers. Returnees would find the tropics full, and there would be a catastrophic struggle for the smaller supply of food. So, having once learned to use fire, people either endured its discomforts—or did away with them by further technological advance. They learned better ways of making fire, heating dwelling places, handling smoke.

No fundamental technological advance has ever been given up willingly by any society. There has been no way to do it.

About 8000 B.C., mankind invented agriculture. Again it made possible an increase in numbers. People had never eaten so well, but it meant they had to give up the free, nomadic life and remain bound to the soil. It meant hard labour. It meant banding together to fight off surrounding tribes who, still food gathering, might help themselves to your crops. It also meant the risk of crop failures.

Where irrigation was introduced to make harvest more dependable, it meant the formation of a large political unit, the social tyranny of a king, an aristocracy, a priesthood. And, even if the land grew prosperous and populous, any infectious disease that got started ran through the crowded population like wildfire.

Why not, then go back to the wilder, freer ways of hunting and food gathering? Wouldn't that mean less work and worry, less war, less pestilence?

But you can't! Abandon agriculture and, out of every 10,000 people, only 100 survive. No, the problems to which agriculture gave rise could be solved only by moving forward with additional advances in technology—the use of oxen in place of men, horses in place of oxen, crop rotation, fertilizer, etc.

So it is now with our industrial age, which has once again increased man's numbers and his range—and brought new problems. If there is a shortage of gasoline, can't we in a pinch abandon our automobiles and go back to horse-and-wagon? Give up our oil furnaces for the fireplace? Give up electric lights and use candles?

No, we can't. There are no longer enough horses to move us about, or enough wood to warm us, or enough candles to light our way. Besides, if we try it for long, we will quickly find that the simple life just won't do.

In 1800, when the earth was still supported almost entirely by nonindustrial methods, the population of the planet was 900 million. Now it is pushing four billion. Where does the food come from to support the extra three billion? It comes from the industrialization of the farm: from the use of high-energy machinery to plow and seed and weed and reap. It comes from fertilizers and insec-

*Isaac Asimov*

ticides produced by sophisticated high-energy chemical factories.

We can't abandon industrialization, if only because our food supply depends on it. You can talk about "natural" food all you want, but if everyone decided to grow food without chemical fertilizers or insecticides or machinery, it would mean that only one quarter of the world population could be fed.

Can we abandon some of our industrial technology and hold onto the rest? That would be very difficult, since it all hangs together.

We can save, conserve, cut out waste, but what we have we must keep. The only solution, as always in the history of mankind, is to solve problems by still further advances in technology.

Originally published in 1975.

### Questions and Topics for Discussion and Writing

1. Why, according to Asimov, can human society not abandon any major aspect of its technology? What are the predicted results? Give some examples.
2. Suppose yourself to be in a debate with Asimov on this topic. Supposing his article to be familiar to the reader, make a counter-argument to it, that is, take a position opposite to his and make the best case you can for it.

# Technology with a Human Face

## E. F. Schumacher

Radical economist E.F. Schumacher takes issue with the view that more technology is the only answer to world problems in this excerpt from his book, *Small is Beautiful.*

As Gandhi said, the poor of the world cannot be helped by mass production, only by production by the masses. The system of mass production, based on sophisticated, highly capital-intensive, high energy-input dependent, and human labour-saving technology, presupposes that you are already rich, for a great deal of capital investment is needed to establish one single workplace. The system of production by the masses mobilises the priceless resources which are possessed by all human beings, their clever brains and skilful hands, and supports them with first-class tools. The technology of mass production is inherently violent, ecologically damaging, self-defeating in terms of non-renewable resources, and stultifying for the human person. The technology of production by the masses, making use of the best of modern knowledge and experience is conducive to decentralization, compatible with the laws of ecology, gentle in its use of scarce resources, and designed to serve the human person instead of making him the servant of machines. I have named it intermediate technology to signify that it is vastly superior to the primitive technology of bygone ages but at the same time much simpler, cheaper, and freer than the super-technology of the rich. One can also call it self-help technology, or democratic or people's technology—a technology to which everybody can gain admittance and which is not reserved to those already rich and powerful. Although we are in possession of all requisite knowledge, it still requires a systematic, creative effort to bring this technology into active existence and make it generally visible and available. It is my experience that it is rather more difficult to recapture directness and simplicity than to advance in the direction of ever more sophistication and complexity. Any third-rate engineer or researcher can increase complexity; but it takes a certain flair of real insight to make things simple again. And this insight does not come easily to people who have allowed themselves to become alienated from real, productive work and from the self-balancing system of nature, which never fails to recognize measure and limitation. Any activity which fails to recognise a self-limiting principle is of the devil. In our work with the developing countries we are at least forced to recognise the limitations of poverty, and this work can therefore be a wholesome school for all of us in which, while genuinely trying to help others, we may also gain knowledge and experience of how to help ourselves.

I think we can already see the conflict of attitudes which will decide our future. On the one side, I see the people who think they can cope with our three-fold crisis by the methods current, only more so; I call them the people of the forward stampede. On the other side, there are people in search of a new life-style, who seek to return to certain basic truths about man and his world; I call the home-comers. Let us admit that the people of the forward stampede, like the devil, have all the best tunes or at least the most popular and familiar tunes. You cannot stand still, they say; standing still means going down; you must go forward; there is nothing wrong with modern technology except that it is as yet incomplete; let us complete it. Dr. Sicco Mansholt, one of the most prominent chiefs of the European Economic Community, may be quoted as a typical representative of this group. "More, further, quicker, richer," he says, "are the watchwords of present-day society." And he thinks we must help people to adapt, "for there is no alternative." This is the authentic voice of the forward stampede, which talks in much the same tone as Dostoyevsky's Grand Inquisitor: "Why have you come to hinder us? " They point to the population explosion and to the possibilities of world hunger. Surely, we must take our flight forward and not be fainthearted. If people start protesting and revolting, we shall have to have more police and have them better equipped. If there is trouble with the environment, we shall need more stringent laws against pollution, and faster economic growth to pay for antipollution measures. If there are problems about natural resources, we shall turn to synthetics; if there are problems about fossil fuels, we shall move from slow reactors to fast breeders and from fission to fusion. There are no insoluble problems. The slogans of the people of the forward stampede burst into the newspaper headlines every day with the message, "a breakthrough a day keeps the crisis at bay."

And what about the other side? This is made up of people who are deeply convinced that technological development has taken a wrong turn and needs to be redirected. The term "home-comer" has, of course, a religious connotation. For it takes a good deal of courage to say "no" to the fashions and fascinations of the age and to question the presuppositions of a civilization which appears destined to conquer the whole world; the requisite strength can be derived only from deep convictions. If it were derived from nothing more than fear of the future, it would be likely to disappear at the decisive moment. The genuine "home-comer" does not have the best tunes, but he has the most exalted text, nothing less than the Gospels. For him, there could not be a more concise statement of his situation, of our situation, than the parable of the prodigal son. Strange to say, the Sermon on the Mount gives pretty precise instructions on how to construct an outlook that could lead to an Economics of Survival.

- How blessed are those who know that they are poor; the Kingdom of Heaven is theirs.
- How blessed are the sorrowful; they shall find consolation.
- How blessed are those of a gentle spirit; they shall have the earth for their possession.
- How blessed are those who hunger and thirst to see right prevail; they shall be satisfied;
- How blessed are the peacemakers; God shall call them his sons.

It may seem daring to connect these beatitudes with matters of technology and economics. But may it not be that we are in trouble precisely because we have failed for so long to make this connection? It is not difficult to discern what these beatitudes may mean for us today:

- We are poor, not demigods.
- We have plenty to be sorrowful about, and are not emerging into a golden age.
- We need a gentle approach, a non-violent spirit, and small is beautiful.
- We must concern ourselves with justice and see right prevail.
- And all this, only this, can enable us to become peacemakers.

The home-comers base themselves upon a different picture of man from that which motivated the people of the forward stampede. It would be very superficial to say that the latter believe in "growth" while the former do not. In a sense, everybody believes in growth, and rightly so, because growth is an essential feature of life. The whole point, however, is to give to the idea of growth a qualitative determination; for there are always many things that ought to be growing and many things that ought to be diminishing.

Equally, it would be very superficial to say that the home-comers do not believe in progress, which also can be said to be an essential feature of all life. The whole point is to determine what constitutes progress. And the home-comers believe that the direction which modern technology has taken and is continuing to pursue—towards ever-greater size, ever-higher speeds, and ever-increased violence, in defiance of all laws of natural harmony—is the opposite of progress. Hence the call for taking stock and finding a new orientation. The stocktaking indicated that we are destroying our very basis of existence, and the orientation is based on remembering what human life is really about.

In one way or another everybody will have to take sides in this great conflict. To "leave it to the experts" means to side with the people of the forward stampede. It is widely accepted that politics is too important a matter to be left to experts. Today, the main content of politics is economics, and the main content of economics is technology. If politics cannot be left to the experts, neither can economics and technology.

The case for hope rests on the fact that ordinary people are often able to take a wider view, and a more "humanistic" view, than is normally being taken by experts. The power of ordinary people, who today tend to feel utterly powerless, does not lie in starting new lines of action, but in placing their sympathy and support with minority groups which have already started. I shall give two examples relevant to the subject here under discussion. One relates to agriculture, still the greatest single activity of man on earth, and the other relates to industrial technology.

Modern agriculture relies on applying to soil, plants, and animals ever-increasing quantities of chemical products, the long-term effect of which on soil fertility and health is subject to very grave doubts. People who raise such doubts are generally confronted with the assertion that the choice lies between "poison or hunger." There are highly successful farmers in many countries who obtain excellent yields without resort to such chemicals and without raising any doubts about long-term soil fertility and health. For the last twenty-five years, a private, voluntary organization, the Soil Association, has been engaged in exploring the vital relationships between soil, plant, animal, and man; has undertaken and assisted relevant research; and has attempted to keep the public informed about developments in these fields. Neither the successful farmers nor the Soil Association have been able to attract official support or recognition. They have generally been dismissed as "the muck and mystery people," because they are obviously outside the mainstream of modern technological progress. Their methods bear the mark of non-violence and humility towards the infinitely subtle system of natural harmony, and this stands in opposition to the life-style of the modern world. But if we now realise that the modern life-style is putting us into mortal danger, we may find it in our hearts to support and even join these pioneers rather than to ignore or ridicule them.

On the industrial side, there is the Intermediate Technology Development Group. It is engaged in the systematic study of how to help people to help themselves. While its work is primarily concerned with giving technical assistance to the Third World, the results of its research are attracting increasing attention also from those who are concerned about the future of the rich societies. For they show that an intermediate technology, a technology with a human face, is in fact possible; that it is viable; and that it reintegrated the human being, with his skilful hands and creative brain, into the productive process. It serves production by the masses instead of mass production. Like the Soil Association, it is a private, voluntary organization depending on public support.

I have no doubt that it is possible to give a new direction to technological development, a direction that shall lead it back to the real needs of man, and that also means: to the actual size of man. Man is small, and, therefore, small is beautiful. To go for giantism is to go for self-destruction. And what is the cost of a reorientation? We might remind ourselves that to calculate the cost of survival is perverse. No doubt, a price has to be paid for anything worth while: to redirect technology so that it serves man instead of destroying him requires primarily an effort of the imagination and an abandonment of fear.

From E. F. Schumacher, *Small is Beautiful* (New York: Harper & Row, 1973).

### Questions and Topics for Discussion and Writing

1. Describe Schumacher's concept of "intermediate technology." What is it intermediate between?
2. Compare the points of view of Schumacher in this article, with that of Asimov in the previous article. What are their points of agreement and disagreement?

# * Technology and the Tragic View

## Samuel C. Florman

In this insightful article the American engineer and writer Samuel C. Florman urges us to take a constructive but mature attitude toward technology—neither a silly naïve optimism, nor a bickering pessimism—and learn to live with technology, recognizing both the good and the bad inherent in it.

The blaming of technology starts with the making of myths—most importantly, the myth of the technological imperative and the myth of the technocratic elite. In spite of the injunctions of common sense, and contrary to the evidence at hand, the myths flourish.

False premises are followed by confused deductions—a maligning of the scientific view; the assertion that small is beautiful; the mistake about job enrichment; an excessive zeal for government regulation; the hostility of feminists toward engineering; and the wishful thinking of the Club of Rome. These in turn are followed by distracted rejoinders from the technological community, culminating in the bizarre exaltation of engineering ethics.

In all of this it is difficult to determine how much is simple misunderstanding and how much is willful evasion of the truth, a refusal to face up to the harsh realities that underlie life, not only in a technological age, but in every age since the beginning of civilization.

Out of the confusion has come a dialogue of sorts, shaped around views that are deemed "pro"-technology or "anti-," "optimistic" or "pessimistic." I believe that we should be thinking in different terms altogether.

*House & Garden* magazine, in celebration of the American Bicentennial, devoted its July 1976 issue to the topic "American Know-How." The editors invited me to contribute an article, and enticed by the opportunity to address a new audience, plus the offer of a handsome fee, I accepted. We agreed that the title of my piece would be "Technology and the Human Adventure," and I thereupon embarked on a strange adventure of my own.

I thought that it would be appropriate to begin my Bicentennial-inspired essay with a discussion of technology in the time of the Founding Fathers, so I went to the library and immersed myself in the works of Benjamin Franklin, surely the most famous technologist of America's early days. Remembering stories from my childhood about Ben Franklin the clever tinkerer, I expected to find a pleasant recounting of inventions and successful experiments, a cheering tale of technological triumphs. I found such a tale, to be sure, but along with it I found a record of calamities *caused by* the technological advances of his day.

In several letters and essays, Franklin expressed concern about fire, an ever-threatening scourge in Colonial times. Efficient sawmills made it possible to build frame houses, more versatile and economical than log cabins—but less fire-resistant. Advances in transport made it possible for people to crowd these frame houses together in cities. Cleverly conceived fireplaces, stoves, lamps, and warming pans made life more comfortable, but contributed to the likelihood of catastrophic fires in which many lives were lost.

To deal with this problem, Franklin recommended architectural modifications to make houses more fireproof. He proposed the licensing and supervision of chimney sweeps and the establishment of volunteer fire companies, well supplied and trained in the science of firefighting. As is well known, he invented the lightning rod. In other words, he proposed technological ways of coping with the unpleasant consequences of technology. He applied Yankee ingenuity to solve problems arising out of Yankee ingenuity.

In Franklin's writings I found other examples of technological advances that brought with them unanticipated problems. Lead poisoning was a peril. Contaminated rum was discovered coming from distilleries where lead parts had been substituted for wood in the distilling apparatus. Drinking water collected from lead-coated roofs was also making people seriously ill.

The advancing techniques of medical science were often a mixed blessing, as they are today. Early methods of vaccination for smallpox, for example, entailed the danger of the vaccinated person dying from the artificially induced disease. (In a particularly poignant article, Franklin was at pains to point out that his four-year-old son's death from smallpox was attributable to the boy's *not* having been vaccinated and did not result, as rumor had it, from vaccination itself.)

After a while, I put aside the writings of Franklin and turned my attention to American know-how in the nineteenth century. I became engrossed in the story of the early days of steamboat transport. This important step forward in American technology was far from being the unsullied triumph that it appears to be in our popular histories.

Manufacturers of the earliest high-pressure steam engines often used materials of inferior quality. They were slow to recognize the weakening of boiler shells caused by rivet holes, and the danger of using wrought-iron shells together with cast iron heads that had a different coefficient of expansion. Safety valve openings were often not properly proportioned, and gauges had a tendency to malfunction. Even well-designed equipment quickly became defective through the effects of corrosion and sediment. On top of it all, competition for prestige led to racing between boats, and during a race the usual practice was to tie down the safety valve so that excessive steam pressure would not be relieved.

From 1825 to 1830, 42 recorded explosions killed upward of 270 persons. When, in 1830, an explosion aboard the *Helen McGregor* near Memphis killed more than 50 passengers, public outrage forced the federal government to take action. Funds were granted to the Franklin Institute of Philadelphia to purchase apparatus needed to conduct experiments on steam boilers. This was a notable event, the first technological research grant made by the federal government.

The institute made a comprehensive report in 1838, but it was not until 14 years later that a workable bill was passed by Congress providing at least minimal safeguards for the citizenry. Today we may wonder why the process took so long, but at the time Congress was still uncertain about its right, under the interstate commerce provision of the Constitution, to control the activities of individual entrepreneurs.

When I turned from steamboats to railroads I found another long-forgotten story of catastrophe. Not only were there problems with the trains themselves, but the roadbeds, and particularly the bridges, made even the shortest train journey a hazardous adventure. In the late 1860s more than 25 American bridges were collapsing each year, with appalling loss of life. In 1873 the American Society of Civil Engineers set up a special commission to address the problem, and eventually the safety of our bridges came to be taken for granted.

The more I researched the history of American know-how, the more I perceived that practically every technological advance had unexpected and unwanted side effects. Along with each triumph of mechanical genius came an inevitable portion of death and destruction. Instead of becoming discouraged, however, our forebears seemed to be resolute in confronting the adverse consequences of their own inventiveness. I was impressed by this pattern of progress/setback/renewed-creative-effort. It seemed to have a special message for our day, and I made it the theme of my essay for *House and Garden*.

No matter how many articles one has had published, and no matter how much one likes the article most recently submitted, waiting to hear from an editor is an anxious experience. In this case, as it turned out, I had reason to be apprehensive. I soon heard from one of the editors who, although she tried to be encouraging, was obviously distressed. "We liked the part about tenacity and ingenuity," she said, "but, oh dear, *all those disasters*—they are so depressing."

I need not go into the details of what follows: the rewriting, the telephone conferences, the re-rewriting—the gradual elimination of accidents and casualty statistics, and a subtle change in emphasis. I retreated, with some honor intact I like to believe, until the article was deemed to be suitably upbeat.

I should have known that the Bicentennial issue of *House & Garden* was not the forum in which to consider the dark complexities of technological change. My piece was to appear side by side with such articles as "A House that Has Everything," "Live Longer, Look Younger," and "Everything's Coming Up Roses" (devoted to a review of Gloria Vanderbilt's latest designs).

In the United States today magazines like *House & Garden* speak for those, and to those, who are optimistic about technology. Through technology we get better dishwashers, permanent-press blouses, and rust-proof lawn furniture. "Better living through chemistry," the old du Pont commercial used to say. Not only is *House & Garden* optimistic, that is, hopeful, about technology; it is cheerfully optimistic. There is no room in its pages for failure, or even for struggle, and in this view it speaks for many Americans, perhaps a majority. This is the lesson I learned—or I should say, relearned—in the Bicentennial year.

Much has been written about the shallow optimism of the United States: about life viewed as a Horatio Alger success story or as a romantic movie with a happy ending. This optimism is less widespread than it used to be, particularly as it relates to technology. Talk of nuclear warfare and a poisoned environment tends to dampen one's enthusiasm. Yet optimistic materialism remains a powerful force in American life. The poll-takers tell us that people believe technology is, on balance, beneficial. And we all know a lot of people who, even at this troublesome moment in history, define happiness in terms of their ability to accumulate new gadgets. The business community, anxious to sell merchandise, spares no expense in promoting a gleeful consumerism.

Side by side with what I have come to think of as *House & Garden* optimism, there is a mood that we might call *New York Review of Books* pessimism. Our intellectual journals are full of gloomy tracts that depict a society debased by technology. Our health is being ruined, according to this view, our landscape despoiled, and our social institutions laid waste. We are forced to do demeaning work and consume unwanted products. We are being dehumanized. This is happening because a technological demon has escaped from human control or, in a slightly different version, because evil technocrats are leading us astray.

It is clear that in recent years the resoluteness exhibited by Benjamin Franklin, and other Americans of similarly robust character, has been largely displaced by a foolish optimism on the one hand and an abject pessimism on the other. These two opposing outlooks are actually manifestations of the same defect in the American character. One is the obverse, the "flip side," of the other. Both reflect a flaw that I can best describe as immaturity.

A young child is optimistic, naively assuming that his needs can always be satisfied and that his parents have it within their power to "make things right." A child frustrated becomes petulant. With the onset of puberty a morose sense of disillusionment is apt to take hold. Sulky pessimism is something we associate with the teenager.

It is not surprising that many inhabitants of the United States, a rich nation with seemingly boundless frontiers, should have evinced a childish optimism, and declared their faith in technology, endowing it with the reassuring power of a parent—also regarding it with the love of a child for a favorite toy. It then follows that technological setbacks would be greeted by some with the naive assumption that ill would turn out for the best and by others with peevish declarations of despair. Intellectuals have been in the forefront of this childish display, but every segment of society has been caught up in it. Technologists themselves have not been immune. In the speeches of nineteenth-century engineers, we find bombastic promises that make us blush. Today the profession is torn between a blustering optimism and a confused guilt.

The past 50 years have seen many hopes dashed, but we can see in retrospect that they were unrealistic hopes. We simply cannot make use of coal without killing miners and polluting the air. Neither can we manufacture solar panels without worker fatalities and environmental degradation. (We assume that it will be less than with coal, but we are not sure.) We cannot build highways or canals or airports without despoiling the landscape. Not only have we learned that environmental dangers are inherent in every technological advance, but we find that we are fated to be dissatisfied with much of what we produce because our tastes keep changing. The sparkling, humming, paved metropolises of science fiction—even if they could be realized—are not, after all, the home to which humankind aspires. It seems that many people find such an environment "alienating." There can never be a technologically-based Utopia because we discover belatedly that we cannot agree on what form that Utopia might take.

To express our disillusionment we have invented a new word: "tradeoff." It is an ugly word, totally without grace, but it signifies, I believe, the beginning of maturity for American society.

It is important to remember that our disappointments have not been limited to technology. (This is a fact that the antitechnologists usually choose to ignore.) Wonderful dreams attended the birth of the New Deal, and later the founding of the United Nations, yet we have since awakened to face unyielding economic and political difficulties. Socialism has been discredited, as was laissez-faire capitalism before it. We have been bitterly disappointed by the labor movement, the educational establishment, efforts at crime prevention, the ministrations of psychiatry, and most recently by the abortive experiments of the so-called counterculture. We have come face to face with *limits* that we had presumed to hope might not exist.

Those of us who have lived through the past 50 years have passed personally from youthful presumptuousness to mature skepticism at the very moment that American society has been going through the same transition. We have to be careful not to define the popular mood in terms of our personal sentiments, but I do not think I am doing that when I observe the multiple disenchantments of our time. We also have to be careful not to deprecate youthful enthusiasm, which is a force for good, along with immaturity, which is tolerable only in the young.

It can be argued that there was for a while good reason to hold out hope for Utopia, since modern science and technology appeared to be completely new factors in human existence. But now that they have been given a fair trial, we perceive their inherent limitations. The human condition is the human condition still.

To persist in saying that we are optimistic or pessimistic about technology is to acknowledge that we will not grow up.

I suggest that an appropriate response to our new wisdom is neither optimism nor pessimism, but rather the espousal of an attitude that has traditionally been associated with men and women of noble character—the tragic view of life.

As a student in high school, and later in college, I found it difficult to comprehend what my teachers told me about comedy and tragedy. Comedy, they said, expresses despair. When there is no hope, we make jokes. We depict people as puny, ridiculous creatures. We laugh to keep from crying.

Tragedy, on the other hand, is uplifting. It depicts heroes wrestling with fate. It is man's destiny to die, to be defeated by the forces of the universe. But in challenging his destiny, in being brave, determined, ambitious, resourceful, the tragic hero shows to what heights a human being can soar. This is an inspiration to the rest of us. After witnessing a tragedy we feel good, because the magnificence of the human spirit has been demonstrated. Tragic drama is an affirmation of the value of life.

Students pay lip service to this theory and give the expected answers in examinations. But sometimes the idea seems to fly in the face of reason. How can we say we feel better after Oedipus puts out his eyes, or Othello kills his beloved wife and commits suicide, than we do after laughing heartily over a bedroom farce?

Yet this concept, which is so hard to grasp in the classroom, where students are young and the environment is serene, rings true in the world where mature people wrestle with burdensome problems.

I do not intend to preach a message of stoicism. The tragic view is not to be confused with world-weary resignation. As Moses Hadas, a great classical scholar of a generation ago, wrote about the Greek tragedians: "Their gloom is not fatalistic pessimism but an adult confrontation of reality, and their emphasis is not on the grimness of life but on the capacity of great figures to adequate themselves to it."[1]

[1] Moses Hadas, *History of Greek Literature*, 75 (New York: Columbia University Press, 1950).

It is not an accident that tragic drama flourishes in societies that were dynamic: Periclean Athens, Elizabethan England, and the France of Louis XIV. For tragedy speaks of ambition, effort, and unquenchable spirit. Technological creativity is one manifestation of this spirit, and it is only a dyspeptic antihumanist who can feel otherwise. Even the Greeks, who for a while placed technologists low on the social scale, recognized the glory of creative engineering. Prometheus is one of the quintessential tragic heroes. In viewing technology through a tragic prism we are at once exalted by its accomplishments and sobered by its limitations. We thus ally ourselves with the spirit of great ages past.

The fate of Prometheus, as well as that of most tragic heroes, is associated with the concept of *hubris*, "overweening pride." Yet pride, which in drama invariably leads to a fall, is not considered sinful by the great tragedians. It is an essential element of humanity's greatness. It is what inspires heroes to confront the universe, to challenge the status quo. Prometheus defied Zeus and brought technological knowledge to the human race. Prometheus was a revolutionary. So were Gutenberg, Watt, Edison, and Ford. Technology is revolutionary. Therefore, hostility toward technology is antirevolutionary, which is to say, it is reactionary. This charge is currently being leveled against environmentalists and other enemies of technology. Since antitechnologists are traditionally "liberal" in their attitudes, the idea that they are reactionary confronts us with a paradox.

The tragic view does not shrink from paradox; it teaches us to live with ambiguity. It is at once revolutionary and cautionary. *Hubris*, as revealed in tragic drama, is an essential element of creativity; it is also a tragic flaw that contributes to the failure of human enterprise. Without effort, however, and daring, we are nothing. Walter Kerr has spoken of "tragedy's commitment to freedom, to the unflinching exploration of the possible." "At the heart of tragedy," he writes, "feeding it energy, stands godlike man passionately desiring a state of affairs more perfect than any that now exists."[1]

This description of the tragic hero well serves, in my opinion, as a definition of the questing technologist.

An aspect of the tragic view that particularly appeals to me is its reluctance to place blame. Those people who hold pessimistic views about technology are forever reproaching others, if not individual engineers, then the "technocratic establishment," the "megastate," "the pentagon of power," or some equally amorphous entity. Everywhere they look they see evil intent.

There is evil in the world, of course, but most of our disappointments with technology come when decent people are trying to act constructively. "The essentially tragic fact," says Hegel, "is not so much the war of good with evil as the war of good with good."

Pesticides serve to keep millions of poor people from starving. To use pesticides is good; to oppose them when they create havoc in the food chain is also good. To drill for oil, and to transport it across oceans is good, since petroleum provides life-saving chemicals and heat for homes. To prevent oil spills is also good. Nuclear energy is good, as is the attempt to eliminate radioactivity. To seek safety is a worthy goal; but in a world of limited resources, the pursuit of economy is also worthy. We are constantly accusing each other of villainy when we should be consulting together on how best to solve our common problems.

Although the tragic view shuns blame, it does not shirk responsibility. "The fault, dear Brutus, is not in our stars, but in ourselves…" We are accountable for what we do or, more often, for what we neglect to do. The most shameful feature of the antitechnological creed is that it so often fails to consider the consequences of not taking action. The lives lost or wasted that might have been saved by exploiting our resources are the responsibility of those who counsel inaction. The tragic view is consistent with good citizenship. It advocates making the most of our opportunities; it challenges us to do the work that needs doing.

Life, it may be said, is not a play. Yet we are constantly talking about roles—role-playing, role models, and so forth. It is a primordial urge to want to play one's part. The outlook I advocate sees value in many different people playing many different

[1] Walter Kerr, *Tragedy and Comedy* (New York: Simon and Schuster, 1967), p. 107.

parts. A vital society, like a meaningful drama, feeds on diversity. Each participant contributes to the body social: scientist, engineer, farmer, craftman, laborer, politician, jurist, teacher, artist, merchant, entertainer.... The pro-growth industrialist and the environmentalist are both needed, and in a strange way they need each other.

Out of conflict comes resolution; out of variety comes health. This is the lesson of the natural world. It is the moral of ecological balance; it is also the moral of great drama. We cannot but admire Caesar, Brutus, and Antony all together. So should we applaud the guardians of our wilderness, even as we applaud the creators of dams and paper mills. I am a builder, but I feel for those who are afraid of building, and I admire those who want to endow all building with grace.

George Steiner, in *The Death of Tragedy* (1961), claimed that the tragic spirit was rendered impotent by Christianity's promise of salvation. But I do not think that most people today are thinking in terms of salvation. They are thinking of doing the best they can in a world that promises neither damnation nor transcendent victories, but instead confronts us with both perils and opportunities for achievement. In such a world the tragic spirit is very much alive. Neither optimism nor pessimism is a worthy alternative to this noble spirit.

We use words to communicate, but sometimes they are not as precise as we pretend, and then we confuse ourselves and each other. "Optimism," "pessimism," "tragic view"—these are mere sounds or scratches on paper. The way we feel is not adequately defined by such sounds or scratches. René Dubos used to write a column for *The American Scholar* that he called "The Despairing Optimist." I seem to recall that he once gave his reasons for not calling it "The Hopeful Pessimist," although I cannot remember what they were. What really counts, I suppose, is not what we say, or even what we feel, but what we want to do.

By saying that I espouse the tragic view of technology I mean to ally myself with those who, aware of the dangers and without foolish illusions about what can be accomplished, still want to move on, actively seeking to realize our constantly changing vision of a more satisfactory society. I mean to oppose those who would evade harsh truths by intoning platitudes. I particularly mean to challenge those who enjoy the benefits of technology but refuse to accept responsibility for its consequences.

Earlier in this chapter I mentioned the problems I encountered in preparing an article for *House & Garden*, and I would like to close by quoting the last few lines from that much-rewritten opus. The prose is somewhat florid, but please remember that it was written in celebration of the American Bicentennial:

> For all our apprehensions, we have no choice but to press ahead. We must do so, first, in the name of compassion. By turning our backs on technological change, we would be expressing our satisfaction with current world levels of hunger, disease, and privation. Further, we must press ahead in the name of the human adventure. Without experimentation and change our existence would be a dull business. We simply cannot stop while there are masses to feed and diseases to conquer, seas to explore and heavens to survey.

The editors of *House & Garden* thought I was being optimistic. I knew that I was being tragic, but I did not argue the point.

---

From Samuel C. Florman, *Blaming Technology: The Irrational Search for Scapegoats,* 181–193 (New York: St. Martins Press, 1981).

---

### Questions and Topics for Discussion and Writing

1. Why does Florman say that the tragic view is uplifting while the comic view is despairing?
2. Take a major technology of your choice and outline a "tragic view" of it in Florman's sense. (Examples of major technologies: nuclear energy, the automobile, television, computers.)

*Unit 3*

# Humans in the Environment

- World Models
- Analyses of the World Environmental Problem
- Proposed Solutions to Environmental Problems
- Atmospheric and Climatic Problems
- Public Health Issues

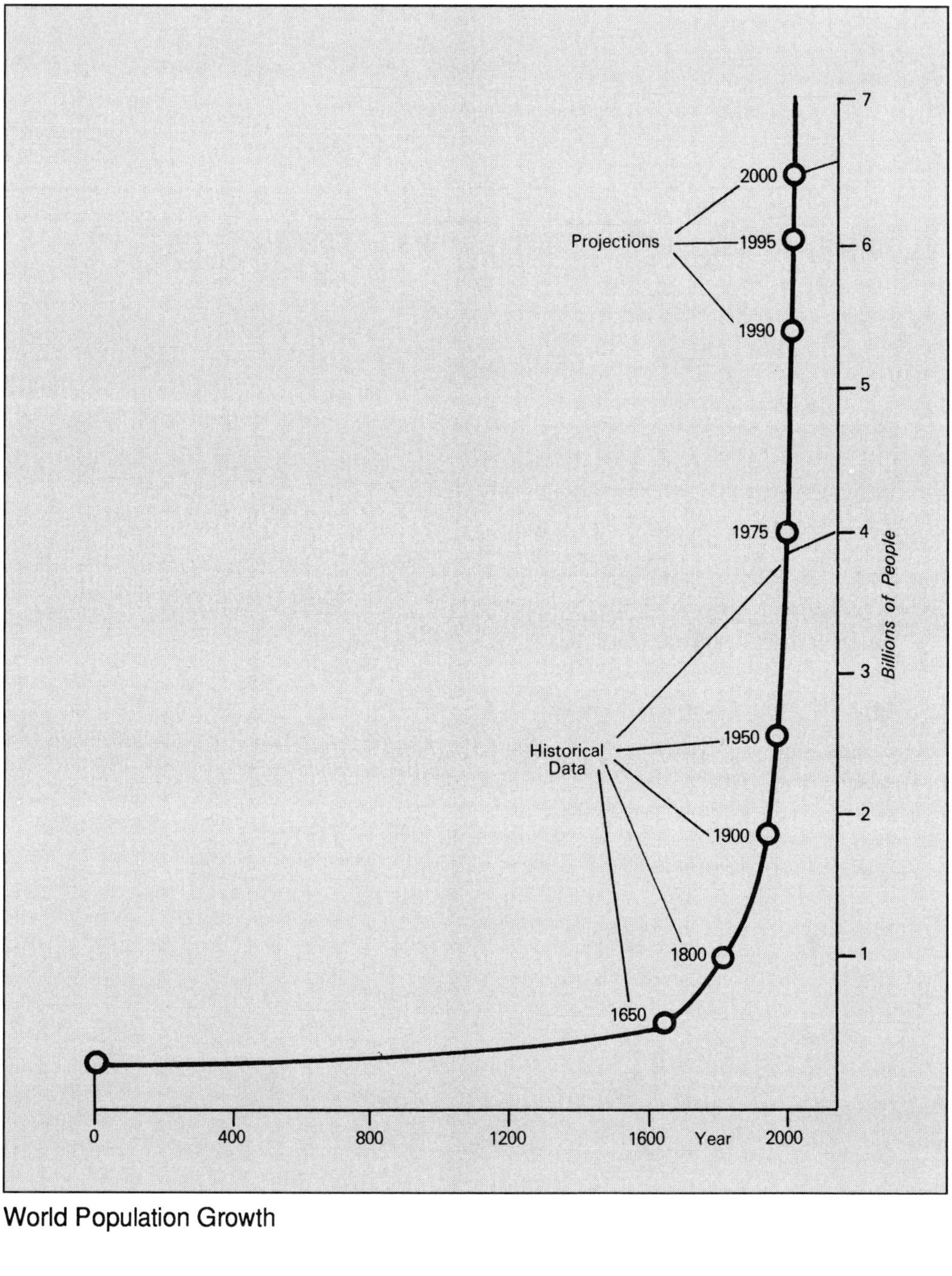


World Population Growth

*Issue*

# World Models

Except for unusual, mostly artificial situations, the environment can be viewed as a closed system only if the whole world is considered at once. To make predictions about the general environment at any considerable time in the future, some way must be found to model the relevant interactions of the world as a whole and then calculate the future effects. Models of this kind are world models.

World models of any complexity having reasonable predictive value have really only been possible in the last few years as high powered computers have become available. Even now, computers must work with a very simplified model of the world that is sure to omit some important influences or estimate interactions badly, so that all predictions from computer simulations are subject to a large amount of uncertainty.

Nevertheless, world models have been around for a long time, in some sense. And even though early models were of necessity very simple, they were still powerful enough to draw attention to phenomena about the world that might otherwise have escaped attention.

The first world model to have caused a great stir among the scientific community is the *Essay on the Principle of Population* by Thomas Malthus, from which the first selection in this section is excerpted. Malthus does not provide an actual simulation, as modern computer models do, but he does do enough calculations to show that, given his assumptions, the rate of growth of the food supply cannot possibly keep up with the natural rate of growth of the population, and hence hard times are inevitably ahead. From Malthus, in 1798, we skip to the present with Michael A. Arbib's "World Simulation," which reviews the current state of computer world models and their use.

# An Essay on the Principle of Population, as it Affects the Future Improvement of Society

## Thomas Robert Malthus

Published in London in 1798, the *Essay on the Principle of Population* by Thomas Malthus has had a tremendous influence on the development of several seemingly disparate subjects. Darwin and Wallace both read the *Essay* in the mid-nineteenth century and got from it the basic ideas of natural selection and struggle for existence that was the key factor in their theory of evolution. Countless social philosophers read the *Essay* and subsequently began to develop theories of the interaction of social classes and to advocate programs to aid the plight of the poor.

What Malthus does in the *Essay* might be called the first *world simulation.* That is, he takes a model of the world with two variables only, food and human population. Then, with a few assumptions about how these variables behave, he shows that the increase in food supply cannot possibly keep up with the increase in population if it were not constrained. Hence future misery is inevitable because there will not be enough to go around.

The *Essay* is a full-length book. The main argument is contained in the first two chapters, which are reproduced here.

### Chapter 1

The great and unlooked for discoveries that have taken place of late years in natural philosophy, the increasing diffusion of general knowledge from the extension of the art of printing, the ardent and unshackled spirit of inquiry that prevails throughout the lettered and even unlettered world, the new and extraordinary lights that have been thrown on political subjects which dazzle and astonish the understanding, and particularly that tremendous phenomenon in the political horizon, the French revolution which, like a blazing comet, seems destined either to inspire with fresh life and vigour, or to scorch up and destroy the shrinking inhabitants of the earth, have all concurred to lead able men into the opinion that we were touching on a period big with the most important changes, changes that would in some measure be decisive of the future fate of mankind.

It has been said that the great question is now at issue, whether man shall henceforth start forwards with accelerated velocity towards illimitable, and hitherto unconceived improvement, or be condemned to a perpetual oscillation between happiness and misery, and after every effort remain still at an immeasurable distance from the wished-for goal.

Yet, anxiously as every friend of mankind must look forwards to the termination of this painful suspense, and eagerly as the inquiring mind would hail every ray of light that might assist its view into futurity, it is much to be lamented that the writers on each side of this momentous question still keep aloof from each other. Their mutual arguments do not meet with a candid examination. The question is not brought to rest on fewer points, and even in theory scarcely seems to be approaching to a decision.

The advocate for the present order of things is apt to treat the sect of speculative philosophers either as a set of artful and designing knaves who preach up ardent benevolence and draw captivating pictures of a happier state of society only the better to enable them to destroy the present establishments and to forward their own deep-laid schemes of ambition, or as wild and mad-headed enthusiasts whose silly speculations and absurd paradoxes are not worthy the attention of any reasonable man.

The advocate for the perfectibility of man, and of society, retorts on the defender of establishments a more than equal contempt. He brands him as the slave of the most miserable and narrow prejudices; or, as the defender of the abuses of civil society, only because he profits by them. He paints him either as a character who prostitutes his understanding to his interest, or as one whose powers of mind are not of a size to grasp any thing great and noble, who cannot see above five years before him, and who must therefore be utterly unable to take in the views of the enlightened benefactor of mankind.

In this unamicable contest the cause of truth cannot but suffer. The really good arguments on each side of the question are not allowed to have their proper weight. Each pursues his own theory, little solicitous to correct or improve it by an attention to what is advanced by his opponents.

The friend of the present order of things condemns all political speculations in the gross. He will not even condescend to examine the grounds from which the perfectibility of society is inferred. Much less will he give himself the trouble in a fair and candid manner to attempt an exposition of their fallacy.

The speculative philosopher equally offends against the cause of truth. With eyes fixed on a happier state of society, the blessings of which he paints in the most captivating colours, he allows himself to indulge in the most bitter invectives against every present establishment, without applying his talents to consider the best and safest means of removing abuses and without seeming to be aware of the tremendous obstacles that threaten, even in theory, to oppose the progress of man towards perfection.

It is an acknowledged truth in philosophy that a just theory will always be confirmed by experiment. Yet so much friction, and so many minute circumstances occur in practice, which it is next to impossible for the most enlarged and penetrating mind to foresee, that on few subjects can any theory be pronounced just, that has not stood the test of experience. But an untried theory cannot fairly be advanced as probable, much less as just, till all the arguments against it have been maturely weighed and clearly and consistently refuted.

I have read some of the speculations on the perfectibility of man and of society with great pleasure. I have been warmed and delighted with the enchanting picture which they hold forth. I ardently wish for such happy improvements. But I see great, and, to my understanding, unconquerable difficulties in the way to them. These difficulties it is my present purpose to state, declaring, at the same time, that so far from exulting in them, as a cause of triumph over the friends of innovation, nothing would give me greater pleasure than to see them completely removed.

The most important argument that I shall adduce is certainly not new. The principles on which it depends have been explained in part by Hume, and more at large by Dr. Adam Smith. It has been advanced and applied to the present subject, though not with its proper weight, or in the most forcible point of view, by Mr. Wallace, and it may probably have been stated by many writers that I have never met with. I should certainly therefore not think of advancing it again, though I mean to place it in a point of view in some degree different from any that I have hitherto seen, if it had ever been fairly and satisfactorily answered.

The cause of this neglect on the part of the advocates for the perfectibility of mankind is not easily accounted for. I cannot doubt the talents of such men as Godwin and Condorcet. I am unwilling to doubt their candour. To my understanding, and probably to that of most others, the difficulty appears insurmountable. Yet these men of acknowledged ability and penetration, scarcely deign to notice it, and hold on their course in such speculations, with unabated ardour and undiminished confidence. I have certainly no right to say that they purposely shut their eyes to such arguments. I ought rather to doubt the validity of them, when neglected by such men, however forcibly their truth may strike my own mind. Yet in this respect it must be acknowledged that we are all of us too prone to err. If I saw a glass of wine repeatedly presented to a man, and he took no notice of it, I should be apt to think that he was blind or uncivil. A juster philosophy might teach me rather to think that my eyes deceived me and that the offer was not really what I conceived it to be.

In entering upon the argument I must premise that I put out of the question, at present, all mere conjectures, that is, all suppositions, the probable realization of which cannot be inferred upon any just philosophical grounds. A writer may tell me that he thinks man will ultimately become an ostrich. I cannot properly contradict him. But before he can expect to bring any reasonable person over to his opinion, he ought to shew, that the necks of mankind have been gradually elongating, that the lips have grown harder and more prominent, that the legs and feet are daily altering their shape, and that the hair is beginning to change into stubs of feathers. And till the probability of so wonderful a conversion can be shewn, it is surely lost time and lost eloquence to expatiate on the happiness of man in such a state; to describe his powers, both of running and flying, to paint him in a condition where all narrow luxuries would be contemned, where he would be employed only in collecting the necessaries of life, and where, consequently, each man's share of labour would be light, and his portion of leisure ample.

I think I may fairly make two postulata.

First, That food is necessary to the existence of man.

Secondly, That the passion between the sexes is necessary and will remain nearly in its present state.

These two laws, ever since we have had any knowledge of mankind, appear to have been fixed laws of our nature, and, as we have not hitherto seen any alteration in them, we have no right to conclude that they will ever cease to be what they now are, without an immediate act of power in that Being who first arranged the system of the universe, and for the advantage of his creatures, still executes, according to fixed laws, all its various operations.

I do not know that any writer has supposed that on this earth man will ultimately be able to live without food. But Mr. Godwin has conjectured that the passion between the sexes may in time be extinguished. As, however, he calls this part of his work a deviation into the land of conjecture, I will not dwell longer upon it at present than to say that the best arguments for the perfectibility of man are drawn from a contemplation of the great progress that has already made from the savage state and the difficulty of saying where he is to stop. But towards the extinction of the passion between the sexes, no progress whatever has hitherto been made. It appears to exist in as much force at present as it did two thousand or four thousand years ago. There are individual exceptions now as there always have been. But, as these exceptions do not appear to increase in number, it would surely be a very unphilosophical mode of arguing to infer merely from the existence of an exception, that the exception would, in time, become the rule, and the rule the exception.

Assuming then, my postulata as granted, I say, that the power of population is indefinitely greater than the power in the earth to produce subsistence for man.

Population, when unchecked, increases in a geometrical ratio. Subsistence increases only in an arithmetical ratio. A slight acquaintance with numbers will shew the immensity of the first power in comparison of the second.

By that law of our nature which makes food necessary to the life of man, the effects of these two unequal powers must be kept equal.

This implies a strong and constantly operating check on population from the difficulty of subsistence. This difficulty must fall somewhere and must necessarily be severely felt by a large portion of mankind.

Through the animal and vegetable kingdoms, nature has scattered the seeds of life abroad with the most profuse and liberal hand. She has been comparatively sparing in the room and the nourishment necessary to rear them. The germs of existence contained in this spot of earth, with ample food, and ample room to expand in, would fill millions of worlds in the course of a few thousand years. Necessity, that imperious all-pervading law of nature, restrains them with the prescribed bounds. The race of plants and the race of animals shrink under this great restrictive law. And the race of man cannot, by any efforts of reason, escape from it. Among plants and animals its effects are waste of seed, sickness, and premature death. Among mankind, misery and vice. The former, misery, is an absolutely necessary consequence of it. Vice is a highly probable consequence, and we therefore see it abundantly prevail, but it ought not, perhaps, to be called an absolutely necessary consequence. The ordeal of virtue is to resist all temptation to evil.

This natural inequality of the two powers of

population and of production in the earth and that great law of our nature which must constantly keep their effects equal form the great difficulty that to me appears insurmountable in the way to the perfectibility of society. All other arguments are of slight and subordinate consideration in comparison of this. I see no way by which man can escape from the weight of this law which pervades all animated nature. No fancied equality, no agrarian regulations in their utmost extent, could remove the pressure of it even for a single century. And it appears, therefore, to be decisive against the possible existence of a society, all the members of which should live in ease, happiness, and comparative leisure, and feel no anxiety about providing the means of subsistence for themselves and families.

Consequently, if the premises are just, the argument is conclusive against the perfectibility of the mass of mankind.

I have thus sketched the general outline of the argument, but I will examine it more particularly, and I think it will be found that experience, the true source and foundation of all knowledge, invariably confirms its truth.

## Chapter II

I said that population, when unchecked, increased in a geometrical ratio, and subsistence for man in an arithmetical ratio.

Let us examine whether this position be just.

I think it will be allowed, that no state has hitherto existed (at least that we have any account of) where the manners were so pure and simple, and the means of subsistence so abundant, that no check whatever has existed to early marriages, among the lower classes, from a fear of not providing well for their families, or among the higher classes, from a fear of lowering their condition in life. Consequently in no state that we have yet known has the power of population been left to exert itself with perfect freedom.

Whether the law of marriage be instituted or not, the dictate of nature and virtue seems to be an early attachment to one woman. Supposing a liberty of changing in the case of an unfortunate choice, this liberty would not affect population till it arose to a height greatly vicious; and we are now supposing the existence of a society where vice is scarcely known.

In a state therefore of great equality and virtue, where pure and simple manners prevailed, and where the means of subsistence were so abundant that no part of the society could have any fears about providing amply for a family, the power of population being left to exert itself unchecked, the increase of the human species would evidently be much greater than any increase that has been hitherto known.

In the United States of America, where the means of subsistence have been more ample, the manners of the people more pure, and consequently the checks to early marriages fewer than in any of the modern states of Europe, the population has been found to double itself in twenty-five years.'

This ratio of increase, though short of the utmost power of population, yet as the result of actual experience, we will take as our rule, and say, that population, when unchecked, goes on doubling itself every twenty-five years or increases in a geometrical ratio.

Let us now take any spot on earth, this Island for instance, and see in what ratio the subsistence it affords can be supposed to increase. We will begin with it under its present state of cultivation.

If I allow that by the best possible policy, by breaking up more land and by great encouragements to agriculture, the produce of this Island may be doubled in the first twenty-five years, I think it will be allowing as much as any person can well demand.

In the next twenty-five years, it is impossible to suppose that the produce would be quadrupled. It would be contrary to all our knowledge of the qualities of land. The very utmost that we can conceive is, that the increase in the second twenty-five years might equal the present produce. Let us then take this for our rule, though certainly far beyond the truth, and allow that by great exertion, the whole produce of the Island might be increased every twenty-five years, by a quantity of subsistence equal to what it at present produces. The most enthusiastic speculator cannot suppose a greater increase than this. In a few centuries it would make every acre of land in the Island like a garden.

Yet this ratio increase is evidently arithmetical.

It may be fairly said, therefore, that the means of subsistence increase in an arithmetical ratio. Let us now bring the effects of these two ratios together.

The population of the Island is computed to be above seven millions, and we will suppose the present produce equal to the support of such a number. In the first twenty-five years the population would be fourteen millions, and the food being also doubled, the means of subsistence would be equal to this increase. In the next twenty-five years the population would be twenty-eight millions, and the means of subsistence only equal to the support of twenty-one millions. In the next period, the population would be fifty-six millions, and the means of subsistence just sufficient for half that number. And at the conclusion of the first century the population would be one hundred and twelve millions and the means of subsistence only equal to the support of thirty-five millions, which would leave a population of seventy-seven millions totally unprovided for.

A great emigration necessarily implies unhappiness of some kind or other in the country that is deserted. For few persons will leave their families, connections, friends, and native land, to seek a settlement in untried foreign climes, without some strong subsisting causes of uneasiness where they are, or the hope of some great advantages in the place to which they are going.

But to make the argument more general and less interrupted by the partial views of emigration, let us take the whole earth, instead of one spot, and suppose that the restraints to population were universally removed. If the subsistence for man that the earth affords was to be increased every twenty-five years by a quantity equal to what the whole world at present produces, this would allow the power of production in the earth to be absolutely unlimited, and its ratio of increase much greater than we can conceive that any possible exertions of mankind could make it.

Taking the population of the world at any number, a thousand millions, for instance, the human species would increase in the ratio of—1, 2, 4, 8, 16, 32, 64, 128, 256, 512, etc., and subsistence as—1, 2, 3, 4, 5, 6, 7, 8, 9, 10, etc. In two centuries and a quarter, the population would be to the means of subsistence as 512 to 10: in three centuries as 4096 to 13, and in two thousand years the difference would be almost incalculable, though the produce in that time would have increased to an immense extent.

No limits whatever are placed to the productions of the earth; they may increase for ever and be greater than any assignable quantity; yet still the power of population being a power of a superior order, the increase of the human species can only be kept commensurate to the increase of the means of subsistence, by the constant operation of the strong law of necessity acting as a check upon the greater power.

The effects of this check remain now to be considered.

Among plants and animals the view of the subject is simple. They are all impelled by a powerful instinct to the increase of their species, and this instinct is interrupted by no reasoning or doubts about providing for their offspring. Wherever therefore there is liberty, the power of increase is exerted, and the superabundant effects are repressed afterwards by want of room and nourishment, which is common to animals and plants, and among animals, by becoming the prey of others.

The effects of this check on man are more complicated. Impelled to the increase of his species by an equally powerful instinct, reason interrupts his career and asks him whether he may not bring beings in the world, for whom he cannot provide the means of subsistence. In a state of equality, this would be the simple question. In the present state of society, other considerations occur. Will he not lower his rank in life? Will he not subject himself to greater difficulties than he at present feels? Will he not be obliged to labour harder? and if he has a large family, will his utmost exertions enable him to support them? May he not see his offspring in rags and misery, and clamouring for bread that he cannot give them? And may he not be reduced to the grating necessity of forfeiting his independence, and of being obliged to the sparing hand of charity for support?

These considerations are calculated to prevent, and certainly do prevent, a very great number in all civilized nations from pursing the dictate of nature in an early attachment to one woman. And this restraint almost necessarily, though not absolutely so, produces vice. Yet in all societies, even those that are most vicious, the tendency to a virtuous attachment is so strong that there is a constance effort towards an increase of population. This constant effort as constantly tends to subject the lower classes of the society to distress and to prevent any great permanent amelioration of their condition.

The way in which these effects are produced seems to be this.

We will suppose the means of subsistence in any country just equal to the easy support of its inhabitants. The constant effort towards population, which is found to act even in the most vicious societies, increases the number of people before the means of subsistence are increased. The food therefore which before supported seven millions must now be divided among seven millions and a half or eight millions. The poor consequently must live much worse, and many of them be reduced to severe distress. The number of labourers also being above the proportion of the work in the market, the price of labour must tend toward a decrease, while the price of provisions would at the same time tend to rise. The labourer therefore must work harder to earn the same time as he did before. During this season of distress, the discouragements to marriage, and the difficulty of rearing a family are so great that population is at a stand. In the mean time the cheapness of labour, the plenty of labourers, and the necessity of an increased industry amongst them, encourage cultivators to employ more labour upon their land, to turn up fresh soil, and to manure and improve more completely what is already in tillage, till ultimately the means of subsistence become in the same proportion to the population as at the period from which we set out. The situation of the labourer being then again tolerably comfortable, the restraints to population are in some degree loosened, and the same retrograde and progressive movements with respect to happiness are repeated.

This sort of oscillation will not be remarked by superficial observers, and it may be difficult even for the most penetrating mind to calculate its periods. Yet that in all old states some such vibration does exist, though from various transverse causes, in a much less marked, and in a much more irregular manner than I have described it, no reflecting man who considers the subject deeply can well doubt.

Many reasons occur why this oscillation has been less obvious, and less decidedly confirmed by experience, than might naturally be expected.

One principal reason is that the histories of mankind that we possess are histories only of the higher classes. We have but few accounts that can be depended upon of the manners and customs of that part of mankind, where these retrograde and progressive movements chiefly take place. A satisfactory history of this kind, of one people, and of one period, would require the constant and minute attention of an observing mind during a long life. Some of the objects of enquiry would be, in what proportion to the number of adults was the number of marriages, to what extent vicious customs prevailed in consequence of the restraints upon matrimony, what was the comparative mortality among the children of the most distressed part of the community and those who lived rather more at their ease, what were the variations in the real price of labour, and what were the observable differences in the state of the lower classes of society with respect to ease and happiness, at different times during a certain period.

Such a history would tend greatly to elucidate the manner in which the constant check upon population acts and would probably prove the existence of the retrograde and progressive movements that have been mentioned, though the times of their vibration must necessarily be rendered irregular, from the operation of many interrupting causes, such as the introduction or failure of certain manufacturers, a greater or less prevalent spirit of agricultural enterprise, years of plenty, or years of scarcity, wars and pestilence, poor laws, the invention of processes for shortening labour without the proportional extension of the market for the commodity, and particularly, the difference between the nominal and real price of labour, a circumstance which has perhaps more than any other contributed to conceal this oscillation from common view.

It very rarely happens that the nominal price of labour universally falls, but we well know that it frequently remains the same, while the nominal price of provisions has been gradually increasing. This is, in effect, a real fall in the price of labour, and during this period the condition of the lower orders of the community must gradually grow worse and worse. But the farmers and capitalists are growing rich from the real cheapness of labour. Their increased capitals enable them to employ a greater number of men. Work therefore may be plentiful, and the price of labour would consequently rise. But the want of freedom in the market of labour, which occurs more or less in all communities, either from parish laws, or the more general cause of the facility of combination among the rich, and its difficulty among the poor, operates to prevent the price of labour from rising at the natural period, and keeps it down some time longer; perhaps, till a year of scarcity, when the clamour is too loud, and the necessity too apparent to be resisted.

The true cause of the advance in the price of labour is thus concealed, and the rich affect to grant it as an act of compassion and favour to the poor, in consideration of a year of scarcity, and, when plenty returns, indulge themselves in the most unreasonable of all complaints, that the price does not again fall, when a little reflection would show them that it must have risen long before but from an unjust conspiracy of their own.

But though the rich by unfair combinations contribute frequently to prolong a season of distress among the poor, yet no possible form of society could prevent the almost constant action of misery upon a great part of mankind, if in a state of inequality, and upon all, if all were equal.

The theory on which the truth of this position depends appears to me so extremely clear that I feel at a loss to conjecture what part of it can be denied.

That population cannot increase without the means of subsistence is a proposition so evident that it needs no illustration.

That population does invariably increase where there are the means of subsistence, the history of every people that have ever existed will abundantly prove.

And that the superior power of population cannot be checked without producing misery or vice the ample portion of these too bitter ingredients in the cup of human life and the continuance of the physical causes that seem to have produced them bear too convincing a testimony...

From Thomas Robert Malthus, *An Essay on the Principle of Population, as It Affects the Future Improvement of Society* (London, 1798).

## Questions and Topics for Discussion and Writing

1. Why can food supply only increase at an arithmetical rate? What assumptions does Malthus make about how food is produced?
2. Why are "misery and vice" the inevitable consequences of the population outstripping the food supply?
3. Comment on the usefulness of a crude model of the world such as this which only considers two variables. Can we learn anything from it, or are too many important considerations left out?

# World Simulation

Michael A. Arbib

Thomas Malthus's *Essay on the Principle of Population* is the prototype of a world simulation model. He took two variables, food and population, and a few assumptions, and with those set up a model of the future, from which he drew conclusions. His model was very simple and did not directly include any interaction effects between the variables. But that was in 1798. Now, nearly 200 years later, much more sophisticated models have been created, which have many more variables and more complex assumptions, and—the most important difference—are calculated using powerful computers which can take into account complex interactions. Nevertheless, these world simulation models are still very primitive and unreliable. Though they give us few results that we can rely upon, they do point clearly to the magnitude of the ecological and environmental problems looming in the future.

This excerpt from Michael Arbib's book, *Computers and the Cybernetic Society*, discusses the world models of the Club of Rome and the problems inherent in the construction of simulation models.

World models [are] attempts to predict, on the computer, ecological and economic changes affecting the population of the entire world. It should be stressed that there are many models of complex systems—such as weather prediction, for example—that are far better developed than world modeling. However, we develop world simulation in some detail because the studies have often captured headlines in newspapers and magazines. They have thus had an impact on the way many citizens view the world's social problems and the way they view the computer's role in society.

Why do people build world models? It is because many people feel that with mounting population, decreasing reserves of oil and metals, and more and more starvation, we are heading toward a worldwide crisis for the human race. Something must be done to avoid this crisis. But what? Some people feel that we should use computer simulation to explore different courses of action to try to find options for the human race that will make the world a better place for our grandchildren and for their grandchildren in turn. We will examine two models commissioned by the Club of Rome .... While we will learn quite a bit from studying them, we will also see that their models are rough and the data are unreliable. Their predictions of the state of the world 50 years from now are to be trusted much *less* than the weather forecast for a week from today. We are still at the stage where we must put most of our effort into model validation, and little of our trust in the predictions that today's primitive models yield.

### Garbage In/Garbage Out

"Garbage in/garbage out" is an old saying in the computer business. It does not matter how fancy a computer you have—if the data you feed in are garbage, then the results your program prints out will be garbage. This message also applies to computer simulation: you cannot obtain accurate predictions unless you have a good model of the system and good data on its inputs and its parameters.

In spacecraft control, the models and the data are both excellent. For hundreds of years, scientists have been studying laws of motion—they really know how to write down the effect of forces on the movement of an object. When engineers build a spacecraft, they know exactly what each piece weighs, where one piece joins another, and how big a force is obtained by turning on the Number 2 engine for one-tenth of a second.

In traffic control, the models are reasonably satisfactory. It is possible to write down equations that show how many cars will get across each intersection during the time a traffic light is green. The data are probably not as reliable, however. While the town planner may have information on the number of cars using a city street each hour, he or she may not be able to tell how many of these people would take a detour, if available, and how many people must use this street to get to their destination. Nonetheless, the data and the models are accurate enough to let planners change around the one-way streets and reschedule the traffic lights and be pretty sure that the traffic will move better as a result.

When we come to weather prediction, the situation becomes more interesting. Weather prediction is still not a "sure thing," but it is far better than it has ever been before. This is because of two breakthroughs in technology. The first breakthrough is in space technology. We now have satellites that can provide a continual photographic record of cloud patterns over the entire planet. The other breakthrough is in computer technology. The millions of photographs would be of little use if they had to be searched through by hand; but put them all on magnetic disks, and it becomes possible to have the computer search for patterns that no human would have the time or patience to find. The problem is: the data are too numerous. What cloud patterns really matter in predicting the weather? How likely is rain if a cold front meets a warm front over a lake or over a desert?

Meteorologists—the people who study the weather—are continually devising new models to predict the way that weather changes from day to day. Using computers, they can handle more realistic models because they can process far more information than they could with hand computation. Furthermore, because of the huge data base they have on their magnetic disks, they can test their models more completely. They can see how well their models can predict weather in a certain region on a certain day on the basis of information about the weather on earlier days and, then, they can repeat the test for many different days and different regions. One model may work better in winter than another, another may work best for desert regions. By trying to preserve the best features in each model, the meteorologists can build more realistic models...

The cost of making weather predictions is worthwhile because, *if the predictions are accurate enough*, the economic impact can be immense. Day-by-day predictions are now fairly reliable. More challenging—and far more important economically—are long-term predictions, a year or more in advance. Such predictions do *not* have to be precise. It really does not matter whether the temperature will be 22°C or 35°C at 1:00 p.m. on the third Thursday of next summer, but it is very important to know if the total rainfall that summer will be normal or whether it will deviate on the high or low side. If farmers can be sure of very high levels of rainfall, they can start planning for flood control and hold back on crops that die if waterlogged. If, on the other hand, a drought is likely, the farmer must invest in an irrigation system and avoid planting "thirsty" crops. Decisions like these can be worth billions of dollars to a nation's economy and can have a dramatic effect on how well the nation's people are fed. That is why space technology and computer technology are not playthings—they are vital tools to help humans live as best they can on this planet.

Many people talk of *ecology*—the study of the interrelations between different species of plants and animals and their environment—as a movement to restore the world to "a state of nature"; but does this mean nature at the time of the dinosaurs, nature as it was before the agrarian revolution, or nature 1,000 years ago before the rise of modern industrialism? None of these seem satisfactory. It seems that one of our aims in studying ecology must be to ensure that this planet will be a fit home for humanity. We now realize that fresh water, clean air, and decent food are not inexhaustible. We realize that some animals are needed to help maintain the environment that supports us—while others should be preserved because they make the world a richer place. It would be sad if our grandchildren could see neither giraffes nor elephants—as I already regret that I will never see a dodo. This does not mean, however, that the ecological movement should turn away from technology. Rather, it should make wise use of technology—and the technology of weather prediction is an important example—to help us learn how to balance what we consume with what we can work with nature to provide, while

*Michael A. Arbib*

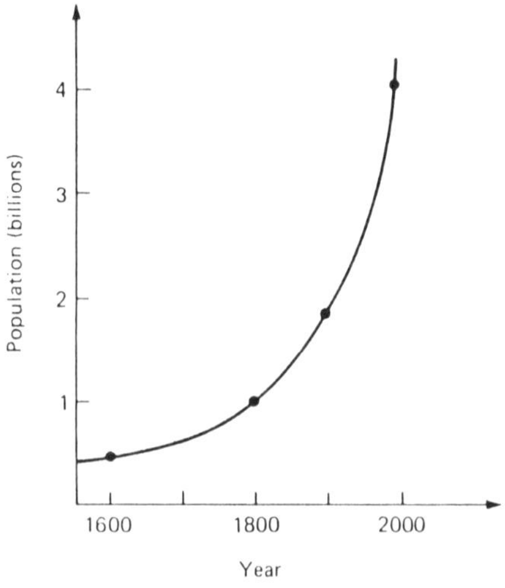


**FIGURE 3.1:** Growth of human population worldwide from 1600 to 2000.

preserving a diversity of animal life and natural scenery to cater to our spiritual needs.

In setting up world models, we must do more than take into account the ecological factors of how populations of organisms interact with one another and their environment. We must also look at the impact of human industry, taking *economics* into account by examining the development and regulation of the material resources and means of production of a community, nation, or group of nations.

Attempts to model the world...[add] many economic, as well as ecological, factors that affect the human population. Some of the numbers in the model are just what the model is meant to predict—like the wolf and rabbit populations. Other numbers are *parameters*—like the breeding rates of rabbits and wolves—and must be ascertained before we can run the model. When we come to the study of the world system and its impact on the human population, we find that some parameters must be determined by careful study—such as the death rate caused by different concentrations of pollutants in the atmosphere. Other parameters—such as the birth rate—may well be dependent on particular government policies (such as those concerning birth control or abortion).

Thus, only some of the important variables can be described in a moderately satisfactory way.... For example, no network of population interactions can explain why the political system in India has not been able to bring about a really effective birth-control program in the Indian villages. Again, these models cannot explain why the approach taken in China has led to a cultural restructuring that has successfully brought about a dramatic lowering of the birth rate. (Of course, political scientists do try to develop models to explain these things—but they are pencil-and-paper models of the way people's attitudes change, not computer models of population interactions.) Consequently, the best strategy for detailed economic planning will use a mixture of the scenario-analysis technique presented in the previous section—to set general courses of action corresponding to such areas as political decision making—and control programs to help specify the numbers—investment rates, food production, and so forth—within that general plan.

### The Limits to Growth

With this general perspective, we can turn to specific world simulations. A graph of the world's total population looks something like the curve of Figure 3.1, with the population doubling in the last 75 years to a figure of around 4,000 million (4 billion).

Recalling the notion of *exponential growth...*, you might suggest that this growth is exponential, with a doubling time of 75 years; but if you look closely, you will see that the growth is *faster* than exponential, with the doubling time getting shorter and shorter. One of my friends, Heinz von Foerster of the University of Illinois, plotted the curve more carefully and found that, if it continued this way, the world population would go to infinity on April 23, 2024! Of course, there are many reasons why the population of the world cannot reach infinity on April 23, 2024, or on any other date....

The first report to the Club of Rome, *The Limits to Growth*, presents an alternative approach.... Taking a curve and extrapolating it gives no idea of the mechanisms involved. Consequently, the authors sought to understand the key variables and their interrelationships. A *small* extract from these relationships is shown in Figure 3.2. At the top of the

*Michael A. Arbib*

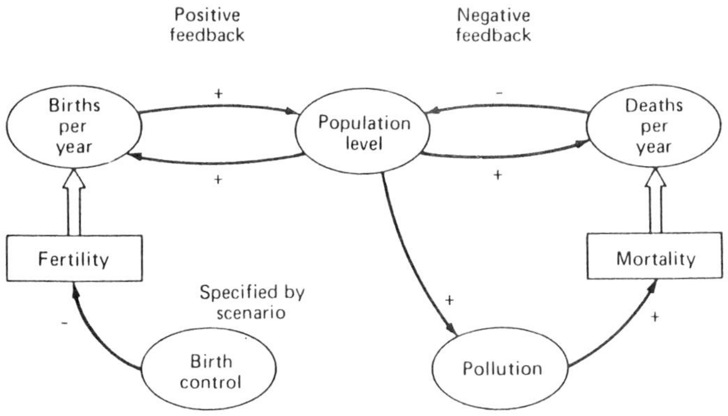


**FIGURE 3.2:** Simplified form of the population-pollution interactions in *The Limits to Growth* world model.

figure, we see a *positive* feedback loop—the higher the population, the more births per year—that serves to *further* increase the population. We also see a *negative* feedback loop—the higher the population, the more deaths per year—that goes *against* increased population.

*Fertility* is the rate at which people reproduce. It can be measured by the average number of children born per year for each thousand people. *Mortality* is the rate at which people die. It can be measured by the average number of deaths per year for each thousand people. The population will increase as long as fertility exceeds mortality....

Figure 3.2 suggests two ways in which fertility and mortality can be brought more into line so as to prevent the population explosion sketched in Figure 3.1. At the left, we see that increased birth control can reduce fertility—lowering the number of children each woman gives birth to—and thus slow down population growth. At the right, we see a more sinister check to population growth. As population increases, so does pollution. People weakened by pollution can die more easily from many different diseases. Thus, increasing pollution increases mortality, and this too counteracts population growth.

The full "Limits" model not only modeled the interaction of population and pollution; it also modeled interactions with natural resources, food production, and industrialization. The model was run with many different scenarios. In one run, natural resources would be depleted, yielding a slowdown in investment rates as more money was tied up in burying scarce and expensive resources. Eventually, the available investment money would be too little to replace worn-out equipment, let alone buy new equipment, As a result, industrial output would collapse, and agricultural collapse would follow as needed fertilizers and equipment became unavailable. Finally, as food resources dwindled, mortality would rise as starvation set in. The human population would drop catastrophically in the middle of the twenty-first century. In another run, the scenario assumed that new technology would maintain a supply of resources. As a result, industrial output would be maintained. In this run, both population and pollution would increase at an alarming rate. In the year 2050, pollution would increase to such a high level that the negative feedback loop of Figure 3.2 would take over, resulting in a catastrophic drop in world population due to widespread deaths from pollution.

The "Limits" team members emphasize that their model cannot predict the *exact* date of catastrophe; but they do suggest that without policy changes of a drastic kind, their projection of population collapse somewhere in the middle of the next century is inevitable. Because they are convinced of the reality of this predicted collapse, the team argues for strong measures to avoid it. They advocate an

*Michael A. Arbib*

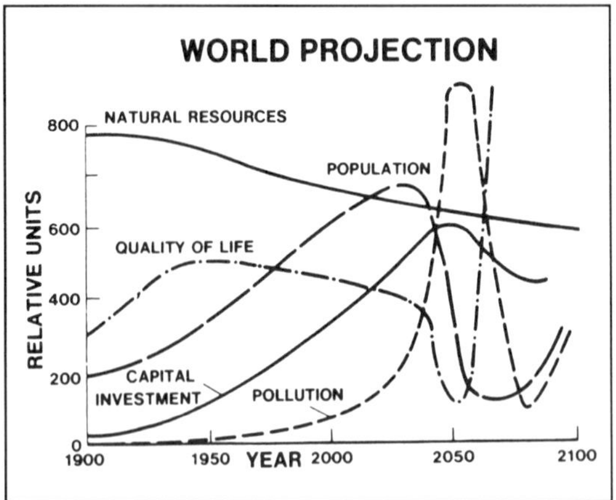


FIGURE 3.3: This typical run from *The Limits to Growth* study shows what might happed if pollution is allowed to increase unchecked. By 2020, pollution would begin to increase the death rate drastically. The population and the quality of life would plunge as pollution skyrocketed until sometime after 2050. By then, the world population would have been so catastrophically reduced, and the industrial base so damaged, that pollution would begin to abate and the population would once more begin to rise.

*end to growth*: population must be held at present levels, they say; the *per capita* industrial output (that is, the average amount of industrial products produced by each person) must be held at the 1975 level; goods must be recycled and pollution controlled; soil restored; and consumer goods made more durable.

Is it really true that our present growth is dangerous—a cancer—or is the model too simple or its data too unreliablel? Perhaps a more careful analysis will allow us different courses of action. We may all argue that recycling goods and controlling pollution and restoring the soil are fine things to do; but must everything we use from now on be a patched-up hand-me-down? Can we grow without inviting disaster in the next 100 years?

The answer to these questions is not clear. In the next subsection, we will see another world model that lets us explore courses of action that could not be included in the "Limits" scenarios. Then, in the final subsection, we will look at data that suggest that we have far more room for growth than the "Limits," team suggests. However, we will also see that there are many inequalities between different nations and between different groups of nations, so that dramatic changes in policy are still required. Inequalities must be reduced even if we reject the call to stop growth as being based on too simple a model of the world…

## A Regional Approach to Global Planning

The "Limits" study modeled the world as a lumped system; consequently, variables such as population, land resources, and so forth, all referred to totals for the entire world. As we have just seen, the "Limits" team predicted that present trends would lead to a worldwide collapse in the middle of the twenty-first century. Their runs showed that even if steps were taken to restrict one variable at a time—such as depletion of natural resources—other variables would reach catastrophic levels. Hence, they recommended an immediate slowdown of economic growth as the only solution to the problem. However, the *second* report to the Club of Rome, *Mankind at the Turning Point*, stressed that one cannot take any single position on growth. While they still agreed that population growth should be halted, they now were willing to see medical services improved—even though this would decrease the mortality rate. While the current rate of depletion of resources by the developed world suggests that their rate of materials consumption should decline, a strong case can nonetheless by made for an increase in materials consumption in the developing nations. The key to the "Turning Point" approach is to model the world as a system of ten distinct regions; including the four regions of the developed world that possess a market economy, namely:

- North America
- Western Europe
- Japan
- The rest of the world that exhibits a market economy, including Australia, South Africa, and Israel.

The next two regions

- Eastern Europe and the USSR
- China

are the two areas of the world with a primarily socialist economy. The next region

- North Africa and the Middle East

has emerged as a region of striking global significance in its own right because of its enormous share of the oil reserves. The other three regions

- Latin America
- Tropical Africa
- South and Southeast Asia

are the regions that we think of as the least developed nations—although here the contrast between Brazil and some of the other Latin American nations is striking.

A quick glance at these ten regions immediately convinces us that the "Limits" model was far too simple, for it is clear that we cannot hope to make meaningful predictions without distinguishing the fate of these different regions. It should be observed, however, that further refinements are indeed possible. For example, lumping Australia and South Africa together in one region immediately rules out an analysis of the impact of African nationalism. Another problem is that the regions are based on existing political entities. A useful refinement would be to find out what states could work together in their planning to form more realistic regions.

Even though the choice of these ten regions is somewhat unsatisfactory, this model did allow the "Turning Point" team to make a number of important projections. In the following discussion, we look at two of them, one concerning *population control* and another concerning *international transfer of development funds.*

We all know that the world population is rising drastically. Some might hope, however, that a *gradual* solution to the problem would be sufficient. They might hope that 50 years is a reasonable period in which to achieve an *equilibrium fertility rate*—one in which the birth rate equals the death rate. However, "Turning Point" computer runs show that—if there is no massive starvation in the meantime—North American will have an additional 4 people per square kilometre by 2024, while South Asia will have 160 more people per square kilometre. This *increase* for South Asia is greater than the *current total* for the already densely populated Western Europe—which now supports 85 people per square kilometre. The message is clear: while North American birth rates are already at a manageable level, the birth rate of South Asia is at a crisis level, a level that must be *speedily* reduced.

To pinpoint the dangers of delay in implementing a population control policy, the "Turning Point" team studied scenarios that project the *population level of the whole underdeveloped world.*

*Scenario 1*: In the first scenario, no population policy is implemented. The simulation shows that if current trends continue unchecked, the population will reach 12 billion by 2010. This number is already so high as to be unsupportable. As a result, famine on a hitherto unimaginable scale will strike the underdeveloped world long before 2010 unless current birth rates are drastically reduced... However, not all food experts agree: some believe that the earth *can* feed many more people.

*Scenario 2*: In the second scenario, a massive birth-control program is immediately initiated. This program succeeds in providing an equilibrium birth rate by 2010. In this case, the computer projects an equilibrium level of 6.27 billion in the underdeveloped nations by 2050.

*Scenario 3*: For a 10-year delay in implementing the birth-control program, the population projection for 2050 rises to 7.97 billion.

*Scenario 4*: For a 20-year delay, the population projection for 2050 rises to 10.17 billion.

Thus a simple 20-year delay in implementing a birth-control program can mean an increase in the *projected* population of the underdeveloped world from 6.27 billion to 10.17 billion. The "Turning Point" team believes that it is unlikely that food could be found for all these extra mouths and that this 20-year delay would result in 500 million additional child deaths. Equally terrifying would be the abject stupor in which many of the survivors would live due to gross malnutrition.

If we accept the results of these scenarios, it is clear that prompt action must be taken. It is also clear that—no matter how quickly the birth rate starts to level off—huge population increases will occur during the rest of this century. Such increases do not simply require more food. Jobs and housing must be found, children must be educated, the sick must be cared for. Tarzie Vittachi, Executive Secretary of the United Nations World Population Year, has estimated that to cope with its population increase, India alone must build 10,000 houses *every day* for the next 20 years. It seems unlikely that such a massive effort can succeed without developmental aid from the richer nations. Thus, we now turn to scenarios that explore the impact of different schedules for delivering that aid.

Let us make the *very optimistic* assumption that people and governments respond to population warnings like those we have just seen, and that each region implements policies that help them achieve an equilibrium fertility rate by the year 2000. On this assumption, current development projections show the per-capita income in the developed world increasing from $3,500 to $14,000 by 2025. The current 5 to 1 ratio of income in the developed world to the average in Latin America would increase to 8 to 1; while the ratio for South Asia would remain at a staggering 20 to 1. The next three "Turning Point" scenarios examine the cost of international aid from the developed countries if all underdeveloped regions are to have at least one-fifth of the developed nations' percapita income by 2025.

*Scenario 1*: Continuous aid is to be supplied from now until 2025 to maintain the growth rate of each underdeveloped nation at a level required to achieve the target ratio of 5 to 1 in the year 2025. The projected total cost is $7,200 billion, with a peak of $500 billion in any one year of the aid project. The cost to individuals in the developed world is a decrease in per-capita income from the projected $14,000 in 2025 to an actual figure of $11,000 in that year—an amount that is till a dramatic increase from $3500.

*Scenario 2*: In this scenario, the cost of delay is once again measured. If the gap between rich and poor nations is allowed to widen until 2000, and if a game of "Operation Catch-up" is then attempted during the remaining 25 years, the projected cost jumps to $10,700 billion. This is a dramatic increase of 50 percent over the cost projected in Scenario 1—and all the more painful because it must be paid out in only half the time.

*Scenario 3:* In the previous scenario, the high cost of delay is once again evident—a cost that is quite apart from the suffering that this delay brings to those who must endure reduced living standards for the extra twenty-five years. Could it be that the humanitarian impulse to provide the aid even more quickly could also prove economically attractive to the developed nations? In Scenario 3, early action is taken to ensure that the 5 to 1 ratio is reached by the year 2000 rather than by 2025. Surprisingly, the total cost is $2500 billion, only a third of that projected for Scenario 1 (and less than a quarter of the Scenario 2 cost), and the peak annual cost is $250 billion per year. Not only is the total cost far less, but the early action strategy yields economic self-sufficiency for the developing regions by the year 2000.

Some readers may still be disturbed by the 5 to 1 gap between the income levels in rich and poor nations—although the political challenges of implementing any of the above three scenarios are large enough as it is. However, it must be stressed that per-capita income is not a true measure of the quality of life. For example, it can be argued that a villager who is within walking distance of his or her fields is better off than a family that must have at least two cars because its members live far from jobs and shops and have no public transportation. There is then no reason for the villager to envy that family the extra $16,000 of "property" that those cars represent. Many thoughtful people have argued that underdeveloped nations would not profit by trying to imitate the inefficient and costly "American way of life," and that they *can* and *should* develop new patterns of housing, employment, transportation, and technology that can yield a highly satisfying life-style on a relatively low per-capita income.

### The Danger of Despair

The reports to the Club of Rome have been widely publicized. Many people are convinced that population growth must be halted at once. Most of us certainly feel that attempts to raise living standards must go hand-in-hand with a reduction in population growth, so that our grandchildren will not only have enough to eat but will still be able to explore a countryside free of high-density human dwellings. If the shock of a possible catastrophe in 2050 can help people cut back the size of their families and use the available resources to enrich their lives and their children's lives, I am all for it.

Unfortunately, though, the vision of catastrophe has also given rise to ugly talk. Some people argue that if resources are so scarce, then we should not waste them on nations like Bangladesh (a country of 70 million people to the northeast of India) that have soaring populations and poor food resources. These people ask us to cut off all aid to such nations, letting them starve to death. Better that some of us get all we need to lead healthy lives, these people say, than that we all lead miserable undernourished lives.

However, it appears that the time for such cruel decisions is not yet upon us. During the last thirty years, the world population has doubled, but total food production has more than doubled. This increase is due to improvements in agriculture, better use of fertilizers, and the "green revolution" that has led to the development of new strains of crops. This growth has led to increased food availability in the developed countries and has kept pace with population increases in the underdeveloped nations—although, unfortunately, it has not led to *increased* food availability there. The latest United Nations reports estimate that 16 percent of the world's population is undernourished. In many cases, however, this undernourishment is *not* a direct result of the lack of food on a nationwide basis. First, there is the problem of distribution: as much as 50 percent of India's food is never eaten because it rots before it reaches the hungry. Again, primitive sanitation leads to a high rate of disease, and many of these diseases impair metabolism and thus block people from getting the full nutritional value of what they do eat. Consequently, investments in transportation networks, running water, sanitation, and education in preventive medicine may be even more important than agricultural advances in lessening the problem of world hunger.

In terms of evaluating our simulations, we have seen that useful actions must be based on hard data on each individual region, not on general predictions made on very rough models. In the case of weather prediction, we have seen the value of building up a massive data base of very accurate information and then using this, as in a process of model validation, to obtain better and better computer simulations. However, the best weather predictions only extend a short way into the future and require accurate data on the region involved, as well as the regions that surround it. In the same way, we can expect the best projections of ecological and economic trends to run at most a few years into the future and to be anchored in solid data on the regions involved.

Having seen the need for better data, let us say something about the need for better models. When...we have to make decisions about the control of a spaceship, we know exactly how to set up a model. Newton's laws of motion provide the tools. Here, there is no controversy. When we turn to world models, however, the tools are still controversial. During the worldwide recession and inflation of the early 1980s, there was no agreed-upon theory to direct the search for economic recovery. President Reagan of the United States put his faith in giving more power to big industry and less to the government. President Mitterand of France took the opposite tack, putting the banks and certain sectors of industry under government control. To date, there is no single framework in which these differences can be resolved. With the aid of computer simulation, this does not have to be the case.

People with different views can use their models to spell out the assumptions that lead to their predictions; and because they spell them out, and because we can determine more of the implications by computer simulation, it should be possible to debate carefully the best features of each model. In this way, political dialogue might contribute to model validation. However, the problem of communicating the findings of such models to the general public may be a formidable one....Since...the behavior of complex systems is counterintuitive, it is all too easy for politicians to resort to slogans that fit into thirty-second television spots, instead of trying to help the public understand.

Even if we agree on which simulation model to use to test how effectively various policies achieve their desired goals, we must still agree on the goals. Here is a very concrete example: In a time of energy crisis, a country's real goal may be to ensure that it has as much energy as possible so that it can maintain its current level of energy usage. This is a very different goal from the "Turning Point" goal of trying to ensure a relatively fair distribution of energy resources throughout the world. Presumably, this will remain a political process—the sort of thing by which political parties will win or lose elections.

Even if we agree on both economic theory and goals, it is still unrealistic to devise fifty-year plans that are to be followed without benefit of revision. The problem with this type of long-range planning

*Michael A. Arbib*

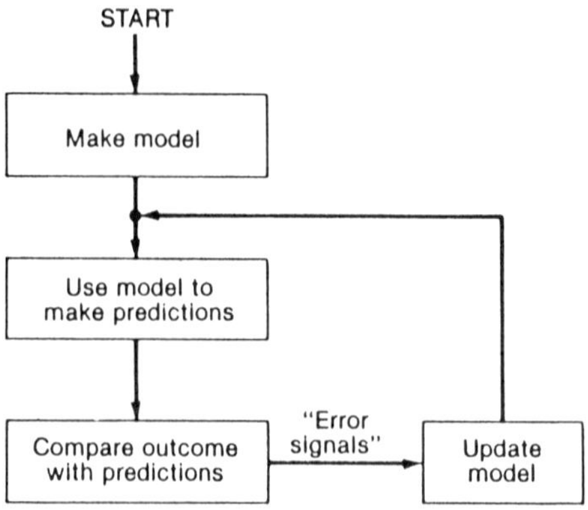


FIGURE 3.4: Model validation as a feedback process.

is twofold. First, it cannot take into account technological and political surprises. Second, models are less and less reliable the longer the range of prediction.

A breakthrough in technology, such as the atom bomb, can greatly change the balance of power or the importance of different types of resources. A political "aberration," such as the rise of Hitler in Germany, will change considerations so drastically that any long-range policy must be sacrificed for awhile so that some form of political stability can be reestablished. Thus, certain scenarios that seemed plausible at the start of a planning period may have to be abandoned simply because they did not foresee these events. While a science fiction writer in 1900 could have "foreseen" the development of an explosive able to wipe out huge cities in a single explosion, or while a writer of historical romances could imagine the impact of a modern Atilla-the-Hun upon Europe and the world, it is quite another thing to include these factors in planning. At the turn of the century, the limits of the available technology ruled out plans for warfare that included explosives of that size. In fact, it was hard to convince the military to take the possibilities of airplanes seriously as late as World War I. If one considers too many "wild" scenarios, one ends up with a prediction that"anything can happen," which is no more help than having no prediction at all. At the other extreme, if one considers scenarios that take no account of possible "surprises," then they cannot help us prepare for a time of rapid change.

We must also continually revise our plans, since the models that we use will always be too crude. How strongly is the rate of production of copper in Zaïre coupled with the growth of the telephone network in Western Europe? Even if a model included an explicit term for such a relationship, it would be very approximate. It would have to be made more accurate numerically and would also have to be expanded to take into account a great many other factors if it were accurately to represent the situation. Thus, while we can expect a model to hold fairly steady for a few years, the odds are that its predictions will become less and less accurate over a period of time. Therefore, we must constantly compare the predictions made a few years ago with the performance of the real world today. We can then use the result of this comparison to go back and analyze the model to find out where it went wrong. In some cases, we will realize that certain relationships that we had ignored are actually very important. Thus, we will add them to the model—so that, hopefully, over the next few years the predictions made by the model will turn out to come closer to the unfolding of reality. *Model validation* is a never-ending process as model builders continually receive feedback (Figure 3.4) from the comparisons of prediction and reality.

## Summary

...We examined two world models that were originally presented as reports to the Club of Rome. The "Limits to Growth" model predicted a worldwide catastrophe in the mid-twenty-first century unless all growth was halted immediately. The "Turning Point" model advocated a leveling off of population but suggested that international aid programs could promote healthy growth in underdeveloped nations. ...[We] closed by stressing the importance of improving modeling and data-gathering techniques before basing drastic action on long-range prediction models.

From Michael A. Arbib, *Computers and the Cybernetic Society*, 2nd ed., 286-300 (Orlando, FL: Academic Press, 1984).

## Questions and Topics for Discussion and Writing

1. What is a world model simulation?
2. What are the differences between the "Limits to Growth" world model and the "Mankind at the Turning Point" world model?
3. Discuss the problems of creating a realistic and useful world model.
4. Sketch a possible world model that you think might be a better predictor than those discussed in the article. Indicate what variables are to be tracked and predicted (such as population, pollution, food supply, etc.), what parameters need to be specified (such as the rates at which one variable affects another), and any special features, such as division of the world into sectors. No details are necessary, but show why you would include each variable specified and what role the parameters would play. (*Note: Even without details, this is a big project.*)

*Michael A. Arbib*

Farm land being stripped for coal.

*Issue*

# Analyses of the World Environmental Problem

World models may show that if the world continues on its present course disaster will befall us. But what is it in the present course that is leading to disaster? The articles in this section are analyses of the problem from a number of different points of view, finding somewhat different culprits, and looking for solutions in different places.

In "Playing Russian Roulette with World Environment," David Suzuki finds the cause of world bio-degradation in our blind adherence to what he calls "sacred truths," such as, that progress equals growth, that more consumption is better, that we are aware of the costs of new technologies. Peter H. Raven, in "We're Killing Our World," is alarmed at the rapid rate of species extinction in the world, reducing the biological diversity of the planet. The main cause, he says, is the rapid rate of deforestation and other destruction of natural resources in Third World countries in their attempt to produce consumer products to pay off their serious debts to the industrialized nations. In "The Problem of Production," E. F. Schumacher applies the accounting concepts of capital and income to what the world produces. A business will soon be bankrupt that consumes its irreplaceable capital instead of only its current income. Likewise the earth's civilization cannot last if it steadily erodes its "capital" in the form of fossil fuels instead of renewable resources such as solar energy.

Finally, Fritjof Capra, in "The Dark Side of Growth," sees the ultimate cause of present world problems as stemming from the mechanistic world view that has been dominant in the West at least since Descartes and Newton, and which leads us to view the world in a fragmented way that makes seeing whole systems very difficult. To Capra, we are blinded by our own concepts.

# Playing Russian Roulette with World Environment

## David Suzuki

David Suzuki, geneticist, columnist, and popular host of Canadian television and radio science programs, deplores the degradation of our environment that is all too visible to any of the older people alive today. He sees part of the reason for it in our blind adherence to "sacred truths" that have no basis in reality.

In the mid-fifties, while I was returning from college in the United States for the summer, I glanced down at the Niagara River as my train passed over. Below, I could see fishermen on the banks flailing away at the water and yanking silver objects from the river as fast as they could cast. The were catching silver bass on a massive annual spawning run.

Those spring runs petered out years ago.

Old fishermen on the east and west coasts of Canada describe an abundance and size of salmon, cod and lobsters when they were starting out that younger fishermen have never seen. It wasn't long ago that we drank from the Great Lakes with confidence in the water's purity, and relished fresh fruits and vegetables without concern about chemical contamination. Only a few decades ago, the quality of our water, soil and air was radically different and there were an abundance and variety of life that now are found only in the most remote parts of the country. The planet has changed almost beyond recognition within the life-time of Canada's elder citizens. Their recollections are not simply old folks' romantic musings on the good old days; they are a living record of the cataclysmic degradation that has taken place around us. In the past, men took canaries into coal mines as biological indicators of the quality of the air; today, our elders are the ones who know that canaries are falling all around us.

Our species boasts the highest ratio of "brain to brawn" of all life forms, and that mental power has gifted us with a conscious strategy for survival. We have invented a notion called the "future," which provided us with options and enabled us to deliberately select a future toward which we aimed. Yet today, with all the amplified brainpower of computers, communications networks, scientists and engineers, we seem unable to use the strategy that got our forebears to where we took over. What has gone wrong?

I believe that we continue to cling to certain "sacred truths" that blind us to many problems and often cause the ones that we do recognize. Let me list some of those sacred truths.

- We equate "progress" with *growth*—growth in the economy, income, consumer goods, material comfort. A need for steady economic growth is repeated like a catechism by every politician, economist and businessperson, and has led us to link *profit* with the goal of society. But human beings today are the most ubiquitous and numerous large mammal in the world, and it is our unrelenting commitment to growth that now has us consuming 40 per cent of the net primary production of energy on the planet. We are only one species among perhaps 30 million. When our population doubles again within 50 years, will we then demand 80 per cent? Nothing on this planet continues indefinitely to grow steadily.

  The current increases in consumption and material wealth are a historical aberration, a blip that will come to a stop within our children's lifetime. The only question is whether we will deliberately bring our demands and consumption under control or allow pestilence, famine or war to do the job.

- We have come to believe in the ability of science to provide us with the knowledge to understand and manage our natural resources. Yet the unique power of science is that its practitioners focus on an isolated part of nature, and thus increase knowledge in fragmented bits and pieces. Modern physicists have learned that it is not possible to put these fragments of knowledge together into a complete picture, because in real life, unlike a jigsaw puzzle, the pieces interact synergistically. Thus, properties emerge from the complex that cannot be anticipated from the known properties of the component parts.

  As a result, there are fundamental reasons why it is not possible to comprehend the behavior of entire ecosystems or even complex components within them. In addition, the degree of our ignorance remains vast. We know very little, for example, about chinook salmon during their four-year stay in the ocean, yet we maintain the illusion that we are managing them scientifically.
- We believe we can manage the effects of new technologies by doing proper cost/benefit analysis to maximize the benefits while minimizing costs. History informs us that every technology, however beneficial, has costs. And invariably the benefits of new technologies are immediate and obvious. That's why we (including me) love technology; it does such wonderful things for us. But the costs of these technologies are almost always hidden, and cannot be anticipated.

  If all technologies exact a price that cannot be foreseen, can we continue to mortgage the future by opting for the immediate benefits of new inventions while postponing the solution to their accompanying costs? Today scientists speak of building machines that can think, tampering with the heredity of babies, manipulating the human mind and releasing genetically engineered organisms into the environment—yet we cannot predict their long-term consequences.
- We believe we can minimize environmental damage from our activity by carrying out environmental assessments. Thus, for example, oil exploration in the high Arctic or off Georges Bank or building a fixed link to Prince Edward Island, depend on the success of an environmental assessment review process (EARP). Of course, there should be such assessments, but in view of how little we know of the constituents of ecosystems and their interactions, fluctuations over time and the tiny window our tests provide, the EARP is far too limited in scope, duration and scale to provide information that is statistically meaningful.

  We cannot assume that once an EARP indicates no hazard, development should have an unconditional green light. Approval should always be provisional, with a continued accumulation of information on which reassessments of that approval are constantly made.
- We believe that in a democracy, we elect people to political office to represent and lead us into the future. Yet most politicians come from two professions, business and law, the two areas whose practitioners have the poorest comprehension of issues scientific and technological.

Today, the most important factor shaping our lives and society is science when applied by industry, medicine and the military. Yet, jurisdictional concerns such as Meech Lake and economic priorities of free trade preoccupy our leaders and subsume the priorities of the planet's ecosystem.

"Conventional wisdom" assumes the truth of the above assumptions. Unless we expose them to the light of critical examination, we will continue to ravage the ecosystem for short-term benefits. We desperately need a new paradigm, a new vision of the human place in nature.

We live in a time when satellites have sent back images of this planet that graphically demonstrate its oneness. I think ecologist Jack Valentyne, who poses as Johnny Biosphere to instruct children about the environment, is on the right track. He reminds us that we all have built into our cells and tissues, atoms and molecules that were once in the bodies of all other people on the planet and of people who lived one or two thousand years ago. Not only that, but we are made of atoms respired by trees and insects and mammals and birds. That's because all life forms share the atmosphere around the world.

Johnny Biosphere tells of an Indian who, on a hot

day hundreds of years ago, swam in Lake Superior. Sodium ions from the sweat of his body are still contained in each drink of water that we take from Lake Ontario. And when one realizes that everything we eat for nutrition was itself once living, we realize that we remain inextricably linked to the rest of life on this planet. Seen in this perspective of sharing and connectedness, we have to behave in a radically different way when we dispose of our wastes or apply new technologies that affect other parts of the ecosystem.

Throughout human history, the boast of our species has been that we love our children and hope that they will have a richer, fuller life than we did. Yet now, for the first time, we know with absolute certainty that our children's lives will be immeasurably poorer in bio-diversity and filled with massive problems that we have foisted on them in our short-sighted pursuit of immediate profit and power.

Can we continue to mortgage our children's future so thoughtlessly? Not if we mean it when we say we love them.

---

David Suzuki, "Playing Russian Roulette with World Environment" *The Globe and Main* (April 23, 1988) D1-D2.

---

## Questions and Topics for Discussion and Writing

1. What are the "sacred truths" that Suzuki says we cling to?
2. What is Suzuki asking us to do?

# We're Killing Our World

## Peter H. Raven

Peter H. Raven, Director of the Missouri Botanical Garden, sums up the world ecological problems with some telling pertinent facts and figures in this address to the American Association for the Advancement of Science in February 1987.

It is with the greatest pleasure that I present this keynote address here to inaugurate the annual meetings of the American Association for the Advancement of Science. By the time I have completed these remarks, I hope you will agree with me that the world that provides our evolutionary and ecological context is in serious trouble, trouble of a kind that demands our urgent attention. By formulating adequate plans for dealing with these largescale problems, we will be laying the foundation for peace and prosperity in the future; by ignoring them, drifting passively while attending to what seem more urgent, personal priorities, we are courting disaster.

Even though we live in a world where far more people are well fed, clothed, and housed today than ever before, we also live in a world in which up to 100,000 people starve to death every day, one in which we consume well over a third of total terrestrial photosynthetic productivity, and one in which our activities are threatening up to a quarter of the other kinds of organisms on earth with extinction in the near future. Since we base our civilization almost completely on our ability to utilize these organisms for out benefit, the loss of so many of them threatens to limit permanently the options that will be available for our children and grandchildren.

Despite these relationships, some of us are apparently so desperate to avoid the need for governmental action of any kind that they attempt to lull us to sleep by pretending that there is no problem, that inaction is best, and that we should simply continue to indulge our selfishness to the fullest extent. To provide such counsel is to offer exceedingly dangerous advice at precisely the wrong time. We scientists, who have the means to know better, must help to inform the public about the facts, and the need for action concerning them.

The global human population, which passed the 5 billion mark for the first time last year, and is growing at an annual rate estimated at 1.7 percent, is a dominant ecological force without precedent. Our numbers have *doubled* since 1950, and will double again in about 40 years if present trends continue. It has recently been calculated that we consume, co-opt, or forego about 40 percent of the total terrestrial photosynthetic productivity, a sure indication of the profound way in which we are affecting the global ecosystem. At the same time, regional climatic problems are becoming increasingly apparent, another sign of wide-scale danger. How will we respond to these threats, and why should those of us who live in the relative comfort of countries such as the United States even care?

Many of our most serious problems are centered in the tropics, where biological diversity is concentrated and is being lost most rapidly and whole ecosystems are being disrupted. In the tropics, three factors are of special importance: (1) the explosive growth of record human populations; (2) widespread and extreme poverty; and (3) an ignorance of the ways in which to carry out productive agriculture and forestry. Let's look at these three factors in turn.

As recently as 1950, about 45 percent of a global population of some 2.5 billion people lived in countries that lie wholly or partly in the tropics; today, the figure is about 55 percent of a global population that is more than twice as large. If present trends continue, nearly two-thirds of the people in the world will be living in these countries (excluding China) by the year 2020. In actual numbers, the 1.1 billion people who inhabited them 36 years ago will have grown to about 5 billion people in another 34 years: a *quadrupling* of the total in 70 years! Meanwhile, the proportion of people living in industrial countries is falling drastically. For each of us living in countries like the United States in 1950, there were approximately 2 other people living elsewhere; by 2020, there will be 5. Population structure suggests that the global population may stabilize in about a century at a level of perhaps 10 or 11 billion people. As its growth slows down, there will be increased opportunities for raising standards of living—a real reason for preserving as many kinds of organisms as possible and taking some of the steps I shall advocate this evening.

The rapidly growing populations of tropical countries include large numbers of poor people. In 1986, for example, the per capita GNP in the United States was estimated at $14,080, whereas that in neighboring Mexico amounted to $2180; in Honduras, it was $670. Overall, the industrial nations with less than a quarter of the global population, control about 80 percent of the wealth, while the largely tropical developing countries, with 54 percent of the population, control about 15 percent of the wealth. Meanwhile, in both Africa and Latin America, per capita income is steadily declining.

Within tropical countries, about 1 billion of the estimated 2.7 billion people live in absolute poverty; they are unable to count on adequate food, shelter, and clothing from one day to the next. Of these, 300 to 400 million people consumed less than 80 percent of UN recommended standards, a diet insufficient to prevent "stunted growth and serious health risks." UNICEF estimates that about 35 million people starve to death every year in the tropics, including more than 14 million children under the age of four. This amounts to nearly 100,000 people, and 40,000 babies, every day. Worse, many millions of additional children exist only in a state of lethargy, their mental capacities often permanently impaired by their lack of access to adequate amounts of food.

In Latin America alone, the 30 million landless rural households, together with those that own less than an acre of land, now represent close to 40 percent of all rural households. The difficulties in improving this situation can be illustrated by Brazil, where President Jose Sarney is attempting to distribute 100 million acres of land to 1.4 million essentially landless families by 1989. These efforts are encountering increasing rural violence; they illustrate how the growing extent of rural landlessness everywhere in the tropics is a serious cause of poverty, malnutrition, and social unrest in many regions.

In addition to large populations and extensive poverty, tropical countries suffer from a lack of information about how to put in place productive agriculture and forestry, and a lack of willingness to apply those facts that are known. As a result, the natural vegetation is often consumed as if it were a non-renewable resource. Many tropical soils are relatively poor, and require careful handling. In the natural ecosystems that develop on such soils, most of the nutrients, except for phosphorus and nitrogen, are held primarily in the vegetation. Cutting and burning the trees releases these nutrients to the soil, fertilizing them and allowing the temporary cultivation of crops; when the excess nutrients are exhausted, usually within a few years, the forests must be given time to recover.

Some traditional systems of cultivation on relatively infertile soils combine trees, which are usually more productive than herbaceous crops, with other plants. *Agro-forestry* systems of this sort will be increasingly important as they are better understood, and as farmers are encouraged to use them by offering them loans for fertilizers, credit for seeds, information about markets, and the like. In addition, relatively fertile soils must be cultivated intensively by the best methods available, with the most suitable crops.

Based on 1981 FAO estimates, about 2.3 million square miles of tropical evergreen forest probably exists in 1987—about half of the original area, now a forest that is roughly three-quarters of the size of the U.S. exclusive of Alaska. In the late 1970s, at least 40,000 square miles of such forest were being cut per year. If that is still the rate of clear-cutting, and if it continues, the forests would last about 60 years—if there were *no* population growth or other increased pressures on the forests from the late 1970s levels.

The decline in tropical forests arises in part from consumer demand in industrialized countries. We obtain much of our timber there, for example, and logging removes about 20,000 square miles of tropical forest—an area nearly the size of West Virginia—every year. Meanwhile, reforestation is proceeding very slowly in the tropics, with 10 trees being cut for each one that is planted, and, in Africa, 29 being cut for each one planted. The developed-world consumption of tropical hardwoods has risen 15 times since 1950, while in-country consumption has increased only three times. Against this background, the inauguration of the International Tropical Timber Agreement in 1986—an agreement that involves all the major producers and consumers of tropical woods—is a development of high importance. In addition, the organization of the Tropical Forest Action Plan by the World Resource Institute (WRI) is of great importance, both for a stable supply of the commodities involved, and for biological conservation as well. The widespread and simultaneous destruction of temperate forests through acid precipitation and other forms of atmospheric pollution compounds the overall problem that we face in securing wood and pulp supplies for the future.

The clearing of tropical forests for pastures affects an area about half the size of that which is logged. For example, the growing imports of beef from southern Mexico and Central America over the past 25 years have been the major factor in the loss of about half of the forests in these regions, for the sake of keeping the price of a hamburger in the U.S. about a nickel less than it would have been otherwise.

In addition, shifting cultivation and fuelwood gathering together are severely damaging an additional area of tropical, evergreen forest about equal to that which is being clear-cut. Only about 200 million people are engaged in these activities worldwide in areas of evergreen, lowland tropical forest. If they were incorporated into the economies of their respective countries, the forests they are cutting would change immediately from a non-renewable to a sustainable resource.

The relentless search for fuelwood is wrecking many tropical forests; about 1.5 billion people—a third of the world population—are cutting firewood faster than it can regrow in the areas where they get it. If plantations for firewood were established on cut over lands, with sufficient economic backing, this trend could be reversed.

Taking both clear-cutting and the kinds of severe disturbance that results from shifting agriculture and firewood gathering together, the current rate of destruction of tropical, evergreen forests may be calculated at roughly 80,000 square miles per year—an area about equal to the size of the state of Kansas. At such a rate, all tropical, evergreen forests would be gone in less than 30 years, even if there were no population growth; this represents roughly the rate of forest destruction that was taking place a decade ago.

Since human populations are in fact growing very rapidly—at the rate of approximately 2.4 percent per year—much of the forest will clearly be gone much sooner, however. Moreover, the actual rates of destruction differ widely from region to region. Three large blocks of lowland, evergreen forest—those in the northern and western Brazilian Amazon, the interior of the Guyanas, and the Zaire Basin of Africa—are larger and less densely populated than the rest, and consequently are being cut more slowly. Their forest might, therefore, last past the middle of the 21st century. The remaining forests, however, will mostly be gone, or at least profoundly altered in nature and composition, much earlier. These include the tropical, lowland forests of Mexico, Central America, and the West Indies, and those of Andean South America, as well as the forests of the southern and eastern Amazon; all similar forests of Africa and Madagascar outside of the Zaire Basin; and all the forests of tropical and subtropical Asia and the Pacific Islands. For the most part, they will have been consumed or severely damaged within the next 15 years or so.

We have been emphasizing lowland, evergreen forests because of their biological richness, and because they are the most extensive tropical forests that have persisted to the present. Dry deciduous forests, and all the other kinds of tropical forests, have mostly been cut over and converted much earlier than tropical, evergreen forests, being better suited to agriculture and cattle grazing. For example, the dry deciduous forests that once occurred along the Pacific Coast of Mexico and Central America covered an area approximately twice the size of Texas 500 years ago. Today, only about 2 percent of this once extensive biome exists in anything resembling its original form. The Guanacaste National Park project in Costa Rica, to which I shall return later, represents an exemplary effort to restore and preserve a significant area of such forest.

The destruction of natural resources in the tropics

and sub-tropics is intimately related to the global economy. The global economy is driving the export of cash crops from many regions of the tropics rapidly upward, often resulting in the displacement of poor farmers who are forced to farm less suitable lands and destroy the forests there. This trend has been evident in the Sahel, for example, for which massive exports of peanuts and cotton are sent to Europe, and in Thailand, which exports huge quantities of cassava.

The international debt is greatly affecting relationships between developing and industrial countries in the 1980s. In 1970, the external debt of Third World countries was about $72 billion; today, it is approximately $1 trillion. The existence of this debt clearly encourages many Third World countries to overexploit their natural resources without the creation of stable, productive alternatives: logging restrictions are eased, poor farmers are displaced to regions that will not support them on the long term, the production of foods that the people can eat (remember that on the average, a fifth of them are malnourished) is decreased in favor of the production of export crops. The associated austerity measures can throw large numbers of people out of work, thus increasing the extent of poverty in the nations involved.

Recently, a New York investment banker was quoted in *The New York Times* as saying: "Somehow the conventional wisdom of 200 million sullen South Americans sweating away in the hot sun for the next decade to earn the interest on their debt so Citicorp can raise its dividend twice a year does not square with my image of political reality." Still more succinctly, Willy Brandt termed the effects of the debt, "A blood transfusion from the sick to the healthy." The existence of the debt is making it increasingly difficult for Third World countries to accept our exports; they must instead export competitively, and to us, to attempt to repay the debt itself. In this light, strenuous efforts to repay the debt, somewhat paradoxically, appears to be weakening, instead of strengthening, our own economy.

The runaway destruction of the natural resources of the tropics is also fueling massive immigration into the industrialized nations of the temperate zone. For example, the U.S. Immigration and Naturalization Service apprehended 1.7 million illegal aliens at the Mexican border alone in 1986, and estimates that more than 3 million might have entered successfully. The hunger and poverty that are responsible for such immigration also underlie economic and political instability in the tropics, and frustrate our efforts to achieve our objectives in these areas.

Another major problem concerns the effects of tropical deforestation on global climates. Such deforestation clearly contributes substantially to the amounts of carbon dioxide in the atmosphere, an issue of increasing concern. In addition, widespread deforestation is impairing the capacities of some tropical systems to recycle rainfall inland, as well demonstrated by the Amazon and possibly related to the past 20 years of drought in Africa also. Combined with the erosion and soil deterioration that accompanies deforestation, these problems are serious ones.

Recent findings suggest that the implications of widespread deforestation may be even more extreme. Last December, at the annual meeting of the American Geophysical Union, it was reported that cutting the evergreen forests in South America would precipitate a regional temperature rise of 3 to 5°C. This would extend the dry season and speed the deterioration of the remaining forests, including reserves, while greatly disrupting agriculture. The effects of such major changes beyond the regions directly impacted can only be guessed at present.

The effects of forest destruction and climatic change have been most severe in sub-Saharan Africa. Here, per capita food production has dropped 20 percent since 1960, and the FAO projects that it will drop another 30 percent over the next 25 years, with population growth greatly exceeding growth in food production. Right now, a majority of people living in the area have too little to eat, and their collective international debt, roughly $200 billion, is equal to 44 percent of gross domestic product or 190 percent of export earnings.

Africa as a whole suffers from overvalued currencies and from historical political situations, such as artificially devalued food prices, that tend to favor urban dwellers over farmers. When Kenya set a higher, more realistic level for the government-controlled price of corn in 1985, for example, production climbed by almost 50 percent the next year. Another negative factor is that most food aid in Africa is made available in the towns, which tends to pull the people out of the countryside, thus causing them to neglect their crops. Development assistance for Africa is falling rapidly, fewer than 5 percent of the couples use modern contraceptive methods, and there are about 1 million more mouths to feed in sub-Saharan Africa every three weeks, a sure prescription for increasing human tragedy.

The most serious, long-term global problem that is resulting from deforestation, however, is the loss of a large portion of the biological diversity on earth within a few decades. Whether this process of extinction is viewed from a scientific, aesthetic, or moral standpoint, or simply as the loss of opportunities that could otherwise have been used for human benefit, it is unquestionably the one problem that will have the most lasting consequences. The extinction rates of the present are a thousand times those of the past tens of millions of years. Despite this, it is all too easy to say, "Goodbye, California condor—but so what? " There has been an unfortunate tendency for some of the media to jump on this bandwagon: it is more stimulating to give the minority (and incorrect) view that extinction presents no problem, or that everything is fine, than it is to document the extent of the problems facing us—besides, it is comforting, and more fun!

In fact, the loss of biological diversity is important to us for many reasons. For example, three species of plants—rice, wheat, and corn—supply over half of all human energy requirements; only about 150 kinds of food plants are used extensively; and only about 5000 have ever been used. Many of these come to us from the tropics. It is estimated, however, that there may be tens of thousands of additional kinds of plants that could provide human food if their properties were fully explored and they were brought into cultivation. There are many uses for plants other than food, too; for example, oral contraceptives for many years were produced from Mexican yams; muscle relaxants used in surgery worldwide come from an Amazonian vine, used traditionally to poison darts; the cure for Hodgkin's disease comes from the rosy periwinkle, a native of Madagascar; and the gene pool of corn has recently been enriched by the finding, on a small area in the mountains of Mexico, of a perennial wild relative.

Among the undiscovered or poorly known plants are doubtless many possible sources of medicines, oils, waxes, fibres, and other commodities of interest to our modern industrial society. Genetic engineering affords us additional possibilities for the transfer of genes from one kind of organism to another, even though the donor may itself be of no economic interest whatever; indeed, as our techniques become more sophisticated, we shall come to depend even more heavily on biological diversity than we do now.

How fast is extinction proceeding? Although only about 500,000 species of tropical organisms have been named, there are at least 3 million, and perhaps ten times that many, yet to be discovered. Investigating the beetle fauna of the canopy of tropical forests in Panama and Peru, for example, Terry Erwin, of the Smithsonian Institution, has fogged the canopies of the trees with insecticides to obtain comprehensive samples for comparative purposes. Based on these samples, he estimates that the true number of species of insects may approach 30 million, or even more.

If we assume that there are 3 million species of tropical organisms, we can use the distributions of the better-known groups as a guide to how many of the total species occur only in those forests that will probably be destroyed or severely damaged in the next 15 years or so. For plants, the figures would be about 120,000 of the estimated total of about 165,000 tropical species; just under half of the total number of species in the world occur only in forests that are rapidly being destroyed. If the distribution of other groups of organisms is roughly similar to that of plants, then about 2.2 million species, or somewhat less that half of the total, would be restricted to these forests. How many of these species can we reasonably expect to survive?

The formation of parks and reserves in the threatened areas would be an important element in the survival of biological diversity; but such protected areas will survive only if they exist in the context of economies that can support them—particularly ones in which the rural poor are supported. The explicit relationship between conservation and development is outlined well in the *World Conservation Strategy*, issued jointly by the International Union for the Conservation of Nature and Natural Resources, the World Wildlife Fund, and the United Nations Environmental Programme in 1980. The future of such reserves will depend in large measure on the functioning of the global economy.

Granted, however, that fragments of natural vegetation will persist in most areas, what level of survival can we expect? The principal basis for making predictions of this sort lies in the theory of island biogeography. This theory rests on the empirical observation that, in general, for islands and mainlands areas alike, the relationship between species number and area is a logarithmic one, such that a tenfold increase in area is associated with a doubling in the number of species of a given group. The reciprocal relationship, and one that is being verified empirically in a number of experiments throughout the world, is that a reduction of area to a tenth or less of its original extent should place half or more of the kinds of organisms that occurred in the larger area at risk.

The biological basis for this relationship is clear. As areas are reduced in size, the population sizes of the organisms in them are also reduced. When a population is small, the chance of survival of the species is reduced on the basis of chance.

An illustration of the way this process works is provided by the history of bird populations on Barro Colorado Island, an island of about 6 square miles that was formed between 1911 and 1914 by the flooding of the Panama Canal. Set aside as a reserve in 1923, Barro Colorado Island was home at that time to 208 species of breeding birds. It has been protected as a reserve continuously. Over the next 60 years, at least 45 of these birds became extinct, despite the fact that Barro Colorado Island is separated from the source areas on the mainland by a distance of approximately 600 feet, a distance that any of the species of birds involved could cross easily.

Hunting and poaching become more intense when surviving patches of vegetation are small. In addition, inbreeding and the consequent loss of genetic variability puts species increasingly at risk. The borders of the patches are affected by sun, wind, and other factors in ways that damage their integrity. Regional climatic changes, such as those affecting precipitation, can change the climate in the patches, leaving the species vulnerable and with nowhere to migrate.

There are many examples of regional extinction on a massive scale. For example, there were at least 88 species of land birds in the Hawaiian Islands when the Polynesians arrived, 43 when Captain Cook explored the islands, and 28 today—two-thirds of which are endangered. The destruction of the lowland forests and the introduction of exotic species have both contributed to the decimation of the native birds and the plants. The process continues rapidly at the present day, because even as the richest nation on earth, we seem unable to protect our natural heritage adequately. The Hawaiian Islands provide a prime example of the way in which ecosystems are simplified, while pests and weedy species become more widespread throughout the world.

One particularly significant example of an area in which large numbers of species are endangered is Madagascar, an island about twice the size of Arizona that lies about 250 miles off the east coast of Africa. Here grow about 8500 species of plants, some 6500 of which are found nowhere else. Madagascar is also the only place where lemurs, one of the major groups of Primates, survive, and there are many other unique and restricted organisms there. All in all, the island is probably home to about 5 percent of the total globe diversity, probably 80 percent of it found nowhere else.

Unfortunately, less than a tenth of the land surface of Madagascar is still covered with natural vegetation. The loss of forests in northeastern Madagascar, the main forested area, has been such that in 1949, only about 66 percent of the original forest was left, and by 1984, only 36 percent, as determined by LANDSAT satellite imagery. Where it once was richly forested in many areas, much of Madagascar is now degraded pasture.

Since the forests where at least half 2.2 million species occur will very probably be reduced to less than a tenth of their total extent over the next 15 years or so, we may assume that over a million species will become extinct during this period or soon afterwards. In western Ecuador, for example, a region that was almost completely forested as recently as 1950 is now almost completely deforested. As extensive deforestation of this kind spreads in many other regions, we can expect the rate of extinction to average more than 100 species a day, with the rate increasing from perhaps a few species a day now to several hundred by the early years of the next century. The great majority of these species will not have been collected, and therefore will never be represented in any scientific collection, preserved, or known in any way. No comparable rate of extinction has occurred since the end of the Cretaceous Period, 65 million years ago; and the background level of extinction is perhaps a thousandth of that we are experiencing now.

From any point of view, the situation that has just been described is extremely serious. Scientifically, we are losing the opportunity to understand the nature of much of the diversity of life on earth. Aesthetically, we are losing the chance to appreciate fully the results of the process of evolution over the billions of years since life appeared on our planet. Economically, we are denying to ourselves, our children, and our grandchildren the opportunity to utilize many of the plants, animals, and microorganisms that exist now for their benefit.

A new kind of global thinking will be necessary to manage the world ecosystem properly for the enormous human populations of the future. To demonstrate this fact, simply recall that we are now using directly, foregoing, or converting about 40 percent of global terrestrial productivity, and that these numbers are projected to double in the next 41 years. Stability everywhere is in the interests of all, yet we are doing relatively little to promote it. Sustainable agriculture and forestry systems must be developed in every country, and their development will benefit everyone. When nations are stable and self-sufficient, they can import and export products, repay debts, provide a decent standard of living for their people, improve their governments, and preserve their biological diversity. When they are unstable and dependent, they cannot afford imports and can organize exports only with difficulty, they will default on debts, cannot provide for their people, who then may emigrate in large numbers, will tend to have unsuitable and unstable governments, and will squander their biological diversity for short-term gain.

Peter H. Raven. "We're Killing Our World," delivered as the Keynote Address, American Association for the Advancement of Science, Chicago, Illinois, February 14, 1987.

## Questions and Topics for Discussion and Writing

1. Discuss the problem of population growth. Does it have a solution?
2. What are the problems associated with deforestation? What are the causes of deforestation?
3. Why should we be concerned about the rate of extinction of species?
4. What kind of action(s) are required to solve the problems discussed by Raven? Compare the actions you propose to those proposed by others (from other articles in this book, or from other sources).

# The Problem of Production

## E. F. Schumacher

In this excerpt from his book *Small is Beautiful,* economist E.F. Schumacher asks us to look at the whole world through the eyes of corporate economics. Viewed as such, he says, the firm, Spaceship Earth, is not living off its *income* but instead is consuming its *capital* at an alarming rate. In other words, we are not consuming only that which is producced on a regular basis and can be produced again as needed, such as crops and energy from the sun, wind, and water, but instead we are using up resources that cannot be replaced when needed, such as fossil fuels, which require thousands of years to form naturally. It is no less alarming to note that the book from which this excerpt is taken was published in 1973, and the problems it warned us about then have only gotten worse since.

One of the most fateful errors of our age is the belief that "the problem of production" has been solved. Not only is this belief firmly held by people remote from production and therefore professionally unacquainted with the facts—it is held by virtually all the experts, the captains of industry, the economic managers in the governments of the world, the academic and not-so-academic economists, not to mention the economic journalists. They may disagree on many things but they all agree that the problem of production has been solved; that mankind has at last come of age. For the rich countries, they say, the most important task now is "education for leisure" and, for the poor countries, the "transfer of technology."

That things are not going as well as they ought to be going must be due to human wickedness. We must therefore construct a political system so perfect that human wickedness disappears and everybody behaves well, no matter how much wickedness there may be in him or her. In fact, it is widely held that everybody is born good; if one turns into a criminal or an exploiter, this is the fault of "the system." No doubt "the system" is in many ways bad and must be changed. One of the main reasons why it is bad and why it can still survive in spite of its badness, is this erroneous view that the "problem of production" has been solved. As this error pervades all present-day systems there is at present not much to choose between them.

The arising of this error, so egregious and so firmly rooted, is closely connected with the philosophical, not to say religious, changes during the last three or four centuries in man's attitude to nature. I should perhaps say: *Western* man's attitude to nature, but since the whole world is now in a process of westernisation, the more generalised statement appears to be justified. Modern man does not experience himself as a part of nature but as an outside force destined to dominate and conquer it. He even talks of a battle with nature, forgetting that, if he won the battle, he would find himself on the losing side. Until quite recently, the battle seemed to go well enough to give him the illusion of unlimited powers, but not so well as to bring the possibility of total victory into view. This has now come into view, and many people, albeit only a minority, are beginning to realise what this means for the continued existence of humanity.

The illusion of unlimited powers, nourished by astonishing scientific and technological achievements, has produced the concurrent illusion of having solved the problem of production. The latter illusion is based on the failure to distinguish between income and capital where this distinction matters most. Every economist and businessman is familiar with the distinction, and applies it conscientiously and with considerable subtlety to all economic affairs—except where it really matters: namely, the irreplaceable capital which man has not made, but simply found, and without which he can do nothing.

A businessman would not consider a firm to have solved its problems of production and to have achieved viability if he saw that it was rapidly consuming its capital. How, then, could we overlook this vital fact when it comes to that very big firm, the economy of Spaceship Earth and, in particular, the economies of its rich passengers?

One reason for overlooking this vital fact is that we are estranged from reality and inclined to treat as valueless everything that we have not made ourselves. Even the great Dr. Marx fell into this devastating error when he formulated the so-called "labour theory of value." Now we have indeed laboured to make some of the capital which today helps us to produce—a large fund of scientific, technological, and other knowledge; an elaborate physical infrastructure; innumerable types of sophisticated capital equipment, etc.—but all this is but a small part of the total capital we are using. Far larger is the capital provided by nature and not by man—and we do not even recognize it as such. This larger part is now being used up at an alarming rate, and that is why it is an absurd and suicidal error to believe, and act on the belief, that the problem of production has been solved.

Let us take a closer look at this "natural capital." First of all, and most obviously, there are the fossil fuels. No one, I am sure, will deny that we are treating them as income items although they are undeniably capital items. If we treated them as capital items, we should be concerned with conservation; we should do everything in our power to try and minimize their current rate of use; we might be saying, for instance, that the money obtained from the realisation of these assets—these irreplaceable assets—must be placed into a special fund to be devoted exclusively to the evolution of production methods and patterns of living which do *not* depend on fossil fuels at all or depend on them only to a very slight extent. These and many other things we should be doing if we treated fossil fuels as capital and not as income. And we do not do any of them, but the exact contrary of every one of them: we are not in the least concerned with conservation; we are maximising, instead of minimising, the current rates of use; and, far from being interested in studying the possibilities of alternative methods of production and patterns of living—so as to get off the collision course on which we are moving with ever-increasing speed—we happily talk of unlimited progress along the beaten track, of "education for leisure" in the rich countries, and of "the transfer of technology" to the poor countries.

The liquidation of these capital assets is proceeding so rapidly that even in the allegedly richest country in the world, the United States of America, there are many worried men, right up to the White House, calling for the massive conversion of coal into oil and gas, demanding ever more gigantic efforts to search for and exploit the remaining treasures of the earth. Look at the figures that are being put forward under the heading "World Fuel Requirements in the Year 2000." If we are now using something like 7000 million tons of coal equivalent, the need in twenty-eight years' time will be three times as large—around 20,000 million tons! What are twenty-eight years? Looking backwards, they take us roughly to the end of World War II, and, of course, since then fuel consumption has trebled; but the trebling involved an increase of less than 5000 million tons of coal equivalent. Now we are calmly talking about an increase three times as large.

People ask: Can it be done? And the answer comes back: It must be done and therefore it shall be done. One might say (with apologies to John Kenneth Galbraith) that it is a case of the bland leading the blind. But why cast aspersions? The question itself is wrong-headed, because it carries the implicit assumption that we are dealing with income and not with capital. What is so special about the year 2000? What about the year 2028, when little children running about today will be planning for their retirement? Another trebling by then? All these questions and answers are seen to be absurd the moment we realise that we are dealing with capital and not with income: fossil fuels are not made by men; they cannot be recycled. Once they are gone they are gone for ever.

But what—it will be asked—about the income fuels? Yes, indeed, what about them? Currently, they contribute (reckoned in calories) less than four per cent to the world total. In the foreseeable future they will have to contribute seventy, eighty, ninety per cent. To do something on a small scale is one thing: to do it on a gigantic scale is quite another, and to make an impact on the world fuel problem, contributions have to be truly gigantic. Who will say that the problem of production has been solved when it comes to income fuels required on a truly gigantic scale?

Fossil fuels are merely a part of the "natural capital" which we steadfastly insist on treating as expendable, as if it were income, and by no means the most important part. If we squander our fossil fuels, we threaten civilisation; but if we squander the capital represented by living nature around us, we threaten life itself. People are waking up to this threat, and they demand that pollution must stop. They think of pollution as a rather nasty habit indulged in by careless or greedy people who, as it were, throw their rubbish over the fence into the neighbour's garden. A more civilised behaviour, they realise, would incur some extra cost, and therefore we need a faster rate of economic growth to be able to pay for it. From now on, they say, we should use at least some of the fruits of our ever-increasing productivity to improve "the quality of life" and not merely to increase the quantity of consumption. All this is fair enough, but it touches only the outer fringe of the problem.

To get to the crux of the matter, we do well to ask why it is that all these terms—pollution, environment, ecology, etc.—have *so suddenly* come into prominence. After all, we have had an industrial system for quite some time, yet only five or ten years ago these words were virtually unknown. Is this a sudden fad, a silly fashion, or perhaps a sudden failure of nerve?

The explanation is not difficult to find. As with fossil fuels, we have indeed been living on the capital of living nature for some time, but at a fairly modest rate. It is only since the end of World War II that we have succeeded in increasing this rate to alarming proportions. In comparison with what is going on now and what has been going on, progressively, during the last quarter of a century, all the industrial activities of mankind up to, and including, World War II are as nothing. The next four or five years are likely to see more industrial production, taking the world as a whole, than all of mankind accomplished up to 1945. In other words, quite recently—so recently that most of us have hardly yet become conscious of it—there has been a unique quantitative jump in industrial production.

Partly as a cause and also as an effect, there has also been a unique qualitative jump. Our scientists and technologists have learned to compound substances unknown to nature. Against many of them, nature is virtually defenceless. There are no natural agents to attack and break them down. It is as if aborigines were suddenly attacked with machine-gun fire: their bows and arrows are of no avail. These substances, unknown to nature, owe their almost magical effectiveness precisely to nature's defencelessness—and that accounts also for their dangerous ecological impact. It is only in the last twenty years or so that they have made their appearance in *bulk*. Because they have no natural enemies, they tend to accumulate, and the long-term consequences of this accumulation are in many cases known to be extremely dangerous, and in other cases totally unpredictable.

In other words, the changes of the last twenty-five years, both in the quantity and in the quality of man's industrial processes, have produced an entirely new situation—a situation resulting not from our failure but from what we thought were our greatest successes. And this has come so suddenly that we hardly noticed the fact that we were very rapidly using up a certain kind of irreplaceable capital asset, namely the *tolerance margins* which benign nature always provides.

Now let me return to the question of "income fuels" with which I had previously dealt in a somewhat cavalier manner. No one is suggesting that the world-wide industrial system which is being envisaged to operate in the year 2000, a generation ahead, would be sustained primarily by water or wind power. No, we are told that we are moving rapidly into the nuclear age. Of course, this has been the story for quite some time, for over twenty years, and yet, the contribution of nuclear energy to man's total fuel and energy requirements is still minute. In 1970, it amounted to 2.7 per cent in Britain; 0.6 per cent in the European Community; and 0.3 per cent in the United States, to mention only the countries that have gone the furthest. Perhaps we can assume that nature's tolerance margins will be able to cope with such small impositions, although there are many people even today who are deeply worried, and Dr. Edward D. David, President Nixon's Science Adviser, talking about the storage of radioactive wastes, says that "one has a queasy feeling about something that has to stay underground and be pretty well sealed off for 25,000 years before it is harmless."

However that may be, the point I am making is a very simple one: the proposition to replace thousands of millions of tons of fossil fuels, every year, by nuclear energy means to "solve" the fuel problem by creating an environmental and ecological problem of such a monstrous magnitude that Dr. David will not be the only one to have "a queasy feeling." It means solving one problem by shifting it to another sphere—there to create an indefinitely bigger problem.

From E. F. Schumacher, *Small is Beautiful* (New York: Perennial Books, 1973).

## Questions and Topics for Discussion and Writing

1. What is the "problem of production" and what does it mean to "solve" it?
2. What is the distinction between "capital fuels" and "income fuels"?

# * The Dark Side of Growth


## Fritjof Capra


Environmentalists have many different concerns, varying with their special interests and areas of expertise, and with the problems each of them consider to be paramount. Yet all environmentalists share one common viewpoint—they view the environment as a holistic *system* which can be affected by an enormous variety of factors, some large, some small; some near at hand, some remote; some with immediate consequences, others which manifest their effects much later. In their efforts to explain environmental change, environmentalists search for events that shift the entire system from a previous position of stability to another state, which may be either stable or unstable.

The philosophically minded physicist, Fritjof Capra, endeavors to see the world and all of society holistically as a great system. He has written thought-provoking analyses of science and culture and its problems, notably in two books, *The Tao of Physics* and *The Turning Point: Science, Society, and the Rising Culture*. This selection is from *The Turning Point*, Capra's analysis of the implications of the scientific world view of the western world and its realized consequences and possible future course.

In the chapter preceding this selection, Capra develops a far-reaching analysis of the mechanistic point of view of Cartesian-Newtonian science, its triumphs and failures. Here, he concentrates on the inability of that mechanistic, building block viewpoint to anticipate the systemic, holistic problems presented by the world environment. He shows how the self-interests of powerful groups such as corporations, and even countries, can be detrimental to the interests of the world as a whole and for the individuals in it. This is an extremely pessimistic and discouraging selection because the problems he discusses seem so intractable. Yet this is far from the end for Capra. As the title suggests, *The Turning Point* is an optimistic book that envisages a new culture, with a wider, more holistic, more environmentally minded viewpoint, rising from the ashes of the bankrupt mechanistic view. This he sees rising into the ascendancy over the older culture which is unable to cope with the world's problems—that is, if the older culture does not first destroy the whole environment.

The mechanistic Cartesian world view has had a powerful influence on all our sciences and on the general Western way of thinking. The method of reducing complex phenomena to basic building blocks, and of looking for the mechanisms through which these interact, has become so deeply ingrained in our culture that it has often been identified with the scientific method. Views, concepts, or ideas that did not fit into the framework of classical science were not taken seriously and were generally disdained, if not ridiculed. As a consequence of this overwhelming emphasis on reductionist science our culture has become progressively fragmented and has developed technologies, institutions, and life styles that are profoundly unhealthy.

That a fragmented world view should also be unhealthy is not surprising in view of the close connection between "health" and "whole." Both these words, as well as "hale," "heal," and "holy," derive from the Old English root word *hal,* which means sound, whole, and healthy. Indeed, our experience of feeling healthy involves the feeling of physical, psychological and spiritual integrity, of a sense of balance among the various components of the organism and between the organism and its environment. This sense of integrity and balance has been lost in our culture. The fragmented, mechanistic world view that has become all-pervasive, and the one-sided, sensate and "yang-oriented" value system that is the basis of this world view, have led to a profound cultural imbalance and have generated numerous symptoms of ill health.

Excessive technological growth has created an environment in which life has become physically and mentally unhealthy. Polluted air, irritating noise, traffic congestion, chemical contaminants, radiation hazards, and many other sources of physical and psychological stress have become part of everyday life for most of us. These manifold health hazards are not just incidental byproducts of technological progress; they are integral features of an economic system obsessed with growth and expansion, continuing to intensify its high technology in an attempt to increase productivity.

In addition to the health hazards we can see, hear, and smell there are other threats to our well-being which may be far more dangerous because they will affect us on a much larger scale, both in space and in time. Human technology is severely disrupting and upsetting the ecological processes that sustain our natural environment and are the very basis of our existence. One of the most serious threats, almost totally ignored until recently, is the poisoning of water and air by toxic chemical wastes.

The American public became acutely aware of the hazards of chemical waste several years ago, when the tragedy of Love Canal was reported in front-page stories. Love Canal was an abandoned trench in a residential area of Niagara Falls, New York, that was used for many years as a dumpsite for toxic chemical wastes. These chemical poisons polluted surrounding bodies of water, filtered into adjacent back yards, and generated toxic fumes, causing high rates of birth defects, liver and kidney damage, respiratory ailments, and various forms of cancer among the residents of the area. Eventually an emergency was declared by the State of New York and the area was evacuated.

The story of love Canal was first pieced together by Michael Brown, a reporter for the *Niagara Gazette*, who then went on to inspect similar hazardous-waste dumps throughout the United States. Brown's extensive investigations have made it clear that Love Canal was only the first of many similar tragedies that are bound to unfold during the coming years and will seriously affect the health of millions of Americans. The U.S. Environmental Protection Agency estimated in 1979 that there are more than 50,000 known sites where hazardous materials are stored or buried, and less than 7 percent has received proper disposal.

These enormous amounts of hazardous chemical waste are a result of the combined effects of technological and economic growth. Obsessed with expansion, increasing profits, and raising "productivity," the United States and other industrial countries have developed societies of competitive consumers who have been induced to buy, use, and throw away ever increasing quantities of products of marginal utility. To produce these goods—food additives, synthetic fibers, plastics, drugs, and pesticides, for example—resource-intensive technologies were developed, many of them heavily dependent on complex chemicals; and as production and consumption increased, so did the chemical wastes that are inevitable byproducts of these manufacturing processes. The United States produces a thousand new chemical compounds a year, many of them more complex than their predecessors and more alien to the human organism, while the amount of annual hazardous waste has increased from ten to thirty-five million tons over the past decade.

As production and consumption accelerated at this hectic pace, technologies appropriate for dealing with the unwanted byproducts were not developed. The reason for this negligence was simple: while the wasteful production of consumer goods was highly profitable for the manufacturers, the proper treatment and recycling of the residues was not. For many decades the chemical industry dumped its wastes into the ground without adequate safeguards, and this irresponsible practice has now resulted in thousands of dangerous chemical dumpsites, "toxic time bombs" that are likely to become one of the most serious environmental threats of the 1980s.

Faced with the grim consequences of its methods of production, the chemical industry has shown the typical corporate response. As Brown shows in case after case, the chemical companies have tried to conceal the dangers of their manufacturing processes and of the resulting chemical wastes; they have also covered up accidents and pressured politicians to avoid full inquiry. But thanks in part to the tragedy of Love Canal, public awareness has increased dramatically. While manufacturers proclaim in slick advertising campaigns that life would be impossible without chemicals, more and more people are realizing that the chemical industry has become life-destroying instead of life-sustaining. We can hope that public opinion will exert increasing pressure on the industry to develop proper technologies for treating and recycling waste products, as is already being done in several European countries. In the long run the problems generated by chemical waste will become manageable only if we can minimize the production of hazardous substances, which will involve radical changes in our attitudes as producers and consumers.

Our excessive consumption and our strong emphasis on high technology not only create massive quantities of waste but also require huge amounts of energy. Nonrenewable energy, derived from fossil fuels, powers most of our production processes, and with the decline of these natural resources energy itself has become a scarce and expensive resource. In their attempts to maintain, and even increase, their current levels of production, the world's industrialized countries have ferociously exploited the available resources of fossil fuels. These processes of energy production have the potential to cause unprecedented ecological disturbances and human suffering.

Exorbitant use of petroleum has led to heavy tanker traffic with frequent collisions, in which huge amounts of oil are spilled into the seas. These oil spills have not only polluted the most beautiful shores and beaches of Europe, but are also seriously disrupting the marine food cycles and thus creating ecological hazards that are still poorly understood. The generation of electricity from coal is even more hazardous and more polluting than energy production from oil. Underground mining causes severe damage to miners' health, and strip mining creates conspicuous environmental consequences, since the mines are generally abandoned once the coal is exhausted, with huge areas of land left devastated. The worst damage of all, both to the environment and to human health, comes from the burning of coal. Coal-burning plants emit vast quantities of smoke, ash, gases, and various organic compounds, many of which are known to be toxic or carcinogenic. The most dangerous of the gases is sulfur dioxide, which can severely impair the lungs. Another pollutant released in the burning of coal is nitrogen oxide, which is also the main ingredient in air pollution from automobiles. A single coal-burning plant can emit as much nitrogen oxide as several hundred thousand cars.

The sulfur and nitrogen oxides emanating from coal-fired plants not only are hazardous to the health of people living in the plant's vicinity but also generate one of the most insidious and completely invisible forms of air pollution, acid rain. The gases thrown up by power plants mix with the oxygen and water vapor in the air and, through a series of chemical reactions, turn into sulfuric and nitric acids. These acids are then carried by the wind until they collect at various atmospheric gathering points and are washed down to the earth as acid rain or snow. Eastern New England, eastern Canada, and southern Scandinavia are heavily affected by this type of pollution. When acid rain falls on lakes it kills fish, insects, plants, and other forms of life; eventually the lakes die completely from acidity they can no longer neutralise. Thousands of lakes in Canada and Scandinavia are dead or dying already; entire fabrics of life that took thousands of years to evolve are rapidly disappearing.

At the heart of the problem, as usual, lies ecological shortsightedness and corporate greed. Technologies to reduce the pollutants that cause acid rain have already been developed, but the corporations that own the coal plants are vigorously opposed to environmental regulation, and they have the political power to prevent stringent controls. Thus American utility companies have forced the Environmental Protection Agency to relax emission standards for the old coal-fired plants in the Midwest, which continue to send vast amounts of pollutants downwind and are expected to be the source of 80 percent of U.S. sulfur emissions by 1990. These actions are based on the same irresponsible attitudes that bring the hazards of chemical waste. Rather than neutralizing their polluting waste products, the industries simply dump them somewhere else, without caring, that in a finite ecosystem there is no such place as "somewhere else."

* * *

During the 1970s the world became acutely aware of a global shortage of fossil fuels, and with the inevitable decline of these conventional sources of energy in sight, the leading industrialised countries embarked on a vigorous campaign for nuclear power as an alternative energy source. The debate about how to solve the energy crisis usually focuses on the costs and risks of nuclear power, as compared to those of energy production from petroleum, coal, and oil shale. The arguments advanced by government and corporate economists, and by other representatives of the energy industry, are usually heavily biased in two ways. Solar energy—the only energy source that is abundant, renewable, stable in price and environmentally benign—is said to be "uneconomical" or "not yet feasible," in spite of considerable evidence to the contrary, and the need for more energy is assumed unquestioningly.

Any realistic discussion of the "energy crisis" has to start from a broader perspective than this, one that takes into account the roots of the present shortage of energy and their connections to the other critical problems facing us today. Such a perspective makes apparent what may at first sound paradoxical: What we need, to overcome the energy crisis, is not *more* energy but *less*. Our ever increasing energy needs reflect the general expansion of our economic and technological systems; they are caused by the patterns of undifferentiated growth that deplete our natural resources and contribute significantly to our multiple symptoms of individual and social illness. Energy, then, is a significant parameter of social and ecological balance. In our present, highly unbalanced state, more of it would not solve our problems but would make them worse. It would not only accelerate the depletion of our minerals and metals, forests and fish, but would also mean more pollution, more chemical poisoning, more social injustice, more cancer, more crime. What we need to overcome our multifaceted crisis is not more energy but a profound change of values, attitudes, and life styles.

Once these basic facts are perceived, it becomes evident that the use of nuclear power as an energy source is sheer folly. It surpasses the ecological impact of large-scale energy production from coal, which is already devastating, by several orders of magnitude, threatening not only to poison our natural environment for thousands of years but even to extinguish the entire human species. Nuclear power represents the most extreme case of a technology that has got out of hand, propelled by an obsession with self-assertion and control that has reached highly pathological levels.

In describing nuclear power in such terms, I am referring to both nuclear weapons and reactors. It is an intrinsic property of nuclear technology that these two applications cannot be separated. The term "nuclear power" itself has two linked meanings. "Power" has not only the technical meaning of "source of energy," but also the more general meaning of "possession of control or influence over others." In the case of nuclear power, these two kinds of power are inseparably connected, and both of them represent today the greatest threat to our survival and well-being...

* * *

The potential of global destruction through nuclear war is the greatest environmental threat of nuclear power. If we are unable to prevent nuclear war, all other environmental concerns will become purely academic. But even without a nuclear holocaust, the environment impact of nuclear power far exceeds all other hazards of our technology. At the beginning of the so-called peaceful use of atomic energy, nuclear power was advocated as cheap, clean, and safe. In the meantime we have become painfully aware that it is none of those. The construction and maintenance of nuclear power plants is becoming ever more expensive owing to the elaborate safety measures imposed on the nuclear industry by public protest; nuclear accidents have threatened the health and safety of hundreds of thousands of people; and radioactive substances continually poison our environment.

The health hazards of nuclear power are of an ecological nature and operate on an extremely large scale, both in space and in time. Nuclear power plants and military facilities release radioactive substances that contaminate the environment, thus affecting all living organisms, including humans. The effects are not immediate but gradual, and they are accumulating to more dangerous levels all the time. In the human organism these substances contaminate the internal environment with many medium- and long-term consequences. Cancer tends to develop after ten to forty years, and genetic diseases can appear in future generations.

Scientists and engineers very often do not fully grasp the dangers of nuclear power, partly because our science and technology have always had great difficulty dealing with ecological concepts. Another reason is the great complexity of nuclear technology. The people responsible for its development and application—physicists, engineers, economists, politicians, and generals—are all used to a fragmented approach and each group concerns itself with narrowly defined problems. They often ignore how these problems interrelate and how they combine to produce a total impact on the global ecosystem. Besides, most nuclear scientists and engineers suffer from a profound conflict of interest. Most are employed by the military or by the nuclear industry, both of which exert powerful influences. Consequently the only experts who can provide a comprehensive assessment of the hazards of nuclear power are those independent of the military-industrial complex and able to adopt a broad ecological perspective. Not surprisingly, they all tend to be in the antinuclear movement.

In the process of producing energy from nuclear power, both the workers in the nuclear industry and the whole natural environment are contaminated with radioactive substances at every step of the "fuel cycle." This cycle begins with the mining, milling, and enrichment of uranium, continues with the fabrication of fuel rods and the operation and maintenance of the reactor, and ends with the handling and storage or reprocessing of nuclear waste. The radioactive substances that escape into the environment at every stage of this process emit particles—alpha particles,[1] electrons, or photons—that can be highly energetic, penetrating the skin and damaging body cells. Radioactive substances can also be ingested with contaminated food or water and will then do their damage from within.

When considering the health hazards of radioactivity, it is important to note that there is no "safe" level of radiation, contrary to what the nuclear industry would have us believe. Medical scientists now generally agree that there is no evidence of a threshold below which radiation may be said to be harmless; even the smallest amounts can produce mutations and diseases. In everyday life we are continually exposed to low-level background radiation, which has been impinging on the earth for billions of years and which also comes from natural sources present in rocks, in water, and in plants and animals. The risks from this natural background are unavoidable, but to increase them means to gamble with our health.

The nuclear reaction that takes place in a reactor is known as fission. It is a process in which uranium nuclei break into fragments—most of which are radioactive substances—plus heat, plus one or two free neutrons. These neutrons are absorbed by other nuclei which, in turn break up, thus setting in motion a chain reaction. In an atomic bomb, this chain reaction ends in an explosion, but in a reactor it can be controlled with the help of control rods, which absorb some of the free neutrons. In this way the rate of fission can be regulated. The fission process releases a tremendous amount of heat that is used to boil water. The resulting steam drives a turbine that generates electricity. A nuclear reactor, then, is a highly sophisticated, expensive, and extremely dangerous device for boiling water.

The human factor involved in all stages of nuclear technology, military and nonmilitary, makes accidents unavoidable. These accidents result in the release of highly poisonous radioactive materials into the environment. One of the worst possibilities is the melt-down of a nuclear reactor, in which the whole mass of molten uranium would burn through the container of the reactor and into the earth, possibly triggering a steam explosion that would scatter deadly radioactive materials. The effects would be similar to those of an atomic bomb. Thousands of people would die from immediate radiation exposure; more deaths would occur after two or three weeks from acute radiation illness; large areas of land would be contaminated and made uninhabitable for thousands of years.

Many nuclear accidents have already happened, and major catastrophes have often been narrowly avoided. The accident at the Three Mile Island nuclear power plant near Harrisburg, Pennsylvania, in which the health and safety of hundreds of thousands of people were threatened, is still vivid in our memory. Less known, but not less frightening, are accidents involving nuclear weapons, accidents that have become more and more frequent as the number and capacity of those weapons have increased. By 1968 there were more than thirty major accidents involving American nuclear weapons that came close to an explosion. One of the most serious

[1] Alpha particles are compounds of two protons and two neutrons.

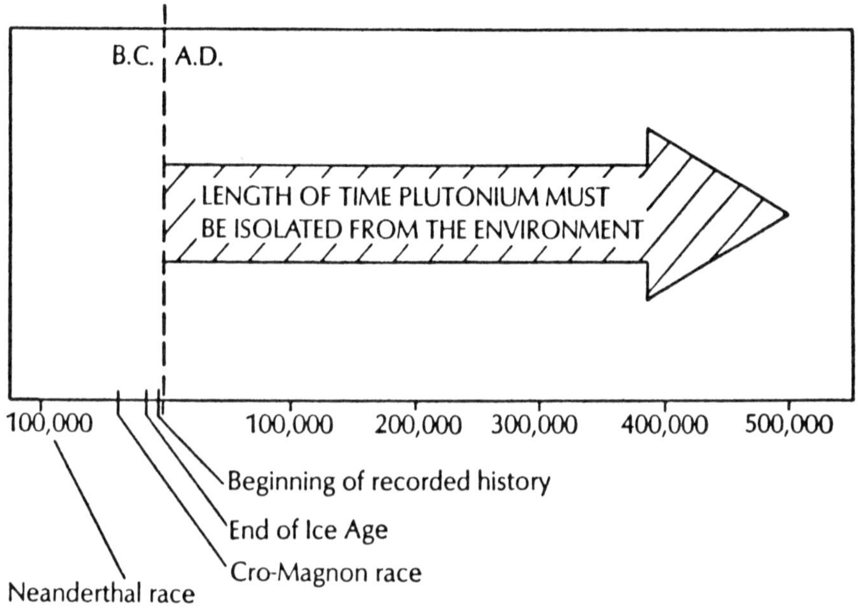


FIGURE 3.5: Length of time plutonium must be isolated from the environment.

occurred in 1961, when an H-bomb was accidentally dropped over Coldsboro, North Carolina, and five of its six safety devices failed. That one safety device protected us from a thermonuclear explosion of twenty-four million tons TNT, an explosion one thousand times more powerful than that of the Nagasaki bomb and, in fact, more powerful than the combined explosions of all the wars in human history. Several of these twenty-four-million-ton bombs have been dropped accidentally over Europe, the United States, and other parts of the world, and these accidents are going to occur even more frequently as more and more countries build nuclear weapons, probably with much less sophisticated safety devices.

Another major problem of nuclear power is the disposal of nuclear waste. Each reactor annually produces tons of radioactive waste that remains toxic for thousands of years. Plutonium, the most dangerous of the radioactive byproducts, is also the most long-lived; it remains poisonous for at least 500,000 years.[1] It is difficult to grasp the enormous length of this time span, which far exceeds the length of time we are used to contemplating within our individual lifetimes, or within the lifetime of a society, nation, or civilisation. Half a million years, as shown on [Figure 3.5] is more than one hundred times longer than all of recorded history. It is a time span fifty times longer than that from the end of the Ice Age to the present day, and more than ten times longer than our entire existence as humans with our present physical characteristics.[2] This is the length of time that plutonium must be isolated from the environment. What moral right do we have to leave

[1] The half-life of plutonium (Pu–239)—the time after which one-half of a given quantity has decayed—is 24,400 years. this means that if one gram of plutonium is released into the environment, about one-millionth of a gram will be left after 500,000 years, a quantity which is minute but still toxic.

[2] The ancestors of the European races are usually identified with the Cro-Magnon race, which appeared about 30,000 years ago and possessed all modern skeletal characteristics, including the large brain.

such a deadly legacy to thousands and thousands of generations?

No human technology can create safe containers for such an enormous time span. Indeed, no permanent safe method of disposal or storage has been found for nuclear waste, in spite of millions of dollars spent during three decades of research. Numerous leaks and accidents have shown the shortcomings of all current devices. In the meantime, nuclear waste keeps piling up. Projections by the nuclear industry anticipate a total of 152 million gallons of intensely radioactive, "high-level" waste by the year 2000, and while the precise amounts of military radioactive waste are kept secret, they can be expected to be enormously larger than those from industrial reactors.

Plutonium, named after Pluto, the Greek god of the underworld, is by far the most deadly of all nuclear waste products. Less than one millionth of a gram—an invisible dose—is carcinogenic. One pound, if uniformly distributed, could potentially induce lung cancer in every person on earth. Given these facts, it is truly frightening to know that each commercial reactor produces four hundred to five hundred funds of plutonium per year. Moreover, tons of plutonium are routinely transported along American highways and railroads and are flown into airports.

Once created, plutonium must be isolated from the environment virtually forever, since even the tiniest amounts will contaminate it for eons to come. It is important to realize that plutonium does not simply vanish with the death of a contaminated organism. A contaminated dead animal, for example, may be eaten by another animal, or it may decay and rot away, its dust scattered by the winds. In any case the plutonium will remain in the environment and will continue its lethal action, on and on, from organism to organism, for half a million years.

Since there is no hundred-percent-safe technology, some plutonium inevitably escapes when it is handled. It has been estimated that if the American nuclear industry expands according to projections made in 1975, and if it contains its plutonium with 99.99 percent perfection which would be something of a miracle—it will be responsible for 500,000 fatal lung cancers per year for about fifty years following the year 2020. This will amount to a 25 percent increase in the total death rate in the United States. In view of these estimates, it is difficult to understand how anybody can call nuclear power a safe source of energy.

Nuclear power also creates many other problems and hazards. They include the unsolved problem of disassembling, or "decommissioning" nuclear reactors at the end of their useful lives; the development of "fast breeder reactors," which use plutonium as a fuel and are far more dangerous than ordinary commercial reactors; the threat of nuclear terrorism and the ensuing loss of basic civil liberties in a totalitarian "plutonium economy"; and the disastrous economic consequences of the use of nuclear power as a capital- and technology-intensive, highly centralized source of energy. The total impact of the unprecedented threats of nuclear technology should make it abundantly clear to anyone that it is unsafe, uneconomical, irresponsible, and immoral: totally unacceptable.

If the case against nuclear power is so convincing, why is nuclear technology still promoted so heavily? The deep reason is the obsession with power. Of all the available energy sources nuclear power is the one that leads to the highest concentration of political and economic power in the hands of a small elite. Because of its complex technology it requires highly centralized institutions, and because of its military aspects it lends itself to excessive secrecy and extensive use of police power. The various protagonists of the nuclear economy—the utilities, the manufacturers of reactors, and the energy corporations[1]—all benefit from a source of energy that is highly capital-intensive and centralized. They have invested billions of dollars in nuclear technology and continue to promote it vigorously in spite of its steadily growing problems and hazards. They are not prepared to abandon that technology, even if they are forced to ask for mas-

[1] "Energy corporations" is an appropriate term used by Ralph Nader to describe the oil companies which have extended their business into all branches of the nergy industry, including the supply of uranium and plutonium, in an attempt to monopolize energy production.

sive taxpayer subsidies and to use a large police force to protect it. As Ralph Nader says, nuclear power has become in many ways America's "technological Vietnam."

* * *

Our obsession with economic growth and the value system underlying it have created a physical and mental environment in which life has become extremely unhealthy. Perhaps the most tragic aspect of this social dilemma is the fact that the health hazards created by the economic system are caused not only by the production process but by the consumption of many of the goods that are produced and heavily advertised to sustain economic expansion. To increase their profits in a saturated market, manufacturers have to produce their goods cheaper, and one way of doing this is to lower the quality of the products. To satisfy customers in spite of these low-quality products, vast sums of money are spent on conditioning consumer opinion and tastes through advertising. This practice, which has become an integral part of our economy, amounts to a serious health hazard because many of the goods produced and sold in this way have a direct effect on our health.

The food industry represents an outstanding example of health hazards generated by commercial interests. Although nutrition is one of the most important influences on our health, this is not emphasized in our system of health care and doctors are notoriously ignorant when it comes to questions of diet. Still, the basic features of a healthy diet are well known. To be healthy and nutritious, our diet should be well balanced, low in animal protein and high in natural, nonrefined carbohydrates. This can be achieved by relying on three basic foods—whole grains, vegetables, and fruits. Even more important than the detailed composition of our diet are the following three requirements: our foods should be *natural*, consisting of organic food elements in their natural, unaltered state; they should be *whole*, complete and unfragmented, neither refined nor enriched; and they should be *poison-free*, organically grown, free from poisonous chemical residues and additives. These dietary requirements are extremely simple, and yet they are almost impossible to follow in today's world.

To improve their business food manufacturers add preservatives to food to increase its shelf life; they replace healthy organic food with synthetic products, and try to make up for the lack of nutritious content by adding artificial flavoring and coloring agents. Such overprocessed, artificial food is heavily advertised on billboards and television, together with alcohol and cigarettes, those other two major health hazards. We are subject to barrages of commercials for "junk food"—soft drinks, sweet snacks, high-fat foods—which have all been proved unhealthy. A recent study in Chicago that analyzed advertising by food companies on four television stations concluded that "on weekdays over 70 percent, and on weekends over 85 percent of food advertising is negatively related to the nation's health needs." Another study found that more than 50 percent of the money spent for food advertising on television is used to promote items closely linked to the most significant risk factors in the American diet.

For a large number of people in our culture the problems of an unhealthy diet are compounded by excessive use of drugs, both medical and nonmedical. Although alcohol continues to cause more problems for individual and social health than all other drugs combined, other kinds of drug abuse have become a significant threat to public health. In the United States aspirin alone is now consumed at the rate of 20,000 tons per year, which amounts to almost 225 tablets per person. But the biggest problem today is the excessive use of prescription drugs. Their sales have soared at an unprecedented rate, especially during the last twenty years, with the strongest increase occurring in the prescription of psychoactive drugs—tranquilizers, sedatives, stimulants, and antidepressants.

Medical drugs can be extremely helpful if used intelligently. They have alleviated a great deal of pain and suffering and have helped many patients with degenerative diseases who, even ten years ago, would have been much more miserable. At the same time countless people have been victims of overuse and misuse. The overuse of drugs in contemporary medicine is based on a limited conceptual model of illness and is perpetuated by the powerful pharmaceutical industry. The biomedical model of illness and the economic model on which drug manufacturers base their business reinforce each other because they both reflect the same reductionist approach to reality. In both cases a complex pattern of phenomena and values is reduced to a single overriding aspect. The pharmaceutical industry is one of the largest industries whose rate of profit has remained very high throughout the past two decades, outranking the rates of other manufacturing industries by significant margins. One of the outstanding characteristics of the drug industry is an excessive emphasis on differentiation of basically similar products. To a large extent research and marketing are devoted to developing drugs that are perceived as distinctive and superior, no matter how closely they resemble competing products, and huge sums are spent on advertising and promotion to establish a drug's distinctiveness far beyond any scientific justification. As a consequence the market has been inundated with thousands of redundant medical drugs, many of them only marginally effective and all with harmful side effects.

To study the ways in which the drug industry sells its products is highly instructive. In the United States the industry is controlled by the Pharmaceutical Manufacturers Association, a policy-making body that influences nearly every facet of the medical system. The PMA has close ties with the American Medical Association, and a large amount of the AMA's revenues comes from advertising in its medical journals. The largest of these periodicals is the *Journal of the American Medical Association*, whose apparent purpose is to keep physicians informed about new developments in medicine but which in fact is largely dominated by the interests of the pharmaceutical industry. The same is true for most other medical journals, which according to reliable estimates receive about half their income from advertising accounts with drug companies.

This strong financial dependence of professional journals on an industry—a unique characteristic of the medical profession—is bound to affect the editorial politices of these journals. Indeed, numerous examples of conflict of interest have been observed. One involved a certain hormone called Norlutin, which was found to have harmful effects on some fetuses when taken during pregnancy. According to a report in the March 1960 issue of the JAMA, these side effects of Norlutin occurred "with sufficient frequency to preclude its use or advertisement as a safe hormone to be taken during pregnancy." Yet in the same issue and for the next three months the journal continued to run a full-page advertisement for Norlutin with no reference to its possible side effects. Eventually the drug was taken off the market. This was not an isolated event. The AMA has consistently neglected to inform physicians sufficiently about the adverse effects of antibiotics, which may well be the drugs most abused by physicians and most dangerous to patients. Unnecessary or negligent prescription of antibiotics has resulted in thousands of deaths, yet the AMA provides unlimited advertising space for antibiotics with no attempts at a disclaimer. This irresponsible advertising is certainly not unrelated to the fact that, next to sedatives and tranquilizers, antibiotics provide the largest advertising income for the American Medical Association.

Pharmaceutical advertising is designed specifically to induce doctors to prescribe ever more drugs. Naturally, then, these drugs are described as the ideal solution to a wide variety of everyday problems. Stressful life situations with physical, psychological, or social origins are presented as diseases amenable to drug treatment. Thus tranquilizers are advertised as remedies for "environmental depression" or "not fitting in," and other drugs are suggested as convenient means to "pacify" elderly patients or unruly school children. The tone of some of these ads, which are addressed to doctors, is absolutely horrifying to the lay person, especially when they advertise the treatment of women. Women suffer disproportionately from drug treatments; they take over 60 percent of all psychoactive drugs prescribed and over 70 percent of all antidepressants. Advertisements often advise doctors in blatantly sexist language to get rid of female patients by giving them tranquilizers for vague complaints, or to prescribe drugs for women who are unhappy with their roles in society.

The influence of drug manufacturers on medical care extends far beyond journal advertising. In the United States the *Physician's Desk Reference* is the most popular reference book on drugs and is consulted regularly by more than 75 percent of the doctors. It lists every drug on the market, with its uses, recommended dosages, and side effects. Yet this standard work amounts to little more than straight advertising, since all its contents are prepared and paid for by the drug companies and it is distributed free to every physician in the country.

For most doctors education about current drugs does not come from independent and objective pharmacologists but is provided almost exclusively by the highly media-conscious and manipulative producers. We can gauge the strength of this influence by noticing how rarely doctors use the proper technical terms when they refer to drugs; instead they use, and thus promote, the brand names made up by the drug companies.

Even more influential than its manual and journal advertising is the pharmaceutical industry's sales force. To sell their goods these "detail men" saturate doctors not only with smooth sales talk but also with briefcases full of drug samples, plus every imaginable promotional ploy. Many companies offer physicians prizes, gifts, and premiums in proportion to the amount of drugs prescribed—tape recorders pocket calculators, dishwashers, refrigerators, and portable TV sets. Others offer week-long educational seminars in the Bahamas with all expenses paid. It has been estimated that the drug companies collectively spend an average of $4,000 a year *per physician* on promotional ploys, which amounts to 65 percent more than they spend on research and development.

* * *

The influence of the pharmaceutical industry on the practice of medicine has an interesting parallel in the influence of the petrochemical[1] industry on agriculture and farming. Farmers, like doctors, deal with living organisms that are severely affected by the mechanistic and reductionist approach of our science and technology. Like the human organism, the soil is a living system that has to remain in a state of dynamic balance to be healthy. When the balance is disturbed there will be pathological growth of certain components—bacteria or cancer cells in the human body, weeds or pests in the fields. Disease will occur, and eventually the whole organism may die and turn into inorganic matter. These effects have become major problems in modern agriculture because of the farming methods promoted by the petro-chemical companies. As the pharmaceutical industry has conditioned doctors and patients to believe that the human body needs continual medical supervision and drug treatment to stay healthy, so the petro-chemical industry has made farmers believe that soil needs massive infusions of chemicals, supervised by agricultural scientists and technicians, to remain productive. In both cases these practices have seriously disrupted the natural balance of the living system and thus generated numerous diseases. Moreover, the two systems are directly connected, since any imbalance in the soil will affect the food that grows in it and thus the health of the people who eat that food.

A fertile soil is a living soil containing billions of living organisms in every cubic centimeter. It is a complex ecosystem in which the substances that are essential to life move in cycles from plants to animals, to soil bacteria, and back again to plants. Carbon and nitrogen are two basic chemical elements that go through these ecological cycles, in addition to many other nutrient chemicals and minerals. Solar energy is the natural fuel that drives the soil cycles, and living organisms of all sizes are necessary to sustain the whole system and keep it in balance. Thus bacteria carry out various chemical transformations, such as the process of nitrogen fixation, which make nturients accessible to plants; deep-rooted weeds bring trace minerals to the soil surface where crops can make use of them; earthworms break up the soil and loosen its texture; and all these activities are interdependent and combine harmoniously to provide the nourishment that sustains all life on earth.

The basic nature of living soil requires agriculture, first and foremost, to preserve the integrity of the great ecological cycles. This principle was embodied in traditional farming methods, which were based on a profound respect for life. Farmers used to plant different crops every year, rotating them so that the balance in the soil was preserved. No pesticides were needed, since insects attracted to one crop would disappear with the next. Instead of using chemical fertilizers, farmers would enrich their fields with manure, thus returning organic matter to the soil to reenter the biological cycle.

This age-old practice of ecological farming changed drastically about three decades ago, when farmers switched from organic to synthetic prod-

[1] Petrochemicals are chemicals isolated or derived from petroleum.

ucts, which opened up vast markets for the oil companies. While the drug companies manipulated doctors to prescribe ever more drugs, the oil companies manipulated farmers to use ever more chemicals. Both the pharmaceutical industry and the petrochemical industry became multibillion-dollar businesses. For the farmers the immediate effect of the new farming methods was a spectacular improvement in agricultural production, and the new era of chemcial farming was hailed as the "Green Revolution." Soon, however, the dark side of the new technology became apparent, and today it is evident that the Green Revolution has helped neither the farmers nor the land nor the starving millions. The only ones to gain from it were the petrochemical corporations.

The massive use of chemical fertilizers and pesticides changed the whole fabric of agriculture and farming. The industry persuaded farmers that they could make money by planting large fields with a single highly profitable crop and controlling weeds and pests with chemicals. The results of this practice of single-crop monocultures were great losses of genetic variety in the fields and, consequently, high risks of large acreages being destroyed by a single pest. Monocultures also affected the health of the people living in the farming areas, who were no longer able to obtain a balanced diet from locally grown foods and thus became more disease-prone.

With the new chemicals, farming became mechanized and energy-intensive, with automated harvesters, feeders, waterers, and many other labor-saving machines performing the work that had previously been done by millions of people. Narrow notions of efficiency helped to conceal the drawbacks of these capital-intensive farming methods, as farmers were seduced by the wonders of modern technology. Even as late as 1970 an article in the *National Geographic* presented the following enthusiastic and utterly naive vision of future agriculture:

> Fields will be larger, with fewer tree, hedges, and roadways. Machines will be bigger and more powerful... They'll be automated, even radio-controlled, with closed circuit TV to let an operator sitting on a front porch monitor what is going on...Weather control may tame hailstorm and tornado dangers...Atomic energy may supply power to level hills or provide irrigation water from the sea.

The reality, of course, was far less encouraging. While American farmers were able to triple their corn yields per acre and, at the same time, cut their labor by two-thirds, the amount of energy used to produce one acre of corn increased fourfold. The new style of farming favored large corporate farmers with big capital and forced most of the traditional single-family farmers, who could not afford to mechanize to leave their land. Three million American farms have been eliminated in this way since 1945, with large numbers of people forced to leave the rural areas and join the masses of the urban unemployed as victims of the Green Revolution.

Those farmers who were able to remain on the land had to accept a profound transformation of their image, role, and activities. From growers of edible foods, taking pride in feeding the world's people, farmers have turned into producers of industrial raw materials to be processed into commodities designed for mass marketing. Thus corn is converted to starch or syrup; soybeans become oils, pet food, or protein concentrates; wheat flour is made into frozen dough or packaged mixes. For the consumer the tie of these products to the land has almost disappeared, and it is not surprising that many children today grow up believing that food comes from supermarket shelves.

Farming as a whole has been turned into a huge industry, in which key decisions are made by "agriscientists" and passed on to "agribusinessmen" or "farming technicians"—the former farmers—through a chain of agents and sales people. Thus farmers have lost most of their freedom and creativity, and have become, in effect, consumers of production techniques. These techniques are not based on ecological considerations but are determined by the commodity market. Farmers can no longer grow or breed what the land indicates, nor what people need; they have to grow and breed what the market dictates.

In this industrialized system, which treats living matter like dead substances and uses animals like machines, penned in feedlots and cages, the process of farming is almost totally controlled by the petrochemical industry. Farmers get virtually all their information about farming techniques from the industry's sales force, just as most doctors get their information about drug therapy from the drug industry's "detail men." The information about chemical farming is almost totally unrelated to the real needs of the land. As Barry Commoner has noted, "One can almost admire the enterprise and clever salesmanship of the petrochemical industry. Somehow it has managed to convince the farmer that he should give up the free solar energy that drives the natural cycles and, instead, buy the needed energy—in the form of fertilizer and fuel—from the petrochemical industry."

In spite of this massive indoctrination by the energy corporations, many farmers have preserved their ecological intuition, passed down from generation to generation. These men and women know that the chemical way of farming is harmful to the land, but they are often forced to adopt it because the whole economy of farming—tax structure, credit system, real estate system, and so on—has been set up in a way that gives them no choice. To quote Commoner again, "The giant corporations have made a colony of rural America."

Nevertheless, a growing number of farmers have become aware of the hazards of chemical farming and are turning back to organic, ecological methods. Just as there is a grass-roots movement in the health field, there is a grass-roots movement in farming. The new organic farmers grow their crops without synthetic fertilizers, rotating them carefully and controlling pests with new ecological methods. Their results have been most impressive. Their food is healthier and tastes better, and their operations have also been shown to be more productive than those of conventional farms. The new organic farming has recently sparked serious interest in the United States and in many European countries.

* * *

The long-term effects of excessive "chemotherapy" in agriculture have proven disastrous for the health of the soil and the people, for our social relations, and for the entire ecosystem of the planet. As the same crops are planted and fertilized synthetically year after year, the balance in the soil is disrupted. The amount of organic matter diminishes, and with it the soil's ability to retain moisture. The humus content is depleted and the soil's porosity reduced. These changes in soil texture entail a multitude of interrelated consequences. The depletion of organic matter makes the soil dead and dry; water runs through it but does not wet it. The ground becomes hard-packed, which forces farmers to use more powerful machines. On the other hand, dead soil is more susceptible to wind and water erosion, which are taking an increasing toll. For example, half of the topsoil in Iowa has been washed away in the last twenty-five years, and in 1976 two-thirds of America's agricultural counties were designated drought disaster areas. What is often called "drought," "wind breaking down the land," or "winter-kill," are all consequences of sterile soil.

The massive use of chemical fertilizers has seriously affected the natural process of nitrogen fixation by damaging soil bacteria involved in this process. As a consequence crops are losing their ability to take up nutrients from the soil and becoming more and more addicted to synthetic chemicals. Because their efficiency in absorbing nutrients this way is much lower, not all the chemicals are taken up by the crop but leach into the ground water or drain from the fields into rivers and lakes.

The ecological imbalance caused by monocropping and by excessive use of chemical fertilizers inevitably results in enormous increases in pests and crop diseases, which farmers counteract by spraying ever larger doses of pesticides, thus fighting the effects of their overuse of chemicals by using even more. However, pesticides often can no longer destroy the pests because they tend to become immune to the chemicals. Since World War II, when massive use of pesticides began, crop losses due to insects have not decreased; on the contrary, they have almost doubled. Moreover, many crops are now attacked by new insects that were never known as pests before, and these new pests are becoming increasingly resistant to all insecticides.

Since 1945 there has been a sixfold increase in the use of chemical fertilizers and a twelvefold increase in the use of pesticides on American farms. At the same time increased mechanization and longer transport routes have contributed further to the energy dependence of modern agriculture. As a result, 60 percent of the costs of food are now costs of petroleum. As the farmer Wes Jackson puts it succinctly: "We have literally moved our agricultural base from soil to oil." When energy was cheap, it was easy for the petrochemical industry to persuade farmers to change from organic to chemical farming, but since the costs of petroleum began their steady climb, many farmers have realized that they can no longer afford the chemicals they now depend upon. With every enlargement of farming technology the indebtedness of farmers increases as well. Even in the 1970s an Iowa banker remarked quite frankly, "I occasionally wonder whether the average farmer will ever get out of debt."

If the Green Revolution has had disastrous consequences for the well-being of farmers and the health of the soil, the hazards for human health have been no less severe. Excessive use of fertilizers and pesticides has sent great quantities of toxic chemicals seeping through the soil, contaminating the water table and showing up in food. Perhaps half the pesticides on the market are mixed with petroleum distillates that may destroy the body's natural immune system. Others contain substances which are related specifically to cancer. Yet these alarming results have barely affected the sale and use of fertilizers and pesticides. Some of the more dangerous chemicals have been outlawed in the United States, but the oil companies continue to sell them in the Third World, where legislation is less strict, as the drug companies sell dangerous prescription drugs. In the case of pesticides all populations are directly affected by this unethical practice because the toxic chemicals come back on fruits and vegetables imported from Third World countries.

One of the principal justifications for the Green Revolution has been the argument that the new agricultural technology is needed to feed the world's hungry. In an age of scarcity, so the argument goes, only increased production will solve the problem of hunger, and only large-scale agribusiness is able to produce more food. This argument is still used, long after detailed research has made it quite clear that the problem of world hunger is not at all a technical problem; it is social and political. One of the most lucid discussions of the relation between agribusiness and world hunger can be found in the work of Frances Moore Lappé and Joseph Collins, founders of the Institute for Food and Development Policy in San Francisco. Extensive research has led these authors to conclude that scarcity of food is a myth and that agribusiness does not solve the problem of hunger but, on the contrary, perpetuates and even aggravates it. They point out that the central question is not how production can be increased, but rather what is grown and who eats it, and that the answers are determined by those who control the food-producing resources. Merely to introduce new technologies into a system marred by social inequalities will never solve the hunger problem; on the contrary, it will make it worse. Indeed, studies of the impact of the Green Revolution on hunger in the Third World have confirmed the same paradoxical and tragic result again and again. More food is being produced, yet more people are hungry. As Moore Lappé and Collins write, "In the Third World, on the whole there is more food and less to eat."

Research codirected by Moore Lappé and Collins has shown that there is no country in the world in which people could not feed themselves from their own resources, and that the amount of food produced in the world at present is sufficient to provide about eight billion people—more than twice the world population—with an adequate diet. Nor can scarcity of agricultural land be considered a cause of hunger. For example, China has twice as many people per cultivated acre as India, yet in China there is no large-scale hunger. Inequality is the main stumbling block in all current attempts to fight world hunger. Agricultural "modernization"—mechanized large-scale farming—is highly profitable for a small elite, the new corporate "farmers," and drives millions of people off the land. Thus fewer people are gaining control over more and more land, and once these large landholders are established they no longer grow food according to local needs but switch to more profitable crops for export, while the local population starves. Examples of this vicious practice abound in all countries of the Third World. In Central America at least half of the agricultural land—and precisely the most fertile land—is used to grow cash crops for export while up to 70 percent of the children are undernourished. In Senegal vegetables for export to Europe are grown on choice land while the country's rural majority goes hungry. Rich, fertile land in Mexico that previously produced a dozen local foods is now used to grow asparagus for European gourmets. Other landowners in Mexico are switching to grapes for brandy, while entrepreneurs in Colombia are changing from growing wheat to growing carnations for export to the United States.

World hunger can be overcome only by transforming social relations in such a way that inequality is reduced at every level. The primary problem is not the redistribution of food but the redistribution of control over agricultural resources. Only when this control is democratized will the hungry be able to eat what is produced. Many countries have proved that social changes of this kind can be successful. In fact, 40 percent of the Third World population now lives in countries where hunger has been eliminated through common struggle. These countries do not use agriculture as the means to export income but rather use it to produce food first for themselves. Such a "food first" policy requires, as Moore Lappé and Collins have emphasized, that industrial crops should be planted only after people have met their basic needs, and that trade should be seen as an extension of domestic need rather than being determined strictly by foreign demand.

At the same time, we who live in industrialized countries will have to realize that our own food security is not being threatened by the hungry masses in the Third World, but by the food and agricultural corporations that perpetuate this massive starvation. Multinational agribusiness corporations are now in the process of creating a single world agricultural system in which they will be able to control all stages of food production and to manipulate both food supply and prices through well-established monopoly practices. This process is now well under way. In the United States almost 90 percent of the vegetable production is controlled by major processing corporations, and many farmers have no choice but to sign up with them or go out of business.

World-wide corporate control of food production would make it impossible ever to eliminate hunger. It would, in effect, establish a Global Supermarket in which the world's poor would be in direct competition with the affluent and thus would never be able to feed themselves. This effect can already be observed in many Third World countries, where people go hungry although food is grown in abundance right where they live. Their own government may subsidize its production and they themselves may even grow and harvest it. Yet they will never eat any because they are unable to pay the prices resulting from international competition.

In its continual efforts to expand and increase its profits, agribusiness not only perpetuates world hunger but is extremely careless in the way it treats the natural environment, to the extent of creating serious threats to the global ecosystem. For example, giant multinational companies such as Goodyear, Volkswagen, and Nestlé are now bulldozing hundreds of millions of acres in the Amazon River basin in Brazil to raise cattle for export. The environmental consequences of clearing such vast areas of tropical forest are likely to be disastrous. Ecologists warn that the actions of the torrential tropical rains and the equatorial sun may set off chain reactions that could significantly alter the climate throughout the world.

Agribusiness, then, ruins the soil on which our very existence depends, perpetuates social injustice and world hunger, and seriously threatens global ecological balance. An enterprise that was originally nourishing and life-sustaining has become a major hazard to individual, social, and ecological health.

* * *

The more we study the social problems of our time, the more we realize that the mechanistic world view and the value system associated with it have generated technologies, institutions, and life styles that are profoundly unhealthy. Many of these health hazards are further aggravated by the fact that our health care system is unable to deal with them appropriately because of its adherence to the same paradigm that is perpetuating the causes of ill health. Current health care is reduced to medical care within the biomedical framework, that is, centered on acute, hospital-based, and drug-oriented medicine. Health care and the prevention of illness are perceived as two different problems and, accordingly, health professionals are not very active in supporting environmental and social policies that are directly related to public health.

The shortcomings of our present system of health care result from the subtle interplay of two tendencies... One is the adherence to the narrow biomedical framework in which the relevance of non-biological aspects to the understanding of illness is systematically denied. The other, no less important, is the pursuit of economic and institutional growth and of political power by the health industry, which has heavily invested in the technologies that emerged from the reductionist view of illness. The American health care system consists of a vast conglomeration of powerful institutions, motivated by economic growth and lacking any effective incentives to hold down health costs. The system is dominated by the same financial and corporate forces that have shaped the other sectors of the economy, forces that are not primarily interested in public health but that control virtually all facets of health care—the structure of health insurance, the management of hospitals, the manufacture and promotion of drugs, the orientation of medical research and education, and recognition and licensing of nonmedical therapists. The dominance of corporate values in this system is evident in the current debates on national health insurance, in which the basic patterns of power are never questioned. That is why none of the schemes currently under discussion is likely to satisfy the health needs of the American people. As one study of health care in the United States has noted, "Just as federal defense appropriations subsidize the military-industrial complex, national health insurance will subsidize the medical-industrial complex."

The aim of the health industry has been to turn health care into a commodity that can be sold to consumers according to the rules of the "free market" economy. To achieve this purpose the "health care delivery" system has been structured and organized like the large manufacturing industries. Rather than encouraging health care in small community health centers, where it can be adapted to individual needs and practiced with an emphasis on prevention and health education, the current system favors a highly centralized and technology-intensive approach that is profitable for the industry but expensive and unhealthy for the patients.

The present "health establishment" has heavily invested in the status quo and is vigorously opposed to any fundamental revision of health care. By effectively controlling medical education, research, and practice, the health industry tries to suppress all incentives to change and to make the current approach intellectually and financially rewarding for the medical elite that directs the practice of health care. However, the problems of rising health costs, the diminishing returns from medical care, and the increasing evidence that environmental, occupational, and social factors are the primary causes of ill health will inevitably force change. In fact this change has already begun and is rapidly gaining momentum. The holistic health movement is active both within and outside the medical system, and is supported and complemented by other popular movements—environmental groups, antinuclear organizations, consumer groups, social liberation movements—that have perceived the environmental and social influences on health and are committed to opposing and preventing the creation of health hazards through political action. All these movements subscribe to a holistic and ecological view of life and reject the value system that dominates our culture and is perpetuated by our social and political institutions. The new rising culture shares a vision of reality that is still being discussed and explored but will eventually emerge as a new paradigm, destined to eclipse the Cartesian world view in our society.

---

From Fritjof Capra, *The Turning Point: Science, Society, and the Rising Culture* (New York: Simon & Schuster, 1982).

---

## Questions and Topics for Discussion and Writing

1. Why does Capra call this chapter "The Dark Side of Growth"? What has "growth" to do with the problems he discusses?
2. Are the problems discussed by Capra inevitable, from his point of view?
3. Why does Capra say that the environmental impact of nuclear power far exceeds all other hazards of our technology?
4. Capra discusses chemical waste, atmospheric pollution, nuclear power, junk food, the pharmaceutical industry, and the Green Revolution all in one chapter. What do these have to do with one another?

Testing unknown contents for toxicity.

*Issue*

# Proposed Solutions to Environmental Problems

Granted that there are some serious environmental problems, can they be solved, and if so how? The articles in this section argue the viability of proposed and tried solutions.

The first selection, "Can Science Save the Future?" is a part of a debate between two scientists carried on in their respective newspaper columns: Russell McNeil arguing that appropriate science and technology can be developed to cope with admittedly serious problems, and David Suzuki rejoining that science will be useful but cannot by itself avert the impending disaster. John Todd's "A Modest Proposal" decries the loss of bio-diversity and proposes the institution of a variety of alternative technologies that are in harmony with the environment. Two articles follow by Craig McInnes. The first, "They Thought River Was Beyond Saving," is a report on the efforts in Cleveland, Ohio, to clean up the terribly polluted Cuyahoga River. The second, "Biodegradable: Does Nature Win or Lose?" reviews the controversy over a new attempt to combat the problem of disposing of plastic wastes—biodegradable plastic garbage bags.

# Can Science Save the Future

## Russell McNeil and David Suzuki

If it is agreed that the world environment is in danger and if trends are not checked that the future looks grim, then the question arises whether the problems can be overcome by human ingenuity. In the following two articles, both by scientists, each side of the question is argued. Both articles appeared in Toronto newspapers in 1989. In the first article, Russell McNeil acknowledges that the "planet is sick" due to human activity, but he holds to the view that the ecological problems are solvable and will be solved whenever we buckle down and develop the science and technology necessary. McNeil's article was in part a comment on a view previously expressed by David Suzuki that the only solution to the problem of the "sick planet" was to reduce world population. A month later Suzuki replied, arguing the McNeil's proposed remedy is simplistic.

## Science Has Ability to Save the Future

### Russell McNeil

Canadian science writer David Suzuki has advocated what he calls "stunningly simple" solutions to the problems that face our planet. "The planet," Suzuki says, "is sick, and the major cause of the disease is thoughtless human activity."

In essence, he advocates cutting our numbers, drastically reducing consumption in the industrialized world, reducing toxic emissions, and incorporating ecological thinking into all of our activities.

Suzuki's premise is correct, the planet is sick, but such solutions are not always so "stunningly simple."

We have been hearing the cry to "reduce our numbers" for 30 years. Our global population today stands at 5 billion. Over the next 40 years it will grow to 8 billion. Internationally co-ordinated energy and resource studies indicate that, with proper management, our planet can sustain and support a population that high.

#### Unheard Cry

In any event, there may be little—short of a global catastrophe we all hope to avoid—that anyone can do to stop this growth. The cry for curbing our numbers has been heard. But it has been heard only in the industrialized West—Canada, the U.S., and Europe.

In much of the Third World, where 75 per cent of the human family lives, the cry has not been heard—and may not—for many decades. Further reductions in our population in the West could produce a skewed and aging population which could impair Western civilization. Reductions in the Third World now do little to address the problem of over-consumption in general—because it is we in the West who consume the most.

What will reducing our numbers achieve? Our planet is going through an enormous growth phase that will probably level out after the year 2030. It seems we may have to live with that. Perhaps it's time we learn to celebrate it.

We are coming to the end of a distressing phase of industrial expansion which began over a century ago. Most of today's environmental problems are caused by dirty industrial fuels and dirty industrial processes—not by too many human beings.

However, we now have the technological potential to develop clean and sustainable fuels and processes. The untenable global sickness we are experiencing will end.

It will end because large and growing forces for social, political and economic change are offering alternatives that are becoming too important for us to ignore. It will end because old style industrial thinking will not solve the problems we now face. It will end because the grass roots movements that demand change are starting to be heard. It will end because governments and multinational corporations must respond to these voices for change, or themselves face extinction.

In the final analysis, it will end because the social and ecological costs of the industrial age are becoming more than you and I are willing to pay. Life in the post-industrial epoch—it is just now starting—will be different.

The problems we face now—terrible and worsening poverty and starvation throughout much of the Third World, the unparalleled assault on our air, water and soil, the unforgivable exploitation of precious tropical forests, will not go on indefinitely.

But it will take time and patience to put the new global structure in place. At the end, new tools and systems will replace those we have now. Cleaner fuels—like hydrogen and solar energy—will provide clean alternatives to those we use now.

This move to clean fuels alone will have an enormous impact on the environmental health of this planet.

Information, and the capacity to use it, will spread to all corners of the globe. With that, opportunities for self realization through education will become available to all.

As these changes begin to evolve over the next four decades, the standard of living everywhere will begin to improve. To suggest that it will not, or that it should not, is to condemn the Third World to inexorable and insufferable poverty—while we in the West continue to consume.

As Third World living standards improve, global consumption will increase. However, this consumption will be in resources we can maintain—sustainable and chemical-free agriculture, renewable energy, and enormously more efficient post-industrial processes.

### Critical Problems

Suzuki's solutions could work—and most will be necessary—but population control has proven difficult, and may even be unnecessary in the very long term. It is true that our problems are critical—but not life threatening for the planet as a whole. What we must do will be stunningly difficult. However, the solutions that will evolve—will allow all of us—and more of us—to live in dignity throughout the world.

Rather than advocating negativity and despair, it might be more productive to promote more research and public awareness to encourage faster development of "appropriate" science and technology—to give us the sustainable infrastructure we must create to free all humanity from its century of slavery.

The sooner our governments, industries and universities develop those tool, the sooner this brighter future will arrive.

---

*Toronto Star* January 1, 1989. H6.

## Don't Bank on Science Saving the World

### David Suzuki

Explosive change has taken place in society and in the natural world since the Second World War.

A major factor underlying that change has been *science* when applied by industry, medicine and the military. In this century, there have been greater leaps in communication, transportation, health, entertainment and knowledge than occurred in the rest of history.

While we ride atop a steep curve of change, we continue to invoke ideas more appropriate to times before the current spasm of planetary transformation.

An illustration of this old mindset is an article by physicist and science writer Russell McNeil under the headline Science Has Ability to Save the Future (*Toronto Star*, Jan. 1). Dr. McNeil recognizes that "the planet is sick," but does not agree that the primary cause is our species. He does not see that the solutions are self-evident—industrialized societies have to cut their consumption and production of toxic emissions, Third-World countries have to reduce their numbers and we all have to learn to live in ecological balance.

I believe that while difficulty lies in trying to achieve those goals, we first have to diagnose the illness and identify the causes before tackling the hard part of finding a cure. However, Dr. McNeil objects to this kind of analysis because it leads to "negativity and despair."

So what is Dr. McNeil's vision? Overpopulation is an issue whose solution will prove elusive in the Third World, he said. (He makes no mention of heroic attempts to control birth in China or India.) Dr. McNeil may be right that "we may have to live with it." However, he goes on to conclude, "It's time we learn to celebrate it." His cause for celebration? "Most of today's environmental problems are caused by dirty industrial fuels and dirty industrial processes, not by too many human beings," and "we now have the technological potential to develop clean and sustainable fuels and processes. The untenable global sickness we are experiencing will end."

In this sweeping critique, the sheer impact of human numbers is ignored. Humanity has increased explosively for the past 100 years and every additional person on Earth represents one more set of lungs to inflate with air, another stomach to fill with water and food, a new body to clothe and shelter. All of these demands must be met by a finite and already overburdened planet.

To suggest, as Dr. McNeil does, that we merely have to find non-polluting forms of energy and industrial processes, is to overlook the massive volume of our output (such as sewage and garbage). It is also to ignore our inability to anticipate all of the environmental consequences of our technologies, however beneficial (for example, the unanticipated scourge of DDT, plastics and even cars). Who could foresee the real costs of even the most apparently benign innovations and inventions, such as antibiotics and television?

Dr. McNeil cheerily predicts that "information and the capacity to use it will spread to all corners of the globe. With that, opportunities for self-realization through education will become available to all." Yet in industrialized countries, numerous studies document the fact that too much information overwhelms and confuses as much as it enlightens. Educators and politicians alike provide corroboration.

Nevertheless, Dr. McNeil is optimistic that with increased population and global enlightenment, "over the next four decades, the standard of living everywhere will begin to improve." (Obviously this definition of standard of living does not include extinction, which is estimated to be the fate of two species an hour.)

What gives him hope that all of this will occur? Why, *science*, of course. "The sooner our governments, industries and universities develop these tools, the sooner this brighter future will arrive." And so he repeats a standard article of faith that is a part of our dilemma today.

Dr. McNeil, the same as many people, simply doesn't want to face up to the reality of what is happening, because it conflicts with deeply held beliefs and demands radical change.

Instead, he prefers to "be positive," thereby perpetuating comfortable beliefs as truth. However, these "sacred truths" obscure the dimensions of the problems and preclude serious solutions.

The twentieth century has been the Age of Science. The industrialized world has benefited beyond imagination in health, longevity, wealth and comfort as a direct result of scientific discovery and invention. But if, as Dr. McNeil believes, science is so full of promise to improve our lives, why are we also beset with species extinction, poverty and starvation, atmospheric degradation, global pollution, desertification and deforestation?

Part of the answer is that science yields an incomplete picture of the world, one that is shattered by scientific methodology into bits and pieces.

In applying insights gained, we are able to bludgeon nature into temporary submission. But our "control" over nature is an *illusion* because scientific perspectives can't encompass intact systems. Tiny facets of nature are studied piecemeal, so what we do know is very incomplete. Therefore, science exacerbates as much as it alleviates global problems. Solutions to the environmental crisis will not come from science but will be political and socio-economic.

Of course, we need more research on cleaner fuels, more environmentally friendly technology and more public awareness, as Dr. McNeil says. But in pushing such a superficial and rosy analysis of the global crisis and failing to debunk some sacred truths, he ensures that we will go on with business as usual and remain locked in a destructive path.

Globe & Mail, February 4, 1989. p. D4

## Questions and Topics for Discussion and Writing

1. On what issues do McNeil and Suzuki agree? On what issues do they disagree?
2. McNeil says that the planet's growth phase "will probably level out after the year 2030." What reasons does he give for this conclusion? What reasons could you suggest to explain this?
3. Why does Suzuki believe that science cannot provide the solution to the world's problems?

# A Modest Proposal

## John Todd

John Todd is the co-founder of the New Alchemy Institute, an organization devoted to producing viable solutions to world ecological problems through alternatives to the methods and materials of modern science and technology. In this article, he describes the problems inherent in the dominant technological systems and the kinds of alternatives that the New Alchemists seek to develop.

> I have been assured by a very knowing *American* of my Acquaintance in *London;* that a young healthy Child, well nursed, is, at a Year old, a most delicious, nourishing, and wholesome Food… I grant this Food will be somewhat dear, and therefore very *proper for landlords;* who, as they have already devoured most of the Parents, seem to have the best Title to the Children.
>
> *Jonathan Swift,* 1729

There is a single overview that is increasingly dominating human affairs, while diversity and indigenous approaches are being set aside with the flourishing of modern science and technology. If the present trend continues, the world community will be shaped into a series of highly planned megalopolises, which are regulated by an advanced technology and fed by a mechanized and chemically sanitized agriculture. This future course is countered largely by the tenacity of many peoples throughout the world, including some of the Indians of North America, marginal and peasant farmers, traditional craftsmen, and many of the young seeking alternatives to the modern industrial state. However, national and international agencies and business enterprises are vigorously attempting to raise the standard of living of most of these peoples and incorporate them into the framework of the dominant societies. The rapid influx of populations into urban areas indicates that these attempts are successful in at least one respect, namely that the numbers and impact of those who live apart from the mainstream of society is constantly being reduced. The world is rapidly becoming more homogeneous and therein may lie one of the most serious problems confronting modern man.

It is my contention that we are in danger of losing an important amount of social variability in the human community at the same time as we are losing the required amount of biological variability in our life-support bases. If we continue on our present path at the present rate, then our chances of maintaining healthy communities and environments will be dramatically reduced before the year 2000, perhaps beyond a point where society as we know it will bc incapable of functioning.

Most current solutions to the immense problems facing us utilize the latest techniques of systems engineering and involve resource and social management on a previously unattainable scale. Increasingly, governments and international agencies are coping with the future by planning and acting on a worldwide basis. FAOs ambitious plan known as the "Indicative World Plan for Agricultural Development" exemplifies this approach and will strongly influence, if not dictate, agricultural development in many of the poorer nations over the next quarter century. There are a number of dangers built into top-down management at national or supranational levels as progressively fewer people are going to be making more and more of the recommendations. This could lead to a lessening of the representation of people and points of view involved in shaping society, particularly if systems specialists take it upon themselves to select the inputs and come up with the answers to future planning. Unfortunately, there is no guarantee that the methods currently in vogue will do any more than identify the crises that are piling up, and if important social or environmental variables are omitted, then these plans may actually aggravate our problems. Equally sobering is the very real possibility that most of the solutions to problems now confronting us will lead to a further loss of social and biological diversity. Since we cannot afford to risk survival in this way, the time has come for science to begin to carve out alternatives, even if they are antithetic to society's dominant emphases.

A few years ago the New Alchemists began their search for ways in which science and the individual could come to the aid of humanity and the stressed planet. We shared the uneasy feeling that modern science and technology have created a false confidence in our techniques and abilities to solve problems. We were also disturbed that most futurology seemed to jeopardize the continued human survival and displayed a real ignorance of biology. It was clear from the outset that social and biological diversity needed to be protected and, if at all possible, extended.

We felt that a plan for the future should create alternatives and help counter the trend toward uniformity. It should provide immediately applicable solutions for small farmers, homesteaders, native peoples everywhere, and the young seeking ecologically sane lives, enabling them to extend their uniqueness and vitality. Our ideas could also have a beneficial impact on a wider scale if some of the concepts were incorporated into society-at-large. Perhaps they could save millions of lives during crisis periods in the highly developed states. This modest and very tentative proposal suggests a direction that society might well consider.

At the foundation of the proposal is the creation of a biotechnology, which by its very nature would:

- function most effectively at the lowest levels of society.
- be comprehensible and utilizable by the poorest of peoples.
- be based upon ecological rather than economic or efficiency considerations, although this would not negate the development of local eco-economies.
- permit the evolution of small decentralist communities, which in turn might act as beacons for a wiser future for many of the world's population.
- be created at local levels and require relatively small amounts of financial support. (This would enable poorer regions or nations to embark upon the creation of indigenous biotechnologies.)

### Unnatural Selection: Loss of Diversity

It is necessary before describing a way of reviving diversity at all levels to evaluate how its loss threatens our future. If an extremely wise ecologist-philosopher from another planet were commissioned to investigate earth he would be dismayed at the outset by our nuclear weaponry and our inability to reach a genuine arms agreement. In fact, he might drop the assignment and leave immediately for fear of his life. If he was daring enough to remain, his confidence in our future would be further shattered by the tendency of the dominant societies, whether "communist" or "capitalist," to be constantly selecting the most efficient or profitable ways of doing things, since the bases for judgment within both political systems are too narrow and short-lived. Our ecologist would ascertain clearly that our narrow approaches are reducing our options. The problem is compounded further by the tendency to condition people to the options that remain. To him it would represent an evolutionary trap and, after his survey of energy use and agriculture was completed, he would confidently predict a major catastrophe. There would be no need to go on to industry, the university, or government, despite the fact that much ecological insanity resides in them.

Examples of unnatural selection are everywhere:

For hundreds of years prior to the industrial revolution a wide variety of energy resources was used. Besides animal power and human toil there was a subtle integration of resources such as the wind and water power and a variety of fuels including peat, wood, coal, dung, vegetable starches, and animal fats. This approach of integration through diversity in providing the energy for society has been replaced by an almost exclusive reliance on fossil fuels and nuclear power. Energy sources are often linked together into huge transmission grids, which provide electric power over large sections of the country. The industrial revolution took place only where there was a large scale shift to fossil fuels as an energy source. The costs to the environment and to the sustained human welfare may yet overshadow its benefits. The production of air pollutants and highly dangerous radioactive wastes continues to increase rapidly, and no downward trend is immediately in sight, despite an increased environmental awareness. Modern society, by reducing the variety of its basic energy sources while increasing its per capita energy needs, is now vulnerable to disruption on an unprecedented scale. It would be foolhardy to disregard the very real possibility of a small group of people destroying our power transmission systems. Tragically there are no widely disseminated backup sources of power available to help the majority of people in a nation hooked on massive amounts of electricity. Society was not as precariously based as this in 1776.

On the country's farmlands, changes have taken place over the last fifty years that have not yet had their full impact on the nation. The majority of the population has been displaced from relatively self-sufficient farms to be replaced by large monoculture industries capable of producing crops efficiently and relatively inexpensively. That many of the displaced farm people are on welfare or adding to the ghettos' problems is not usually considered by agricultural planners. Unfortunately, the trend is worldwide as former colonial regimes and present economic involvement of the powerful industrial nations have created a climate of uncontrolled urbanization in underdeveloped countries. There is a contemporary theory that contends that the industrial powers have contributed directly to the conditions that led to their dangerously high population levels.

Proselytizers on behalf of modern agribusiness rarely consider the key role of numerous small farms as a social buffer during periods of emergency or social breakdown. This oversight could well be the result of a lack of civilian research into the needs of a major industrial nation under the stress of severe crises, despite the fact that a disaster could occur. A depression of the magnitude of the one that befell the country in 1929 could well take place, but if one should happen in the 1970s the social consequences would be much more severe. In 1929, a large percentage of Americans had friends or relatives on farms that could operate on a self-sufficient basis during lean periods. Today the situation is alarmingly different as the rural buffer is largely gone and far fewer people have access to the land. The problem is compounded by the fact that today's farms have little resemblance to those of forty years ago, as the modern farm is in no way independent and like other businesses it requires large amounts of capital, machinery, and chemicals to maintain its operations.

The replacement of large rural populations, their unique social organizations, and the many small farms by agribusinesses operated primarily on the basis of short-term incentives rather than as legacies for future generations, is resulting in a tremendous increase in homogeneity in the countryside. When the land and landscapes become just another commodity, society as a whole suffers. It might not be too serious if the loss of a viable countryside was all that was threatened by modern agriculture, but a close look at present agricultural methods suggests that many of them are damaging to the environment and a loss of biological variability is rapidly taking place.

### The Green Revolution: Unnatural Selection

Over the past several decades the agricultural sciences have created a number of major advances in food raising and the widely acclaimed green revolution has come to symbolize the power of applied science and technology working on behalf of humanity. Our confidence has been renewed that the mushrooming populations can be fed if only western agriculture can be spread rapidly enough throughout the world. But the green revolution has not been shaped by an ecological ethic and its keenest enthusiasts are usually manufacturers of chemicals and agricultural implements backed by government officials rather than farmers and agricultural researchers, many of whom are aware of the immense complexity of stable agricultural systems. A cursory examination of the ads in a wide variety of American journals and magazines would lead one to believe that the agricultural revolution is a chemical revolution, and perhaps it basically is.

A number of biologists and agricultural authorities are cautious about the future, since they foresee environmental decimation that will offset the agricultural gains before the turn of the century. Among some of them there is a disquieting feeling that we are witnessing the agricultural equivalent of the launching of the *Titanic*, only this time there are several billion passengers.

The modernization of agriculture has resulted in the large scale use of chemical fertilizers upon which many of the new high-yielding strains of grains depend. Coupled with this is a basic emphasis on single cash crops, which are grown on increasingly large tracts of land. The dependency on fertilizers for successful crops has created depressed soil faunas and an alarming increase in nitrates in the ground waters in some areas. The nitrate levels are often above the safety limits set by the U.S. Public Health Service for infants' drinking water.

Along with the widespread use of chemical fertilizers has been the rapid increase of biocides to control pests and weeds. These substances, particularly the chlorinated hydrocarbons, but also the shorter-lived organophosphates, are altering ecosystems and have the potential to threaten their stability. The use of weed killers and pesticides has reduced the number of species of soil animals in many farm fields, with subsequent reductions in the quality of humus that is essential to the sustained health of soils. Unfortunately, these changes are occurring just as we are beginning to discover how much the soil fauna, particularly the earthworms, contribute to plant growth and health. The use of biocides has triggered a vicious cycle; soils decline in quality, which in turn makes crops more vulnerable to attack by pests or disease organisms. This creates a need for increasingly large amounts of pesticides and fungicides for agricultural production to be sustained.

The full impact of biocides has yet to come. It is as if ecology and agriculture represent a modern Janus in their antithetic stances. While a team of ecologists has recently announced that the full impact of DDT often does not show up in long-lived birds, predatory animals, and man for 25 years after application, agricultural planners confidently predict a 600 percent increase in the use of pesticides in the underdeveloped world over the next few years. By the year 2000 the developing nations, as the beneficiaries of an uncontrolled experiment, will have reason to resent the blessings of modern technology.

The most notable achievement of the green revolution has been the creation of new, high-yield strains of rice, wheat and corn. World agriculture has in the space of a few years been made more efficient, and in the short run, more productive because of these supergrains, particularly the Mexican semi-dwarf varieties of wheat. They represent a triumph of the modern plant breeder's art, but they are in no way a panacea to the world food shortage. The grain revolution has an Achilles heel; the new varieties, grown on increasingly vast acreages, are causing the rapid extinction of older varieties and a decline in diversity of the germ plasm in nature. The genetic variability, which initially enabled the new types to be created, is threatened, and the very foundation of the new agriculture is being eroded. In Turkey and Ethiopia thousands of local wheats have become extinct over the last several decades and the phenomenon is widespread. It is possible that the genetic variability of wheats could be irreplaceably lost and Erna Bennett of FAO has stated recently that "the world is beleaguered as far as its genetic resources are concerned." Some of the most influential agricultural experts are deeply aware of the problem and are attempting to create the necessary "gene banks" before it is too late. It has been suggested that the race to save our genetic resources may be hampered by another biological fact of life, namely that seed storage may not be enough as "reserves" of the original micro-climates, and ecosystems may be required if the viability of the local strains are to be maintained.

The tend away from cultivating local varieties to a few higher yielding forms is placing much of the world's population out on a limb. If the new varieties are attacked by pathogens the consequences could be worldwide, rather than, local, and plant breeders may not be able to create new strains before it is too late. Such events are not without precedent. An earlier counterpart of the green revolution occurred in Ireland in the eighteenth century with the introduction of the Irish potato from the western hemispheres. Production of food dramatically increased and by 1835 a population explosion had taken place as a result of the land's increased carrying capacity. During the 1840s a new fungal plant disease appeared, destroying several potato crops, and one-quarter of the Irish people died of starvation. The recent devastation of coffee plants in Brazil is partly the result of their narrow genetic base and their consequent vulnerability to leaf rust disease. The present corn leaf blight in the U.S. is caused by a fungus that attacks plants that carry the T gene for male sterility, and 70 to 90 percent of the corn hybrids carry this gene. Despite heavy applications of fungicides, corn blight is spreading with heavy crop losses, and blight-resistant varieties may not be available in quantity until 1973 at the earliest. A modern agriculture, racing one step ahead of the apocalypse, is not ecologically sane, no matter how productive, efficient, or economically sound it may seem.

There are other hidden perils associated with the modernization of agriculture, but the loss of genetic diversity is perhaps the most readily analyzable example of analogous changes taking place at every level of society. Since a scientific or technological advance on one level (e.g., the super grains) may be pushing us closer to disaster on another, it is time to look carefully at the alternatives before these avenues have disappeared behind us.

## The Future: An Alternative

The environmental dilemma is mirrored by comparable changes in the human situation. Unnatural selection is causing a loss of diversity in the human sphere and this loss may be leading toward social instability. Unlike the biosphere, society as a whole may have increased in complexity or remained fairly constant. However, the roles of most contemporary men and women grow ever more simplified as they relinquish the various tasks of living to myriads of machines and specialist. The brilliant anthropologist Lévi-Strauss has shown just how far this narrowing of roles has progressed. Unlike our ancestors we have little control over the creation of our power, food, clothing, or shelter and this loss may be harmful to the human psyche. This oversimplification and impoverishment of the lives of most of us could lie close to the roots of much of the chaos, violence, and disintegration that threatens modern society. Erich Fromm has suggested that violence, especially, is related to boredom and it seems probable that ineffectualness and boredom are allied in this case.

Function, particularly in the twentieth century, has been replaced by social dither. The immense popularity of snowmobiles, garden "tractors," motorboats, hunting, and contact sports indicates that we have a strong need to recapture the roles of explorer, farmer, navigator, hunter-provider, and warrior. In part, these are displacement activities, attempts to regain equilibrium with the progressive specialization of most people's lives.

The individual's retreat from function and diminution of roles is a negative trend as it removes him or her from the totality of the world. By restoring and extending a genuine interaction with the life processes upon which people depend, it may be possible to reverse this course. This task should begin at the lowest functional levels, at the level of the individual or the small group and their life-spaces.

A real knowledge of microcosms is essential to understanding higher levels. The basic tenet of this proposal is that to build a viable future for humankind we must begin to place emphasis on restoring microcosms in a meaningful way—be they forests, fields, small farms, or communities. If they are healthy, society as a whole will respond, and if they are ill, the human community cannot be well, no matter how much legislation, time, and money are directed toward saving it.

Fraser Darling, in his perceptive studies of remote Scottish peoples, showed how self-sufficiency was a positive force in their lives. The most independent communities were far more diverse and socially alive than the single-industry towns and those heavily dependent on a lifeline to the outside. He also found that they coped far better in their dealings with the world at large. Equally important, the independent communities cared for their environment and were less prone to despoil it for short term monetary gain.

The first step toward countering homogeneity would be to create a biotechnology based upon an ecological ethic. This biotechnology would function at the lowest levels of society, providing inexpensive life-support bases for individual families, small farmers or communities who desire more independence and a way of life that restores rather than destroys this fragile planet. It would not be founded upon profit or efficiency considerations but on the philosophical view that all things are interconnected and interdependent, and that the whole cannot be defined in monetary terms. Energy production, agriculture, landscapes, and communities must be tied together within individual research programs and each area should be considered as a unique entity worthy of study. From indigenous research projects would evolve a biotechnology that reflects the needs of each region and peoples. In this way it will be possible to have fantastically varied communities and landscapes, as each develops its own integration with the world around them.

The New Alchemists have begun studies to shape the skills needed to establish modern, relatively self-contained communities that capture their own power, grow their own foods, and utilize their wastes. Smaller systems, which could be adopted by suburban dwellers or any peoples interested in significant personal changes on behalf of the environment, are in the early stages of development. It is our feeling that the adoption of such support systems would increase individual independence and reestablish a much-needed link with the organic world. If such a bond were created, people may begin to relate more realistically with the larger world around them. It should not be overlooked that the findings of this new biotechnology, if widely adapted, would permit a nation to function more normally during periods of hardship.

The first task of this science is to explore the little-used, ancient, nonpolluting forms of energy such as the sun and the wind, and perhaps even the waves and water currents. On a small scale it will be possible to harness and create new potentials for "old-fashioned" forms of energy. Recent technological advances in energy conversion and storage ensure the likelihood of success. It will not be the task of this biotechnology to create energy systems capable of handling the needs of large cities or industries. However, its discoveries might make decentralization or a return to the countryside more appealing.

Research on new uses for the clean energy sources should be closely linked to an ecologically based agriculture and aquaculture. With this orientation there would be no imitation of the present agricultural sciences, which are basically committed to extending the large, unstable, monoculture agricultural enterprises and increasing specialization in food production. This bioscience should focus on the creation of rich soils, and the raising of high-quality plants and animals together in sophisticated polyculture schemes. Its goal would be the founding of stable and beautiful agricultural environments. Rather than being geared toward displacing people from the land, it would attempt to reverse the trend by emphasizing intensive forms of culture, more varied diets on a local basis, and nurture a love of food producing and agronomy for its own sake. Noncommercial varieties of nutritious foods would be investigated, such as many of the nuts, weeds, and berries, as each area of the country has its own set of neglected wild crops that would benefit from selective breeding.

Besides borrowing many of the research methods and insights of orthodox agriculture and ecology, food research programs should be tied directly with the community's needs and require little capital for adoption. Planting and tending of ecologically sophisticated agricultural plots will require far more training, knowledge, and labor than is needed on contemporary farms. Yet this fact should prove an asset by raising the status of those involved. Persons highly skilled in the creating of good soils and raising nourishing foods will be highly respected and emulated. To work toward restoring the landscapes should become a major intellectual and physical pursuit of the present generation.

Those who teach, investigate, and adopt the findings of this science will be deeply involved in most phases of raising of foods and the generation of power to meet their needs. Diverse multicrop food projects will add to community life by shaping exquisitely beautiful landscapes, which will be a great pleasure to live and work in. These microcosms may contain within themselves some of the seeds of change for the larger society around them.

It is proposed that throughout the country centers of research and education be established and modestly funded. They should not be controlled by the universities, although affiliation in some instances might add to the variety of inputs into their development. Apart from research funding, these centers should be relatively self-sufficient and the knowledge gained be made available to all through community extension on the part of the centers' residents.

Our experiences in attempting to create a holistic biotechnology for small communities has convinced us that many outstanding young scientists would gladly commit themselves to its development and success, for no less than our future may be at stake.

*John Todd*

John Todd, "A Modest Proposal," in *The Book of the New Alchemists*, ed. Nancy Jack Todd (New York: Dutton, 1977).

## Questions and Topics for Discussion and Writing

1. Why is loss of diversity in power technology and in agriculture a problem in and of itself?
2. The title of Todd's article is taken from a satirical essay of that title by Jonathan Swift. A quotation from Swift's "A Modest Proposal" appears at the beginning of Todd's article. What is the point that Todd wishes to make by choosing his title and printing the quotation from Swift?

# They Thought River was Beyond Saving

## Craig McInnes

*It is easy to point to environmental damage and wring one's hands in despair. It is much harder to find a way to combat it that really works. This story about Cleveland's cleanup of the Cuyahoga River underscores both the magnitude of the problems and the viability of real solutions.*

The fire that broke out on the surface of the Cuyahoga River on the morning of June 22, 1969, lasted only about 25 minutes.

But by the time the fireboat Antony J. Calabrezze had the burning oil and debris under control, Cleveland had become an international joke, and public indignation over pollution had reached the boiling point.

Twenty years later, that indignation and the political action it inspired have transformed the Cuyahoga. Although still polluted—it is one of the 42 hot spots around the Great Lakes singled out for special attention by the International Joint Commission—the river is as remarkable now for how much it has been cleaned up as for the problems that remain.

Pleasure boats now cruise waterfront bars and restaurants on waters once piled only by freighters, fireboats, floating sewage and oil slicks. There are four marinas near the river mouth, and traffic has become so heavy the U.S. coast guard has had to establish safety zones to keep runabouts from being squashed by freighters.

"Back in 1973, if somebody had told me we were going to have problems with recreational boating on the mouth of the Cuyahoga they would have been laughed out of the office," said Robert Wysenski, an environmental scientists at the Ohio Environmental Protection Agency who is co-ordinating cleanup plans for the river.

Although the Cuyahoga fire made Cleveland synonymous with pollution, people who live here share with other denizens of the Great Lakes region the dubious distinction of being subjected to more toxic chemicals in their food and drinking water than anyone else in North America, according to a report by the Royal Society of Canada and the U.S. National Research Council.

And the Cuyahoga shares with other waterways in this heavily populated region the burden of chemicals, heavy metals and sewage delivered by industry, farms, homes and runoff from city streets. Before the cleanup began, those pollutants had formed a toxic cocktail so potent that virtually nothing lived in the river from Akron to its mouth at Lake Erie.

Pumping life back into the corpse of the Cuyahoga has not been easy.

"It's been a long expensive road to get to where we are now," Mr. Wysenski said.

The river's revival started with the formation in 1972 of federal and state environmental protection agencies. And for the first time, legislation was passed setting quality standards for water and air.

Many people believe the fire on the Cuyahoga provided the impetus for legislators to overcome objections of the industrial and municipal lobby, which until then been able to ward off any meaningful controls.

The Cuyahoga is a small river, but it is subject to incredible stress. Two of its largest tributaries are sewage treatment plants, one in Akron and one in Cleveland.

Effluent from the Akron plant contributes as much as two-thirds of the flow of the river, and so in its turn does the larger Cleveland plant farther downstream. A single industrial complex, LTV Steel Co. in Cleveland, uses up to 750-million litres a day, at times more than half the river's minimum flow, before returning the water to the channel.

When the cleanup began, Congress passed legislation requiring action, but it also provided the means. Washington paid 75 per cent of the capital cost of the sewage treatment infrastructure.

Hundreds of millions of dollars were invested in sewage treatment plants. But industries along the river had to pay the cost of curtailing their emissions.

Industrial discharges were required to meet standards based on how much life the Ohio EPA expected the river to be able to sustain. In the shipping channel, water quality is only supposed to meet "nuisance prevention criteria." Above the shipping channel it must meet the needs of a "warm water habitat," a designation that means fish, plants, insects and bottom dwellers should be able to survive.

Everyone agrees that the river needed to be cleaned up, but the definition of clean has been contentious.

"The lower section of the Cuyahoga is probably the most litigated piece of river in the state," Mr. Wysenski said. The latest battle is between the Ohio EPA, LTV Steel and the Cleveland and Akron sewage plant operators, who are arguing that the "warm water habitat" designation is unreasonable—fish will never be able to survive in the section immediately above the ship canal.

At stake are millions of dollars. The more stringent the regulations, the greater the cost of cleaning up. Ohio EPA won the first round, but the polluters are appealing.

The question of how clean is clean has become central to much of the debate concerning toxic substances in the environment.

"It's a question of resources, that's what it comes down to every time," Mr. Wysenski said. "When you do a survey, everybody says, 'Yes, we want cleaner water.' What you need to include in that survey is, 'Are you willing to pay...a 20–30 per cent increase in your utility rate for sewer and water? '"

In Ontario, where enforceable water pollution standards are being developed under the Municipal Industrial Strategy for Abatement, the standard has been defined as the best available technology that is economically achievable.

But is cleaning up toxic pollution just a question of money and political will?

There are seemingly intractable problems. Six years ago, for example, Howard Ferguson, then senior Environment Canada official in Ontario, said Hamilton Harbor might never be purified because there was no safe way of dealing with the highly toxic bottom sediments.

Today, however, most observers believe that the problems are solvable.

"There is no longer a technological barrier to clean up," said Paul Muldoon, co-author of a report on reducing toxic substances in the Great Lakes for the Canadian Institute for Environmental Law and Policy. "It's a question of political will and economic priorities."

Ultimately, whether clean water is a priority will depend on public response. Acceptable pollution levels are still being set based as much on what the public nose will stand as on scientific evidence.

For Mr. Wysenski, who has spent much of the past 17 years trying to bring back the Cuyahoga, settling those levels is a key first step to further cleanup.

"I think we need to come up with realistically attainable, socially acceptable and economically affordable standards," he said.

"We're not going to have a trout stream flowing through downtown Cleveland."

*The Globe and Mail*, April 15, 1989. D2

## Questions and Topics for Discussion and Writing

1. Why was the Cuyahoga River polluted?
2. What does the term "clean water" mean? Who is to decide what standards for environmental controls should be.
3. Who should pay the costs of environmental cleanup?

*Craig McInnes*

# Biodegradable: Does Nature Win or Lose?

## Craig McInnes

One of the most intractable environmental problems caused by modern industrial society is the problem of waste disposal. The sheer bulk of urban garbage is a vast problem in itself, but it is further complicated by the portion of that garbage which is not biodegradable, in particular, plastics. The latest attempt to address this problem is the development of "biodegradable plastics." This article reviews the mixed reception of biodegradable plastics among environmentalists.

The recent development of plastics that are designed to break down once discarded has given shoppers a chance to decide between competing products on the basis of their effects on the environment.

However, there are warning that these "environmentally friendly" products may actually do more harm than good.

For many, plastics have long been a symbol of everything that is wrong with the way modern society treats the environment.

They are made from non-renewable resources, they are disposable, and when discarded they offend the landscape for decades. None of the microorganisms that act as natural janitors have been able to clean up plastics.

But within the past five years, science and industry have come up with ways of making plastic disintegrate. Products have come on the market that are supposed to be biodegradable (designed to break down in dumps) and photodegradable (designed to break down in sunlight).

Finally, it seemed, industry had produced environmentally friendly forms of plastic. And certainly that is the way they have been promoted.

The additives used to make them degrade have names such as Ecostar and Ecolyte.

The Loblaws grocery chain calls its biodegradable garbage bags Nature's Choice. and the Liquor Control Board of Ontario uses plastic bags that are "photodegradable to help our customers contribute to a clean and healthy environment."

But critics say the new plastics won't help alleviate the garbage crisis or reduce litter. And worst of all, they will impede recycling programs that are taking shape.

It is recycling and reuse that are part of the real solution to the garbage crisis, environmentalists say.

"That's the No. 1 thrust; it shouldn't be going into the garbage dump whether it's biodegradable or not," said Colin Isaacs, executive director of the environmental group Pollution Probe.

"It is the throwaway society that represents much of the environmental problem that we are facing today....Every tonne of plastic that goes into the garbage represents a tremendous waste of oil and gas."

Dow Chemical Co. and Domtar Inc. have announced a joint venture to recycle about 23 million kilograms of mixed plastics a year.

In manufacturing products using some recycled plastic, "they've indicated that they will tolerate (no more than) one half of 1 per cent contamination (foreign material in the plastic), unless it is degradable plastic," in which case "they will tolerate absolutely none at all," said John Hanson of the Recycling Council of Ontario.

Manufacturers are concerned that products made from recycled plastic made from degradable stock will end up disintegrating.

"Some of the end products they manufacture are roofing materials, wallboard, conduit for plumbing and ground piping and consumer products. Obviously you don't want those types of things falling apart," Mr. Hanson said.

But the developers of a cornstarch additive used to make plastic biodegradable deny that their products will impede recycling.

"Four years ago when we first got into this there was virtually no recycling" said Wayne Maddever, manager of business development at St. Lawrence Starch Co. Ltd.

"Now that recycling has taken off, particularly here in Ontario, we've taken an active step back...and are saying this is appropriate technology for those things that are not going to be recycled, one-way items such as garbage bags and diapers."

To put biodegradable plastics in widespread use and to put them into pop bottles, that would be irresponsible, and "we are not trying to do that."

In addition to the question of whether biodegradable plastics are useful in principle, there is the question of whether they actually degrade as promised.

Critics say it is not clear that they will break down in landfill sites, and if they do, it is not clear they will break down into benign substances.

In theory, the cornstarch in biodegradable products will be attacked by micro-organisms and eaten away, leaving the plastic that remains in small pieces. Then other additives are supposed to speed the process of oxidation, which ultimately will break down the plastic. (Plastic does eventually break down, but no one knows how long it takes).

"Under the correct conditions, starch, which is a naturally occurring material, is biodegradable. It is easy to get these conditions to apply in a laboratory. In a landfill site or a garbage dump, it is rather more difficult," said Dr. David Wiles, director of the chemistry, division of the National Research Council.

"There isn't a shred of evidence that these things will work inside a landfill," Mr. Isaacs said.

Even naturally biodegradable materials don't always degrade quickly in landfill sites.

"I've got pictures in my office of carrots that have (not decayed) in landfill sites for 10 years," said Fred Edgecombe, technical director for the Society of the Plastics Industry of Canada.

If biodegradable plastic products don't break down quickly, they will do nothing to alleviate the landfill crisis. Plastics make up about 7 per cent by weight of garbage going into dumps and a larger amount by volume.

Even if they do break down quickly, the 95 per cent of the biodegradable product that is plastic will not follow suit, Dr. Wiles said, so the result will be a lot of tiny pieces of plastic in the landfill site.

In some cases, biodegradable products actually contain more plastic than the products they are replacing because the plastic-starch mix is not as strong as pure plastic.

"They have to add 10 per cent more plastic (10 per cent more for each individual product) to the biodegradable ones in order to maintain strength," said David Sparling of the Ontario Ministry of the Environment. "Well, that's adding 10 per cent more plastic to the waste site, even though the film itself may disappear in six months."

The ministry hasn't formed a position yet on degradable plastics, Mr. Sparling said. Along with the Society of the Plastics Industry, the ministry is commissioning a study on the issue.

Photodegradable plastic, which degrades in the sun, has been promoted as an effective means of dealing with litter.

Several states have passed laws requiring that certain items, including the rings that hold six-packs of beer and soft drinks together, be made with the material.

But environmentalists in Canada fear that people will see degradable plastics as a licence to litter.

"I would certainly strongly oppose sticking a label on a plastic container that says this thing is photodegradable, because that would encourage at least some percentage of the public...to throw it out of the window because they would be given the impression that is would magically disappear," Mr. Isaacs said.

"If we want to tackle litter...it is time to start developing programs to encourage the public to stop littering."

For companies that have been trying to respond to environmental problems, the controversy over degradable plastics has been frustrating.

Last fall, the Globe and Mail switched to biodegradable plastic bags to keep newspapers dry. The bags, which are more expensive than the ones they replaced, were ordered in response to readers' environmental concerns.

*Craig McInnes*

Now some readers have been calling to complain about the degradable bags.

"You are basically damned if you do and damned if you don't," said Dimitri Chrus, Toronto circulation manager for The Globe.

For Mr. Isaacs, at least, the choice between degradable and regular plastic is clear.

"As a consumer, I still buy the cheapest garbage bags."

*Craig McInnes*

*The Globe and Mail* (March 4, 1989): D4.

## Questions and Topics for Discussion and Writing

1. What is a "biodegradable plastic"? How is it made?
2. Outline the major objections of environmentalists to biodegradable plastics.

**Photograph on next page:**
Smog in a German industrial town.

*Issue*

# Atmospheric and Climatic Problems

Damage to the environment takes many forms, including deterioration of soil, adulteration of the water table, poisons in the food chain, chemical spills, etc., but a special place must be reserved for atmospheric and climatic problems. The air we breathe is not and cannot be confined to one location (excepting totally controlled environments specially constructed for people with serious medical problems). Air circulates freely around the globe, and in time mixes completely. It has been said, by way of illustration, that in each and every breath of air we take, the probability is greater than one-half that we will have inhaled a molecule of air that was exhaled by Shakespeare when he first said "To be or not to be, that is the question."

The atmosphere is what makes life as we know it possible in the first place. The atmosphere collects and holds just enough heat to make life possible; a little more heat or a little less heat makes a very large difference to the climate and to the balance of nature. The atmosphere also protects us from harmful rays from the sun.

The articles in this section discuss the very alarming evidence that the atmosphere has already been damaged and that the damage will increase whatever we do about it now. The consequences for the climate are likely to be severe. Jovan V. Jovanovich in "Is Man Meddling?" cites evidence of changes in the global climate and discusses the adverse consequences that will result from the changes, pointing to destruction of rain forests and industrial release of chlorofluorocarbons (CFCs) as contributing causes. Lydia Dotto, in "Dangerous Holes in the Sky Above Us," reports on the vast holes in the ozone layer, which protects us from damaging ultraviolet rays. In "Endless Summer: Living with the Greenhouse Effect," Andrew C. Revkin gives a detailed analysis of the effects we may expect from the warming of the climate that is the virtually certain consequence of pollutants now found in the atmosphere.

# Is Man Meddling?

## Jovan V. Jovanovich

Thomas Malthus predicted disaster purely on the basis of the rates of increase of food versus population. He did not take into consideration the additional problems that are becoming obvious today, caused by the ever-growing population meddling with the delicately balanced atmosphere that regulates our climate. This article by Professor Jovan Jovanovich from the University of Manitoba explains what is at stake.

In the past few years there has been frequent news of recurring droughts and famines in Ethiopia and Africa below the Sahara. In 1986, drought swept the southern part of the United States. In 1987 came a killing heat wave in Greece. Last year, most of the North American grain belt had unusually warm and dry weather—one of the worst droughts on record. By the time the summer ended, record forest fires and a record hurricane had struck.

Droughts are, of course, nothing new. They come and go, regularly. The rains always return. But this time, there is something new.

Man has been changing the atmosphere. By burning fossil fuels, by using CFC (chloro-fluoro-carbon) gases in refrigerators, spray cans and foam coffee cups and by cutting down the forests, along the Amazon or in British Columbia, we may be affecting the global climate. Is man's effect on the atmosphere responsible for the recent droughts?

The importance of changes in global climate cannot be overemphasized. To give one example: suppose that the rains shift from Ethiopia's highlands to somewhere else and that the Nile River dries up to a trickle. What happens to the people living along its banks? There are 50 million of them.

The current popular phrase for the possible shifting of the world's climate is the greenhouse effect.

By burning fossil fuels and destroying forests at an ever-increasing rate, we have increased the amount of carbon dioxide in the atmosphere by about one-quarter since the beginning of the Industrial Revolution. We are also releasing CFC gases into the atmosphere at an accelerating rate—gases that weren't there before and will stay there for centuries. The carbon dioxide and other gases warm up the Earth by creating the greenhouse effect.

The Earth receives energy from the sun, mainly in the form of visible sunlight, and radiates exactly the same amount of energy into outer space, in the form of invisible infra-red (thermal) radiation. Physicists say the Earth is in a thermal equilibrium with the sun and outer space.

The laws of physics tell us that this equilibrium is reached when the Earth's temperature is, on average, about minus 20 Celsius. How is it, then, that the average surface temperature is about 15 degrees above zero?

The reason is that the Earth has just the right combination of the "greenhouse" gases (carbon dioxide, water vapor, ozone and nitrogen oxides) and the "non-greenhouse" gases (oxygen, nitrogen and argon) in the atmosphere. Acting together, these gases trap just the right amount of solar heat near the surface of the Earth and raise its temperature to plus 15 degrees. At the top of the atmosphere, however, it is very cold, which is necessary to maintain the thermal equilibrium with the sun and outer space.

Venus, for some reason, has an atmosphere loaded with greenhouse gases and a surface temperature of about 500 degrees, hotter than fire. On Mars, the atmosphere contains too few greenhouse gases and the temperature is about 40 below zero.

Without greenhouse gases in the Earth's atmosphere, there would have been no life on the planet. The whole of Earth would be as cold as Mount Everest. If there were excess greenhouse gases, Earth would be too hot to support life. We owe our existence to the very delicate temperature balance maintained by just the right mixture of gases in the atmosphere.

By changing the level of these gases, however, we interfere with this delicate balance. The result is an inevitable change in the average temperature on Earth.

There is no doubt about this, despite some claims to the contrary. It should be pointed out that a temperature increase on the average does not exclude the possibility of one relatively small part of the Earth becoming cooler and another part correspondingly much warmer, providing the over-all average is warmer.

Unfortunately, at present, science cannot tell reliably how much and how fast the Earth will warm up or cool off if we change the balance of greenhouse gases by a certain amount. Determining the effect on the climate in each part of the world is even more complicated.

But there has been progress. Climatologists have developed better climate models and computer programs, which indicate that, if we double the amount of carbon dioxide in the atmosphere (having already increased it one-quarter), the average global temperature may increase somewhere between 1.5 and 4.5 degrees Celsius. Rains should increase north of the 35th latitude, but decrease to the south, except in a narrow zone near the equator.

The Prairies will likely grow hotter and drier. So will most of the mountains in the West. The North should become warmer and wetter.

Will the Prairies become as dry as Sahara is today? Will forests in the mountains become so dry that they will eventually burn up? We do not know—but they may.

Eighty-four per cent of the water in the Nile River originates in the Ethiopian highlands. Ethiopia has been hit with a series of severe droughts. Could the Nile dry up? We do not know. What we do know, and this is very disturbing, is that the water levels in the Nile, as well as in the Niger and Lake Chad, have fallen appreciably in the past decade. And the Sahara Desert has been spreading southward.

Are these the first major signs that we are changing the climate? Again, we do not know, but they may be. The weather patterns are consistent with those predicted by the climatologists' computer models. However, because of natural climatic variabilities, this short experience doesn't necessarily prove that the models are correct.

So what do we do? Do we wait until climatologists present us with the scientific proof that we *are* changing the global climate before we decide that it is justified to act, before we decide to reduce the burning of fossil fuels and the release of CFC gases?

Climate experts have no way of giving us scientific proof until their models are tested—that is, until the climate has actually changes. And who wants to change the global climate?

Unfortunately, we are doing just that. We are, according to a statement from a recent conference on the atmosphere, "conducting an unintended, uncontrolled, globally pervasive experiment" with the climate. We are doing so by following the path of least resistance: we want the cheapest energy, cheap refrigerators, foam plastic and convenient spray cans.

There is no use waiting for scientific proof. We will not get it for a long time, if ever. We have to make our decisions on the basis of incomplete information, on the basis of our perception of future risks, on the basis of our willingness to give up something today, for the sake of our children tomorrow.

Before it is too late.

*Jovan V. Jovanovich*

*The Globe and Mail* (January 10, 1989): A7.

## Questions and Topics for Discussion and Writing

1. What are the factors given by Jovanovich through which we have been affecting the atmosphere?
2. What evidence does Jovanovich cite that the climate may be changing?
3. Why should climactic change concern us?

# Dangerous Holes in the Sky Above Us

## Lydia Dotto

Environmental damage is particularly insidious in that the effects of preventable human actions often do not show up for decades. By then, much damage has already been done and more, much more, will inevitably follow as a result of human activities that are already in the past. An example is damage to the ozone layer. Scientists have been predicting it for years and urging the banning of clorofluorocarbons (CFCs), which deplete the ozone layer. Now that the damage is visible, the task *still* remains to convince governments—around the world—to act *now* to prevent more damage in the next century. It is a foregone conclusion that much more damage will occur in the next few decades that we have no way of reversing. This article by Lydia Dotto summarizes the nature and extent of the problem, both scientific and political, as analyzed by scientists in mid-1988.

Human beings are conducting a huge, uncontrolled experiment on the earth's atmosphere. No one knows exactly what the result will be, but the early warning signs suggest that our children will inherit a planet with a markedly changed climate. Scientists say this experiment could cause global environmental disruptions second only to a massive nuclear war—and that these disruptions are more likely to happen than nuclear war. Indeed, the occurrence of large holes in the *ozone* layer and extreme weather events all over the world suggest that these have already begun.

Many scientists are so convinced that a serious problem exists, despite uncertainties about its exact dimensions, that they've stopped pussyfooting around with hedged statements larded with scientific jargon. Displaying a rare degree of consensus and a willingness to step beyond cautious, narrow interpretations of their findings, they're urging political leaders and the public to act now to protect the atmosphere, or face serious economic and social problems that could threaten world peace.

A case in point is the statement issued at an international conference, "The Changing Atmosphere: Implications for Global Security," held in Toronto in June [1988]—in the midst of a record-breaking heat wave. Sponsored by the Canadian government, it attracted more than 300 scientists and policy makers who concluded that

> the earth's atmosphere is being changed at an unprecedented rate by pollutants resulting from human activities, inefficient and wasteful fossil fuel use and the effects of rapid population growth in many regions. These changes represent a major threat to international security and are already having harmful consequences over many parts of the globe.... The best predictions available indicate potentially severe economic and social dislocations for present and future generations, which will worsen international tensions and increase the risk of conflicts among and within nations. It is imperative to act now.... If rapid action is not taken now by the countries of the world, these problems will become progressively more serious, more difficult to reverse and more costly to address.
> Humanity is conducting an unintended, uncontrolled, globally pervasive experiment whose ultimate consequences could be second only to a global nuclear war.

Nothing has contributed to the sense of urgency more than the discovery of large and unexpected "holes" in the ozone layer which protects life on earth from the sun's damaging ultraviolet radiation. There's a hole over Antarctica that encompasses an area about half the size of Canada and appears to be spreading. A smaller, but still extensive hole was found over the Arctic in 1986. Scientists believe that industrial chemicals known as chloroflurocarbons (CFCs) are major culprits in this massive destruction of ozone over the Antarctic, but they don't know yet if the same fate will befall the rest of the ozone layer. The circumstantial evidence against CFCs is "pretty clear, but whether the hole will spread is anyone's guess," says Alex Chisholm, director general of research at Environment Canada.

"A great majority of working scientists agree that CFCs are implicated," says Kenneth Hare, University Professor Emeritus of [the University of Toronto] and chairman of the Canadian Climate Planning Board, who was a speaker at the Toronto conference. Discovery of the Antarctic hole had important political ramifications because, he says, "it was taken as the smoking gun." Last year—13 years after the first warnings about CFCs were sounded—an international plan to cut back production of the chemicals was hammered out in Montreal.

Unfortunately, even if we stop using all CFCs tomorrow, the ozone problem will get worse before it gets better. Nearly all of the millions of tonnes of CFCs already released remain in the lower atmosphere and they will continue slowly making their way into the upper atmosphere to destroy ozone well into, and perhaps beyond, the next century.

Equally worrying is the fact that CFCs, like carbon dioxide, contribute to the "greenhouse effect"—the atmospheric warming that occurs when certain chemicals trap heat from the earth's surface and prevent it from escaping into space. (Ironically, it's a natural greenhouse effect that makes this planet liveable in the first place, maintaining the average global temperature about 35 degrees Celsius higher than it would otherwise be.) Several of the warmest years in recorded history have occurred during this decade and many scientists believe that greenhouse warming has already begun to affect global weather patterns. They project an increase in the earth's mean surface temperature of 1.5 to 4.5 degrees C. by the middle of the next century—a change that is far from trivial when you consider that just five degrees separates the earth's present temperature from the last ice age.

The likely consequences include a global rise in sea level, major changes in wind patterns and rainfall distribution and greater climatic variability—more severe and more frequent storms, floods and droughts and greater temperature swings. Scientists warn that major climatic disruptions are likely to occur within the next 50 to 100 years that could exceed anything experienced in human history. "Not since the abrupt end of glacial climates a little over 10,000 years ago have temperatures changed as much or so rapidly," says Hare. "I have no doubt that we are discussing the central environmental problem of our times."

That problem stems directly from our profligate use of energy and various technologies which produce chemicals that alter the nature of the atmosphere—notably carbon dioxide and CFCs. Carbon dioxide exists in nature, but we are steadily increasing the amounts in the atmosphere by burning fossil fuels and cutting down large tracts of trees. CFCs are synthetic industrial chemicals widely used as coolants in refrigerators and air conditioners, as solvents for cleaning microchips and other electronic products, and in manufacturing foam insulation and other foam products such as food containers and automobile fittings. They were also used as spray-can propellants until the early 1980s, when these uses were banned or curtailed in many countries.

Invented in 1928, CFCs were being produced in the hundreds of tonnes annually by the 1960s. They grew at a rate of nearly 10 per cent a year, peaking in 1974 at more than 800,000 tonnes. Because they are non-toxic, non-flammable and chemically inert (they don't react with anything), they are less directly hazardous to human health than many alternatives and they were, with reason, thought to be environmentally benign. But it is precisely because they do not interact with anything near the ground that they can hang around in the atmosphere for decades, finally making their way into the stratosphere where ozone is located.

Most of the ozone is in a layer about 20 km thick, between 15 km and 35 km above the earth's surface. The concentrations are low; if the entire layer were compressed to ground-level pressure, it would form a band only three millimetres thick. This fragile barrier is all that stands between us and the sun's biologically damaging ultraviolet radiation, known as UV-B. (Another irony—when located at ground level, ozone is a serious pollutant and a threat to human health.)

It's estimated that a one per cent depletion in

stratospheric ozone would cause a two per cent increase in UV-B at the earth's surface. This would increase the incidence of sunburn, skin cancers and cataracts, a leading cause of blindness; it would also cause skin aging. Studies indicate that a one per cent ozone depletion would cause a four to six per cent increase in non-melanoma skin cancers, which generally are not fatal but can be disfiguring and painful and expensive to treat. A five per cent decrease in ozone would be expected to increase the incidence of these cancers by 22 to 33 per cent, which would translate into an annual worldwide increase of tens of thousands to hundreds of thousands of new cases among light-skinned populations.

Although the relationship between UV-B and malignant melanoma, an often fatal form of skin cancer, is less clear, studies suggest a link and experts say we should assume that ozone depletion would result in an increase in the incidence of melanomas, perhaps at a rate of about one to 1.5 per cent for every one per cent of ozone depletion.

Animal studies indicate that UV-B exposure can cause impairment of the immune system, increasing susceptibility to infections. There's some evidence that UV-B has similar effects on humans. Though this aspect needs further study, Lester Grant of the U.S. Environmental Protection Agency says we should expect ozone depletion to result in increased suppression of the human immune system. This, in turn, would cause a higher incidence of all infectious diseases. Grant adds that immune system effects could affect more diverse populations than the skin cancer effects (less likely to occur in dark-skinned populations) and have an even greater impact on the health care system.

Increased levels of UV-B would also affect agriculture and natural ecosystems. The most important food crops—wheat, rice, corn and soya beans—are sensitive to UV-B. Studies on soya beans indicate that a one per cent ozone depletion would result in a one per cent reduction in crop yield. Organisms that live in the top layers of the oceans including fish species important as food sources, would also be susceptible to increased UV-B. Harvard University scientist Michael McElroy says that in 1987 the Antarctic ozone hole resulted in exceptionally high ultraviolet levels at the surface, causing concern about the impact on the very productive polar ocean ecosystems, including the marine food chain.

CFCs destroy ozone in a particularly efficient way: they produce chemicals that become involved in chain reactions with ozone. With each reaction, an ozone molecule is destroyed but the destructive chemical emerges intact. This is why even tiny concentrations of CFCs can have a huge effect: every CFC molecule can destroy thousands of ozone molecules.

Nor are CFCs the only culprits; halons, methane and nitrous oxide are all precursors of chemicals that similarly destroy ozone. Halons are used as fire retardants. Methane and nitrous oxide are produced by both natural biological processes and human activities, particularly agriculture. Atmospheric concentrations of these chemicals are increasing at rates ranging from 0.2 to about five per cent per year and some CFCs and halons are growing at even faster rates. Although there was a drop in global CFC production between 1975 and 1982 because of the spray can ban—to a low of about 600,000 tonnes in 1982—production has since increased by five to six per cent a year and had reached 750,000 tonnes by 1986.

"There is now compelling observational evidence that the chemical composition of the atmosphere is changing at a rapid rate on a global scale," says Robert Watson, head of the upper atmosphere research program of the U.S. National Aeronautics and Space Administration. Any climate or ozone changes caused by these long-lived chemicals would be very persistent. "Full recovery of the system will take several tens to hundreds of years after the emission of these gases into the atmosphere is terminated."

Because there are large natural variations in stratospheric ozone, it's difficult to determine exactly what is going on up there, but the evidence to date indicates that between 1978 and 1985 there was about a 2.5 per cent drop in the annual average amount of ozone in mid-latitudes in the northern and southern hemispheres (after allowing for natural variability). It has been calculated that continued growth of CFCs and halons at three per cent a year would deplete the ozone layer by about six per cent by 2040 and more thereafter, while a freeze on emissions at or below the levels projected for 1990 would reduce ozone less than one per cent by the year 2015 and less thereafter.

The events in the Arctic and Antarctic are currently the most worrying phenomena. In 1986, Environment Canada scientists found that ozone over the Arctic thinned for about six weeks during March and April. This hole appeared to be similar to though less extensive than the one in the Antarctic—about one-third the size and one-third the depth—and it moved around the North Pole, shifting from northern Europe to northern Canada. Studies done in March 1987, however, indicated that Arctic ozone was back to normal.

In the Antarctic, the ozone hole occurs during the southern spring, late August to early November, and it appears to be getting deeper, lasting longer and covering more territory each year. It now covers a huge area ranging from the South Pole up to about 45 degrees S. latitude. In 1987, the largest depletion—fully 50 per cent—was measured and it persisted longer than ever before—until early December. In October 1987, ozone concentrations in the Antarctic hole fell to less than five per cent of their August levels between the altitudes of 15 and 20 km. Last year, ozone was lower at all latitudes south of 60 degrees S. than at any time since measurements began. While the depletion is worst during the Antarctic spring, ozone has decreased at all latitudes south of 50 degrees S. throughout the year.

Three theories were advanced to explain the Antarctic hole: it was being caused by CFCs and halons; changes in the circulation of the atmosphere were causing ozone-poor air to be carried into Antarctica; and increased solar activity was causing ozone-destroying chemistry to occur in the region. Watson says that recent studies indicate that, while wind patterns definitely play a critical role, "the chemical composition of the Antarctic stratosphere is highly perturbed compared to elsewhere on the planet." These unusual chemical conditions are linked to CFCs. However, scientists cannot yet determine the relative importance of the two factors because the hole is still a poorly understood phenomenon. "It is not yet evident whether the behaviour of ozone above the Antarctic is an early warning of future changes in global ozone or whether it will always be confined to the Antarctic because of the special geophysical conditions that exist there."

Ozone depletion and climate change, each extremely complicated problems in themselves, are intricately woven together into an even more complex tapestry. CFCs, though far less abundant than carbon dioxide, have a greenhouse effect 10,000 times more powerful and it's estimated they now account for more than 15 per cent of the greenhouse effect resulting from human activates. Other ozone destroyers, such as halons, methane and nitrous oxide, are also greenhouse gases and, together with CFCs, are expected to cause warming nearly equal to that of carbon dioxide. "What has been previously thought of as the carbon dioxide/climate problem should be more properly thought of as the trace gas/chemistry/climate problem," says Watson.

The growing evidence that something really is happening to the earth's atmosphere has been alarming enough to prompt political action. In 1987, at an international meeting in Montreal, 24 nations pledged to reduce the use of CFCs by 50 per cent by 1999. The agreement has since been ratified by four countries, including Canada. It requires ratification by 11 countries that produce two-thirds of the world's CFCs. Chisholm says there is "every expectation" the agreement will be ratified and come into effect at the beginning of 1989.

Many scientists are already urging stronger measures; delegates to the Toronto climate conference recommended that the protocol be amended in 1990 to ensure "nearly complete elimination" of ozone-destroying CFCs by the year 2000 and that additional measures to limit other destroyers should also be considered. "The ozone hole will persist indefinitely if we simply adhere to the original schedule of emission controls outlined in the Montreal protocol," says McElroy. "It should be amended as soon as possible." The Canadian government has already announced its intention to ban all non-essential uses of CFCs and halons.

McElroy says the CFC cutbacks will have the added benefit of slowing the pace of the greenhouse effect, but it will not be enough. "Ultimately, we must strive to drastically curtail use of fossil fuel." The statement issued at the Toronto conference recommended the establishment of a world atmosphere fund, financed in part by a levy on fossil fuel, to support efforts to protect the global environment. To stabilize atmospheric concentrations of carbon dioxide, current emission levels must be cut by more than half and the statement proposed an initial goal of reducing carbon dioxide emissions by 20 per cent of the 1988 levels by 2005.

*Lydia Dotto*

The question inevitably arises: is this a case of "too little, too late"? Not necessarily, says Hare, who thinks the scientific warnings are finally sinking in. The public often takes action before governments do and "sooner or later the politicians will budge, and budge rapidly." (It was a consumer boycott of spray cans, in advance of legislation, that prompted both industry and governments to take action to reduce the use of CFCs in the 1970s.) But he admits that some political leaders still are not convinced of the urgency of the problem. "The heads of many nations simply don't believe it, won't believe it," he says, citing specifically the leaders of the United States, Britain and France.

Politicians are reluctant to act because the actions required—conservation, restraint in our use of resources and energy, and reducing or eliminating the use of technologies we consider essential—are politically unpalatable and they can always point to the scientific uncertainties in our understanding of these problems as a reason for delay. But Hare says that the predictions of scientists are "much sounder than the socio-economic indicators governments use and they're more likely to be true. If decision-makers are willing to listen to economists, they should be at least as ready to listen to natural scientists. Complex as it is, the natural system is probably more predictable than the human economy."

While Hare believes that action is needed, he admits he has little patience with the "technological fix" approach. Proposals that we develop technologies to regenerate the ozone layer are enough to provoke him to a rare display of temper. "Nature is a colossal system with immense energy; it is extremely complex and quasi-chaotic. To monkey about with the atmosphere with a machine—that's just not on."

And yet, monkeying about with the atmosphere seems to be pretty much what we're doing.

---

*University of Toronto Magazine* (Autumn, 1988): 8–11.

---

## Questions and Topics for Discussion and Writing

1. Explain the importance of the ozone layer, what it is and how it functions.
2. Explain the particular hazards of chlorofluorocarbons (CFCs) and why we have so much of them in the atmosphere.
3. Discuss the problems of curtailing the further depletion of the ozone layer. What must be done? How can an effective program of change become established? What manner of agreements must be made and enforced?

# * Endless Summer: Living with the Greenhouse Effect


## Andrew C. Revkin



The composition of the Earth's atmosphere is affected by such human activities as industrialization and agriculture, which release gases into the atmosphere that traps the warmth from the sun. The resulting general warming trend is the "greenhouse effect"—a perceptible changing of the world climate which will have dramatic and in general devastating consequences. This article, by science writer Andrew Revkin in the American science magazine, *Discover*, explains the scope of the problem.


On June 23 [1988] the United States sizzled as thermometers topped 100 degrees in 45 cities from coast to coast: 102 in Sacramento; 103 in Lincoln, Nebraska; 101 in Richmond, Virginia. In the nation's heartland the searing heat was accompanied by a ruinous drought that ravaged crops and prompted talk of a dust bowl to rival that of the 1930s. Heat waves and droughts are nothing new, of course. But on that stifling June day a top atmospheric scientist testifying on Capital Hill had a disturbing message for his senatorial audience: Get used to it.

This wasn't just a bad year, James Hansen of the NASA Goddard Institute for Space Studies told the Senate committee, or even the start of a bad decade. Rather, he could state with "99 percent confidence" that a recent, persistent rise in global temperature was a climatic signal he and his colleagues had long been expecting. Others were still hedging their bets, arguing there was room for doubt. But Hansen was willing to say what no one had dared say before. "The greenhouse effect," he claimed, "has been detected and is changing our climate now."

Until this year, despite dire warnings from climatologists, the greenhouse effect has seemed somehow academic and far off. The idea behind it is simple: gases accumulating in the atmosphere as by-products of human industry and agriculture—carbon dioxide, mostly, but also methane, nitrous oxide, ozone, and chlorofluorocarbons—let in the sun's warming rays but don't let excess heat escape. As a result, mean global temperature has probably been rising for decades. But the rise has been so gradual that it has been masked by the much greater, and ordinary, year-to-year swings in world temperature.

Not any more, said Hansen. The 1980s have already seen the four hottest years on record, and 1988 is almost certain to be hotter still. Moreover, the seasonal, regional, and atmospheric patterns of rising temperatures—greater warming in winters than summers, greater warming at high latitudes than near the equator, and a cooling in the stratosphere while the lower atmosphere is warmer—jibe with what computer models predict should happen with greenhouse heating. And the warming comes at a time when, by rights, Earth should actually be cooler than normal. The sun's radiance has dropped slightly since the 1970s, and dust thrown up by recent volcanic eruptions, especially that of Mexico's El Chichon in 1982, should be keeping some sunlight from reaching the planet.

Even though most climatologists think Hansen's claims are premature, they agree that warming is on the way. Carbon dioxide levels are 25 percent higher now than they were in 1860, and the atmosphere's burden of greenhouse gases is expected to keep growing. By the middle of the next century the resulting warming could boost global mean temperatures from three to nine degrees Fahrenheit. That doesn't sound like much, but it equals the temperature rise since the end of the last ice age, and the consequences could be devastating.

Weather patterns could shift, bringing drought to once fertile areas and heavy rains to fragile deserts that cannot handle them. As run-off from melting glaciers increases and warming seawater expands, sea level could rise as much as six feet, inundating low-lying coastal areas and islands. There would be dramatic disruptions of agriculture, water resources, fisheries, coastal activity, and energy use.

"Average climate will certainly get warmer," says Roger Revelle, an oceanographer and climatologist at the University of California at San Diego. "But what's more serious is how many more hurricanes we'll have, how many more droughts we'll have, how many days above one hundred degrees." By Hansen's reckoning, where Washington now averages one day a year over 100 degrees, it will average 12 such scorchers annually by the middle of the next century.

Comparable climate shifts have happened before, but over tens of centuries, not tens of years. The unprecedented rapid change could accelerate the already high rate of species extinction as plants and animals fail to adapt quickly enough. For the first time in history humans are affecting the ecological balance of not just a region but the entire world, all at once. "We're altering the environment far faster than we can possibly predict the consequences," says Stephen Schneider, a climate modeler at the National Center for Atmospheric Research in Boulder, Colorado. "This is bound to lead to some surprises."

Schneider has been trying to generate interest in the greenhouse effect since the early 1970s, although largely unsuccessfully. Frightening as the greenhouse effect is, the task of curbing it is so daunting that no one has been willing to take the necessary steps as long as there was even a tiny chance that the effect might not be real. Since greenhouse gases are chiefly the result of human industry and agriculture, it is not an exaggeration to say that civilization itself is the ultimate cause of global warming. That doesn't mean nothing can be done; only that delaying the effects of global warming by cutting down on greenhouse-gas emissions will be tremendously difficult, both technically and politically. Part of the problem is that predicting exactly what will happen to the local climate, region by region, is a task that's still beyond the power of even the most sophisticated computer model.

Some parts of the world could actually benefit from climate change, while others could suffer tremendously. But for the foreseeable future the effects will be uncertain. No nation can *plan* on benefiting, and so, says Schneider, we must all "hedge our global bets," by reducing emissions of greenhouse gases. "The longer we wait to take action," he says, "and the weaker the action, the larger the effect and the more likely that it will be negative." Says meteorologist Howard Ferguson, assistant deputy minister of the Canadian Atmospheric Environment Service, "All the greenhouse scenarios are consistent. These numbers are real. We have to start behaving as if this is going to happen. Those who advocate a program consisting only of additional research are missing the boat."

* * *

While the greenhouse effect threatens to make life on Earth miserable, it is also part of the reason life is livable in the first place. For at least the last 100,000 years atmospheric carbon dioxide, naturally generated and consumed by animals and plants, was in rough equilibrium, at a couple of hundred parts per million. Without this minute but critical trace to hold in heat, the globe's mean temperature would be in the forties instead of a comfortable 59 degrees. The amount of carbon dioxide has risen and fallen a bit, coinciding with the spread and retreat of glaciers as ice ages have come and gone. But until the Industrial Revolution, atmospheric carbon dioxide levels never rose above a manageable 280 parts per million.

Then, beginning early in the nineteenth century, the burning of fossil fuels, especially coal, took off. By 1900, carbon dioxide levels in the atmosphere had begun to rise steadily, reaching 340 parts per million last year.

Levels of the other greenhouse gases have also risen. Methane, for example, is generated primarily by bacterial decomposition of organic matter—particularly in such places as landfills, flooded rice paddies, and the guts of cattle and termites—and by the burning of wood. Methane concentration in the atmosphere has grown steadily as Earth's human population has grown, rising one percent a year over the last decade. Levels of chlorofluorocarbons, which are used as refrigerants, as cleaning solvents, and as raw materials for making plastic foam, have climbed 5 percent annually.

The amount of nitrous oxide in the atmosphere has quickly increased as well, with about a third of the total added by human activity—much of that emitted by nitrogen-based fertilizers, and half of *that* from just three nations: China, the Soviet Union, and the United States. This gas is also released by the burning of coal and other fossil fuels, including gasoline. And ozone, which forms a beneficial shield against ultraviolet radiation when high in the stratosphere, is an efficient greenhouse gas when it appears at airliner altitudes—as it increasingly does, since it too is a by-product of fossil fuel burning.

All these gases are far more efficient at absorbing infrared energy (the invisible radiation that ordinarily carries Earth's excess heat into space) than is carbon dioxide. Indeed, atmospheric chemists have estimated that the combined warming effect of these trace gases will soon equal or exceed the effect from carbon dioxide. And even as growth has slowed in the industrialized nations, the Third World is rushing full tilt into development. All told, billions of tons of greenhouse gases enter the atmosphere each year.

* * *

The big question is, given the inexorable buildup of these gases—a growth that even the most spirited optimists concede can only be slowed, not stopped—what will the specific effects be? It's hard to say, because the relationship between worldwide climate and local weather is such a complex phenomenon to begin with. The chaotic patterns of jet streams and vortices and ocean currents swirling around the globe and governing the weather still confound meteorologists; in fact, weather more than two weeks in the future is thought by some to be inherently unpredictable.

So far, the best answers have come from computer models that simulate the workings of the atmosphere. Most divide the atmosphere into hundreds of boxes, each of which is represented by mathematical equations for wind, temperature, moisture, incoming radiation, outgoing radiation, and the like. Each mathematical box is linked to its neighbors, so it can respond to changing conditions with appropriate changes of its own. Thus, the model behaves the way the world does—albeit at a very rough scale. A typical model divides the atmosphere vertically into nine layers and horizontally into boxes that are several hundred miles on a side.

Climate modelers can play with "what if" scenarios to see how the world would respond to an arbitrary set of conditions. Several years ago, for example, computer models were used to bolster the theory of nuclear winter, which concluded that smoke and dust lofted into the atmosphere in a nuclear war would block sunlight and dangerously chill the planet. To study the greenhouse effect, climatologists first used models to simulate current conditions, then instantly doubled the amount of carbon dioxide in the atmosphere. The computer was allowed to run until conditions stabilized at a new equilibrium, and a map could be drawn showing changes in temperature, precipitation, and other factors.

But Hansen's latest simulations—the ones he used in his startling congressional testimony—are more sophisticated. In them he added carbon dioxide to the atmosphere stepwise, just as is happening in the real world. The simulations, begun in 1983, took so much computer time that they were not completed and published until this summer [1988].

Even the best climate model, however, has to oversimplify the enormous complexity of the real atmosphere. One problem is the size of the boxes. The model used at the National Center for Atmospheric research, for example, typically uses boxes 4.5 degrees of latitude by 7 degrees of longitude—about the size of the center's home state of Colorado—and treats them as uniform masses of air. While that's inherently inaccurate—the real Colorado contains such fundamentally different features as the Rocky Mountains and the Great Plains—using smaller boxes would take too much computing power.

Another problem is that modelers must estimate the influence of vegetation, ice and snow, soil moisture, terrain, and especially clouds, which reflect lots of sunlight back into space and also hold in surface heat. "Clouds are an important factor about which little is known," says Schneider. "When I first started looking at this in 1972, we didn't know much about the feedback from clouds. We don't know any more now than we did then."

* * *

So it is not surprising that while the more than a dozen major global climate models in use around the world tend to agree on the broadest phenomena, they differ wildly when it comes to regional effects. And, says Robert Cess, a climate modeler at the State University of New York at Stony Brook, "The smaller the scale, the bigger the disagreement."

That makes it extremely hard to get national and local governments to take action. Says Stephen Leatherman, director of the Laboratory for coastal Research at the University of Maryland, "Unless you can put something down on paper and show the effects on actual locations—even actual buildings—then it's just pie in the sky."

There are, however, some consequences of a warming Earth that will be universal. Perhaps the most obvious is a rise in sea level. "If we went all out to slow the warming trend, we might stall sea level rise at three to six feet," says Robert Buddemeier of Lawrence Livermore National Laboratory, who is studying the impact of sea-level rise on coral reefs. "But that's the very best you could hope for." And a six-foot rise, Buddemeier predicts, would be devastating.

It would, for one thing, render almost all low coral islands uninhabitable. "Eventually," Buddemeier says, "a lot of this real estate is going to go underwater." For places like the Marshall Islands in the Pacific, the Maldives off the west coast of India, and some Caribbean nations, this could mean nothing less than national extinction. "You're really looking at a potential refugee problem of unprecedented dimensions," says Buddemeier. "In the past, people have run away from famine or oppression. But they've never been physically displaced from a country because a large part of it has disappeared."

Coastal regions of continents or larger islands will also be in harm's way, particularly towns or cities built on barrier islands and the fertile flat plains that typically surround river deltas. Bangladesh, dominated by the Ganges-Brahmaputra-Meghna Delta, is the classic case, says Buddemeier. "It's massively populated, achingly poor, and something like a sixth of the country is going to go away."

Egypt will be in similar trouble, according to a study by economist James Broadus and several colleagues at Woods Hole Oceanographic Institution. Like the Ganges-Brahmaputra-Meghna, the soft sediments of the Nile Delta are subsiding. Given even an intermediate scenario for sea-level rise by the year 2050, Egypt could lose 15 percent of its arable land, land that currently houses 14 percent of its population and produces 14 percent of its gross domestic product.

One mitigating factor for some coastal nations that are still developing, such as Belize and Indonesia, is that they generally have committed fewer resources to the coastline than their developed counterparts—Australia, for example, or the United States, with such vulnerable cities as Galveston and Miami. "Developed countries have billions invested in a very precarious, no-win situation," Buddemeier says. "The less developed countries will have an easier time adapting."

Indeed, the impact on coastal cities in developed countries may be enormous. The Urban Institute, a non-partisan think tank, is completing a study for the Environmental Protection Agency on what a three-foot sea level rise would do to Miami. Miami is particularly vulnerable. Not only is it a coastal city, but it is nearly surrounded by water, with the Atlantic to the east, the Everglades to the west, and porous limestone beneath—"one of the most permeable aquifers in the world," says William Hyman, a senior research associate at the institute. "The aquifer in Miami is so porous that you'd actually have to build a dike down one hundred fifty feet beneath the surface to keep water from welling up." In an unusually severe storm nearby Miami Beach would be swept by a wall of water up to 16 feet above the current sea level.

Storms are an even greater danger to Galveston, which Leatherman has studied extensively. Given just a couple of feet in sea-level rise, a moderately bad hurricane, of the type that occurs about once every ten years, would have the destructive impact of the type of storm that occurs once a century. And Galveston is typical of a whole range of resort areas on the eastern and Gulf coasts, such as Atlantic City, New Jersey ("almost the whole New Jersey coast, really," says Leatherman); Ocean City, Maryland; and Myrtle Beach, South Carolina. "The point is, all these cities have been built on low-lying sandy barrier islands, mostly with elevations no higher than ten feet above sea level," Leatherman says. "Just a small rise in sea level will result in a lot of complications."

Even as cities become more vulnerable to moderate storms, the intensity of hurricanes may increase dramatically, says Kerry Emanuel, a meteorologist at MIT. Hurricane intensity is linked to the temperature of the sea surface, Emanuel explains. According to his models, if the sea warms to predicted levels, the most intense hurricanes will be 40 to 50 percent more severe than the most intense hurricanes of the past 50 years.

James Titus, director of the Environmental Protection Agency's Sea Level Rise Project, says communities will have two choices: build walls or get out of the way. For cities such as New York or Boston the answer may well be to build walls. But for most other coastal regions, picking up and moving may work out better. One of the first examples of a regional government making a regulation based on the greenhouse effect took place in Maine last year. The state approved regulations allowing coastal development with the understanding that if sea level rises enough to inundate a property, the property will revert to nature, with the owner footing the bill for dismantling or moving structures.

Another worldwide consequence of global warming is increased precipitation: warmer air will mean more evaporation of ocean water, more clouds, and an overall rise in rain and snow of between 5 and 7 percent. But it won't be evenly distributed. One climate model at Princeton University's Geophysical Fluid Dynamics Laboratory predicts that central India will have doubled precipitation, while the centers of continents at middle latitudes—the midwestern United States, for example—will actually have much drier summers than they have now (this summer's drought could, in other words, be a foretaste). Some arid areas, including southern California and Morocco, will have drier winters; and winters are when such areas get most of their precipitation. Moreover, the effect may be self-perpetuating: drier soil, says Syukuro Manabe, the climatologist who developed the model, leads to even hotter air.

The changes could be political dynamite for nations that already argue over water resources. A prime example is Egypt and Sudan, both of which draw their lifeblood from the north-flowing Nile. Sudan has been trying to divert a bigger share of the river's water; but downstream, Egypt is experiencing one of Africa's fastest population explosions and will need every drop of water it can get. A string of droughts in the Sudan could make the conflict far worse. The same situation occurs in many other parts of the world.

Not all the tensions will be international. Within nations, local effects of global warming will cause internecine fights for increasingly scarce water. In the United States, for example, western states have long argued over who won what fraction of the water in such rivers as the Colorado. In California 42 percent of the water comes from the Sacramento and San Joaquin river basins, which are fed by runoff from the Sierra Nevada and other mountain ranges. Most of the water falls as snow in the winter, which melts in the spring to feed the rivers, reservoirs, and subterranean aquifers. The state's normal strategy for water management calls for keeping the reservoirs low in winter, to provide protection against floods, and keeping them as high as possible in summer, to ensure an adequate supply for the giant farming operations in the Central Valley (one of the most productive agricultural regions in the world) and for arid southern California.

Peter Gleick of the Pacific Institute for Studies in Development, Environment and Security, in Berkeley, California, has devised a widely praised model that predicts a dramatic disruption of the state's water supply in the event of global warming, even if total precipitation remains unchanged. It focuses on the Sacramento River basin, which alone provides 30 percent of the state's water and almost all the water for agriculture in the Central Valley.

According to the model, higher temperatures will mean that what falls in winter will increasingly be rain, not snow, and that more of it will run off right away. California may get the same amount of total annual run-off, but the water-distribution system won't be able to deal with it. "California will get the worst of all possible worlds—more flooding in the winter, less available water in the summer<170 Gleick says. "This will reverberate throughout the state." San Francisco Bay will feel a secondary effect. As freshwater supplies shrink in the summer, seawater, which has already infiltrated freshwater aquifers beneath the low-lying Sacramento Delta, will continue its push inland. Rising sea level will just compound the effect.

* * *

Food is another crucial resource that will be affected by the global greenhouse. Taken by itself, a rise in atmospheric carbon dioxide might not be so bad. For many crops more carbon dioxide means a rise in the rate of photosynthesis and, therefore, in growth; and with increased carbon dioxide some plants' use of water is more efficient, according to studies done in conventional glass greenhouses. Also, as the planet gets warmer, crops might be cultivated farther north. But as usual, things are not so simple. A temperature rise of only 3.5 degrees in the tropics could reduce rice production by more than 10 percent.

In temperate regions also, the picture is mixed. Cynthia Rosenzweig, a researcher based at Goddard, has been using crop-growth computer models to predict effects of carbon dioxide buildup and climate change on wheat, the most widely cultivated crop in the world. Plugging in temperature changes derived from the Goddard climate model, Rosenzweig tested a world with doubled carbon dioxide levels. Because the Goddard model is bad at predicting precipitation, she did separate runs for normal and dry conditions. She found that in normal years the wheat grew better, thanks to the extra carbon dioxide. But in dry years there was a marked increase in crop failures, because of excessive heat. Given the likelihood that heat waves and droughts are increasing, she says, no one should count on better yields in years to come.

The nations most likely to reap the benefits of warmer climate are Canada and the Soviet Union, much of whose vast land area is too cold for large-scale crop cultivation. There has even been speculation that these countries might go slowly on controlling the greenhouse effect, or even oppose such control; anyone who has spent the winter in Moscow or Saskatoon would be sorely tempted by the prospect of better weather.

But again, atmospheric scientists stress that no nation can count on benefits. "The models suggest that ecological zones will shift northward," says planetary scientist Michael McElroy of Harvard. "The southwestern desert to the Grain Belt; the Grain Belt to Canada. There might be winners and losers if this shift occurs slowly. But suppose it shifts so fast that ecosystems are unable to keep up?" For example, he says, there is a limit to the distance that a forest can propagate in a year. "If it is unable to propagate fast enough, then either we have to come in and plant trees, or else we'll see total devastation and the collapse of the ecosystem."

According to Irving Mintzer, a senior associate with the Energy and Climate Project of the World Resources Institute in Washington, there is another reason to be leery of projections for regional agricultural benefits. Just because climatic conditions conducive to grain cultivation move north, that doesn't mean that other conditions necessary for agricultural superpowerdom will be present. Much of Canada, for example, does not have the optimum type of soil for growing wheat and corn.

* * *

Wildlife will suffer, too. In much of the world, wilderness areas are increasingly hemmed in by development, and when climate shifts, these fragile ecosystems won't be able to shift with it. Plants will suddenly be unable to propagate their seeds, and animals will have no place to go. Species in the Arctic, such as caribou, may lose vital migratory routes as ice bridges between islands melt.

In the United States the greatest impact will likely be on coastal wetlands: the salt marshes, swamps, and bayous that are among the world's most diverse and productive natural habitats. James Titus of the Environmental Protection Agency estimates that a five-foot rise in sea level—not even the worst-case scenario—would destroy between 50 and 90 percent of America's wetlands. Under natural conditions marshes would slowly shift inland. But with levees, condominiums, and other man-made structures in the way, they can't. The situation is worst in Louisiana, says Titus, which has 40 percent of U.S. wetlands (excluding those in Alaska); much of the verdant Mississippi River delta may well vanish.

In many parts of the tropics, low forests of mangrove trees thrive in the shallow waters along coastlines. Their dense networks of roots and runners are natural island-building systems, trapping sediment and cushioning the damaging effects of tropical storms. But rising sea levels will flood the mangroves; the natural response would be for them to shift with the tide, spreading their roots farther inland. But in places where development has encroached on the shore, the mangrove forests will feel the same squeeze that will threaten marshes.

* * *

The only way to eliminate the greenhouse problem completely would be to return the world to its pre-industrial state. No one proposes that. But researchers agree that there is plenty that can be done to at least slow down the warming. Energy conservation comes first: using less coal, finding more efficient ways to use cleaner-burning fossil fuels, and taking a new look at nonfossil alternatives, everything from solar and geothermal energy to—yes, even some environmentalists are admitting it—nuclear power.

Getting the world's fractious nations to agree to a program of remedial measures sounds extremely difficult, but Stephen Schneider sees signs that it may not be impossible. Schneider was one of more than 300 delegates from 48 countries who attended the International Conference on the Changing Atmosphere, which took place in Toronto, coincidentally, just a week after Hansen's congressional testimony. It was, says Schneider, the "Woodstock of $CO_2$" (an obvious reference to the "Woodstock of Physics" meeting held last year, during which news of the high-temperature superconductors exploded into the public consciousness).

The meeting was the first large-scale attempt to bridge the gap between scientists and policymakers on a wide range of atmospheric problems, including not just the greenhouse effect but also acid rain and the depletion of the protective layer of ozone in the stratosphere. Four days of floor debates, panel discussions, and closed-door sessions produced an ambitious manifesto calling for, among other things, the following:

- A 20 percent reduction in carbon dioxide emissions by industrialized nations by the year 2005, using a combination of conservation efforts and reduced consumption of fossil fuels. A 50 percent cut would eventually be needed to stabilize atmospheric carbon dioxide.
- A switch from coal or oil to other fuels. Burning natural gas, for example, produces half as much carbon dioxide per unit of energy as burning coal.
- Much more funding for development of solar power, wind power, geothermal power, and the like, and efforts to develop safe nuclear power.
- Drastic reductions in deforestation, and encouragement of forest replanting and restoration.
- The labeling of products whose manufacture does not harm the environment.
- Nearly complete elimination of the use of chlorofluorocarbons, or CFCs, by the year 2000.

* * *

Of all the anti-greenhouse measures, the last should prove easiest to achieve. Although CFCs are extremely persistent, remaining in the upper atmosphere for decades, and although they are 10,000 times more efficient than carbon dioxide at trapping heat, the process of controlling them has been under way for years, for reasons having nothing to do with the greenhouse effect. Since the early 1970s atmospheric scientists have known that CFCs could have destructive effects on ozone. CFCs were banned from spray cans in the United States and Canada in the late 1970s, and the appearance of a "hole" in the ozone layer over Antarctica in the early 1980s created an international consensus that CFCs must go. Last year 53 nations crafted an agreement that will cut CFC production by 50 percent over the next decade; the chemicals may well be banned altogether by the turn of the century.

CFCs are a special case, however. Since they are entirely man-made, and since substitutes are available or under development, control is straightforward. "There are only thirty-eight companies worldwide that produce CFCs," says Pieter Winsemius, former minister of the environment of the Netherlands. "You can put them all in one room; you can talk to them. But you can't do that with the producers of carbon dioxide—all the world's utilities and industries."

Also, there is a lack of basic information on the flow of carbon dioxide and the other greenhouse gases into and out of the atmosphere and biosphere. Just as one example, there is no good estimate of how much carbon dioxide, methane, and nitrous oxide are produced by fires, both man-made and naturally occurring. "We need to better assess global biomass burning as a source of greenhouse gases," says Joel Levine of the NASA Langley Research Center in Hampton, Virginia. "We have to understand what we're actually doing when we burn tropical forests and when we burn agricultural stubble after harvest. We don't know on a global basis what the contribution is."

Remarkably, the conference spurred some specific promises from political leaders rather than just vague platitudes. Standing before a 40-foot-wide photorealist painting of a cloud-studded skyscape, prime ministers Brian Mulroney of Canada and Gro Harlem Brundtland of Norway pledged that their countries will slow fossil fuel use and forgive some Third World debt, allowing developing countries to grow in a sustainable way. Says Schneider, "In the fifteen years that I've been trying to convince people of the seriousness of the greenhouse effect, this is the first time I've seen a broad consensus: First, there is a consensus that action is *not* premature. Second, that solutions have to occur on a global as well as a national scale."

In the end, the greatest obstacle facing those who are trying to slow the output of greenhouse gases is the fundamental and pervasive nature of the human activities that are causing the problem: deforestation, industrialization, energy production. As populations boom, productivity must keep up. And even as the developed nations of the world cut back on fossil fuel use, there will be no justifiable way to prevent the Third World from expanding its use of coal and oil. How can the developed countries expect that China, for example, which has plans to double its coal production in the next 15 years in order to spur development, will be willing or even able to change course?

* * *

And then there is poverty, which contributes to the greenhouse effect by encouraging destruction of forests. "Approximately seventy-five percent of the deforestation occurring in the world today is accounted for by landless people in a desperate search for food," says José Lutzenberger, director of the Gaia Foundation, an influential Brazilian environmental group. Commercial logging accounts for just 15 percent of tropical forest loss worldwide. Unfortunately for the atmosphere and the forests themselves, working out an agreement with the tropical timber industry will be far easier than eliminating rural poverty.

Industrialized nations, which created most of the greenhouse problem, should lead the way to finding solutions, says State Department official Richard Benedick, who represented the United States during negotiations for cuts in CFCs and who was a conference attendee. The first priority, he says, should be strong conservation efforts—an area in which the United States lags far behind such countries as Japan. The effect of such measures, Benedick feels, can only be positive and the cost is not great. "Certain things make sense on their own merits," he says. Technology can be transferred to developing countries. In some Third World nations a partial solution can be as simple as modernizing energy production and distribution. Upgrading India's electric-power distribution system, Benedick says, could double the effective energy output of existing coal-fired power plants.

Addressing the conference Canadian minister of energy Marcel Masse noted that there is cause for optimism. One need look no further than the energy crisis of a decade ago. From 1979 to 1985, thanks primarily to conservation, substantial cuts were made in the use of fossil fuels by industrialized nations. Only since 1986 and the current oil glut, said Masse, has there been a resurgence in oil use and coal burning.

Michael McElroy concluded, "If we choose to take on this challenge, it appears that we can slow the rate of change substantially, giving us time to develop mechanisms so that the cost to society and the damage to ecosystems can be minimized. We could alternatively close our eyes, hope for the best, and pay the cost when the bill comes due."

*Discover* (October 1988): 50–61

## Questions for Discussion and Writing

1. List some of the predicted probable consequences of the greenhouse effect.
2. What are the causes of the greenhouse effect?
3. What can be done about the greenhouse effect?

Rural Health Education in Venezuala.

*Issue*

# Public Health

Even when food and water are abundant, the atmosphere clean, the climate benign, and we think we are in good health, we can be felled by poisons that we unwittingly ingest or by infectious diseases that are spread from person to person. When whole communities are at risk from toxins in the environment or epidemic diseases, the problems become issues of public health. All modern governments have public health departments that remain ever on guard against the spread of disease and toxic substances through the population. The actions taken by public health departments are for the good of the community as a whole, and sometimes must take precedence over the convenience and comfort of individuals.

The three articles in this section illustrate different aspects of public health issues. S. C. Gilfillan's "Lead Poisoning and the Fall of Rome" discusses the disastrous cumulative effect of exposure to lead used in cookware, water pipes, paint, cosmetics, and other uses. The Romans were not aware of the danger of such cumulative exposure. Similarly, we have only recently become aware of the deadly effects of exposure to PCBs, asbestos, and other substances previously thought to be harmless. Geoffrey Bilson's "Science, Technology and 100 Years of Canadian Quarantine" details how one country, Canada, with ports of entry for travellers from all over the earth attempted to control the spread of infectious diseases through quarantines. Bilson well illustrates just how difficult it is to achieve the concerted action required by the whole community to prevent the spread of disease, especially when there is disagreement over the nature of the disease to be controlled and how it is spread. Those two historical articles are good mental preparation for reading about today's incurable epidemic in Harvey V. Fineberg's "The Social Dimensions of AIDS." Fineberg's article should be enough to make anyone stop viewing AIDS as someone else's problem.

# Lead Poisoning and the Fall of Rome

## S.C. Gilfillan

How serious are environmental concerns? Have we not lived through countless scares that predicted dire consequences from breathing the air and drinking the water? Well, maybe we have; maybe we need only be really concerned if the present problems are of a totally different magnitude from those of the past (which they doubtless are). But on the other hand, maybe if we knew more about what actually happened in the past we would see that we who survive now are merely the descendants of those who were not killed off by past human bungling.

As a case in point, consider the great and powerful Roman Empire. Why did it decline? How could it possibly go from its position as the most powerful empire in the western world to degradation and ruin. Many historians have addressed the issue, pieced together seemingly significant events and developments, and proffered their explanations. In this article, S.C. Gilfillan offers a rather different explanation: the Romans were poisoned by lead in their own environment.

Our novel theory concerns what mattered most about the fall of Rome. This was not the empire's military defeat and break-up. Those events might even have been of benefit to mankind. What concerns us here is the decay of Rome, the decline of every aspect of her civilization, except technology; the debasement of her arts; the cessation of novelty and progress; the complete disappearance of science after the second century; the retreat from almost all that had spelled "the glory that was Greece, and the grandeur that was Rome." There are many indications that Greece had been smitten in the same way a few centuries earlier, when her genius suddenly declined and her population shrank, and that it was the importation of Greek winemaking and cookery into the Roman mansions that chiefly ruined the Roman aristocracy; but the proof as to Greece is left for better Greek scholars.

We present here a completely new explanation for the Roman decay. It is derived not from the realm of philosophy or history, but from such firm sciences as toxicology, vital statistics, archeology, and evidence such as old bones and old Roman recipes and lead-lined pots for brewing poisons esteemed as delicious by the ancient well-to-do. We are well aware that the fall of Rome has been a favorite theme not only for professional philosophers of history, but for every curbstone amateur, and that each tends to ascribe that downfall to whatever evil he most opposes today. And we know that the fall of Rome was really a net result of many causes, such as the rising civilization of the barbarians. It included salutary as well as destructive forces, so that a true answer must include them all, and not just lead poisoning.

### Nature and Origins of Lead Poisoning

Lead poisoning, although still a potent industrial disease, is unfamiliar to most. Continued daily intake of lead or any of its compounds by mouth or lungs in an amount greater than about 1 mg per day may be dangerous—and 1 mg is only 1/28,300 of an avoirdupois ounce.

The varying symptoms of chronic plumbism [i.e. lead poisoning] include at first, most conspicuously, an obstinate and painful constipation and colic that can be relieved by pressure around the navel and by opium. These were symptoms the Romans knew and recorded. Anemia, pallor, emaciation, a metallic taste in the mouth, dark stools, loss of appetite, and pains in the joints are other signs. Further stages include paralysis of the extremities, starting with a peculiar wrist drop, headache, insomnia, blindness, and mental disturbances extending to insanity. "Crazy as a painter" was a common phrase not long ago when those craftsmen usually had lead poisoning. A wide range of further baleful effects occur in extreme cases, including death. Those most interesting to us are various interferences with the fertility of men and especially of women, whom sufficient lead brings to sterility, miscarriage, stillbirth, premature labor; children born to such women are apt to die shortly after birth.

Children also seem to be especially susceptible to plumbism, and the results are usually, or perhaps always, some degree of permanent physical or mental damage. Premature birth also often produces mental retardation. American city hospitals treat a succession of lead-poisoned children, usually from the slums, where they developed pica, eating flecks of lead paint from old walls. (Lead is banned from modern interior paint.) Also, infants crawl on the floor, then put their dusty hands in their mouths, whence it is likely the high death rate of children of lead workers originates. The well-to-do Romans had lead paints, and the favorite color of their rich walls was the famous Pompeian red, produced by minium, a salt of lead or of mercury—also poisonous.

## The Argument

Our reasoning involves the chain of facts listed below. Some have already been proved by familiar and conclusive evidence and will be but briefly treated. Some of the points are not well known, but have been proved by Hofmann, Kobert, and other specialist authorities and have been set forth with adequate documentation and argued with sufficient cogency. The major ideas that have been personally added, and that give the argument its import are shown below in italics and are developed as needed.

1. Roman culture declined and lost progressiveness, except in technology.
2. Rome's cultural progress, except in technology, depended upon the upper class.
3. The Roman upper class, beginning about the second or first century B.C., died out with extreme rapidity; each generation being perhaps a fourth of the previous one, largely from rearing few children. (All possible causes for this extermination are examined.)
4. Lead poisoning of the wives and mistresses of upper-class Romans occurred chiefly through their diet, after the introduction of Greek cookery around 150 B.C. and the relaxation of the old rule against wives drinking wine. The most significant sources of the lead were wine, grape syrup, and preserved fruit.
5. Lead in women produces sterility, miscarriage, stillbirth, heavy child-mortality; it brings permanent mental impairment in children.
6. *This lead poisoning was sufficient to produce the observed decimation of the upper class; nothing else was, venereal disease being then unimportant.*
7. *The Roman free poor and the slaves had much less lead in their diet.*
8. They did have other checks to their reproduction, especially poverty, and positive interferences with slave reproduction.
9. Despite these they must have maintained their number without much decline, because the Roman population as a whole did.
10. *It follows from items 3–6) that whatever qualities enabled Roman individuals to make money, or to marry or mate with money, were rigorously bred out of the race and culture by lead and lesser forces.*
11. *Technical invention (aside from architecture, civil engineering, and the art of war) was in ancient times confined to the artisan class, so invention continued, albeit retarded.*

Belief in eugenics is not necessary for acceptance of our thesis, for whatever personal qualities come not from heredity, must come from education and other environment, and the upper-class environment was extinguished along with the aristocratic scions. The son of a slave or peasant might succeed to a vacant chiefship, but he could not have had the education, family traditions, nor prestige of the scion. In short, Roman culture lost its source of progress, when and because its elite, endowed by heredity and/or upbringing with the powers to make or mate with money, was exterminated by various factors, above all by lead in the diet of the upper-class women...

Why were the ancient rich so reckless as to drink lead compounds daily? They had scarcely an idea of their toxicity. They knew plumbism (since Nicander in the second century B.C.) only as a disease of lead workers, the first or second industrial disease to be discovered. Also their physicians knew about acute poisoning by lead, and were cautious about lead in medicines. But as to lead poisoning from water or food, we can find hardly more than a reference or two, chiefly that of Vitruvius in his first-century A.D. treatise on water supply, echoed by Galen in the second century. It was certainly not a well-known danger.

But were not the poor dying out at the same time, for the same reason? In Greece, to some extent, they were, but hardly in Rome. In both areas the poor were constantly reinforced in some degree by war-captured slaves, although the reproduction of these—especially that of the Northern captives and other gang laborers in the Roman world—was much interfered with in various ways. The diet of the poor was not so badly poisoned with lead under the Romans. It is safe to say that the poor had not much wine, olive oil, grape syrup, other drinks made from fruit or honey, or preserved fruits—all the chief distinctive dietary sources of lead for the best people. Nor did they have the luxuries of cosmetics, or lead paint. They drank the same water, which would be leaded if it came through lead pipes and was acid from decaying vegetation or from volcanic, basaltic soil, or was neutral, like rainwater. Hard, alkaline water takes up no lead. But lead pipes were presumably less used in the country, where more of the poor and the unintellectual always live. The diet of the poor was mostly porridge, made from barley and other grains and perhaps acorns, but it also included bread, legumes, and other vegetables, greens, vinegar, and a little meat, fish, milk, fresh or dried fruit, and wine. Less cooking might be done than for the rich, since wood fuel was expensive, the nearby forests being denuded, and transportation bad from lack of horseshoes and suitable harness. For cooking and liquid food storage, the poor would be likely to use cheap earthenware, everywhere locally made, rather than expensive lead-lined bronze carried from afar.

Cooking involving heat greatly heightens the take-up of lead, especially if the pot be scraped from time to time to prevent burning, as is needed with thick liquids. The ancient writer Columella said this scraping had to be done in making grape syrup (used for sweetening and preserving), which the well-to-do prepared by boiling down grape juice to as little as one-third the original volume. Sugar, imported from India, hardly existed, leaving for sweetening only costly honey and the various fruits.

### Ancient Uses of Lead

To understand further how lead could be ingested in so many ways by the ancient aristocracy, we should consider its position among the metals and other substances that might have been substituted for it, especially as containers of acid food.

Lead was a metal the ancients produced in superabundance, and hence used wholesale. It is the easiest of all metals (save sometimes iron) to mine, smelt, and fashion, and needs only moderate heat. It was used in Anau I (4200 B.C.). Later, especially with the invention of the cupellation process in the third millennium B.C., lead became usually a byproduct of silver mining. Today the reverse is true; silver usually is a byproduct of mining lead and other metals. In the great silver mines of Laurion, the fountain of Athens' prosperity, 300 oz. of lead were produced for every 1 oz. of silver. Sometimes they threw the lead away. It has been estimated that 2,100,000 tons of it were produced there in 3 centuries. A Greek wine-ship of about 150 B.C., recently salvaged, had tacked on it a rather useless 200 tons of cast lead-sheathing. Again, 1200 tons were used to lay an aquaduct under one valley. Most of this dangerous metal was consumed in a great variety of domestic ways, often in contact with food or water, since there was so little idea of its poisonous nature. They had easily learned about copper poisoning from bronze, because this is detectable at once by color, taste, and sickness. Lead poisoning is far less evident and immediate. So the bronze vessel was replaced by a leaden or lead-lined one, save when by rare luck they used expensive pure tin. Most often plating was done with a mixture of the two metals. Lead was used for water pipes and containers, cups, toys, statuettes, tokens, corns, lids, sieves, household repairs, solder, paint, cosmetics, external medicines, boxes, markers, writing tablets, coffins, and roofing, and many alloyed it with copper (lead bronze) and with tin (pewter) and used it in many other ways.

Romans put lead into their wine in as many as 14 different ways, as we shall describe below. The final, saturnine touch of perfection was to warm the wine before serving it in cold weather. Since the houses were scarcely heated we can imagine they reckoned many cold days. The warming was naturally done in a lead-lined bronze apparatus with a charcoal brazier. In early Republican times, up to the Gracchi, wine had been scarce and precious; but after the conquest of Greece it became abundant for the wealthy, along with other luxuries.

The aristocracy's most significant source of lead was the grape syrup described above, which they called, irregularly, *sapa* or *defrulum,* according to the degree of boiling down of the unfermented grape juice (must), or sometimes by its Greek names, *car(o)enum, hepsema,* or *sireion (siraeum).* Such syrup is still used for flavoring and sweetening in some Mediterranean countries and in American food manufacturing, so that we have all consumed it as "grape concentrate." Pliny gives directions for making it: "Leaden and not bronze pots should be used." And the like said Cato, Palladius and Columella. We have cited the latter's injunction for stirring. Pliny evidently envisioned a solid leaden vessel, for he warns against leaving a pebble in it, lest this melt a hole through the pot.

This slow poison, this delicious syrup, was frequently put into wine to give it bouquet, strength, and keeping quality. The Romans often boiled the must before fermenting it to wine, since with their insanitary methods the wine often would not last more than a year, if that. The lead had two virtues, only one of which they understood correctly—that it gave a slightly sweetish taste. More importantly, it poisoned the microorganisms that cause fermentation and souring. Not having the modern art of sterilization by canning, they knew not that they were also sterilizing themselves. Grape syrup was further used for preserving olives, plums, cherries, apples, pears, peaches, and quinces—all cooked in lead. The syrup figures in a large number of the luxury dishes in Apicius' cookbook. Grapes were preserved for a year simply by placing them in rainwater boiled down to one-third. Why such a virtue in boiled down rain-water, which should be only pure water, we should think? Evidently because it would take up lead in long boiling, not being hard water—and perhaps it had also been collected from a lead roof and stored in a leaden cistern.

A weak drink for invalids, *adynamon*, was made by boiling down grape syrup with sea-water. It was a favorite of courtesans. With myrrh the syrup made a special drink for women. Wine boiled with wormwood might be used for a hang-over. Grape syrup with onion made an abortifacient.

Democritus said that if the must is too sour its acidity must be blunted by minium, lead oxide. Some minor amounts of lead in wine were from storage or drinking in leaden vessels, from the water always mixed with it, from the grape syrup used to cement the amphora's stopper, from mending an amphora with leaden clamps inside and out (but there was a coating of pitch), and perhaps leaden parts in the wine-press. Similar additions of lead, deliberate or ignorant, have continued occasionally almost until present times, although under legal attack as early as the 1100s.

Some recognition of the toxicity of the wine, though not of the lead in it, perhaps appears in the rule of the earlier Republic, that wives must not drink wine. People may have observed that winebibbing wives had fewer children. The rule ceased to be observed in later Republican times, when wine had become no longer a rare luxury, but an everyday drink of the well-to-do. There was also an ancient tradition that fat women were more fertile than thin ones. This would be the reverse of the attitude today, but would be true in ancient times, since emaciation is one of the symptoms of lead poisoning. A fat woman would be one who had somehow avoided lead. There were also ancient statements that certain Greek wines produced sterility, miscarriage, constipation, headache, or insomnia—all symptoms of plumbism.

That man has a sweet tooth is evidenced by the American consumption of sugar–nearly 100 lb. per capita per annum. So the ancients who could afford grape syrup or the limited supply of honey, used them freely. Pliny and Celsus said that raw honey was a good laxative, but that cooked, it was antidiarrheic. This could only happen through the action of lead. Dioseurides (v. 16) said that hydromel, *aqua mulsa*, was honey mixed with twice its volume of spring water, then boiled down to a third to prevent fermentation. A sour honey preparation called oxymel, Pliny says was made by heating honey with sea-water 10 times. At any rate, two of its uses were as a contraceptive and an abortifacient.

Another drink, whose name we might translate as wine-mead, was composed of must, water, and honey, fermented. Celsus and Pliny say again that drunk cold it was laxative, but when hot, stopped diarrhea. Such opposite effects after cooking an acid food show that lead was involved: the possible escapes through cooking in tin-plate or pottery were not utilized.

Kobert gathered the bones of 22 people from the ancient world. His student, Rosenblatt, found lead in 2 from Roman Carthage, and 2 from the same city before the Roman conquest, including a child in a stone coffin, hence rich. The remaining 18 came from outside the Roman Empire in time or place and showed no lead, although 5 early ones showed tin, and 1, copper. The chemistry of 1906 hardly permitted determinations in bone of more than the presence or absence of a metal. Later the dithizone and other methods enabled much more sensitive and accurate analyses for lead. So, in 1959, necromancers Specht and Fischer called up Pope Clement II, who had died 9 centuries before, and was buried in a stone coffin in Bamberg. The amount of lead they found in a rib was so great that he must have died of plumbism. So I have set out to gather bones from the Roman Empire and Greece before, which can be identified from the conditions of burial as definitely rich or poor, and with further data desirable. My first samples are being analyzed by the Kettering Laboratory in the University of Cincinnati, which has been carrying out accurate studies of the occurrence of lead in bones today. Distinction will be made according to which bone is analyzed, since they store lead unequally, and also according to the subjects' age and sex, and whether circulating ground waters would have deposited any lead found, and how much. If it shall appear that, on the average, the rich of the Imperial period were markedly poisoned by lead, and the poor not too much for reproduction, clinical necromancy will have added a final seal of proof to evidence logically sufficient without it. Returns from the tests so far received on the bones of 40 persons are consistent with the theory here presented. The complete returns will lead to a subsequent report perhaps devoted to that single item.

### Birth and Death Rates

Sociologists always want census statistics. We shall try to oblige, with what little can be found from the ancient world, viz., for an aristocratic quarter of Roman Troy (Ilion) and for Miletus—both Greek. In the Trojan quarter, aside from the widowed and the new citizens, there were 101 men over age 19, of whom 66 were unmarried. Of the 35 married, 18 were childless, 10 had 1 child, and 7 had 2 or 3—a total of 27 children from 101 men counted. They were doubtless younger than those making up our adult population, and their families not yet complete in some cases, but the deceased, departed and illegitimate children would not be counted. However we refigure it, there was evidently a low rate of surviving children, and an extraordinary amount of celibacy. The Milesian data are from the third to first century B.C. and show that the naturalized men, an old group probably, had 1.36 children apiece.

From the top of Roman society, emperors and such, we have abundant data on the few legitimate children. Few of the emperors had a son to succeed him. So the Antonines established the excellent system of adopting a fit man for heir. But at last Marcus Aurelius had a son, and by folly appointed him, the fool Commodus. If he had been mentally damaged by lead in childhood, it would not be surprising. So then the adoptive system was wrecked, and thereafter the "system" was a strong invitation to civil war every time an emperor died or someone chose to assassinate him. Here lies some explanation of Rome's political decay, and a contrast with the medieval and modern routine of thrones descending almost always automatically to the king's eldest son or another close relative.

The Romans recognized what was happening. The bachelor Horace persuaded Augustus, who had only a daughter, to pass laws penalizing aristocratic bachelors, and granting special privileges, the *Jus trium liberorum*, to fathers of 3 or more children. But these laws were little enforced, despite long continuance on the books and efforts of various emperors to revive them. There were also payments to poor men with children. In time the *Jus trium liberorum* came to be awarded even to childless men and bachelors. Ovid, the "singer of tender loves," as he rightly called himself, provides a case typical of the Augustan age. Of equestrian birth [i.e. from the order of *equites*, or knights], with one brother who died young, he had four marriages and one surviving daughter.

For a group to survive, its birthrate must equal or exceed its death rate. So we seek the death rate of the aristocratic Romans and find some statistics based on tombstone inscriptions—their presence evidence of a class able to afford that modest luxury, though they were often freedmen. Indicated is an expectation of life at birth of 22–25 years comparing with 72 for white Americans today. Expectation was longest about 5 years after birth. Such a high death rate suggests two things: First, that a very high birthrate would have been requisite to preserve such a race, as opposed to the very few births indicated by our literary and census data. (Perhaps the aristocratic birthrate was one-fourth of what would have been necessary to maintain their number. Multiply one-fourth by itself 18 times—for the 18 generations of our history—and one obtains the result that of the original aristocracy there would remain but a trace, and about as little from most of the successive leadership cadres promoted as replacements from the most capable of the plebs.) Second, the high death rate as well as the low birthrate strongly suggests lead poisoning. Many must have died of it.

Three factors with which we must also reckon were deliberate restriction of offspring by birth control and abortions; homosexuality; and above all the extraordinary frequency of celibacy. First, as to contraception—the science of the process was so little understood, and the worthless and the uncertain methods so mixed with the effective, that I do not think voluntary restriction was the major factor in the infertility. A method might work 9 times out of 10, and still plenty of children be conceived. Abortion was not rare and, like contraception, sometimes involved lead. Homosexuality was certainly frequent, though not so much as in Greece. But so long as people had heterosexual relations even occasionally—e.g., being married, there would still have been many children, just as with uncertain birth control. Celibacy, including intervals between the frequent divorces, was certainly an important fact. But why did so many men prefer celibacy? We may find lead a cause among others, thus: The chief purpose of marriage was to rear children; otherwise the ancient men would prefer full freedom, as Metellus Macedonicus said in 131 B.C. in an often quoted speech. Children were wanted both for patriotic and religious motives—to continue the family. As with Hindus and Chinese, this religious motive declined. In addition, as they looked about them they would have perceived that none of their prosperous fellows had many children, that most who wanted to beget a child could not do so by the wife, and that a child born would like as not never grow up. So when the desire for children decreased, and also the possibility of having them, men largely gave up marrying: Why try to have a son and heir? Why marry?

The Roman moralists often declaimed against selfishness as the cause of celibacy and the rearing of few offspring. But consider that a moralist must blame people for immoral choices, or he is no moralist. He does not deplore the facts of nature, or what seem to be such—e.g., the infertility of old women, or the high death rate of children, or the fact that the wealthy eat this and that which is tasty and seems harmless.

The continued decimation of the Roman aristocracy is a historic fact, variously attested. But decimation is not the right word, since it properly means the killing of one-tenth, whereas it appears about three-quarters were eliminated in each generation on the higher hills of Roman society. One line of evidence is the disappearance of the old family of "gentile" names of Republican days, like Julius and Cornelius. With the families all gone about as early as the Republic, the Romans changed the basis of their naming system from strict inheritance to personal choice or parental bestowal, so that the name test cannot be used later. By Nero's time, Tacitus said that nearly all the knights and most of the senators betrayed a servile origin.

Still other possible causes, indeed all possibilities anent this aristothanasia, or death of the best, must be inquired into. Political proscriptions were certainly one factor, especially those of Sulla, Marius, and Octavius, who slaughtered their prominent political opponents by thousands. The later emperors were less wholesalers in massacre of leaders. But it was always the prominent, the leaders by brains, devotion, energy, and money, who were most often politically proscribed. Still, such leaders are usually men of middle age or older, whose consorts were probably so hampered by age and accumulated lead that they would have had no more children anyway. And the Roman proscriptors did not slaughter the wife and children with the father, as savage tyrants have done.

### Binding up the Argument

By this time the reader should be thoroughly confused, as to what happened and why, among the many reasons advanced for the aristothanasia that occurred. But though the items be somewhat conjectural, their total sum should be clear—viz., that the Roman upper class came to have a birthrate far below its death rate, resulting in a six-centuries-continued aristothanasia, killing off the brightest and winsomest from the poorer class, who in each generation were promoted to wealth and extinction, like the prize animals garlanded with flowers as they were led to slaughter at an altar. Whatever the relative strength of the causes, the result, aristothanasia, stands established.

And a second fact should be clear—that among all those causes one new one stands proved as certainly important: that the well-to-do were poisoned by lead, and thereby their fertility must have been much diminished, their death often hastened, and their surviving children often damaged mentally. This fact, in its class aspect, was not known before.

Thirdly, the aristothanasia entailed that the ultimately surviving race, which in time had to fill all the posts from lowest to highest, was the screened-out remnant, composed of those who had least capacity to get money, whether by merit or rapacity, or by virtue of the beauty and charm that would enable a poor girl to marry money.

The beatitude said, "Blessed are the meek, for they shall inherit the earth." They did, and the result was the dysgenic and cultural disaster of the Dark Ages and Byzantine stagnation.

*Journal of Occupational Medicine* 7 (1965): 53–60.

## Questions and Topics for Discussion and Writing

1. Why did lead poisoning affect the Roman aristocracy more than the peasants?
2. If lead poisoning was ultimately fatal or caused sterility, why did the Roman population not decline?
3. Explain why Gilfillan believes that systematic aristothanasia (death of the aristocracy) would lead to the decline of Rome?

# * Science, Technology and 100 Years of Canadian Quarantine


## Geoffrey Bilson



Governments and their medical advisers have been aware of public health issues since the time of the Black Death in the Middle Ages. Epidemics were matters of special concern because their causes were often unknown and cures were often not available. The only effective prevention was to separate those infected from those not infected. For those travelling across national borders, this could be accomplished with a quarantine. But quarantines had their own problems and were not entirely effective. This analysis of Canadian quarantines established to combat yellow fever, cholera, and other highly communicable diseases reveals the complex interdependence of medical knowledge, technology, and public opinion involved in administering this aspect of public social policy.


Cholera reached the British Isles in 1831. Because it seemed certain to cross the Atlantic a quarantine was established in Canada at the end of that year, as had been done in earlier years when yellow fever threatened. Early in 1832 a decision was made to buy Grosse Isle as a quarantine station. For more than a hundred years, Grosse Isle would serve as the quarantine station in the St. Lawrence. Similar stations were set up at other ports of entry in the Maritimes by the colonial governments. After Confederation these stations passed into dominion jurisdiction and later were joined by one at William Head in British Columbia.

The need for a quarantine station arose out of a combination of technical, medical and political needs. The political need was plain. Whatever the complaints of shipping companies and traders might be about interruption to their business, governments could not allow thousands of potentially disease-carrying immigrants into cities without inspection. Quarantine was intended to stop the sick before they reached the local population. The fact that quarantines had failed to stop the passage of cholera from India to Europe and had little success against other diseases did not permit governments to ignore public fears.

The technology of mass travel and the state of medical knowledge in the 1830s dictated the form which quarantine took. Mass travel was nasty, brutish but long with passengers crowded together in the holds of vessels. Some ships were able to carry only fifteen passengers, but all took six weeks or more to make the crossing. It was almost impossible to keep clean under these conditions and a slow passage in a dirty ship undermined the health of passengers, many of whom were undernourished before the voyage began. There was no clear agreement amongst medical men on the mechanism by which disease was transmitted, but it was agreed generally that there was some link between dirt and disease.

Quarantine regulations were designed to remove the sick and to force passengers and ships to be cleaned before they came into contact with Canadian cities. Vessels with sickness on board were detained at the station for a period after the last case appeared. Other vessels landed their passengers so they could clean themselves and their baggage while the ship was cleaned before proceeding to port. This policy had the effect, in 1832, of mixing passengers from ships with sickness on board with those from healthy craft and thus spreading the diseases it was meant to check. In later years, efforts were made to keep the passengers from different vessels apart. The quarantine did not prevent disease from entering Canada, but it was retained because, as the Quebec Board of Health said in 1832, it was "adequate to the removal of a particular and known cause of evil...it will enable us to disinfect a ship...whereas if there be no such place as Grosse Isle or yet any substitute for it that evil is without remedy.

The essential features of the quarantine system laid down in the 1830s endured for nearly fifty years. Although modified in detail from time to time, the regulations required vessels with infectious disease on board to stop at the quarantine station and land all steerage passengers while the ship was disinfected. Cabin passengers continued to be exempted from this requirement. In those fifty years, the technology of ocean travel changed radically and the condition of the passengers changed for the better. By the early 1860s, nearly half of the Atlantic immigrants travelled by steamer, and before the end of the decade, nearly all did. Immigrants were prepared to pay a premium of nearly fifty per cent for the speed that steam travel offered. The steamers were healthy and carried surgeons, and the majority of cases of sickness were taken from the slow sailing vessels. Improvements in travel conditions continued to be made into the twentieth century, at which point a commentator remarked on the "wave of increasing luxury which has reached the steerage of the transatlantic steamships." It outstripped improvements in the accommodation at Grosse Isle which remained "rather in keeping with the previous condition of accommodation for steerage passengers" and led to bitter complaints by immigrants and their relatives.

The speed of passage meant that passengers arrived healthier but it also encouraged ship's masters to ignore the regulations. It was better to pay the fine than lose time by stopping at the station. In 1882 no vessel stopped at Grosse Isle for inspection. The desire for speed also led the authorities to exempt the mail steamers from the regulations. The result was a continual series of complaints from the residents of the ports that sick immigrants were being landed. Frederick Montizambert, a doctor who spent fifty years in the public service rising from assistant to the Superintendent at Grosse Isle to Director General of Public Health, proposed changes in the regulations in the early 1880s which would reflect the new technology of ocean shipping. The need was to combine protection with the desire for speed. He suggested that all ships be required to stop at Grosse Isle, but that in cases of minor infectious disease, only the sick and their immediate contacts should be landed. All steerage passengers would land if the ship had cases of serious disease on board. The new system would be imposed on mail steamers and enforced by heavy penalties.

Montizambert's distinction between minor disease and serious disease reflected one of the major developments in the century. The attempt to control the spread of infectious disease was becoming an international effort. In the years after 1851, international sanitary conferences were held in a number of European cities in an effort to coordinate action against disease. Improved communication made such international action possible and the attention of the conferences was focused on major infectious diseases (cholera, smallpox, yellow fever, plague and typhus). Montizambert's suggestions would bring Canadian practice into line with this developing movement, although in later years he preferred to take a firmer line against minor diseases than did other nations.

The technology of transport made traditional quarantine even more cumbersome than it had been in the days of sail. The passengers, travelling in smaller groups in separate compartments, were less exposed to infection than earlier voyagers in their stinking holds. Public perception of them, however, had not kept pace with these changes. It was not the 29,777 immigrants who arrived in good health at Quebec in 1881 but the one case of smallpox, two cases of diphtheria and sixteen cases of measles which caught public attention. Medical knowledge would have to be exploited if a new system of quarantine was to be worked out which would recognize changes in the nature of ocean travel while at the same time calming public fears.

The 1880s were years which saw new knowledge and new technology applied to the problems of quarantine. Dr. Montizambert was becoming recognized as a leading figure in Canadian public health and he attended conferences in the United States and toured facilities in quarantine stations there. He returned from these tours in 1885 convinced of a medical truth. The germ theory was fact. From that fact flowed a series of consequences. The work of Dr Joseph Holt in New Orleans convinced Montizambert that "maritime sanitation" should replace quarantine detention. The new medical knowledge could be combined with a new technology to bring speed and increased safety to the quarantine. If the Canadian government hesitated at the cost they should consider the fact that the USA was interested especially in seeing the St. Lawrence made safe. United States public health officials reminded Canada that cholera and smallpox both had entered the USA via the St. Lawrence. The new idea of maritime sanitation required inspection of vessels and the rapid disinfection of travellers and of the vessel with a variety of proven disinfecting methods using chemicals and dry and wet heat. Vaccination against smallpox, either before leaving home, on the voyage or at the station would be an essential part of the system.

By 1886, these methods had been adopted at Grosse Isle. A steam yacht, the *Hygeia*, was stationed there to inspect vessels and carry off the sick. The yacht was equipped with apparatus designed to pump steam into the holds of ships and "displace the dangerous atmosphere and replace it by one intensely germicidal." Passengers, even in cabin class, were examined for evidence of vaccination. The delegates of the American Public Health Association, meeting in Toronto that year, declared the new regulations "sufficient for the protection of Canada and the people of the United States if they are properly enforced." That was a large qualification, for in 1886 only thirty-seven of 800–900 ships entering the St. Lawrence stopped at Grosse Isle.

The next year marked "an entirely new era in the history of the quarantine system of the St. Lawrence," according to Dr Montizambert writing in his annual report. Partly from fear of a cholera epidemic in Europe, and partly as general legislative housekeeping while preparing a new Immigration Act, new quarantine regulations were made which were the first major revisions since the early years of the century. All vessels had to stop for inspection. A ship could not enter customs without showing pratique [i.e., papers giving permission for a ship to enter port, load and unload, after passing quarantine]. The quarantine station operated twenty-four hours a day and delays were slight. The regulations required ships to provide hospitals to ensure the isolation of passengers with infectious diseases. The regulations provided for the detention of people suffering from minor infectious diseases and most of those detained were victims of scarlet fever or measles. That did not make sense to the shipping lines, who argued that there was little point in protecting Canadian communities from diseases already treated in the local hospitals. Montizambert replied that the minor diseases were killers (in the USA in 1880 smallpox had killed 871, scarlet fever 16,416 and diphtheria 38,398) and that it was sensible to exclude fresh foci of infection entering the country. On one question, medical science yielded to social and political power. The requirement that cabin passengers be vaccinated, if they had not been, was dropped. They had protested strongly against the indignity. Montizambert sympathized that the "ladies…find themselves suddenly and unexpectedly confronted with the requirement that they must show their arms to, and possibly submit to vaccination by, a ship's surgeon, or a quarantine officer, about whom they know nothing and in whom and the purity of whose vaccine, they may have no confidence." They could save themselves these troubles by being vaccinated at home. The shipping companies, however, feared the loss of trade to US shippers for whom there was no similar requirement. Steerage passengers still were required to be protected.

The debate over vaccination is an interesting example of how a medical question is shaped by other considerations. The question of the incubation period for smallpox appears to be a medical question, yet it was set in Canada by political compromise. Until 1898 it was assumed that smallpox would show itself within fourteen days after exposure and that was the period used to quarantine contacts. Then three cases of smallpox developed among passengers arriving in Winnipeg from the S.S. *Pisa* after they had served a fourteen day quarantine. Manitoba authorities demanded that the quarantine period be extended to twenty-one days. Montizambert investigated the cases, found that the medical authorities had been negligent in handling the *Pisa* passengers and recommended no change. An order-in-council, however, was issued raising the period of detention to eighteen days and so it remained, politically determined, until a new order-in-council restored the fourteen day rule in 1913.

Political pressure was brought to bear by the Anti-Vaccination League of Great Britain in the years before the first World War, but they were ignored more easily than the government of Manitoba. The League pointed out the discrepancy between treatment of the cabin and steerage passengers, and between Canadian practice and that of the other Dominions which did not require vaccination. They warned of the loss of immigrants repelled by compulsory vaccination. The Canadian reply was that vaccination was not compulsory, the alternative being to remain at quarantine for the period of observation. One distinction was rejected by Canadian authorities—returning Canadians were treated in the same way as immigrants. If they came third class, they had to be vaccinated or serve the observation period. Smallpox it seems did not respect nationality, even if it appeared to respect class lines.

To vaccinate large numbers of passengers, the quarantine service required a reliable vaccine. Canada could not provide it. Experience showed that the best form was to have the lymph on a glass point. No one in Canada manufactured in that way and consequently the service had to buy its supply from a company in Pennsylvania. When a company began to manufacture vaccine in Ontario, the Minister of Agriculture was advised that the product was unreliable in strength, unsupervised in manufacture, and available only in tubes. The quarantine doctors preferred glass points and so did the patients. Points were convenient to use and each patient was treated with a fresh point. "The throwing of that point down on the desk and the taking of another leads to the removal of many objections to vaccination on the part of the immigrants." Some thought was given to preparing vaccines in government laboratories "free from the temptation that must accompany their manufacture as a commercial product," but the service continued to import vaccine in the early twentieth century.

Smallpox was not the only matter in which political considerations affected medical questions. On the west coast, it became standard practice to insist on disinfecting oriental passengers, crew and their baggage on arrival. Montizambert, as General Superintendent of Canadian Quarantines, objected to the practice because of its disruptive effect on travel and because it was not necessary. The same rules applied to diseases brought in from the Orient as from Europe. Each case of illness could be traced to a specific person and medical inspection and vaccination offered protection. The practice was continued, however, and a few years later Montizambert was obliged to insist that the Superintendent of Quarantine carry out the procedures laid down by the Minister. "Such matters as the inconvenience of the saloon passengers in having cooks and waiters taken off the ship and the consequent delay to meals of which you speak do not constitute an adequate reason for omitting any matter that you consider of importance," he advised Dr Watt. Routine disinfection was ended in 1903.

The quarantine service made a similar concession a few years later when Montizambert agreed to issue each Chinese immigrant with a Bill of Health. This was done at the insistence of the Chief Controller under the Chinese Act, even though it was medically and legally unnecessary, to "reduce friction" between quarantine officers and controllers. Thus another unnecessary humiliation was added to immigration from China. Montizambert, himself, was alert to the dangers the Orient might pose to Canada. On morning, in 1912, he read of the revolutionary decree in China abolishing the pigtail. Anticipating a flood of pigtails into the "hair markets of this and other countries" he hurried to secure an Order-in-Council to amend the regulations for disinfecting unmanufactured or uncleaned hair. Montizambert also recommended that Canada not adhere to international sanitary conventions drawn up in 1903 and 1912 for a number of reasons including their provision for the free entry of mail. He did not wish to see "mail matter from China and other places in the Orient during a Plague epidemic...admitted into the China-towns of British Columbian cities without disinfection."

The new quarantine of inspection, disinfection and limited detention required more sophisticated medical techniques. By 1890, progress in identifying the specific organisms responsible for each disease had made the bacteriological laboratory an "essential requirement" for quarantine stations. Within a few years a doctor appointed as medical assistant at Grosse Isle was being told in February that, if he had no bacteriological knowledge, he should acquire some by April and be able to prepare culture media to carry on the study and practice of bacteriology at the Station. The bacteriological laboratories began to prepare serum and prophylactic fluid against plague and other diseases. The work of preparing the anti-plague fluid stopped after the death in Germany of a scientist. Some quarantine doctors took to laboratory work happily. One had to be reminded, in 1905, that the purpose of the bacteriological laboratory was to diagnose doubtful cases and not to support original research. Not all stations had laboratories, and they were expected to use local facilities or rely on clinical diagnosis. They were not to expect laboratory work to be done in Ottawa. After sending one package to the capital, Dr J. Macdonald of Chatham, N.B., was told by Montizambert "do not send rats to Ottawa."

The system was elaborate and required a battery of technical devices. By 1892 Grosse Isle had a 100-bed hospital, a shed for cholera and smallpox cases, accommodation for 2000 healthy passengers, bath houses, hot air forges, fumigating rooms, a steam disinfecting house under construction, two steamers always ready to sail to meet incoming craft and support services. Vessels could be disinfected by steam, by Dioxide Sulphuric Blast and by Bichloride Mercuric Drench. Other stations in the service were being brought up to these standards. The new system of quarantine won the praise of public health organizations and was taken as a model by delegates to an International Conference in the United States in 1893 which considered methods "aimed at...obtaining efficient disinfection scrupulously carried out, but avoiding such extreme measures as to do harm to individuals and, being superfluous, bring science into disrepute."

Even as it was developing, the Canadian system was beginning to fall behind practices in some other countries. Canada, unlike most other nations which cooperated in International Sanitary Conventions, continued to include minor diseases in the list of quarantinable diseases. The cumbersome machinery was directed against chickenpox as carefully as it was against smallpox until 1924. This won popular support, for when the policy was changed it brought fresh protests from communities that they were being laid open to danger. In some other countries, such as Great Britain, the detention of contacts was being given up in the 1890s in favour of surveillance. Canada rejected this method of control, which depended on good communications and a degree of centralized authority, on the grounds that it did not suit local conditions. The distance immigrants travelled within Canada, the size of the country and the divided jurisdiction over health matters all made surveillance difficult. Detention served Canada's needs better. The same problems emerged under the new system as under the old and shippers began to protest its inconvenience. One concession made on the St. Lawrence was to pass vessels at night without inspection on the sworn statement of the ship's master or surgeon that they were free of infectious disease. Dr Pagé estimated that, while he was in charge of Immigrant Medical Services at Quebec between 1904–20, forty-five per cent of the ships passed Grosse Isle without inspection, many of them with minor infectious disease on board.

While popular opinion might be served by a quarantine of detention, professionals were coming gradually to the conclusion that other methods were better. International cooperation and communication could warn of the approach of major diseases in time to take special precautions, as was done when cholera reappeared in Europe in the early 1890s. When individual vessels began to carry wireless, the control of major disease could be directed precisely at a particular vessel. Minor diseases offered fewer worries as the twentieth century passed. Growing confidence that the mechanisms of disease transmission were understood clearly led to the gradual abandonment of domestic quarantines and the elaborate precautions of disinfectant soaked sheets hung over the sick room door. By the mid 1920s the new technologies of communication and the current state of medical knowledge suggested that elaborate quarantines were unnecessary. The quarantine station at Charlottetown, P.E.I., had retained a salaried officer in all the years from Confederation to the mid 1920s "although no patient coming from a foreign port had been detained there all this time." That was one station which could be closed.

Canada signed the International Sanitary Convention of 1926 and began the process of bringing her quarantine into line with that of other signatories. Quarantine regulations were exchanged with departments in other nations. Regular inspection of all craft would continue but minor diseases generally would be dealt with in immigration or community hospitals not at the quarantine stations. Ships with infectious disease on board were able to warn authorities of that fact by wireless. One agency which developed a short-lived interest in quarantine was the RCMP. They wondered if rum runners' boats could be held in quarantine after making a pickup from a ship which would have been subject to quarantine had it docked. The men administering the system increasingly found it unduly complex. Plague was unlikely to be introduced by ships as deratinization techniques became more sophisticated. Cholera was unlikely to spread in modern cities with clean water supplies. Typhus was unknown in Canada after 1915 and was kept under control by delousing immigrants. Smallpox was seen occasionally, but it was controllable by vaccination. Yellow fever was unknown in Canada. By the early 1930s, which were times of economic hardship, an expensive and elaborate scheme was being maintained to deal with major diseases rarely or never seen in the country. The technological marvels of drench and blast had gone unused since 1926 when routine disinfection was abandoned.

Dr Haegerty pointed out, when recommending changes in the regulations in 1936, that advancing knowledge of how disease was transmitted and of methods of providing for the public health had reduced the need for a maritime quarantine. He had twenty-five years of experience with boarding ships and he thought it useless. New regulations in 1937 reflected Haegerty's views. Ships on the Atlantic coast could obtain pratique by wireless, as could ships approaching the Pacific coast from the Americas, and would be visited on arrival in dock. Vessels without wireless, with disease on board or from infected ports would be boarded in the stream. Grosse Isle and Partridge Island (Saint John, N.B.) would remain open only as emergency detention centres. William Head would continue in operation because of the greater incidence of quarantinable disease on the Pacific Coast. Lawlor's Island (Halifax, N.S.) would be closed down. If Grosse Isle was not yet closed, its establishment was greatly reduced from what it had been fifteen years before when sixty staff and their 170 dependents spent the summer on the island "all with free fuel and electricity, two churches with resident clergymen, baker, carpenter, two plumbers, two teamsters, greenhouses with gardeners, a herd of cows and horses to take care of 15–20 patients a week at the height of the season at a cost of $118,095.41." The pastoral days of the quarantine service were over.

This brief survey of the quarantine in Canada raises some points relevant to the study of critical issues in the history of science, technology and medicine in Canada, for it combines elements of all three. The quarantine was instituted to meet not only medical but also social needs. Changes in quarantine regulations could be made only with regard to public opinion and a closer study of the balance of interests between shippers and general public should yield some insights into the relation between science and public policy. The ability to make change politically acceptable is tied to the state of public knowledge about health questions. The changes of the mid–1930s appear to have provoked fewer questions than did those of the 1920s or 1880s, which perhaps could be related to developments in public health in the mid-twentieth century. The inertia the service showed, particularly in the nineteenth century when changes in shipping techniques made the quarantine obsolete, is related to the political nature of the role it played as well as to the apparently inevitable inertia of any institution.

The 1880s were a period when medical knowledge and technology both offered opportunities to make over the service. The men in charge of the service were confident in their medical knowledge and in their ability to create a technology that was based on that knowledge. The result was a quarantine procedure not much less cumbersome than the one it replaced. Quarantine operated within a developing attempt to create a formal international structure for public health questions. The decision of Canada to go against the prevailing tendencies to concentrate on major disease and to replace detention of contacts by surveillance was a product partly of personality—Montizambert's influence was important here—and of the peculiar nature of Canada—its size, limited communication network and divided political responsibility for health. It would be interesting to know how the particular problems of Canada affected developments in other areas of science and technology.

As with other scientific and medical questions, the quarantine did not operate in a vacuum. Current social and political attitudes affected such questions as the treatment offered first class passengers or Chinese immigrants and led to regulations at odds with medical needs. The service also created needs which Canada had a limited ability to meet, from the provision of public laboratories to the supply of smallpox vaccine in suitable form and quantity. These social restraints and economic and technical shortcomings are unlikely to be peculiar only to the quarantine question. We are reminded once again that the history of medicine cannot be divorced from its social context. While we must be alert to developments elsewhere, we must also ask how the Canadian context influenced the particular problem we choose to study.

From Richard A. Jarrell and Arnold E. Roos, eds., *Critical Issues in the History of Canadian Science, Technology and Medicine,* 89–100 (Thornhill, Ontario: HSTC Publications, 1983).

### Questions and Topics for Discussion and Writing

1. Discuss the different treatments given to passengers of different nationalities, or different economic circumstances upon entering Canada.
2. Discuss the relation between medical and political considerations in the formation and enforcement of quarantine regulations in Canada.

# The Social Dimensions of AIDS

## Harvey V. Fineberg

The world has known many scourges in its history—epidemics that sweep whole nations and rage out of control and out of understanding—the plague, smallpox, tuberculosis, malaria, cholera, typhus. These epidemics are worst when the disease is not understood and no cure is available. The world's present scourge is AIDS, Acquired Immune Deficiency Syndrome, a disease that kills not by directly attacking the body, but by destroying the body's ability to combat infection. The fact that the disease is usually fatal and that there is no known cure is terrible enough. But the horror is compounded by two other factors: the long latency period and the mode of transmission. A person can have AIDS, that is, carry the Human Immunodeficiency Virus (HIV), for years without knowing it and therefore can spread the virus unwittingly to many other people before being diagnosed as suffering from the disease. That, combined with the fact that the main mode of transmission of the disease is through sexual contact, is particularly ghastly because it turns powerful expressions of love and affection into potential preludes to premature death. The only reliable way to prevent the spread of AIDS throughout the population is to stop the virus-spreading contacts between HIV carriers and non-carriers. But this is easier said than done.

This article by Harvey V. Fineberg, Dean of the Harvard School of Public Health, was published in *Scientific American* in October 1988. It reviews the impact of AIDS on American society and the efforts to combat the epidemic up to 1988.

The AIDS epidemic exposes hidden vulnerabilities in the human condition that are both biological and social. AIDS prompts courageous and generous acts, and it provokes mean-spirited and irrational responses. AIDS throws new light on traditional questions of value, compels a fresh look at the performance of the institutions we depend on and brings society to a crossroads for collective action that may, with the passage of years, mark a key measure of our time.

In the seven years AIDS was recognized, the epidemic has touched on almost all aspects of society. Its reach extends to every social institution, from families, schools and communities to businesses, courts of law, the military and Federal, state and local governments. It has also had a profound impact on the way science, medicine and public health are practiced in the world.

Through its association with sex, blood, drugs and death, AIDS evokes basic human fears and inhibitions. In her book *Illness as Metaphor* Susan Sontag writes: "Although the way in which disease mystifies is set against a backdrop of new expectations, the disease itself...arouses thoroughly old-fashioned kinds of dread. Any disease that is treated as a mystery and acutely enough feared will be felt morally, if not literally, contagious...Contact with someone afflicted with a disease regarded as a mysterious malevolency inevitably feels like a trespass; worse, like the violation of a taboo."

Although she was reflecting on cancer, Sontag's words are even more appropriate for AIDS, a condition that is literally as well as morally contagious. The contagion is compounded by the stigma attached to the behaviors most prominently associated with HIV infection in the U.S.: homosexual intercourse and intravenous drug use. Knowledge of HIV and its mode of spread, convincing as it is to scientists and epidemiologists, is not powerful enough to fully dissolve the public sense of mystery and old-fashioned dread. The protective garb needlessly donned by workers transporting a person with AIDS is reminiscent of the costume worn by physicians treating plague victims in 18th-century France. People known to be infected with HIV have lost jobs, homes and friends. Children with AIDS have been denied access to public schools and in

1987 a major air carrier temporarily refused to transport patients with AIDS. People with AIDS have even been denied transportation to the grave, as some funeral directors have refused to handle their corpses.

AIDS is a modern affliction. The AIDS epidemic was fomented by changes in social mores and lifestyle that are unique to the latter part of the 20th century: urbanization in Africa, gay consciousness and liberation in the U.S., development of technologies for the preservation and shipment of blood-clotting factors for hemophiliacs, and modern air travel. Unlike some other infectious diseases, the AIDS virus is carried and transmitted by the human host; there is no apparent insect or other animal vector and the virus has no special climatic requirements. Because AIDS spreads directly from one person to another, the disease is—at least potentially—a universal problem. It is the one contemporary disease that is keenly felt as an urgent problem in both industrialized and less developed countries.

HIV is insidious. It corrupts vital body fluids, turning blood and semen from sources of life into instruments of death. The virus insinuates itself into the genetic material of selected cells, where it may remain quiescent for prolonged periods of time. When it is active, the virus gradually undermines the body's immune system, eventually rendering it vulnerable to opportunistic infections. During the latency period, which may average eight years or longer, the patient feels perfectly well yet is capable of transmitting the virus to others. HIV infection remains at the present time incurable, a pointed reminder of humanity's thrall to the tyranny of nature.

* * *

Because of its association with sex and its long latency period, AIDS has altered our thinking and prompted much discussion about human relations, love and sexuality. The AIDS epidemic has heightened awareness of homosexuality in our society, promoting understanding and tolerance in some and reinforcing aversion in others. The ease and readiness with which many now speak in public about homosexuality, sexual practices, the use of condoms and similar matters could hardly have been foreseen 10 years ago. The willingness of so many to see formerly taboo subjects presented in the media testifies to the extent to which AIDS has affected the standards of public discourse. The National AIDS Awareness Test presented on U.S. television in September, 1987, was introduced with the warning that some viewers might be offended; viewers were assured that nearly all those surveyed during the program's preparation believed the subject should be aired. Soon such reassurances about the need to discuss AIDS candidly will seem superfluous, because it will be obvious that we can no longer afford to live according to old inhibitions in discussing sexual practices and other risk factors that relate to this disease.

The HIV epidemic is marked by sharp variation in geographic, racial and gender composition. Globally three disparate patterns in the distribution of AIDS have been discerned. In the U.S. and other industrialized countries with large numbers of cases, the predominant modes of spread have been through homosexual activities and intravenous drug use, and the ratio of male to female cases is approximately 10 to one. In central, eastern and southern Africa and in parts of the Caribbean, heterosexual spread predominates, with a male to female ratio of about one to one. In these economically disadvantaged parts of the world perinatal transmission is high and blood-borne spread continues to be significant problem because of inadequate or absent screening procedures. In some parts of the world, such as eastern Europe, the Middle East and Asia, very few cases have been reported. Officials in these countries tend to ascribe most of their cases to travel, or to contacts with travelers from endemic areas, much as cases of heterosexual transmission in the U.S. are mainly attributed to contact with individuals who are bi-sexual or intravenous drug users. In neither setting should the current pattern offer much reassurance about the future.

* * *

Within the U.S. the geographic distribution of AIDS is highly uneven, minorities are disproportionately represented and intravenous drug use plays an increasing role in transmission. By mid–1988 the U.S. had counted 65,000 cases. More than half the states have reported fewer than 400 cases each, with a range of from fewer than 10 in each of the Dakotas to more than 16,000 in New York. The distribution is expected to be lopsided for some time, but the rest of the country is tending to catch up with the epicenters of the epidemic in New York City and San Francisco. In 1984 these cities had half of all the AIDS cases in the U.S.; in 1987 they had only 25 percent of new cases.

In San Francisco 85 percent of all reported cases of AIDS are among homosexual men who deny use of intravenous drugs; in contrast, 36 percent of cases in New York City are related to intravenous drug use. The majority of infected women in the U.S., who constituted more than 10 percent of the new cases of AIDS in the first half of 1988, are exposed by intravenous drug use, and an estimated 70 percent of HIV infection in newborns is related to intravenous drugs. The epidemic has hit minority communities particularly hard. Blacks and Hispanics constitute about 20 percent of the U.S. population yet make up 40 percent of AIDS cases.

The principal means by which the spread of HIV infection can be stemmed—education and altered behavior patterns—are at once clear and elusive. Behavior related to sex and drugs is biologically based, socially conditioned and resistant to change. In some of the homosexual communities most severely affected by AIDS, particularly those in San Francisco, sustained and intensive educational efforts have been rewarded by striking changes in behavior and arrested transmission of the HIV.

Yet the gap between knowledge and personal action remains wide. In a national poll conducted in August, 1987, more than 90 percent of Americans knew they could contract AIDS from having sex or sharing needles with an infected person. Yet when they were asked about the possibility of contracting AIDS themselves, 90 percent of all respondents said they viewed their own risk as low or nonexistent. Surveys taken after a 1987 New York City advertising campaign for AIDS prevention showed that 80 percent of the respondents agreed that sexually active people should carry condoms and women should tell their sexual partner to use a condom. Yet the reported numbers and frequency of sexual contacts in the preceding month had not changed and more than 60 percent of those surveyed said they had failed to use a condom more than just some of the time. If the effectiveness of education is to be measured by behavioral change, success will not come easily.

* * *

Health officials are particularly concerned about the increase in HIV infection among intravenous drug users. In 1987 they represented 16 percent of new AIDS cases; in the first half of 1988 that number had grown to 21 percent. Serum surveys reveal that 50 percent or more of the intravenous drug users in New York City have antibodies to HIV. Of the more than 1.2 million intravenous drug users in the U.S., fewer than 250,000 are estimated to be in treatment at any one time. In some cities the waiting period for those who seek treatment is longer than six months.

Such bleak statistics led President's Commission on the Human Immunodeficiency Virus Epidemic to call for 2,500 new treatment sites and an additional annual investment of $1.5 billion in drug-control programs. At the community level, street workers in a number of cities are attempting to protect drug users from HIV by showing them how to clean their needles and syringes with dilute bleach solution. Following the lead of European cities, Portland, Ore., recently undertook a trial program of providing drug users with sterile needles in exchange for dirty ones. Similar proposals have been made in Boston and New York, where they have met with considerable controversy. Critics oppose any appearance of state-sanctioned drug use and doubt the efficacy of exchange programs; advocates hold the preservation of life as a higher value and argue in favor of trial programs.

Another controversial proposal to stem the spread of AIDS, considered by legislatures in more than 30 states, is mandatory premarital screening for antibodies to HIV. Public-health officials and others have argued strenuously against such measures, saying that universal premarital screening would be counterproductive at this time. They point out that such tests would yield few truly positive results in a low-risk population, yet would overwhelm test sites, produce needless anxiety among those tested and waste resources. For the most part the arguments against screening have been convincing, although several states, including Illinois, adopted such legislation last year. Early experience in Illinois, however, bears out the predictions of public-health officials, suggesting that there are many problems and few benefits associated with universal premarital screening.

As with many other public-health measures, decisions about HIV screening tests should be reexamined in light of the changing dynamics of the AIDS epidemic. As the prevalence of the disease increases, the ration of false-positive results to true positives will decrease. (The reasons for the decrease are technical and somewhat beyond the scope of this article, but basically they stem from the increase of the fraction of the population that is infected.) Technical advances in testing and improvements in the quality assurance of laboratories where the tests are conducted will also enhance performance. If there are advances in therapy, such as development of an effective and safe treatment for the asymptomatic HIV carrier, then increased emphasis on screening would be more desirable.

Containing the spread of HIV infection in the U.S. today requires special attention to minority communities. Some black leaders have been understandably reluctant to add the stigma of AIDS to the burden of racism. An increasing number of them, however, are now prepared to take up the challenge of stopping the spread of HIV. The singer Dionne Warwick, for example, who was appointed Ambassador of Health by the U.S. Department of Health and Human Services in 1987, had made AIDS one of her highest priorities. She has focused her efforts on the minority community, enlisting the support of other celebrities to raise money for education, research and patient services. In fiscal 1988 the U.S. Centers for Disease Control spent $10 million on state programs to combat HIV in minority communities, with $3 million earmarked for community organizations. Some private foundations are also giving special attention to community-based programs. With the support of the Kaiser Family Foundation in Menlo Park, Calif., the School of Medicine at Morehouse College in Atlanta is managing a health-promotion program that includes AIDS education for minority communities in 15 eastern states.

Intravenous drug use flourishes in areas that are burdened by unemployment, homelessness, welfare dependency, prostitution, crime, school dropout and teen-age pregnancy. These conditions are so intertwined that no one of them can be solved in the long term without providing the fundamental infrastructure—jobs, schools and housing—needed by any community. Such an infrastructure would go a long way toward creating the individual self-respect, dignity and hope for the future that can forestall the turning to drugs in the first place.

* * *

Perhaps the specter of AIDS will arouse the nation's determination to face up to those realities. The darker possibility is that racial discrimination will become camouflaged under the delusion that AIDS is a problem for poor blacks and Hispanics and need not concern white, middle-class America. It is as dangerous and shortsighted for whites to view AIDS as a minority disease as it has been for black and Hispanics to view AIDS as a white homosexual disease. Anyone who engages in risky activities, including heterosexual sex outside of a monogamous relationship, stands a chance of becoming infected. The risk in some geographic areas and some population groups is now exceedingly low, but no one can foresee with confidence the course of the HIV epidemic over the next 25 years.

The advent of AIDS has indelibly marked the practice of medicine in the U.S. The adoption of universal precautions by many hospitals means that the blood and certain body fluids of all patients are to be regarded as potentially infectious to health-care workers. Some hospitals in cities with large numbers of AIDS patients have established dedicated clinical units to care for hospitalized AIDS patients. At the other extreme, a private pediatric hospital recently announced that it was not going to admit HIV-infected children. If an admitted child is found to harbor the AIDS virus, then the child will be transferred to another hospital. The same hospital also began a systematic testing program for its employees to ensure that the entire institution would remain free of HIV infection. Many hospitals frankly do not want AIDS to drive away their "real" patients: those who can most easily pay.

The stress of AIDS on health-care workers can be tremendous. Doctors and nurses face young and desperately ill patients suffering from a disease for which there is at present no cure. The medical and insurance systems around them resist the kind of counseling, home treatment and hospice care that the patient may need most. The doctor may be caught in conflicts between patients, lovers, family and friends; other AIDS patients may have no evident social support at all.

What is more, health-care workers have legitimate concerns about occupational exposure to HIV, although that risk is low. (Available data suggest that the risk of transmission from a single needle stick is less than half of 1 percent). Some may harbor prejudice or moral judgments about the behavior of their patients. Fewer physicians today are choosing to pursue careers in internal medicine, and it may be that AIDS is part of the reason. No other disease in modern times has engendered such frustration, resentment and anxiety or demanded more compassion, intelligence, selflessness and integrity on the part of health professionals.

* * *

A disease such as AIDS drains an economy in many ways. AIDS imposes an economic toll on every business, school, public agency, church congregation and community group responsive to the epidemic. In direct costs (those covering medical, scientific and other social expenditures) AIDS will cost the American public tens of billions of dollars over the next decade; indirect costs (such as lost wages from premature death and disability) will add several hundred billion more.

U.S. Public Service expenditures on AIDS have grown from approximately $60 million in fiscal 1984 to more than $900 million in fiscal 1988. The budget request for 1989 exceeds $1.2 billion, including $400 million for the Centers for Disease Control and $600 million for the National Institutes of Health. These sums cover scientific research, disease surveillance, prevention and control efforts. Total Federal expenditures for AIDS in fiscal 1989 are projected to exceed $2 billion, including $600 million for the Federal share of patient care through Medicaid.

At the state level, expenditures on AIDS have also risen dramatically, from less than $10 million in 1984 to more than $150 million in 1988. Much of that is spent by California and New York; in fiscal 1988 the two states accounted for 46 percent of all AIDS cases and more than 60 percent of state expenditures. On the level of the cities, expenditures are also great: New York City spent more than $130 million on AIDS in 1988 and has budgeted $170 million for the fiscal year 1989, mainly because most of the city's AIDS patients are cared for in public hospitals.

One of the most remarkable and hearting by-products of the HIV epidemic in the U.S. has been the development of grass-roots organizations dedicated to serving the needs of people with AIDS. An early and sustained commitment to building these organizations has come from the homosexual community. Groups such as the Shanti Project in San Francisco, the Gay Men's Health Crisis, Inc., in New York City and the AIDS Action Committee in Boston were begun in an effort to reach out and relieve the suffering of patients the world seemed to have turned against. As nonprofit, community-based organizations, they have provided a way for thousands of volunteers to give countless hours of assistance and comfort to patients, their loved ones and families. These organizations developed AIDS telephone hotlines and created specific educational materials for various cultural groups at high risk of HIV infection. They have also been outspoken and effective advocates for all those touched by the epidemic. In a larger social sense, these groups have served as bridges between the gay and lesbian and the straight communities, bringing together individuals who share a commitment to humanitarian goals and a refusal to give in to a lethal enemy.

AIDS has attracted the support of celebrities, business leaders and private foundations. Following the death of her friend Rock Hudson, Elizabeth Taylor became the National Chair of the American Foundation for AIDS Research (AmFAR), the only national foundation dedicated solely to combating AIDS, AmFARhas raised and devoted millions of dollars to scientific research and has recently broadened its agenda to include innovative educational and community-based service programs.

In 1986 the Robert Wood Johnson Foundation of Princeton, N.J., committed more than $20 million to projects for developing comprehensive and coordinated care for patients with AIDS. More recently the Johnson Foundation has invited applications for support of community-based prevention and service programs. In 1987 the Ford Foundation announced a collaborative, $4.5-million AIDS prevention and service program. By mid–1987 more than 150 foundations were providing support to AIDS-related projects.

Several major insurance companies have spent millions of dollars sponsoring AIDS education programs. Metropolitan Life underwrote the 1987 broadcast of the National AIDS Awareness Test on U.S. television. The New York Life Insurance Co. provided support for the New York City Department of Health's initial advertising campaign to prevent AIDS, a campaign that was developed pro bono by the advertising firm Saatchi & Saatchi Compton, Inc. Scores of prominent individuals in the arts, sports and business communities have lent their time, names and dollars to the struggle against AIDS.

* * *

Medical care for those suffering from AIDS is expensive. Estimates of the average lifetime medical costs per patient in the U.S. have ranged from less than $30,000 to more than $140,000, with more recent figures in the vicinity of from $50,000 to $60,000 per patient. These costs do not, of course, include the many thousands of hours that volunteers, family members and friends have contributed to the care of AIDS patients in cities across the U.S. It should be noted that although the cost of treating a patient with AIDS is high, it is well within the range of costs for severely ill patients with other conditions. Patients who require liver transplants, for example, have lifetime medical costs that are three to four times higher on the average than those of an AIDS patient.

Advances in medical care have begun to lower the cost of high-quality services to AIDS patients. The Kaiser-Permanent Medical Group of northern California, for example, established an outpatient center in 1986 to administer various AIDS therapies, including the life-prolonging drug AZT. During its first 18 months of operation the outpatient treatment center saved an estimated 3,500 hospital days. Although pharmacy costs nearly doubled between 1986 and the first half of 1987, average costs for the care of an AIDS patient declined 20 percent because of a 36 percent drop in overall hospital expenses during that period. New treatments in the future may further reduce and possibly eliminate some costs. On the other hand, new and expensive treatment could increase the average cost of care. The uncertain cost of future therapy is another hazy segment in the crystal ball foretelling the future of AIDS.

Total personal medical costs for AIDS will depend both on the average cost per patient and the number of patients. Uncertainty about the future size of the epidemic increases with the distance of the projection. The U.S. Public Health Service recently predicted that 450,000 cases will have been diagnosed by the end of 1993, extending its earlier estimate of 270,000 by the end of 1991. Personal medical costs for AIDS patients during 1991 have been projected to reach levels of between $4.5 and $8.5 billion.

Other costs associated with AIDS patients are subtler. When a hospital adopts universal precautions requiring frequent use of disposable gloves, gowns, masks and protective eyewear, hires additional infectious-disease specialists and infection-control personnel, follows special blood-screening and laboratory procedures and undertakes education and counseling programs for its staff, such costs are spread over the patients and are not found on the bills of those having a diagnosis of AIDS.

The HIV epidemic also results in medical expenditures for patients who are not infected with the virus. The "worried well" who experience general symptoms of fatigue, anxiety or poor appetite may seek medical care and testing because they are concerned about HIV infection. Family members and friends of patients with HIV infection may appropriately seek psychological counseling. HIV infection can also indirectly contribute to the rise of other infections in the community. After declining for many decades, tuberculosis has begun to increase in the U.S. Between 1984 and 1986 reported cases jumped 36 percent in New York City. Today these new cases are found mainly in patients with HIV infection, but as more people in the community develop active tuberculosis the risk of spread to those not infected with HIV will increase. As if to taunt progress in the life sciences in the 20th century, HIV not only has caused the disease most feared in America near the end of the century but also has fueled a resurgence of tuberculosis, the disease most feared at the beginning of the century.

* * *

Between 10,000 and 20,000 children in the U.S. are expected to have symptomatic HIV infection in 1991, most of them infected at birth by their mothers. This will represent a 10-to–20-fold increase over the number of newborns afflicted by the end of 1988. In New York City at present between 1 and 2 percent of all women giving birth are infected with HIV, and the proportion rises to more than 5 percent in some areas. The circumstances of life for many of these mothers—poor, ill, unwed and dependent on drugs—prevent them from caring for their infants. Half of the babies may escape infection, yet they have no place to go; many remain in the hospital, where the cost for their care can exceed $250,000 per year. Both the newborns and those who pay their bills would benefit from expanded nursery care outside the hospital, and if the projected number of infected newborns is even remotely correct, the need for many more nonhospital nursery facilities is acute.

The HIV epidemic exposes and exacerbates shortcomings in the system of paying for health care in the U.S. One in five AIDS patients has no insurance; 40 percent of AIDS patients are covered by Medicaid (more than four times the proportion in the general population). Medicaid, a program designed to cover the medical costs of the indigent, is a partnership between the Federal and state governments. Because of the variability in state rules for eligibility, Medicaid covers only 40 percent of those with incomes below the poverty line and frequently pays less than the cost of care. Even private insurers often do not cover the kinds of services—patient, home and hospice care—most needed by people infected with HIV. Private health insurance, for example, generally covers only 15 percent of the cost of drugs prescribed outside the hospital.

These problems can be solved by adopting a number of different strategies: state-based insurance risk pools or subsidies for uninsured patients; reliance on case managers to determine whether insurers should pay for services normally not covered; adjustments in the national standards of Medicaid eligibility and payment; simplification of procedures for states to request flexibility in Medicaid coverage; further extension of insurance coverage for employees who lose their jobs; mandated employer health insurance, and broadened Federal and state support of health insurance. In the case of AIDS, medicine and economy go hand in hand: failure to bring the means of payment into line with patient needs is dime wise and dollar foolish.

Although the costs associated with AIDS are undeniably great, those costs must be put into perspective. The U.S. spends more than half a trillion dollars per year on medical care. Even allowing for a relatively pessimistic projection of the AIDS epidemic, the billions spent on AIDS over the next five years will amount to a small fraction of the country's total health-care expenditures. In the cities and states hardest hit by the epidemic, however, the financial toll will be much heavier. Medical care for the additional cases expected in New York City by 1991 are estimated to cost $100 per resident, in San Francisco $350 per resident.

In parts of Africa and the Caribbean, where HIV infection is more prevalent than in the U.S., weakened economies are much less able to sustain the onslaught of the disease. In the city of Kinshasa in Zaire between 4 and 8 percent of the total population and more than a fourth of the hospitalized patients are thought to be infected with HIV. Many of those infected are in the educated middle class and are business people and professionals. In a dramatic press conference held in October, 1987, the president of Zambia announced that his son had died of AIDS. Demographic projections suggest that the long-term impact of AIDS on these populations may be similar to a prolonged war. In countries where the per capita national product is measured in hundreds of dollars and annual per capita expenditures on health care are $5 or less, a drug such as AZT that costs $8,000 per year might as well be nonexistent.

* * *

The World Health Organization's Global Program on AIDS convened an extraordinary World Summit of Ministers of Health in London in January, 1988. Their conference concluded with a declaration on AIDS prevention that emphasized broadening the scope of education, promoting worldwide exchange of information and reinforcing the importance of nondiscriminatory policies. The 41st World Health Assembly in Geneva in May adopted a formal resolution endorsing confidentiality of HIV testing and urging member states to avoid discrimination against AIDS patients in the provision of services, in employment and in travel. A similar call for antidiscrimination laws was the first recommendation of the June 1988 report of the U.S. Presidential Commission on the HIV Epidemic.

Public-health officials should have three primary objectives in coping with the AIDS epidemic. They are, first, to provide compassionate, effective and cost-sensitive care to the people who have the disease; second, to prevent further transmission of the disease, and third, to aggressively pursue scientific research that may lead to more effective prevention, diagnosis and treatment. The first objective requires committed and well-trained health professionals, increased services and a responsive health-care financing system. The second requires a sustained and unprecedented educational effort, judicious use of available public-health measures and special attention to minority communities, intravenous drug users and other populations at high risk of infection. The third requires building new human and institutional capacities, balancing basic and applied objectives and designing coherent research plans.

A strategy to accomplish these objectives stands on four cornerstones. The first is leadership that will inspire, direct and organize the fight against AIDS at local, state, national and international levels. The second is adequate financial resources to do the job, drawn mostly from public sources but also from some private ones. The third is legal protection against discrimination, on which so much else depends. And the fourth is an accurate and timely surveillance system that can track and project the status of the epidemic. The future course of the HIV epidemic is uncertain and a strategy that takes account of such uncertainty is imperative.

Our world has been made a different place by the human immunodeficiency virus. More profoundly, our society is being shaped by our response to the epidemic. Will AIDS enhance understanding and tolerance of different sexual orientations, or will it harden traditional norms of acceptable and deviant sexual behavior? Will AIDS be perceived as a universal threat to all humanity, or will it be regarded as a problem of the underclass, the poor and uneducated, and the minorities? Will AIDS heighten the tension between moralistic and pragmatic approaches to behavior and health, or can solutions be found that are both effective and morally acceptable? Will AIDS evoke the selfless dedication of physicians, nurses and other health professionals, or will caregivers shun AIDS patients and seek other ways to practice their craft? How we choose to answer such questions, and the society we thus shape, is up to us.

*Scientific American* (October, 1988): 128–134.

## Questions and Topics for Discussion and Writing

1. Describe the AIDS disease and HIV from a medical viewpoint.
2. Describe the different ways in which HIV is spread and the relative proportions of cases attributed to each.
3. How does AIDS differ from other epidemics in terms of transmission, contagiousness, seriousness, and capability of being controlled?
4. Discuss the conflict between the rights of individuals suffering from AIDS and the interests of the community as a whole.

*Unit 4*

# Current Issues in Science

- Chemicals in Society
- Nuclear Energy and Nuclear Warfare
- Biotechnology and Related Issues
- Computers and Automation in Society
- Scientists in Society

A send-up of a well-known advertising slogan of a major chemical company taken in the heart of hippie culture, the Haight-Asbury of San Francisco around 1965.

*Issue*

# Chemicals in Society

Chemicals have played a role in human society from the earliest times. Substances have been extracted, mixed together, boiled down, used in natural form or processed in a number of ways and then used for human purposes in foods, in medicines, in the household environment, and in public utilities. Ingenious applications of the chemical properties of matter have been of phenomenal value to the societies in which they were used. But along with the benefits gained, there have always been costs that were not recognized at first. A case in point is the article on lead poisoning in Rome in the previous section.

With the coming of the synthetic chemical industry, the uses of chemicals in society has expanded enormously. Chemical wonder products are among the distinguishing features of modern industrial society. It would be difficult to imagine what a modern North American city would be like without plastics, dyes and paints, glues, detergents, antibiotics and other wonder drugs, such essential components of business machines as photocopy machine toner and floppy disks—and on and on. There are so many "miracle" chemical products that any attempt to list them appears to be an arbitrary and eccentric selection.

It is no surprise that the rapid pace of new applications of chemistry has engendered some serious harmful side effects. Emphasizing only harmful effects runs the danger of losing sight of the benefits of applied chemistry that we would be loath to give up and suggests that we can point the finger of blame at the chemical and biochemical industries for poisoning the environment and ourselves. This would be foolish and arrogant. On the other hand, as the articles in Unit 3 overwhelmingly show, whatever appears to be a small environmental problem today can grow into a major environmental disaster in the not-too-distant future.

There is some justification for harping on the dangers of chemicals in society and ignoring the benefits for the moment. The chemical manufacturers will do their best to make the benefits of their products known to us. We must make it our business to be alert to the dangers. The articles in this section are all concerned with harmful effects. As those effects are often environmental, many of the articles in Unit 3 would fit equally well in this section.

With the general environmental issues already represented in Unit 3 the four articles in this section concentrate on issues of industrial safety and pharmaceuticals. Harrison S. Martland's "Occupational Poisoning in Manufacture of Luminous Watch Dials" is a physician's report of an unfortunate incident involving careless handling of radioactive materials in the early days of chemical manufacture of radioactive (and thereby luminous) paint causing the death of several workers. A more recent case causing concern among affected workers is discussed in "PCBs—Unanswered Questions," a report of the Ontario Workers' Occupational Safety and Health Centre. Elisabeth Rosenthal's "Madison Avenue Medicine" raises the issue of the independence of judgement of physicians who are bombarded by drug company representatives with sales hype for their products. Taking that a step further is the article "Hazards Implicit in Prescribing Psychoactive Drugs" by Henry L. Lennard et al., which calls attention to the subtle pressure from pharmaceutical manufacturers to characterize psychological and emotional issues as medical problems which can, therefore, be treated by drugs.

# * Occupational Poisoning in Manufacture of Luminous Watch Dials

## Harrison S. Martland, M.D.

It is always a shock to look back in history, sometimes very recent history, and read about products, processes, or customs that we would consider highly dangerous or destructive today, but which were viewed with relative equanimity then. Such a case is the former practice of painting watch dials with luminous paint so that the time could be determined in the dark. That the paint was radioactive was known, but not considered dangerous until some of the workers in the watch factories began to fall ill. This report, published in 1929, reviews the cases of the affected workers and what was known about the hazards at that time. Much of the medical detail of the original report has been omitted from the selection printed here.

Because of recent newspaper publicity, the result of litigation; and the publication of scientific and lay articles, some erroneously explaining and others denying the existence of the so-called radium poisoning in watch dial painters, it seems advisable that a review of the history, symptomatology, pathology, prognosis, treatment, and scientific literature of this industrial hazard, first adequately described by me and my associates, be published at this time...

### Number of Girls Exposed in the New Jersey Cases

During the years from 1917 to 1924, an accumulative total of about 800 girls were employed in a factory in New Jersey painting the dials of watches and clocks with luminous paint. In the busy periods there were never more than 250 girls working at one time. Many of the girls did this work for only a short period. The pay was attractive and the work comparatively easy. It is important to note that the known cases of radium poisoning all affected the girls who had worked steadily in this plant for from one to two or more years.

Among these employees there have been fifteen known deaths in which the so-called radium poisoning is strongly suspected. It was proved to be the cause of death in five cases. That other deaths have occurred is not at all unlikely. The late development of symptoms, often occurring from one to seven years after the patients leave the employment, and their resemblance to various other diseases would make a diagnosis almost impossible if the patients had moved to another locality. Physicians not aware of this condition might treat a victim for sepsis, anemia, Vincent's angina, rheumatism or "God knows what." How soon the true state of affairs would come to light would depend on the physician and the facilities at his disposal.

When one estimates the number of girls who worked steadily at this employment for a considerable period of time and realizes that most of the fatal and severe cases came from this group, it will be seen that the incidence of this poisoning compares with that of most other forms of occupational poisoning. For instance, in phosphorus poisoning among match makers, only 1 per cent of those working in Great Britain developed necrosis of the jaw and from 1 to 3 per cent in France and Switzerland. One of the great medicolegal questions raised against the radium-mesothorium cases being occupational was, Why did not a larger number of the girls become affected? Carelessness in handling poisons, individual susceptibility, individual resistance, habits and length of exposure all play important parts in occupational poisoning, and some are not thoroughly understood. There is no way of ascertaining in these cases how many have lost their lives and how many have been permanently or temporarily harmed.

### Luminous Paint

*General Principles and Theory.* Lind states that while the general principles of the preparation of luminous paint have been described and known for some time, the actual methods of preparing phosphorescent zinc sulphide have not been fully described in the literature, and the methods in commercial use have been carefully guarded. The following quotations are selected from the excellent summary by Lind in his book on the alpha particle:

> Certain substances, under the influence of various kinds of radiation, emit light of visible wavelengths at ordinary and at extremely low temperatures... In the radium luminous paints, radium salts are intimately mixed with phosphorescent zinc sulphide, which mixture is then applied to the dial to be illuminated. These mixtures have come into extensive use on watch and clock dials, electric push buttons, etc., and during the war were widely used for illuminating dials on aeroplanes, battleships, or in any place where it is desired to have a feeble light not visible for more than a few yards.

This luminescence depends on the phenomenon known as scintillation, discovered by Sir William Crookes in 1903. It is due to an alpha particle which consists of a helium nucleus containing two positive charges, striking the crystalline, phosphorescent zinc sulphide and causing a flash.

> Scintillations can be individually observed and counted only when the number of alpha particles falling on a given area of the screen in unit time is limited. When the number is greatly increased the screen appears to the eye to be uniformly illuminated, while under the microscope the number is too great to be counted.
>
> The nature of the reaction furnishing the light is not thoroughly understood. Rutherford proposed the theory that a phosphorescent substance contains initially a large number of "active centers" or molecular aggregates which are disrupted by alpha particles with the emission of light. Rutherford has calculated that a single alpha particle destroys all the centers in its path within a radius of $1.3 \times 10^{-1}$ *cm* for ZnS.

Without going into further detail, it becomes apparent that after the luminous paint has been applied there will be a constant decrease in luminosity as weeks and years go by, owing to a decay of the zinc from constant disruption by the bombarding alpha particles and also to weakening in the radioactive substances used by reason of their natural and uninfluenceable decay. In the average watch this luminosity is greatly impaired in about three years with the radium-mesothorium mixtures...

*Character of the Paint Used.* Of course the detailed preparation of the paint used in these cases has for purely commercial reasons never been published. It is possible, even probable, that its ingredients and formulas change from time to time. According to my information, however, an average paint used during this period on watches consisted mainly of crystalline zinc sulphide (ZnS, Sidot's blende) mixed with various proportions of radium, mesothorium, and radiothorium in order to obtain the desired amount of luminosity...

The paint used by the girls contained...chiefly zinc sulphide rendered luminous by activation with about 20 to 30 per cent of radium and from 80 to 70 per cent of mesothorium containing radiothorium. But doubtless this percentage varied from time to time. This is most important in view of the fact that later, in the extraction from the bones in the fatal cases, mesothorium and its decay products were found greatly in excess of the radium in four of my cases, while in the case recorded by St. George and Gettler only radium was found...

As to whether certain ingredients of the paint other than the radioactive substances were responsible for the symptoms in these cases, it may be stated that there is no previous knowledge or record of chronic poisoning due to the cumulative effects of zinc, copper, cadmium and manganese showing anything like the clinical pictures seen in these cases...

*Mode of Application of Paint.* The paint was mixed with acacia and thinned out with distilled water, and was given to each girl in a porcelain crucible containing about 10 cc. of the mixture.

It was the habit of most of the girls, who were doing piece work, to point the camel hair brushes used in painting with their lips, as quicker and more accurate painting could thus be performed. We had assumed in 1925, as a basis for argument, that a girl painting 250 watches a day and licking her brush once for each dial might possibly ingest a minimum of about 125 mg of the paint. If she pointed her brush fourteen times per dial (one girl stated that she often licked her brush twice for each watch figure), she might ingest about 1.75 gm of paint a day. Working throughout the week, it was possible, therefore, for a girl to swallow from 15 to 215 micrograms of radioactive substances during a week...

Hoffman, in May, 1925, was the first to call attention to this habit of licking brushes. Speaking of twelve cases that he reported, he states:

> There is some question of doubt as to whether they were instructed to do their work this way or whether they worked contrary to instructions. They certainly seem to have copied the habit one from another, and found it decidedly convenient, the brush being better adapted to the purpose when penciled to a fine point with the tip.

He also stated that the paint contained mesothorium.

## Chronological Order of Events in the New Jersey Cases

*Early Deaths.* In 1922, 1923 and 1924, nine girls died who had been employed as watch dial painters in this plant and who had worked steadily for years at this employment.

Absolutely no investigations were made in any of these cases as to the cause of death. The death certificates were signed by the family physicians and some of the following conditions were given as the cause: ulcerative stomatitis, syphilis; primary anemia, Vincent's angina; phosphorus poisoning, necrosis of the jaw; occupational poisoning (character unknown), necrosis of the jaw; anemia, necrosis of the jaw. It is impossible to obtain any data in these cases as to the special symptomatology, or any blood studies on the type of anemia. Autopsies were not performed. In fact, except for slight suspicion on the part of a few family physicians that something was wrong, no medical recognition of the existence of an occupational poisoning was recorded. The material represented by these cases is therefore wasted with the exception of one case. In this case, which was one of the first to terminate fatally in 1922, the body was exhumed five years after death for litigation purposes. Gettler found undisputed evidence of radioactivity in the bones.

*Treatment by dentists.* During this time, local dentists were treating a considerable number of girls employed, or who had been employed, by this plant for various degrees of jaw necroses, many of which were resistant to the usual forms of treatment. There is no way of determining the number of girls treated, but the assumption of about fifty would not be an exaggeration.

*First recorded suspicion of the existence of an occupational disease.* Blum, a New York dentist, is to be credited with the first report indicating that an occupational form of poisoning existed. He reported in September, 1924, a case of unusual and intractable osteomyelitis of the mandible in one of these girls. He knew nothing regarding the nature of the paint or industry, but thought from the unusual clinical behavior that some sort of occupational poisoning existed.

*Investigation by lay and civic bodies.* In 1924, the Consumers' League of the Oranges [a geographical region in New Jersey] called the attention of the state department of labor to the objectionable fumes coming from the plant, and the possibility of an occupational poisoning. An investigation was made but with entirely negative results. The US Public Health Service had briefly considered the matter but did not make an investigation.

*The Hoffman report.* In 1925, Hoffman, whose attention was called to the plant by the New Jersey Consumers' League, found that there was a great similarity in the reading of the death certificates of deceased former employees. Most of them mentioned jaw necroses, anemias, and buccal lesions. He thought that from a statistical standpoint the matter should be investigated. In May and September, 1925, he reported, from the statistician's point of view, five deaths and twelve cases occurring among girls still alive who had developed a resistant infection of the jaw with buccal lesions and a marked anemia. He was of the opinion that they represented an unusual occupational poisoning, and thought that mesothorium in the paint was mainly responsible and that there was an ingestion of this paint as a result of the habit of licking the brushes. This is the first recorded report of the nature of the disease. In the investigation he was greatly aided by the experience of von Sochocky, who was intimately acquainted with conditions at the factory. No description of the types of anemia or other symptomatology was given, and as there had not been any autopsies there were no records of pathologic changes, and the subject was unproved.

*The Castle and Drinker Report.* During this time the factory executives had undertaken an investigation of their own to determine whether any occupational hazard existed. This was made by Castle and the Drinkers in March, 1924. Unfortunately, at the request of the company, this report was not published on account of threatened litigation until August, 1925. According to my information these investigators were asked to ascertain whether there was any evidence of zinc poisoning and to comment on the general hygiene of the plant. Nothing was told them about mesothorium. They inspected the factory and the employees and concluded that there was undoubtedly proof of excessive exposure to radiation. Sealed dental films when placed in any part of the factory showed fogging in from one to two weeks, and in the painting room in from two to three days, showing exposure to gamma radiation. The skin, dresses, and underwear of many girls when examined in a dark room showed the presence of phosphorescent zinc sulphide and appeared luminous. Dust collected from the chandeliers, wall beams and even in the offices, when exposed to light was luminous in the dark room, showing the presence of the zinc sulphide. Samples tested later for radioactivity were positive. While they considered the skin and gastrointestinal tract as probable portals of entry, they concluded that undue exposure to radioactive substances existed, chiefly from the inhalation of particulate material containing a radium salt. They did not ascertain the presence of mesothorium. They stated that the habit of pointing brushes in their mouths had been general among the employees, but was stopped six months before the investigation started. They concluded that the implication was not proved, but the safest course would be to assume that an occupational poisoning existed. They made recommendations for the future protection of workers. The main value of this report is that it came from the factory's side and demonstrated without a doubt that all the girls employed as dial painters were subjected to excessive exposure.

*First Patients Examined.* In May, 1925, I examined two girls with extensive jaw necrosis and severe anemia. Both patients died and I performed autopsies. At the same time I saw in consultation the company's physicist, who was suffering from a profound anemia, apparently of the regenerative type. He died soon afterward and I performed the autopsy. These were the first autopsies made in such cases...

### Death, from Aplastic Anemia, of Originator of Formula Used at New Jersey Plant

Nov. 14, 1928, Dr Sabin A.von Sochocky died of an aplastic or regenerative anemia. Von Sochocky had formerly been a technical director in the company in which these cases occurred. While the general principles of luminous paint are well known, each plant had its own formula which was a commercial secret. The paint used by these girls was mainly von Sochocky's formula. From 1913 to 1921 he had been exposed to intense radiation from radioactive substances, during which time he had personally extracted about 30 gm. of radium from the ore, being exposed continually to heavy penetrative radiation. He was exposed to the inhalation of dust in the crystallizing laboratories. From 1919 to 1920 he was exposed to heavy radiation in a small room from large amounts of emanation by inhalation and ingestion and to highly concentrated dust from radium, mesothorium, and the like. On four occasions he was exposed to explosions of tubes containing high concentrations of radium and mesothorium.

Since 1921 he had occasionally been exposed to external penetrative radiation from radium used for therapeutic purposes.

He had an extensive radium dermatitis of the fingers of both hands. He also had considerable jaw necrosis with buccal lesions, and at times was quite anemic. However, he was able to combat the anemia and buccal sepsis by hygienic measures and by living at times in high altitudes.

In 1925 he gave me considerable advice with regard to measuring the amounts of radioactivity in the bones at autopsy. In June, 1925, while testing some of the living patients and devising a technic for the expiratory air test as a control in one of our experiments, we tested his expiratory air for emanation and found larger amounts than in any of the living patients. Later, when he blew his expiratory air over a screen of phosphorescent zinc sulphide, scintillations were seen, too numerous to be counted. At this time he realized his condition and its apparent hopelessness but was stoical to the end...

He died a horrible death, similar to that seen in chronic benzene poisoning, the acute leukemias and other acute blood dyscrasias. During the time I knew him he gave all that was in him to help and comfort others suffering from this disease. Without his valuable aid and suggestions we would have been greatly handicapped in our investigation.

## Litigation

In 1926 two of the fatal cases in which I had performed the autopsy were settled out of court without the company assuming any legal responsibility. The families of the two girls received a lump sum of money approximately equal to what they would have received under the compensation act, which arrangement was to extend over a period of time. Of course, the attorney extracted his usual fee.

In May, 1927, suit was started in the supreme court by five girls said to be suffering from this disease. This was made known to the public through the newspapers as "The Case of the Five Women Doomed to Die." It is of great medicolegal importance and will therefore be briefly discussed. I had no interest in either side but was subpenaed by the plaintiffs as a county official, and refused to accept more than the sum of 50 cents customarily tendered with the service of a subpena.

I shall try to bring out the shortcomings of the present compensation law and show the impossibility of obtaining compensation for occupational diseases unless the disease is specifically included in the law.

After suit was started the defense pleaded that the statute of limitations[1] barred action and the women's counsel was compelled to transfer the case to the chancery court[2] in July, 1927. The case was continually postponed and no hearings were held until Jan. 12, 1928. After three days the hearing was adjourned to April 25, when it was again continued for three days. The plaintiffs case being in, further hearings were then put off until September 24. Shortly after an editorial appearing in the New York World, which stated that "this is one of the most damnable travesties on justice that has come to our attention," it was decided that the case never belonged in chancery and should go back to the supreme court for trial.

At last everything was ready for the trial, even the sob writers, when suddenly an uninvolved judge, out of the kindness of his heart, believing that there should be some relief for the girls and that the case might drag on for years in the courts, arranged a settlement out of court. The company again did not assume any legal responsibility that the cases arose from occupational poisoning. Great credit was given this judge for his very humanitarian act.

In a recent review, De Laet cites the cases of the five New Jersey girls as the only known instance in which persons injured from the handling of radioactive substances have received compensation. He says that these cases have no medicolegal precedent.

The settlement is unique. The girls received a definite sum of money and will have a yearly pension as long as they live, or as long as a committee of three experts decides that they are still suffering from radium poisoning. One member of the committee was selected by the company, one by the girls and the third by the other two. Two members of the committee will constitute a majority. No mention was made of mesothorium and no explanation of the omission was offered.

[1] A New Jersey statute of limitations says that suit for damages must be brought within two years after inception of the disease. If this law was pleaded and upheld these girls would not have been entitled to bring suit, as most of them did not develop symptoms or realize that they were suffering from this disease until five, six and eight years after its inception.

[2] Literally, the Court of King's Conscience dating from the early eighteenth century. Chancery had jurisdiction when there were no forms of action by which relief could be obtained at law, in respect of rights that ought to be enforced. Said King James, speaking in the Star Chamber: "Where the rigor of the law in many cases will undo a subject, then the chancery tempers the law with equity, and so mixes mercy with justice, as it preserves a man from destruction."

Three of the five girls had been examined by me in 1925 and showed evidence of radioactive deposits in their bones and emanation in their expiratory air. At this time they were in rather good physical condition. In fact, one girl who we predicted might have trouble in the future was reported in our original paper as a "healthy case" on whose life any insurance company would have issued a policy. This girl now has bone lesions of a crippling nature and a moderately severe anemia. Only one of the five girls has had any severe necrosis of the mandible. This patient still has jaw necrosis, which has already necessitated about twenty dental operations...

The clinical picture in these five girls is a little different from that in the early fatal cases. None have developed so far the profound anemias seen in the early cases. All but one have recovered from, or have never had, the necrosis of the mandible and maxillae that was such a prominent feature of the early cases. They seem to be suffering chiefly from the effects of radiation osteitis, more or less localized in such bones as are subject to external trauma or pressure, such as the scaphoid bones of the feet or the bones that bear considerable weight, such as the lower dorsal and lumbar spines, the heads of the femurs, the humerus, and the glenoid cavities. The effect has been to produce chronic bone lesions, with pain, limitation of motion, and a consequent crippling effect. Two of the girls cannot walk and practically have to be carried.

To what, then, is this different clinical picture due? The most feasible explanation would seem to be that a sufficient length of time has elapsed to allow the mesothorium, which forms the largest part of the deposits, to decay in its own natural, uninfluenceable manner, and that these girls may escape the fatal anemias and sepsis seen in the earlier cases.

Before the final hearing in chancery, I had the opportunity of witnessing an examination of these patients by the company's experts, in the presence of experts representing the plaintiffs. At this examination the only test made to determine whether the girls were radioactive was to have them sit in front of a gamma electrometer. The instrument used was said to be unaffected by atmospheric changes. This electrometer was supposed to detect penetrative rays coming from a living body thought to contain deposits of radioactive substances, chiefly in the bones. The total amount distributed throughout the skeleton was probably below 10 micrograms, and 95 per cent of the radiation coming from these deposits was alpha. In addition, important parts of the body, such as the tibia and femurs, which would contain a large amount of the radioactive substances, were from 2 to 31/2 feet away from the instrument, allowing what little gamma rays were given off to be almost entirely absorbed by the body flesh. The readings were made by Schlundt and Flinn for the company, and separate readings were made by Mrs. Hughes and Gettler for the girls. I was informed that as a result of these tests the company's experts concluded that the girls were not radioactive, although Gettler told me that his readings in one case strongly suggested the presence of penetrative gamma rays. It is worth noting that, after the girls had been tested, one of the experts seated himself in the same chair in front of the instrument with 10 micrograms of radium element, or its equivalent, 5 micrograms in each of two vest pockets, and that the readings were no greater than those found in one of the girls...

Shortly after this examination, at the suggestion of von Sochocky, Mrs. Hughes prepared screens of pure phosphorescent zinc sulphide uncontaminated by radioactive substances. After these had been examined in the dark room for scintillations, with negative results, each girl blew her expired air over these screens beneath the microscope. Scintillation from the presence of alpha particles in the girls' expired air was easily demonstrated in all five cases, proving beyond a doubt that they had deposits of radioactive substances in their bodies, which were giving off emanation.

### Author's Interpretation of the Disease

On the basis of the studies made by me and my associates, I am of the opinion that the anemias, jaw necroses, and various forms of crippling bone lesions encountered in these girls were due to the ingestion of insoluble sulphates of radium, mesothorium and radiothorium which occurred when they were employed as dial painters and which were due to their habit of pointing the brushes in their mouths. Most of the paint swallowed passed through the gastrointestinal tract, but there was always a small amount that was absorbed day after day, month after month, and year after year, as they worked. This was eventually deposited as insoluble sulphates of colloidal or particulate size in the phagocytic cells of the sinusoids of the reticuloendothelial system, and especially in the bones, liver, and spleen, with by far the greater amount in the bones.

In none of my patients that died, with the exception of the physicist, were there any appreciable radioactive deposits in the lungs. In this patient the type of exposure was different from that in the girls, consisting chiefly of the inhalation of emanation and radioactive dust and of penetrative external radiation. This observation is against the supposition of Castle and the Drinkers, who thought that the chief mode of entrance in these cases was by inhalation, and supports the original theory of Hoffman that the main danger lay in the licking of the brushes, with the consequent ingestion of the paint.

*General Symptomatology.* A sufficient time has now elapsed. to permit the clinical recognition of two types of cases: the early fatal ones and those in which death has not occurred.

In the early fatal cases in which there have been at least fifteen known deaths, all the patients suffered from a severe leukopenic anemia, and severe necroses of the mandible or maxillae or both, with buccal lesions and often terminal sepsis. Symptoms were first noted in these cases in from one to seven years after the patients ceased working as dial painters.

The later group, in which the five living girls cited belong, and in which there is evidence that perhaps other girls are similarly affected, presents a somewhat different clinical picture. In these cases the necroses of the mandible and the maxillae are not a prominent feature, nor is the severe anemia. It is quite possible that some of the patients may yet develop symptoms similar to those seen in the early cases, but in them the disease seems to follow a more chronic and slower course, characterized chiefly by low-grade crippling lesions of the bones due to a radiation osteitis set up by the deposits in the individual bones...

*Journal of the American Medical Association* (February 9, 1929): 466–473.

## Questions and Topics for Discussion and Writing

1. Why were the effects of the luminous paint on the workers not noticed sooner than they were?
2. What is the role of "pointing the brush" with the lips in the process of the radium poisoning? How could this practice have developed without the knowledge of company officials?
3. Comment on this case as an instance of occupational hazards. What lessons can be learned from this that might be of use to us today?

# PCBs—Unanswered Questions


## Ontario Workers' Occupational Safety and Health Centre


Polychlorinated biphenyls—PCBs for short—are only one of the many groups of chemicals that were introduced as miracles of modern science only to later be found to have some very serious harmful side effects. What happens then, especially when meanwhile the chemicals have been very widely used and are present in society in many forms? Often, what starts then is the legal wrangling as damage is assessed and responsibility is determined. This report concerns PCBs at Stelco, Canada's huge steelworks in Hamilton, Ontario, and the struggle between the labour union and the Ontario Ministry of Labour to investigate it properly.

Perched above Stelco's PCBs storage compound the eye sees row upon row of metal drums. Next to them sit a series of now-defunct transformers—all of course contain dielectric oils laden with PCBs or polychlorinated biphenyls.

Set up after studies of the '70s began identifying PCBs as a dangerous environmental contaminant and a suspected human carcinogen, this Hamilton storage unit has reportedly become one of the largest of its kind.

In Canada, to date there are approximately 6,500 tonnes of concentrated PCB liquids in storage awaiting disposal. A number of provinces including Ontario, have initiated plans to build destruction facilities, but none as yet have been approved.

First used in 1929, PCBs were hailed the miracle chemical. Its properties of stability, fire-resistance, electrical insulating and low volatility at normal temperatures recommended them as desirable components in a gamut of industrial and consumer products.

Although primarily used as a cooling and insulating fluid for industrial transformers and capacitors, PCBs were also used as hydraulic fluids, surface coatings for carbonless copy paper and plasticizers in sealants, synthetic resins, rubbers and paints.

But then the bottom fell out. Now PCBs are allowed only in existing closed electrical and hydraulic systems. Any leaks, and out come the face shields, impervious gloves and disposable clothing, all of which are in turn sealed in a drum and stored away with the contaminant itself.

When the cleanup at Stelco came in 1977, Joe Choscinski, Chief Health and Safety representative for the electrical department, USWA Local 1005, remembers what a shock it was. "PCBs had been everywhere—in all kinds of walkways," he says.

Sometimes, says Joe, workers would spill PCB oil all down the front of them as they tested transformers. Often, he adds, they would eat on the job without cleansing their hands of them.

Gary Knox, Joe's predecessor in the electrical department, says he thought nothing of reaching down into a transformer, PCB solution up to his armpit, and changing the diodes.

But he also recalls in the years just previous to the restrictions of PCBs, how only a few including himself were assigned to work on the transformers. "Here I thought they were taking me into their confidence. All those years I was thinking they liked me," says Gary.

The surge to rid the plant of PCBs though, was a direct result of pressure exerted by the union's health and safety reps. "This union has done an awful lot to get controls around here," says Gary.

Labeling of all transformers that contain PCBs, catch trays around the base for any leaks, personal protective equipment, work procedures for handling transformers, and the enclosed storage compound meeting Ministry of Environment and Labour requirements have all come about because the union forced the hand of the company, says Don

Fraser, Local 1005's health and safety chairman and OWOSHC qualified instructor.

* * *

Then again, trained health and safety labour activists have been making the difference for some time now. "Over the last decade virtually all regulatory actions against carcinogens in the workplace, in consumer products, and in the environment have been developed only at the initiative of public interest groups or organized labour," writes Dr Samuel Epstein, a professor of Occupational and Environmental Medicine, at the University of Illinois' Medical Center (The Politics of Cancer, 1978).

In the wake of renewed concern over PCBs at the Hamilton Stelco plant, union health and safety reps have been requesting an independent epidemiological study to be conducted by a third party and funded by the Ministry of Labour, thus making it acceptable to all parties. So far though, the union has been frustrated in their attempts.

Concern first broke out last fall when a Stelco electrician went to the press telling them his blood test read 38 parts per billion (PPB) of polychlorinated biphenyls.

Attracted by the newspaper reports, field officers from the Ministry's Hamilton branch decided to pay the plant a visit, where they toured three areas and reviewed PCB records. As a result of that visit and the recent upgrading of chemical guidelines to enforceable regulations the Ministry of Labour issued an order on December 19, 1986 which stipulates: "Where a worker has been exposed to PCBs the worker shall at the expense of the employer undergo medical examinations and clinical tests...to determine whether the worker has an occupational illness because of exposure to PCBs."

Plainly, this order falls short of a proper epidemiological study. Members who have retired, quit, or been laid off are all excluded.

And yet it's "the older guys who are upset. Several retired members have called worried about PCBs," says Fraser. These are also the members who would have experienced the highest and longest exposure, he adds.

Since then though, the Ministry has further watered down their position. On February 11th, they rescinded their order requiring the testing of all workers exposed to PCBs, and issued a new one that requires the testing of current power department workers only.

At best this new order encompasses 38 to 40 people, many who have never been exposed to higher levels of PCBs, or are too young to have developed any latent ill health effects.

In response the union held a press conference challenging the competency of the Ministry. Their release read: "We condemn the Chief Physician of Ontario and the Ministry of Labour for their back tracking on this issue and we would question why it took over two months for them to find a so called mistake in their order. The mistake being that it was not the order they meant to write. This order leaves the question in people's minds, did someone influence higher officials in the Ministry of Labour to rescind this order? "

Two days later, Ministry of Labour hierarchy, including the Minister himself, all blew into Hamilton. Justifying their new order, the Ministry argued that the original order—aimed at anyone who had any PCB exposure—was too broad. As such, they said if a worker challenged it, refusing to undergo the mandatory testing, it wouldn't stand up in court.

* * *

Ironically, the union never requested mandatory testing in the first place. They called for an independent epidemiological study from the start—no orders, no mandatory testing, only volunteer worker participation.

When earlier asked about the fate of their independent study request Fraser said: "Stelco is staying silent on it and the Ministry of Labour is determined to wash their hands of it." At this last meeting Stelco refused outright, and the Ministry was as determined as ever.

They did however waffle over the new order. Picking up on the implications of this order, union representatives asked if the Ministry intended to exclude persons with past exposures to every substance that has been newly regulated.

From the five Ministry reps present there were five different responses, says Fraser. He argues: "If they don't apply to past exposures they're pretty toothless."

Backing off the Ministry reps ended the meeting promising an interpretation from their Legal

Branch on the scope of the new regulation and hinting that they may work out a middle ground between the original and new order.

Middle ground or not, by restricting themselves to small studies with no examination of the long-term effects the Ministry offers little or no help to workers, especially in their bid for compensation, says Norm Carriere, Health and Safety Representative for USWA, District 6. "Unless we can get a large group, they probably hurt our cause," he says.

"Testing has to be something well correlated with an epidemiological study," says Dr Allan Okey, a pharmacologist researching dioxins and PCBs at Toronto's Sick Children's Hospital. Testing also has to be followed up into retirement and even death. On their own, tests "don't tell you about risk," says Dr Okey.

* * *

As of February 5 no blood levels taken through Local 1005's health and safety clinic were over 10 PPB, says Fraser. But ten PPB of PCBs in your blood or 38 PPB of PCBs in your blood only means just that, you have 10 or 38 PPB in your blood.

One member, at Local 1005's January membership meeting declared in a frenzy that their members are dying or have died from exposures to PCBs. However the majority of the scientific community answers in a unanimous refrain that "the jury is still out."

In animals though, we do know that certain PCBs and related chemicals cause cancerous tumours, birth defects, liver injury, a depression of white blood cells and changes in blood fat levels (triglyceride).

Although scientists have failed to conclusively link PCBs with cancer in humans, reserving judgement until further testing is done, nonetheless no one suggests PCBs are as harmless as some newspaper columnists have recently claimed.

Certainly those Japanese who experienced a form of poisoning known as "Yusho" (oil disease) are likely none too complacent about PCBs. In the summer of 1968 almost 2,000 of them ingested rice oil which had been contaminated with high concentrations of PCBs during its processing.

As a result the victims suffered an acne-like breakout (chloracne) on their skin and a darkening of their pigmentation: cyst-like swelling on the eyelids and an associated thick discharge; damage to the nervous system including fatigue, headaches, pain in the limbs and swelling of the joints; and respiratory symptoms most notably chronic bronchitis. The list continues: liver damage, increased blood fats, hormonal changes and reproductive damage such as low birth weights and a transmission of many signs present in the parent. Of the 31 Yusho victims who have died, 11 (35.4%) have died of cancer, a rate that is high than expected for a normal population.

Many of these same effects would later show up in a Taiwan PCB poisoning incident, by comparison however as yet, fewer have been noted in occupationally exposed workers. Nonetheless, they have still demonstrated a number of biologic and toxic effects, including chloracne, fatigue, nausea, abdominal pains, a loss of appetite, and liver malfunction accompanied by jaundice.

One study of 31 workers heavily exposed to PCBs (Aroclor 1254) suggests an increased incidence of melanomas or skin cancer, but as is the case with all types of PCBs, a great deal more study needs to be done to confirm these observations.

Along with these facts there are also a number of theories that need further investigation. For instance, many contend that polychlorinated dibenzofurans (PCDFs) also present in the PCB contaminated rice oil, compounded the toxic effects.

Dr Douglas Harding a toxicologist for the Ministry of Labour, Special Studies Branch, claims PCBs may not be a direct carcinogen, but an indirect carcinogen or promoter of cancer when present with other chemicals. Although, if this argument is borne out, Dr Okey suggests that the distinction is a "semantic argument" or "waffling." Surely the distinction would also be immaterial or a moot point to a worker who contracts cancer.

* * *

Finally, a recent preliminary study heralds some PCBs as protectors from the toxic effects of dioxins and PCDFs. The same study also claims most low-level environmental PCB exposures likely have little effect on humans. It does not however rule out other PCBs as possible carcinogens, when used as they have been in the workplace.

Neither does the International Agency for Research in Cancer (IARC). In fact, they conclude that given the "sufficient" evidence in animals, but "inadequate" evidence in humans PCBs are "probably carcinogenic to humans." They continue: "for practical purposes polychlorinated biphenyls should be regarded as if they were carcinogenic to humans."

Taking IARC's conclusion one step further Dr Epstein writes: "Whenever it is announced that a chemical has been shown to be carcinogenic in animals, there is usually an accompanying disclaimer that "the chemical has not been found harmful to humans. Every time you read this kind of statement in the press, or hear it announced on the radio or television, you should mentally insert the word 'yet'."

Very wisely though, Don Fraser of Local 1005 chooses to stay outside the scientific fray. "It's not for me to argue one way or the other. My job now is just to make sure workers won't be exposed to any more PCBs."

Ontario Workers' Occupational Safety and Health Centre, *At the Source* 8, No. 1 (1987): 6–8.

## Questions and Topics for Discussion and Writing

1. What evidence is cited in this report of the harmful effects of PCBS?
2. Why is exposure to PCBS an issue for labour unions and government labour departments to concern themselves with? Who has the responsibility to assure that workers are not knowingly exposed to dangerous substances?
3. Who should bear the responsibility (and pay compensation for damages, if any) for harmful effects from a substance that resulted from exposure *before* the substance was found to be harmful?

# Madison Avenue Medicine

## Elisabeth Rosenthal

Wherever power is exercised over matters of interest to others—matters of, say, political or economic importance—, there will be people trying to influence the control of that power. Whenever you go to your doctor with a complaint and come away with a prescription, someone, and probably quite a few people, have been there before you, trying to persuade your doctor to prescribe *their* company's medications. This article by physician Elisabeth Rosenthal explains just how pervasive and subtle that persuasion is.

I am about to bite the hand that feeds me. Really. Today it fed me and my fellow residents a very pleasant barbecue dinner of chicken and hamburgers, on a roof overlooking a river. The party marked the end of our second year of residency, two years of hard work and exhaustion. Most of the people knew one another extraordinarily well. There was only one stranger at the party, a pretty young woman dressed in blue high heels and a miniskirt.

Ostensibly, she fitted in perfectly, alternately flirting with the men and gossiping with the women. She danced to the reggae that issued from a tape deck, and she pitched in to turn the burgers. Yet I regarded her with suspicion. This young woman was in fact a "detail man" for Wyeth, the pharmaceutical company that was bankrolling our burgers. While we were celebrating, her purpose at our party was to promote an antianxiety drug called Ativan.

The party had been organized by several of my colleagues. As is fairly common practice for hospital department functions, they had approached the drug company for "sponsorship" of the event. The clues that Wyeth had anything to do with our party were few: the popcorn tins were all labeled Orudis, another Wyeth medication, and at one point the young woman balanced herself on a lawn chair and delivered a 30-second speech about Ativan's superiority to its competitors. But the plugs for the drug were easy enough to ignore, or at least not too difficult to rationalize: Ativan, after all, is a good medication. So why did this woman's presence make me squirm?

The seduction of doctors by drug companies begins early. Immediately upon arriving at medical school, I remember, I received a box of reference books courtesy of Upjohn. A year later, just before the start of clinical rotations, a $50 stethoscope appeared from a company whose name I have long since forgotten—every second-year medical student in the country receives one. As a fourth-year student I somewhat guiltily picked up a drug-company pen at a company-sponsored lunch; it would be almost a year before I could return the favor by writing a prescription.

The arrival of the stethoscopes prompted the first and last official ethical discussion during my medical training regarding the wisdom of accepting drug-company gifts. The entire second-year class was invited to a debate that echoed my internal musings. The devil on my left shoulder urged me to take up the tempting offer; although I had already purchased my own stethoscope, it would be nice to have this perfectly adequate instrument as a backup. The angel on my right shoulder warned me that accepting the stethoscope would set a bad precedent. I'm afraid, after endless discussions, the devil won out: I, and everyone else I knew, kept the tainted instrument. The class ahead of mine at least had had the ingenuity to send their free stethoscopes to a clinic in Central America.

After I received my medical degree, and with it the license to write prescriptions, the gifts became more frequent and the sales pitches more shameless. Each morning thousands of drug-company detail men arrive at hospitals around the country to promote their wares. Where I now work there is a detail man I'll call Jerry; he is known as the Zantac man. He's the high priest of a popular antiulcer drug manufactured by Glaxo Pharmaceuticals. As far as I know, this gentleman's one and only job is to patrol the halls where I work and make sure that Zantac is never far from anyone's mind. He is good at it.

* * *

I am sitting in the library with another doctor, discussing the management of a particular patient, and in comes the Zantac man with his briefcase overflowing with samples. Zantac is actually a wonderful drug, but nonetheless he reminds me of a snake-oil salesman.

"Do you have just a few minutes," he implores, "to discuss once-a-day Zantac dosing? Studies have shown that..." My mind starts to wander. He drones on about new studies demonstrating still more ways in which Zantac is infinitely superior to its primary competitor, a drug called Tagamet from Smith Kline & French. Jerry knows a lot about Zantac, and there is probably some useful information in his speeches, but I have long since ceased to hear it. He is like a piano player in a seedy bar where the din of conversation is drowning out the music. "Thanks for your time," he says when he's finished. He begins to pack up, leaving behind Zantac datebooks and Zantac pens and offering free samples of Zantac to each of us.

Every doctor has to make a separate peace with these drug reps. Personally, I find they make me nervous with their affable manners and their attempts to become my pal, and I prefer to keep my distance—but that is a gut reaction rather than a reasoned decision or a moral stance. I am cordial and will listen to them if time permits, but I would never dream of returning their friendly overtures or requesting free Zantac for my Uncle Jack in Cleveland.

Not everyone I know is so aloof. Some of my colleagues, fine doctors whom I generally respect, meet drug reps for lunch. They call the reps up every so often to chat—just to keep the conduits open between the drug-company coffers and the department social fund. When I tease my colleagues about their antics, they shrug their shoulders as if to say it's just good business. They remind me, rightly, that I am not above eating Wyeth hamburgers on the right occasion.

Why do the Jerrys of this world even exist? After all, the medical journals I read are replete with pharmaceutical advertising. The answer is simple: Within most classes of drug there are many brands marketed, all of which are essentially the same from the clinician's standpoint. They may vary a bit in chemical structure, and a little in cost, side effects, and dosing frequency, but they will all do the job.

In the antiulcer department Zantac and Tagemet are the current leaders. To treat minor muscle and joint pain there are dozens of nonsteroidal anti-inflammatory medications: Feldene, Motrin, Naprosyn, Anaprox, Clinoril, Orudis. They will all treat a strained tendon. It is the detail man's job to convince me somehow that Feldene or Motrin or Clinoril is far superior to the competition. Barring that, it is his task to turn my fancy toward the drug, the same way Kellogg's advertising seeks to ensure loyalty to its brand of cornflakes. And barring that, it is his job to keep my world so full of, say, Naprosyn—Naprosyn pens, Naprosyn pads, Naprosyn meals—that the letters *N-A-P-R-O-S-Y-N* flow reflexively from my pen.

To be fair, the detail men can at times be valuable sources of information. They can tell me about the resistance patterns of microbes to their antibiotics. They can find out if my patient's rash might be due to a rare side effect of their drug. But the hype always seems to outweigh the content.

Any number of tactics are used by drug-company reps to ingratiate themselves with the people who prescribe their wares. Judging from the number of meals we're offered, they must feel that the way to a doctor's heart is through his stomach. During the first years of training these are generally free lunches on the hospital premises—we'll bring you sandwiches if you'll listen to a five-minute spiel about Naprosyn. There are two or three of these lunches a week.

To those outside medicine a sandwich may not seem like much of an incentive, but the drug companies know their audience. Internship is memorable for its long hours of sometimes degrading work, few thanks, and even fewer concrete rewards. So when a detail man comes in, treats you with respect, values your opinion (okay, so it's only about antacids), and offers you lunch, you're grateful.

As training progresses, meals become more sumptuous, and the settings more exotic. A dinner

*Elisabeth Rosenthal*

for 20 at an elegant Spanish restaurant sets the scene for a presentation on balloon angioplasty (a procedure to clear blocked blood vessels) and a plug for Cardizem, a cardiac drug made by Marion Laboratories, which picks up the tab for the occasion. In time, pharmaceutical firms actually offer to pay you to eat with them. One company recently invited several of my co-workers out to a very expensive ($100 a head) Italian restaurant and paid them each a $150 consultancy fee to discuss what they thought about the firm's new antibiotic. With very little resourcefulness, young doctors can line up two or three dinners a week and make a handsome second income from offering their opinions over pasta primavera.

Tickets to sports events are also frequent enticements. So I imagine if you play the game really well, you could enjoy dinner at Le Cirque sponsored by one drug company, followed by a Knicks game at Madison Square Garden courtesy of its competitor. To the average resident, who earns somewhere around $25,000 a year and is shouldering twice that much in medical school loans, such luxuries are tempting, and you tend to dismiss the favors they imply. Who bought these seats anyway? Was it Feldene? Or Motrin?

Most doctors I know feel they are above being influenced by drug-company propaganda and laugh at the notion that the care of their patients might in any way be affected. If anyone is swayed, it's always the other guy.

But I'm willing to bet that Syntex and Glaxo wouldn't be paying a huge sales force millions of dollars a year unless their strategies were working. And I suppose that their success is staring us in the face. Thanks to Glaxo's efforts, the hospital where I work is clearly Zantac turf: Zantac is the only antiulcer medication in the pharmacy, and I can recite the dosing of Zantac—by mouth or intravenously—in my sleep. Should I for some reason switch to prescribing Tagamet, I would have to fetch the hefty *Physicians Desk Reference* from the shelf to figure out the proper dose.

Some of the more intellectually pure hospitals in the country have sought to stunt the influence of drug companies by banning detail men from the buildings. But these are persistent suitors, and in my experience this sort of decree generally moves the seduction to restaurants several blocks from the hospital's front door. The struggle for markets is a form of guerrilla warfare, and all tactics are fair.

Drug-company knickknacks are scattered liberally about the hospital and hawked shamelessly at medical conventions. Some of the medical paraphernalia is quite useful: calipers for reading cardiograms, from Squibb (makers of Capoten, a heart medication); reflex hammers courtesy of Syntex; penlights and tourniquets from two antibiotics manufacturers. At conventions, where doctors in a festive mood are gathered in high concentrations, the connection between the gift and the advertisement is often more elusive: umbrellas for an antibiotic called Imipenum, wall clocks from a maker of transdermal nitroglycerin, hard hats advertising Cardizem.

* * *

Some people, I'm sure, will react to my tales of detail men with a big "So what? " Brokers take out their bankers for dinner; layers treat prospective clients to basketball games. So what? Maybe I am overreacting to the appearance of Madison Avenue sales techniques in what I've always considered a helping profession. But the hard sell of the drug companies makes me nervous. Does using that tainted stethoscope from medical school make me somehow more closely allied to the pharmaceutical industry? Does eating Wyeth barbecued chicken make me more likely to prescribe Ativan?

I must admit, I like my nifty pen from Nitro-Dur, the nitroglycerin-patch medication. And I am quite fond of my Amikin antibiotic mug with its inscrutable inscription, TOO MANY VARIABLES, TOO LITTLE TIME. Perhaps these gifts might influence someone—not me, of course, but maybe the other guy.

---

*Discover* (October, 1988): 34–36.

---

### Questions and Topics for Discussion and Writing

1. Why is Dr Rosenthal uneasy about the attention lavished upon her by pharmaceutical companies?
2. Is there an alternative to the way pharmaceutical companies now market their goods? If so describe it. If not explain why.

# Hazards Implicit in Prescribing Psychoactive Drugs

## Henry L. Lennard, Leon J. Epstein, Arnold Bernstein, and Donald C. Ransom

When it comes to doctors prescribing drugs to patients, there are more issues than just *which* of several similar drugs to prescribe—there is also the issue of whether a condition should be treated by drugs at all. This question arises particularly in the case of "psychoactive" drugs, that is, drugs which alter the patients' emotional states. Here, the appropriateness of the "chemical fix" depends largely on whether the problem is one which falls within the purview of medicine and medicinal cures. This article, by a team of prominent psychiatrists and psychologists, examines some of the risks and assumptions of prescribing psychoactive drugs.

In order to extend the potential market for its product, the pharmaceutical industry, in its communications to physicians, all too often practices mystification...in relabeling an increasing number of human and personal problems as medical problems. In the context of current usage, drugs are medical agents whose function is the solution of medical problems. Only to the extent that interpersonal and other human problems can be construed as medical-psychiatric problems can they be considered appropriate targets for drug treatment. As more and more facets of ordinary human conduct, interactions, and conflicts are considered to be "medical" problems, physicians and, subsequently, patients become convinced that intervention through the medium of psychoactive drugs is desirable or required.

It is apparent that the pharmaceutical industry is redefining and relabeling as medical problems calling for drug intervention a wide range of human behaviors which, in the past, have been viewed as falling within the bounds of the normal trials and tribulations of human existence. Much evidence for this position is to be found in the advertisements of drug companies, both in medical journals and in direct mailings to physicians.

A series of examples will be sufficient to illustrate this point. The first involves the potential personal conflict a young woman may experience when first going off to college.

On the inside front cover of one journal an advertisement states: "A Whole New World...of Anxiety"..."to help free her of excessive anxiety...adjunctive Librium." Accompanying the bold print is a full page picture of an attractive, worried looking young woman, standing with an armful of books. In captions surrounding her, the potential problems of a new college student are foretold: "Exposure to new friends and other influences may force her to reevaluate herself and her goals."..."Her newly stimulated intellectual curiosity may make her more sensitive to and apprehensive about unstable national and world conditions." The text suggests that Librium (chlordiazepoxide HCl), together with counseling and reassurance "can help the anxious student to handle the primary problem and to 'get her back on her feet.'" Thus, the normal problems and conflicts associated with the status change and personal growth that accompany the college experience are relabeled medical-psychiatric problems, and as such are subject to amelioration through Librium.

Another journal has an advertisement that advises a physician on how he can help deal with such everyday anxieties of childhood as school and dental visits. This advertisement, in the *American Journal of Diseases of Children*, portrays a tearful little girl, and in large type appear the words: "School, the dark, separation, dental visits, 'monsters.'" On the subsequent page the physician is told in bold print that "The everyday anxieties of childhood sometimes get out of hand." In small print below he reads that "A child can usually deal with his anxieties. But sometimes the anxieties overpower the child. Then, he needs your help. "Your help may include Vistaril (hydroxyzine pamoate)."

The advertisement, in effect, presents an oversimplified conception of behavior and behavior change. Potential anxiety engendered by new and different situations is defined as undesirable, as constituting a medical and psychiatric problem which requires the intervention of a physician and, most particularly, intervention through the prescription of a psychoactive drug.

Physicians and parents with low tolerance for anxiety, or those with limited ability to meet the demands of even a temporarily troubled child, are more prone to believe that the child is disturbed and in need of drug treatment.

There is, however, no substantial evidence for the proposition that the prescribed drug does indeed facilitate children's participation in school situations. What is especially disturbing about advertisements such as this is that they tend to enlist the help of physicians to introduce children to a pattern of psychoactive drug use. Paradoxically, such drug use, at a later date, without a physician's prescription, is deplored both by the medical profession and the community at large.

Finally, we come upon a box used to distribute samples of Tofranil (imipramine hydrocholoride) to physicians. On the box is a picture of an adolescent girl. Above the picture in bold print is the legend, "Missing, Kathy Miller." Below the picture we read, "$500 reward for information concerning her whereabouts." Alongside in white print we read the plea, "Kathy, please come home! " Inside the box is a letter entitled, "Kathy, We love you...Please come home." We quote: "Dear Doctor: For parents, inability to communicate with their children is a significant loss. The "What did I do wrong?" lament of the parent may be accompanied by feelings of incapacity, inferiority, guilt and unworthiness. Many may, in fact, be suffering from symptoms of pathological depression. What can Tofranil, imipramine hydrochloride, do for your depressed patient?"

There are multiple levels of mystification in this advertisement. First, the cover suggests that the problem of a runaway child is a medical or psychiatric problem, rather than a human or family problem to be viewed within the larger framework of conflicts between generational outlooks and values. Second, the insert suggests a parent's grief reaction is pathological and characteristic of depression. And third, once having placed the problem into the psychiatric realm the advertisement recommends the use of the antidepressant Tofranil.

This advertisement is only one of many which...imply "that antidepressant therapy is indicated in the griefs of everyday life" and are "part of a trend to suggest the use, both of antidepressants and tranquilizers, not only for specific mental illness but to soothe life's ordinary woes."

The contemporary trend of increasing prescriptions of psychoactive drugs seems to be contributing to the recruitment of more and more persons into a way of life in which the regulation of personal and interpersonal processes is accomplished through the ingestion of drugs. Thus, when a physician prescribes a drug for the control or solution (or both) of personal problems of living, he does more than merely relieve the discomfort caused by the problem. He simultaneously communicates a model for an acceptable and useful way of dealing with personal and interpersonal problems. The implications attaching to this model and its long-term effects are what concern us...

When a physician prescribes drugs to a middle-aged woman who is upset about her child's rebellion or who is miserable in an unhappy marriage, or to an elderly man or woman who has been isolated from children and community; or when he prescribes drugs to children who cause trouble in schools, he often only masks the problem. In these instances drugs serve to decrease the anxiety or unhappiness of the individual and, more importantly, decrease the amount of trouble his anxiety, misery, or unhappiness may be causing others. Drugs may thus indeed make it more possible for others to manage or cope with the disturbed or disturbing individual.

The drugs do not, however, reach the sources of the anxiety or misery which may, for example, reside in an unhappy marriage, in the unfortunate situation of the elderly in our society, or in the unsuccessful socialization of many youngsters into group settings. In other words, drugs do not remedy the unfavorable social and interpersonal arrangements and personal circumstances which generate anxiety or unhappiness. Through the creation of chemical barriers, and through diminishing gross social deviance, drugs may actually serve to perpetuate malignant patterns and social arrangements. Were drugs not so readily available, pressure for other solutions and the pursuit of alternative options might be encouraged.

In the giving and taking of drugs, one pays for what one gets. The double entries in this ledger are too often ignored and too easily denied by the youthful user of illegal drugs who thinks that he gets something for nothing, who considers only the immediate sensory experience, the ecstatic "high," and freedom from anxiety, without attending to the personal and group costs exacted. This proposition concerning costs is also strenuously denied by those advocates of "legal" drugs—physicians who casually and consistently prescribe tranquilizers and sedatives. It is part of contemporary medical mythology that drugs somehow do not exact the same price from the user when they are prescribed by a physician and that a patient can get relief from his symptoms and escape from his troubles through psychoactive drugs, provided they are duly prescribed, without paying a cost.

One may well ask what costs are involved. Briefly, we see two major kinds: costs at the level of the individual and his personal functioning and experience, and costs at the level of human relatedness in significant social systems within which the drugged person lives.

The first kind range from the multitude of side effects attending drug usage, from the "feeling of being drained" often reported after amphetamine use to a robot-like sensation frequently noted after phenothiazine ingestion—to the alterations in the body's own self-regulating functions, which are assumed by the drugs ingested.

The second kind is less visible, but potentially even more important. Involved here are fundamental alterations at the level of social interaction and group responsiveness to distressed and deviant behavior. The use of drugs erodes a group's ability to make provisions and develop strategies of human relatedness, which serve to regulate particular kinds of psychological processes of its members, such as anxiety, grief, rage, and other more extreme forms of behavior. The regulatory function is delegated to drugs, not to the significant parties in the interpersonal environment. We anticipate that the eventual cost in human group relatedness, of this decision to substitute a drug modality for an interpersonal one, will be a heavy one. The more potent the drug, the more sustained its use, the more likely is there to be an increase in side effects, the unanticipated and unwanted consequences of drug use.

Drug giving and drug taking represent all too brittle and undiscriminating responses, and ultimately, in our view, they will breed only more frustration and more alienation. Changing the human environment is a monumental undertaking. While seeking to change cognitive shapes through chemical means is more convenient and economical, the drug solution has already become another technological Trojan horse.

The ultimate task is to alter the shapes of human relatedness and social arrangements that determine the context and the substance of our existence. To maintain, as do significant groups within the pharmaceutical industry, the medical profession, and the youth culture, that this can be accomplished merely through chemical means is indeed to have fallen victim to mystification.

*Science* 169 (July 31, 1970): 438–441.

## Questions and Topics for Discussion and Writing

1. What kind of conditions are psychoactive drugs used to treat?
2. Make an argument *in support* of the use of psychoactive drugs for some of the conditions discussed in this article.
3. Make an argument *against* the use of psychoactive drugs for some of the conditions discussed in this article.
4. Who should decide whether a problem is medical or not? Doctors? The patient? The community? The pharmaceutical industry?

*Issue*

# Nuclear Energy and Nuclear Warfare

Much of the history of technology is a chronicle of successive discoveries of ways of harnessing the power of nature to do work for human purposes. As the means of harnessing power became known, more uses were found, and hence more power was wanted. So long as the sources of nature's power were renewable resources, the increase in demand was not a matter of concern, but once the sources used were not renewable, or not renewable fast enough, the question of how future generations would cope arose.

The tremendous increase in energy demand has come in this century, when the technology was found to turn any form of energy into electricity, and more and more machines were designed to run on electricity. Electrical demand, combined with the general adoption of the automobile as a means of transportation and central heating in homes and offices, has created a voracious demand for readily available energy in the developed countries. The easiest and generally cheapest sources of energy have been fossil fuels: coal, oil, and gas. But all of these substances are the end product of thousands of years of natural processes. They are therefore limited in extent. Even though many years of reserves remain in the ground, eventually society must turn to other sources. Moreover, the burning of fossil fuels add pollutants to the atmosphere, contributing to the greenhouse effect and reducing the supply of free oxygen.

It was therefore of major importance to the world when, early in this century, scientists discovered that so-called "atoms" were not actually "atomic"—that is, indivisible—but rather were held together by extremely powerful forces concentrated in a central nucleus. When further it was determined that the nucleus of certain atoms could be split apart, releasing some of that powerful energy, it seemed that the answer to society's energy problems might be solved.

Even though only some relatively rare substances could be made to produce nuclear fission, uranium and plutonium in particular, the amount of energy released from even a tiny amount of these substances could produce a tremendous amount of energy. So, despite uranium and plutonium also being non-renewable resources, the problem of when the supply of fuels would run out was at least postponed for many generations.

But, as with so many technologies, there was a cost that was not obvious at first. Nuclear fuels produce radioactive nuclear waste that is very, very difficult to dispose of safely. It is also difficult to handle safely. Thus accidents in a nuclear power plant are potential disasters to the workers in the plant, the residents in the immediate vicinity, and, if the accident is bad enough, to the world at large.

Of course, it doesn't end here. Any technology that produces huge amounts of power can be used in warfare as well as in peacetime. Nuclear weapons are not just bigger and more powerful bombs; they unleash the same highly toxic, environment-destroying radioactive waste that nuclear power plants do, only they do it deliberately, in huge amounts, and right into the atmosphere—affecting not just the target, but the world as a whole. This leads to the paradox that a nuclear arsenal is an effective weapon (and deterrent) only so long as it is never used.

There are now many nuclear power plants in the world, and more are being built, though awareness of the hazards has led to the cancellation of plans to build many more. There are also an incredible number of nuclear weapons that have been built and stand ready to be used, though awareness of the insanity of ever using them has recently led to long awaited steps being taken toward disarmament. Even if no more nuclear power plants and nuclear weapons are ever built, and if all current nuclear power plants and weapons are eventually dismantled, there will still be a very serious problem for the world in how to dispose of the nuclear waste.

The story does not end here. The energy of the nucleus of atoms can be released by other means than by nuclear fission. Another alternative is *fusion*—the melding together of two atomic nuclei to form one larger atom. Fusion also releases energy, and does so without creating anywhere near the amount of hazardous radioactive waste that fission does. Moreover, fusion is possible with very light elements, hydrogen in particular, that are abundant in the world.

The hitch is that fusion seemed to be possible only in the presence of enormous amounts of heat—temperatures that were possible only in an uncontrollable explosion. Hence, fusion bombs (hydrogen bombs) were possible, but not fusion power plants.

That was how it seemed to most scientists until March 1989 when two researchers announced that they had produced fusion reactions in a small apparatus at room temperature. If true this is certainly the most important scientific development of our time, because once again, an answer to the world's energy problems would appear to have been found.

The articles in this section address the major issues raised above: nuclear accidents, nuclear waste disposal, nuclear warfare, and the new fusion research. Geoffrey York's article "Clouds of Doubt Linger Seven Years After Leak at Three Mile Island," written just after the disaster at the Chernobyl nuclear plant, reports on the aftermath of the earlier accident at Three Mile Island in Pennsylvania. Dan Grossman and Seth Shulman, in "A Nuclear Dump: The Experiment Begins," describe the preparations being made in the United States for a massive nuclear waste disposal site. In "The Consequences of Nuclear War," Carl Sagan and Paul R. Ehrlich give a very detailed and thoroughly horrific analysis of what would happen to the world if a nuclear war were ever started. John Phillips and David Michaelis, in "The A-Bomb Kid," show how relatively easy it is for anyone to build a working atomic bomb. In "The New Star Wars," John Polanyi criticizes the latest proposal to protect the world from nuclear war, the so-called "brilliant pebbles." Finally, the excitement and controversy over the new fusion research is captured in "Cold Fusion," a series of newspaper reports.

# Clouds of Doubt Linger Seven Years After Leak at Three Mile Island

## Geoffrey York

In the last forty years nuclear energy has been touted as the answer to the world's energy problems. Nuclear power plants would not be affected by depleting coal and oil supplies and would not pollute the air with the products of combustion. Or so it was thought. Unlike conventional coal or oil power plants, nuclear energy plants *are* cleaner in their day-to-day operation—but also unlike the conventional plants, nuclear plants pose a much greater hazard from malfunction and from disposal of wastes.

Though there have been several nuclear plant malfunctions over the years, two are better known and more discussed than the others: Three Mile Island in Pennsylvania in 1979 and Chernobyl in the Soviet Union in 1986. The latter was the far greater disaster, but the former, being closer to home, is more easily analyzed. This article, about repercussions from Three Mile Island, was written just after the Chernobyl disaster.

On a peaceful country road in the shadow of the massive towers of the Three Mile Island nuclear plant, Jane Lee is pointing to the silent houses and reciting a roll-call of death and disease. As the car climbs to the top of the ridge, she describes the illnesses in each home. It is a devastating litany of brain tumors, leukemia, pancreatic cancer, stillbirths and thyroid sicknesses.

When the nuclear plant came close to a catastrophic meltdown in 1979, plumes of radiation drifted across the hills and farmland on both sides of the Susquehanna River. Equipment malfunctions and human errors had exposed the radioactive uranium core of the plant. The amount of radiation released at Three Mile Island is believed to be hundreds of times smaller than the emissions from the Chernobyl disaster last month in the Soviet Union. Yet the local residents—and many prominent scientists are refusing to accept the official reassurances that "no one died at Three Mile Island."

Ms. Lee, a dairy farmer dear Goldsboro [Pennsylvania], has spent hundreds of hours on her surveys of the health of people who lived in the pathway of radiation emissions from the 1979 nuclear accident. Other residents have conducted similar studies. Their discoveries have sparked a furious debate in the scientific community. A survey by Marjorie Aamodt, for example, found 20 cancer deaths in an area where five would normally be expected. Another resident, Mary Osborn, has collected dozens of plants with bizarre abnormalities, including gigantic maple leaves and dandelions 78 centimeters long. A world authority on the effects of radiation has concluded that the abnormalities are "entirely comparable to those induced by ionizing radiation."

The survey by Mrs. Aamodt has fueled the biggest controversy. In September it was vigorously criticized by the Pennsylvania Department of Health. But officials of the Three Mile Island Public Health Fund—an agency established by court order after the 1979 accident—have confirmed the existence of "a significant excess" in cancer deaths among the 450 people surveyed by Mrs. Aamodt. The state health department conducted an official survey of all cancer cases within 32 kilometres of Three Mile Island, and decided there was no significant increase in the cancer rate. But the department's methodology and its conclusions have been criticized by several experts.

In an effort to resolve the dispute, the TMI health fund has commissioned a $300,000 study by epidemiologists from Columbia University. The Columbia study will be the most sophisticated and detailed analysis of cancer rates ever attempted in the region of Three Mile Island. The researchers will examine 80,000 hospital charts. And so, seven years after the worst nuclear accident in US history, no one is yet certain of the impact of the radiation leak on health and the environment. The Columbia report—expected in 1987 or 1988—might provide the final answers. Yet the study would never have been commissioned if the local people had accepted the official reassurances in 1979. It was the citizens—not the bureaucrats or the scientists—who finally found the evidence that prompted the Columbia study.

Meanwhile, there is increasing hostility between the residents and the government officials. The Pennsylvania health department openly ridicules the research by thc residents. Indeed, the state continues to insist there is nothing unusual at all in the health of the TMI residents. "It may be part of a massive psychological denial of the whole thing," says Jonathan Berger, executive secretary of the TMI health fund. "It could be a coping mechanism. The citizens are cast in ignominy."

The evidence gathered by the residents seems to be overwhelming. In some cases, it is backed by respected scientists. For example:

- The plant abnormalities found by Mrs. Osborn and other residents have included stem tumors, multiple buds and growth stimulation. The residents showed the plants to James Gunckel, a botanist who studied the effects of radiation for 34 years at the Brookhaven National Laboratory and at Rutgers University. According to an affidavit by Mr. Gunckel, the abnormalities could have been caused by radioactive fallout from the 1979 accident. Jan Beyea, a nuclear physicist who has worked for the TMI health fund, says the abnormalities should be investigated.
- Farmers reported a rash of stillbirths and abortions in their livestock after the accident. A local veterinarian, Dr Richard Weber, said there was a sharp increase in stillbirths. Three years after the accident, he noticed a sudden rise in the number of animals with cancer. The Pennsylvania Agriculture Department however, says it has found no abnormalities in the death rate for farm animals near TMI.
- A study in 1985 by US Government scientists found that the functioning of the immune systems of the TMI residents was significantly depressed after the accident. They theorized that the immune systems were affected by "chronic stress" in the residents after the accident. There was an unexplained jump in the infant mortality rate in 1979. From April to June of that year the mortality rate was 19.3 per thousand births within 16 kilometres of Three Mile Island, compared with a rate of 7.6 per thousand a year earlier. The mortality rate returned to normal by the end of 1979. The state health department argues that the accident should not be blamed for the increase because the death rate would have remained high until the end of 1979 if radiation had affected pregnant women.
- Because of inadequate monitoring of radiation emissions during the accident, no one is certain of the radiation exposure among residents around Three Mile Island. Estimates range from 276 to 63,000 person-rems for the region within 80 kilometres of the plant. Based on standard assumptions of the danger from radiation, this would produce as many as 13 extra cancer deaths in the area. But some scientists—including the TMI health fund's advisers—believe the standard risk assumption is too low. These scientists suggest that the number of fatalities could be as high as 130, Mr. Beyea said. Mr. Berger says the maximum number is 190 extra deaths.
- More than 2,000 residents have filed lawsuits against the owners and operators of the Three Mile Island nuclear plant. About 350 have received out-of-court payments to settle their injury claims. The payments included $1.1 million for a Down's syndrome child who was born just nine months after the accident, and $855,000 for an infant who suffered cerebral palsy.

Against this evidence, Pennsylvania points to its study of cancer cases in the area surrounding Three Mile Island. According to the study, there were 7,924 cancer deaths from the start of 1979 to the

*Geoffrey York*

end of 1983, in an area where statisticians would expect 8,177 deaths. But there were a number of weaknesses in this study. It failed to consider the pathways of the radiation plumes, and it failed to focus on the specific types of cancer which can be induced or accelerated by radiation. The study did not consider the terrain or the wind patterns, both of which can affect the level of exposure to radiation. The study also included some areas which were beyond the 32-kilometre radius from Three Mile Island. Moreover, the study confirmed that there was an increase in cancer deaths in the area surveyed by Mrs. Aamodt, though it argued that the increase was not significant.

"The Pennsylvania Department of Health has a propensity for ignoring things that it finds," Mr. Berger said in an interview. "You can assume that the department did everything it could do to use a conservative method of testing ... We're critical of its work."

Maureen Hatch, a Columbia University epidemiologist who is working on the current study, said the Columbia researchers are taking "a more sensitive approach" than the Pennsylvania study. If the local residents are correct in their assessment of the radiation exposure, most of the cancers will begin appearing in the next few years. Many residents feel as if they are waiting for a time bomb.

"There's a feeling of impending disaster," says Joyce Corradi, a mother of five who helped found an association of mothers in the Three Mile Island region. "I don't think we should have to live with this apprehension. I'm praying that my children won't be a statistic in all this." The residents tried to prevent the start-up of Unit One of the nuclear plant, but they lost several court battles and the unit was restarted last October, after being shut down for six and a half years. (Unit Two, damaged in the 1979 accident, is still out of operation. A $1.2-billion repair operation is continuing.)

"There's almost a feeling of hopelessness in this area," says Paola Kinney, a mother of four. Mrs. Kinney and her family are moving away from the region. They sold their house earlier this month. They decided to move when Unit One returned to operation.

Mrs. Corradi, however, plans to remain—even after she heard of the Chernobyl disaster. "We have a bond with the people of Chernobyl," she says. "We are all guinea pigs. The information about us could be invaluable to future generations."

*The Globe and Mail* (May 24, 1986)

## Questions and Topics for Discussion and Writing

1. Why is there such disagreement between the residents and the government on the extent of the radiation damage from Three Mile Island?
2. After an accident like that at Three Mile Island, damage can be extensive, not only in the immediate vicinity but a long distance away, and can show up only after many years. Discuss the problem of who should be appointed to investigate the accident and assess damages. How far afield and for how long should the investigation continue? Who should bear the responsibility for the damages?

# A Nuclear Dump: The Experiment Begins

**Dan Grossman and Seth Shulman**

Even without accidents and malfunctions, nuclear power plants have one very serious and inescapable problem: waste disposal. Unlike the residues of coal or oil power plants, spent nuclear fuel cannot simply be dumped in open places—it must be sealed tightly in materials that will not corrode for thousands of years. It is not a problem that can be ignored. This article describes the elaborate preparations being made for a massive nuclear waste dump in the state of Nevada.

The train ride into the middle of Rainier Mesa is dark and deafening. As the dusty, roofless metal cars clang and screech their way into the blackness, an occasional incandescent bulb exposes the rough-hewn walls; the bulbs mark the closed-off entrances to the many passages branching from the main shaft of G-Tunnel. For an instant as the train rolls past, the dim yellow glow illuminates signs that warn of radioactive contamination in the side tunnels. Twenty-five years ago they were the site of some of the United States' earliest underground nuclear explosions.

A mile into the mesa and 1,400 feet underground the train slows as it reaches Bob Schuch's laboratory. Up an incline the tunnel widens to a large chamber where workers drill and saw into the volcanic rock; the area is cluttered with tools, instruments, and heavy machinery. Beneath the desolate landscape of the Nevada Test site, in this subterranean maze that once thundered with explosions of our burgeoning nuclear arsenal, Schuch and his crew are part of a $2 billion, seven-year effort by the Department of Energy to settle on a final repository for the radioactive refuse of the nuclear age: the overflowing stacks of spent fuel from the nation's nuclear power plants and some of the deadly spoilage from our nuclear weapons factories. The radioactive waste already exceeds 22,500 tons. If Schuch's work pans out, a nearby ridge called Yucca Mountain will be riddled with passages in which the nuclear material will be stored.

As test coordinator at G-Tunnel, Bob Schuch's job is to determine whether the volcanic rock in the area will provide the waste with a stable and contained resting place for at least 10,000 years—that's how long the Department of Energy says it will take for the slowly decaying material to lose its toxicity. Schuch must study the behavior of the rock under various stresses and predict what the fate of the region will be for millennia to come. Before any spent fuel can be buried at Yucca Mountain, every conceivable way that the waste could escape must be considered. Will the waste leak into the water table? Will it escape into the atmosphere? What will the weather be like for the next 10,000 years? What are the odds of a catastrophic earthquake or volcanic eruption?

The bustle of activity in G-Tunnel belies the long road ahead before a repository is finally built. It will take years of cutting and testing, both at Rainier Mesa and Yucca Mountain itself, before enough is known to determine whether Yucca Mountain is an appropriate site. Only then can the actual construction begin, at an eventual cost that current estimates place as high as $35 billion.

The person with overall responsibility for the project is Carl Gertz, who from his Las Vegas office directs an extended staff of more than 1,400 researchers and support personnel in four government laboratories across the West, in addition to many private contractors. Standing by a scale model of a completed repository and a map of the surrounding countryside, Gertz confidently describes the project. "We are right here," he says, aiming a pointer at 1988 on a time line of the project. If all goes according to plan, Gertz will finish the study of the site by 1995, about a foot to the right on the chart. Two feet to the right of Gertz's pointer is 2003, the year the agency hopes to open the repository. If the chart spanned the entire repository life, a marker 10,000 years from the opening of the facility would be located more than a quarter of a mile farther away.

According to Congress's plan, shafts like Schuch's G-Tunnel will honeycomb an area of more than 1,500 acres inside Yucca Mountain. The total length of tunnel will be 112 miles—nearly the size of the New York City subway system, the world's largest. Altogether the tunnels will accommodate some 63,500 tons of high-level nuclear waste—a staggering amount of radioactive material. Nevada officials have estimated that even with trucks arriving at the site every workday at 90-minute intervals, it would take 28 years just to bring all the waste to Yucca Mountain.

Of course, that's all according to the plan as it's now formulated. By the end of those 28 years there could be tens of thousands more tons of waste to be dealt with. The Department of Energy is required to report to congress before 2010 as to whether the repository must be expanded or whether a new site must be added.

If the struggle that led to the tentative choice of Yucca Mountain is any indication of the difficulties that lie between now and 2003, Gertz, Schuch, and their crew have their work cut out for them. The debate over what to do with radioactive waste is as old as the nuclear industry itself. Nearly all 114 existing US commercial nuclear power plants were built long before any real solution to the waste problem was envisioned. And there are 16 more plants planned. It has just always been assumed that sooner or later a storage method and a site would be found. Former Nevada governor Grant Sawyer, chairman of the state's high-level nuclear waste commission, excoriates the nuclear industry and federal regulators for failing to find a solution. He likens the inaction to "sending John Glenn into orbit without figuring out how to bring him down."

About 17 per cent of the waste slated for Yucca Mountain will come from military nuclear reactors, in the form of highly radioactive liquids that have been vitrified—encased in solid blocks of glasslike material. The other 83 per cent of the waste will be in the form of spent fuel rods from nuclear power plants around the country. Fuel rods form the core of a nuclear reactor. They consist of thousands of uranium pellets encased in 12-foot metal tubes that are just twice the diameter of a pencil. A typical US commercial reactor contains 40,000 fuel rods, bundled in "fuel assemblies" of around 200 rods each.

After three to four years in a nuclear power plant the fuel rods are spent; that is, they no longer can sustain a usable nuclear reaction. At this point they are among the most highly radioactive objects on Earth. About 4 per cent of the weight of a rod is composed of extremely radioactive plutonium, strontium, cesium, and other isotopes created during the fission reaction that is the source of a reactor's power. The remainder contains less radioactive but still hazardous uranium isotopes. While there is little chance that the waste can explode like a nuclear bomb, even the briefest human contact with a fuel rod is fatal. The tiniest fleck of plutonium—say, a millionth of a gram—will cause lung cancer. And up to ten years after they're spent, the rods still give off, along with the radiation, 10 kilowatts' worth of energy in the form of heat, adding to the challenge of safe disposal.

Today the spent rods are stored in cooling ponds adjacent to the nation's nuclear reactors. Many of the cooling ponds are already near capacity; the ponds were designed, after all, only as temporary way stations for the fuel until a permanent disposal site was found. And as they've filled up, operators at many reactors have opted to store the rods closer together, at a density that the pools weren't designed to handle. Quite simply, there aren't too many other options. Some nuclear power plants have already applied for permission to build new, off-site cooling ponds to store their continuing accumulation of spent rods.

Initially Yucca Mountain was just one of several sites around the country that were to be studied as a home for the radioactive waste. The Nuclear Waste Policy Act of 1982 specified that several sites be evaluated and that two—one on either side of the Mississippi—ultimately be chosen based on technical suitability. On the last day of its 1987 session, however, Congress amended the law, voting to evaluate only Yucca Mountain for the nuclear dump.

Many Nevadans see that vote as a political decision that made a sacrificial lamb out of a state with one of the smallest delegations in Washington. Although debate on the floor focused on the cost savings of evaluating only one site, they have little doubt that the decision was designed to minimize opposition to the facility. In what Richard Bryan, US senator fromm Nevada, says was "a feeding frenzy," 49 states "ganged up" on Nevada.

State officials and some scientists now worry that the work at Yucca Mountain will be biased toward finding the terrain acceptable. "Congress put all its nuclear eggs in one basket," says Bryan. Others outside the state agree. As Massachusetts congressman Edward Markey warned his colleagues: "If this site proves unsound...we will be left without a nuclear waste disposal program of any kind. As Nevada goes, so goes our nation's nuclear waste policy."

Everyone involved in the project can feel the pressure. Nevertheless, Gertz insists that the scientific conclusions will be technically sound. If the research does not prove the site to be safe, he vows, "then we don't want to build the repository there."

Increasingly, though, some scientists involved in the effort are complaining that the lack of alternatives is making acceptance of Yucca Mountain a foregone conclusion, no matter what the scientific evidence shows. Recently, for instance, some of Gertz's recruits from the US Geological Survey wrote a disgruntled letter claiming that the work on the repository may have already "moved away from objective site characterization and into site construction." The 17 scientists and engineers stated that "in subjugating the technical program to satisfy Department of Energy political objectives, we may succeed in making the program comply with regulations, while being scientifically indefensible." Joe Downey, a hydrologist who signed the letter, says, "I've never seen the Survey involved in such a mess before. There are serious problems with Yucca Mountain that should be studied and fairly presented."

Standing on the summit of Yucca Mountain it is easy to imagine why the site was chosen. The view is an unbroken panorama, extending more than 50 miles in every direction, with the only evidence of civilization the single-lane dirt road leading to the mountain. In fact, very little life of any kind can be seen beyond the sparse, drab green brush and tan tufts of grass that put a slight fuzz on distant hills.

Despite its isolation, there is a long list of questions that must be answered before Yucca Mountain can get final approval. The concerns range from such "low probability but high consequence events" as volcanoes or earthquakes to the many questions concerning water, both deep in the Earth and falling as rain.

The view from the summit also makes it easy to understand why Gertz is concerned about geologically "high consequence events." The moonlike landscape surrounding the ridge bespeaks a violent past. Busted Butte, for example, a small peak only about five miles away, is sheared in half, the eastern side lifted several hundred feet upward. Yucca Mountain itself is bounded by a number of geologic faults, the most obvious being the Solitario Canyon Fault, which created a sheer cliff that plunges nearly 1,000 feet from the summit into Solitario Canyon.

Just to the west a handful of perfectly symmetrical volcanic cones rise above Crater Flat. One of these is the Lathrop Wells Volcano, only 12 miles from Yucca Mountain. Last March, Stephen Wells, a University of New Mexico geologist, reported that research he conducted for the Department of Energy showed this volcano to be fewer than 20,000 years old, not 300,000 as once believed. Wells made his estimate by comparing the physical features of the Lathrop Wells Volcano with a volcano in the Mojave Desert that's known to be 15,000 years old. In Lathrop Wells, he says, "there are no gullies or channels." This lack of erosion implies that the volcano may have appeared only 5,000 years ago, increasing the likelihood of another eruption. He is conducting further studies to pinpoint the age.

While this news could hardly be comforting to proponents of the site, Gertz remains undaunted. He emphasizes that the volcanoes in Crater Flat were formed from oozing lava and not from the kind of explosive eruptions that created Yucca Mountain itself over 9 million years ago. "Another eruption like we discovered at Lathrop Wells," he says, "would not hurt a thing." His office has, however, revised its estimate of the probability of an eruption actually occurring within the repository area: originally given as one in a billion per year, the probability is now said to be within a range that stretches from one in a billion down to one in 10 million.

The Department of Energy is also making plans to test a hypothesis that there may be a chamber filled with magma, or molten rock, deep under Yucca Mountain. Because rock generally reflects and refracts seismic waves differently when it's molten, explosives will be detonated to produce shock waves that travel deep underground; the waves will be distorted by differences between deposits, and if a molten body exists, it should return a distinctive signal.

Earthquakes are another concern, and Yucca Mountain sits between two prominent faults. Since 1857, when the region was settled, there have been eight major earthquakes—with a rating of at least 6.5 on the Richter scale—within about 250 miles of the site. Gertz says that the repercussions on the underground repository of even a sizable earthquake would probably be minimal because tremors diminish in effect that far below the surface. He acknowledges, however, that an earthquake could be devastating to the aboveground facility where the waste would be delivered and prepared for burial. But this would be a concern for only the next 100 years or so; after that the repository would be full and the waste-handling buildings dismantled or sealed in concrete. To lessen the short-term risk, however, the surface structures would be constructed to resist a severe quake.

* * *

Along with geologic disruptions, the other great concern is water. The Yucca Mountain site—which is not far from Death Valley, the driest place in North America—was selected in part because of the lack of rainfall: the area receives only three to six inches of rain each year. This is important because rainwater could leach down through the rock and hasten the corrosion of the canisters that are designed to contain the waste for the first few hundred years. Water reaching the waste could ultimately cause it to seep down to the underground water table. Plans now call for the radioactive material inside Yucca Mountain to sit about a thousand feet above the water table but still well below the surface. Geologists call this the unsaturated zone.

Before it can be shown that a repository at Yucca Mountain can safely contain the waste, much more must be learned about the properties of this zone. Early estimates indicate that the amount of rainwater that would actually reach the depth of a repository is probably less than .02 inch per year, well below the amount that would be a concern. But little is known about the way water can travel through tuff—the volcanic rock that makes up Yucca Mountain—over thousands of years.

The tuff deposits forming the area were produced by volcanoes 13 million to 18 million years ago. Some of the volcanic deposits cooled more quickly than others, creating soft rock formations known as nonwelded tuff. The problem with nonwelded tuff is simple: the spongy material crumbles with just a slight kick. The Department of Energy will attempt to locate the repository in a layer of rock that cooled more slowly, in material known as densely welded tuff. This rock is still porous, but strong, with 12 times the compression strength of concrete. Densely welded tuff also happens to be one of the world's most difficult types of rock to excavate. In G-Tunnel, Schuch and other researchers are conducting a number of experiments on the rock. Some of them will be repeated in an exploratory shaft that will eventually be dug at Yucca Mountain itself.

One of these experiments, known as the heated block, will help predict the performance of a repository filled with hot waste. Spent nuclear fuel continues to decay and generate heat long after it has been removed from a power plant. Schuch notes that tuff has many properties that make it ideal for holding such hot material. It is a good conductor of heat, so it won't melt. And its fractured structure would allow steam to escape if water seeping through pores in the rock reached the boiling point. Without these fractures, steam under high pressure could further weaken the rock.

The heated block is an eight-foot cube of rock cut into the tunnel floor. It looks like a huge square peg in a slightly larger square hole, remaining attached only on the bottom. Schuch's crew needed more than three months to form the block by drilling lines of adjacent holes and breaking the thin walls of stone between them. Since this block was formed a tool has been devised that will greatly ease the preparation of such experiments in the future: an eight-foot, remote-controlled, diamond-tipped chain saw that seems to eat away a slot almost effortlessly, shooting rock and debris against a Plexiglas shield. With the saw, a heated block can be created in less than a month.

Heaters placed in holes drilled in the block simulate the heat of waste canisters. Pressure is applied to the cube by steel chambers cemented between the block and the floor from which it was cut. More than 300 instruments, all connected to a central computer, stud the block and measure temperature, compression, and expansion.

For one experiment already performed at G-Tunnel, researchers applied pressure to the clock in a variety of directions, they heated it to as high as 212 degrees Fahrenheit; and, after driving all of the water out of the block, they even flooded it. The idea was to apply every conceivable condition.

Other experiments will help determine exactly how water flows in unsaturated welded tuff. If water travels only by seeping between the pores of the rock, it will move much slower than if it flows through the tiny fractures found throughout the deposit. One of these tests will measure how quickly water travels between a series of bored holes and a sealed alcove that can be placed under pressure.

* * *

Given the importance of keeping water away from nuclear waste, it comes as a surprise that deep inside G-Tunnel a steady trickle of water can be seen dripping from the uneven ceiling. Schuch is quick to explain that his is only "perched water" and thus of little consequence. What he means is that these streams come from reservoirs deposited in the distant past somewhere in the mountain towering above the tunnel. Construction of the tunnel itself released the water and caused it to begin dripping. Schuch says these deposits are of limited size and would not threaten canisters of waste. Even Downey, the hydrologist who has criticized the Yucca Mountain research, confirms that this water is not a great concern. "It's really a small amount of water in the great scheme of things," he says. Just to be sure, however, the Geological Survey is working on ways to collect and measure the flows.

Much of the work at G-Tunnel is designed to improve the tools and techniques for studying water in the unsaturated zone. For example, most methods for drilling and cutting tunnels use water to cool the tools and to prevent rock powder from becoming airborne, where it can endanger the health of workers. But excessive water usage at Yucca Mountain's exploratory shaft would contaminate the studies of water movement and could even make the deposit unsuitable for waste storage. Researchers from Los Alamos National Laboratory are testing dry cutting methods that remove dust with powerful air blowers, like a vacuum cleaner, rather than jets of water. Another technique, for tools that cannot operate dry, involves recirculating and recovering as much water as possible.

The tests of water content in the rock samples are critical, and so great care is being taken not to contaminate cores of extracted rock. A special drill coats the samples with a plastic tube as fast as they are extracted from the deposit so that the traces of water found at each depth are retained. These "sealed cores" will be catalogued and stored for later study.

Last April, just to ferret out any last "alternative hypotheses" for how waste stored in Yucca Mountain might be disturbed and contaminate the environment, the Department of Energy held an open conference. According to Gertz, the only theory any of the 150 participants could come up with that was not already being studied closely was proposed by Jerry Szymanski, a Department of Energy physical scientist. According to Szymanski's theory, which is detailed in a maverick report that originally reached the public after it was released by the Nevada governor's office, the region around Yucca Mountain is in the midst of a cyclical state of geologic activity that causes the volcanic rock to contract and expand over time.

Every 20,000 to 30,000 years, Szymanski believes, these cycles may cause a large shift in the depth of the region's aquifer, pushing the water table up as much as 1,000 feet. Given the current level of stress measured in the fractures of the rock, Szymanski thinks such a geologic shift may be on its way at Yucca Mountain. If he is right, a repository constructed at Yucca Mountain could someday be flooded. The current level of the water table is, he says, "purely a matter of tectonic consensus, subject to change without notice."

* * *

One of the key clues that led to Szymanski's theory was the discovery of mineral deposits—what Szymanski calls "footprints of the past"—on and just under the surface of the area around Yucca Mountain that indicated that water had been there at one time. If the mineral deposits came from rainwater, says Szymanski, they will peter out with depth. But if they were formed by a rising water table they will extend deep underground. The only way to know for sure would be to conduct further investigations. If the theory is corroborated by new evidence, Szymanski's report concludes, "serious consideration should be given to abandoning the Yucca Mountain site and declaring it as unsuitable for the purpose of permanent disposal of the high-level nuclear wastes."

Szymanski believes that a concerted effort could settle the issue in just three to four months, although he concedes that the characterization program will eventually gather the necessary data in any event. "We're going to study everything he's recommended," says Gertz, but he has no plans to give Szymanski's theory a higher priority than is given to any other.

More doubts about Yucca Mountain were raised in November by researchers from the University of Nevada and Arizona State. They questioned a method used by government scientists to determine the date of geologic disturbances. The technique analyzes a paper-thin coating of minerals that formed on rock surfaces that were exposed to the air by past landslides, earthquakes, or other geologic disturbances. Over time, various elements leach out of such layers, allowing an estimation of the date when the rock was first exposed. The disagreement is over the rate at which the elements dissolve. The university geologists claim the age of geologic disturbances is being overestimated. The project scientist agree that the technique at least needs to be refined.

Even as doubts linger about the suitability of Yucca Mountain and the quality of the research there, growing disclosures of mismanagement of nuclear facilities around the country by the Department of Energy have eroded public confidence in the department's ability to handle such materials. Many of the nation's largest and most important facilities producing nuclear weapons materials have recently been shut down because of contamination and safety violations. Equally disturbing, a long-awaited storage facility for liquid nuclear waste from military reactors, built near Carlsbad, New Mexico, continues to be stalled because water leaks in the deep salt caverns underneath the desert were found even before the facility officially opened.

Frank Clements, office manager at the state-sponsored Nuclear Waste Task Force in Las Vegas, sums up the discontent and unease felt by many: "The Department of Energy maintains a hundred twenty-seven facilities, and it seems that ninety percent of them leak. Coming from a gaming town, those don't sound like good odds to me."

*Discover* (March 1989): 49–56

## Questions and Topics for Discussion and Writing

1. Why is the disposal of spent nuclear fuel such a problem?
2. "NIMBY" stands for "Not in my backyard." Proposals to locate a nuclear waste dump have provoked intense "NIMBY" reactions from the residents of all proposed sites. Who should have the right and responsibility to decide where nuclear dumps should be located?
3. If as wealthy and well-governed a country as the United States has to go to as much trouble and expense to locate and prepare a nuclear dump as it is now doing, what can the world expect will happen to the radioactive wastes of nuclear power plants of the smaller countries? How will decisions be made? How will construction of dumps be financed? Who will police the disposal procedures? Discuss these questions.

# * The Consequences of Nuclear War


## Carl Sagan and Paul R. Ehrlich


Ever since the bombing of Hiroshima and Nagasaki in 1945, the world has known that a war fought with atomic bombs would be far more terrible and destructive than any other war the world has ever known. In spite of that, perhaps even because of that, the world's arsenals of nuclear weapons—especially those of the United States and the Soviet Union—have grown to incredible proportions. Yet no atomic bombs have been used in warfare since World War II. What would happen if nuclear weapons were used? A few years ago this question was extensively studied by two groups of scientists—physical scientists, who studied the atmospheric and climatic consequences, and biological scientists, who studied the biological effects. The general results of their studies are reported in the following two lectures by two of the scientific world's most respected spokespersons: Carl Sagan, who reports on the atmospheric and climatic consequences and Paul R. Ehrlich, who reports on the biological effects.

## The Atmospheric and Climatic Consequences


### Carl Sagan


It is the Halloween preceding 1984, and I deeply wish that what I am about to tell you were only a ghost story, only something invented to frighten children for a day. But, unfortunately, it is not just a story. Our recent research has uncovered the surprising fact that nuclear war may carry in its wake a climatic catastrophe, which we call "nuclear winter," unprecedented during the tenure of humans on Earth.

We stumbled upon these results by accident, by a circuitous route, by one of those circumstances common in science where studying something purely for its intellectual interest leads you to conclusions of surprising practical utility. For me, it began in 1971 with the Mariner 9 exploration of the planet Mars. Mariner 9 was the first spacecraft to orbit another planet. Its engineers guaranteed that it would work only for three months after orbital injection. The spacecraft arrived at Mars to find the planet completely covered with a global dust storm. After a month of photographing an almost entirely featureless disk, we began to worry seriously that by the time the dust would all settle out of the Martian atmosphere the spacecraft would no longer be working. The dust storm in fact took three months to dissipate, but the spacecraft worked far better than the engineers had said—and for the next year we were able to examine the planet pole to pole in the first detailed orbital reconnaissance of another planet.

During those first three months, there was very little to look at except the dust in the atmosphere. There was an instrument on board the spacecraft called an infrared interferometric spectrometer, which had the ability to examine the atmosphere at various wavelengths and therefore to probe to different depths in the atmosphere—from very high altitudes down to the surface. We were able to see the temperature of the atmosphere and that of the surface change with time. The results showed that the atmosphere was considerably warmer than is usually the case on Mars, and the surface considerably colder. As the dust settled out, the atmosphere became cooler and the surface warmer—both approaching their usual, or "ambient," values. It was not difficult to understand the reasons for this. The winds had stirred a great deal of dust off the Martian deserts into the atmosphere. Sunlight was being absorbed by the high-altitude dust, thereby heating the atmosphere. But, by the same token, the sunlight was impeded from reaching the surface, and so the surface was cooled. An observer on Mars would have noticed, after the dust storm stirred, that cold and darkness were spreading over the planet. After many months (the dust storm had started several months before Mariner 9 arrived at Mars), the dust had mainly fallen out of the atmosphere, and conditions had returned to normal.

Such dust storms are a Martian commonplace, and have been noted by ground-based observers for more than a century. They characteristically arise in the same few locations on Mars, spread first in longitude, then in latitude, and in a matter of a few weeks at most typically cross the Martian equator into the other hemisphere. Now, the surface atmospheric pressure on Mars is about the same as that in the stratosphere of the Earth. Mars rotates, as the Earth does, once every twenty-four hours, and its axis of rotation is tilted to its orbital plane by just about the same angle as the Earth's. There are differences between Mars and Earth, of course—including the absence of oceans on Mars, and the fact that it is farther away from the sun. But it seemed to us that the Martian experience might be relevant to Earth.

A number of us, having little before us for the first three months after orbital injection but the dust storm, set to calculating by how much the atmosphere should be warmed and the surface cooled for a given amount of dust put up into the atmosphere. A rough calculation was not very difficult, and several different groups were able to understand not just qualitatively but quantitatively the temperature changes that the dust storm had brought temporarily to Mars. My colleagues (and former students) James B. Pollock and O. Brian Toon, both now at the NASA Ames Research Center, were eager to apply this kind of computational armamentarium to terrestrial problems. We set out trying to understand what happens to the climate of the Earth when a large volcano goes off and distributes stratospheric aerosols worldwide. In some cases, we know how much dust is put into the upper atmosphere, what the particle sizes of the dust are (generally smaller than a micrometer [a ten-thousandth of a centimeter]), and what the composition of the fine particles is (generally sulfuric acid and silicates). Because the stratosphere is very dry, rain does not carry these aerosols out; and because convection is very muted in the stratosphere, atmospheric motions tend not to carry the fine aerosols out. And so they slowly sink by their own weight—slowly because their sizes are so small—taking more than a year for the stratosphere to clear. At the same time, there are, for many volcanic explosions, measurements of a small but definite global temperature decline—for all volcanic explosions in the last few centuries, a cooling of a degree or less. We found that we were able to calculate these temperature declines fairly accurately; the methods developed for Mars, and considerably extended since, worked quite well for Earth.

It was then proposed by Alvarez et al. that the extinction of the dinosaurs and many other species 65 million years ago, at the boundary of the Cretaceous and Tertiary epochs, was due to the collision with the Earth of an asteroid 10 kilometers across, and the subsequent spewing of enormous quantities of fine dust into the atmosphere. Joined by Richard Turco of R&D Associates in Marina del Rey, California, Pollock and Toon calculated that a severe cooling and darkening event might have been attendant to such an asteroidal collision. I wish to stress, however, that our conclusions on the climatic consequences of nuclear war do not depend on this interpretation of the Cretaceous/Tertiary extinctions. The dinosaurs could have died of influenza without affecting the validity of our conclusions.

We had known, of course, that nuclear explosions put large amounts of fine dust into the atmosphere, and had talked on and off for a period of years about calculating what the climatic effects of this dust might be. At a meeting at Ames Research Center (devoted in part to the question of the origin of life) in 1981, we decided to go ahead with the calculations. The effort was further spurred a year later by word of some very interesting work performed by Paul Crutzen of the Max-Planck-Institute for Chemistry in Mainz, Federal Republic of Germany, and John Birks of the University of Colorado. Crutzen and Birks had made a preliminary estimate of the amount of smoke from the burning of forests and cities that might be released into the atmosphere in a nuclear war. Clearly here was an additional important source of fine particles that might attenuate sunlight.

So now I come to the question of the effects of nuclear war. The immediate consequences of a single thermonuclear weapon explosion are well-known and well-documented—fireball radiation, prompt neutrons and gamma rays, blast, and fires. The Hiroshima bomb that killed between 100,000 and 200,000 people was a fission device with a yield of about 12 kilotons (the explosive equivalent of 12,000 tons of TNT). A modern thermonuclear warhead uses a device something like the Hiroshima bomb as the trigger—the "match" to light the fusion reaction. A typical American thermonuclear weapon might have a yield of about 500 kilotons (or 0.5 megaton, a megaton being the explosive equivalent of a million tons of TNT). There are many weapons in the 9- to 20-megaton range in the strategic arsenals of the US and the USSR today. The highest-yield weapon ever exploded is 58 megatons.

Strategic nuclear weapons are those designed for delivery by ground-based or submarine-launched missiles, or by bombers, to targets in the adversary's homeland. Many weapons with yields roughly equal to that of the Hiroshima bomb are today assigned to "tactical" or "theater" military missions, or are designated "munitions" and relegated to ground-to-air and air-to-air missiles, torpedoes, depth charges, and artillery. While strategic weapons often have higher yields than tactical weapons, this is not always the case. Modern tactical or theater missiles (e.g., Pershing 2, SS–20) and aircraft (e.g., F–15, MiG–23) have sufficient ranges to make the distinction between "strategic" and "tactical" or "theater" weapons increasingly artificial. Both categories of weapons can be delivered by land-based missiles, sea-based missiles, and aircraft, and by intermediate-range as well as intercontinental delivery systems. Nevertheless, by the usual accounting, there are around 18,000 strategic and theater thermonuclear weapons and the equivalent number of fission triggers in the American and Soviet strategic arsenals, with an aggregate yield of about 10,000 megatons. The total number of nuclear weapons (strategic plus theater and tactical) in the arsenals of the two nations is close to 50,000, with an aggregate yield near 15,000 megatons. For convenience, we here collapse the distinction between strategic and theater weapons and adopt, under the rubric "strategic," an aggregate yield of 13,000 megatons. The nuclear weapons of the rest of the world—mainly Britain, France, and China—amount to many hundred warheads and a few hundred megatons of additional aggregate yield.

No one knows, of course, how many warheads with what aggregate yield would be detonated in a nuclear war. Because of attacks on strategic aircraft and missiles, and because of technological failures, it is clear that less than the entire world arsenal would be detonated. On the other hand, it is generally accepted, even among most military planners, that a "small" nuclear war would be almost impossible to contain before it escalated to include much of the world arsenals. (Precipitating factors include command and control malfunctions, communications failures, the necessity for instantaneous decisions on the fates of millions, fear, panic, and other aspects of real nuclear war fought by real people.) For this reason alone, any serious attempt to examine the possible consequences of nuclear war must place major emphasis on large-scale exchanges in the 5,000- to 7,000-megaton range—between about a third and a half of the world strategic inventories—and many studies have done so. Many of the effects described below, however, can be triggered by much smaller wars.

The adversary's strategic airfields, missile silos, naval bases, submarines at sea, weapons manufacturing and storage locales, civilian and military command and control centers, attack assessment and early-warning facilities, and the like are probable targets ("counterforce attack"). While it is often stated that cities are not targeted "per se," many of the above targets are proximate to or collocated with cities, especially in Europe. In addition, there is an industrial-targeting category ("countervalue attack"). Modern nuclear doctrines require that "war-supporting" facilities be attacked. Many of these facilities are necessarily industrial in nature, and engage a workforce of considerable size. They are almost always situated near major transportation centers, so that raw materials and finished products can be efficiently transported to other industrial sectors, or to forces in the field. Thus, such facilities are, almost by definition, cities, or near or within cities. Other "war-supporting" targets may include the transportation systems themselves (roads, canals, rivers, railways, civilian airfields, etc.), petroleum refineries, storage sites and pipelines, hydroelectric and nuclear power plants, radio and television transmitters, and the like. A major countervalue exchange therefore might involve almost all large cities in the United States and the Soviet Union, and possibly most of the large cities in the Northern Hemisphere. There are fewer than 2,500 cities in the world with populations of over 100,000 inhabitants, so the devastation of all such cities is well within the means of the world nuclear arsenals.

Recent estimates of the immediate deaths from blast, prompt radiation, and fires in a major exchange in which cities were targeted range from several hundred million to—most recently, in a World Health Organization study in which targets were assumed not to be restricted entirely to NATO and Warsaw Pact countries—1.1 billion people. Serious injuries requiring immediate medical attention (which would be largely unavailable) would be suffered by a comparably large number of people, perhaps an additional 1.1 billion. Thus it is possible that something approaching half the human population on the planet would be killed or seriously injured by the direct effects of a nuclear war. Social disruption; the unavailability of electricity, fuel, transportation, food deliveries, communications, and other civil services; the absence of medical care; the decline in sanitation measures; rampant disease and severe psychiatric disorders would doubtless claim collectively a significant number of further victims. But a range of additional effects—some unexpected, some inadequately treated in earlier studies, some uncovered by us only recently—makes the picture much more somber still.

Destruction of missile silos, command and control facilities, and other hardened sites requires—because of current limitations on missile accuracy—nuclear weapons of fairly high yield exploded as ground bursts or as low air bursts. High-yield ground bursts will vaporize, melt, and pulverize the surface at the target area and propel large quantities of condensates and fine dust into the upper troposphere and stratosphere. The particles are chiefly entrained in the rising fireball; some ride up the stem of the mushroom cloud. Most military targets, however, are not very hard. The destruction of cities can be accomplished, as demonstrated at Hiroshima and Nagasaki, by lower-yield explosions less than a kilometer above the surface. Low-yield air bursts over cities or near forests will tend to produce massive fires, in some cases over a total area of 100,000 square kilometers or more. City fires generate enormous quantities of black smoke which rise at least into the upper part of the lower atmosphere, or troposphere. If firestorms occur, the smoke column rises vigorously, like the draft in a fireplace, and may (the question is still unresolved) carry some of the soot into the lower part of the upper atmosphere, or stratosphere. The smoke from forest and grassland fires would initially be restricted to the lower troposphere.

The fission of the (generally plutonium) trigger in every thermonuclear weapon and the reactions in the (generally uranium–238) casing added as a fission yield "booster" produce a witch's brew of radioactive products, which are also entrained in the cloud. Each such product, or radioisotope, has a characteristic half-life (defined as the time to decay to half of its original level of radioactivity). Most of the radioisotopes have very short half-lives, and decay in hours to days. Particles injected into the stratosphere, mainly by high-yield explosions, fall out very slowly—characteristically in about a year, by which time most of the fission products, even when concentrated, will have decayed to much safer levels. Particles injected into the troposphere by low-yield explosions and fires fall out more rapidly—by coagulation, gravitational settling, rainout, convection, and other processes—before the radioactivity has decayed to moderately safe levels. Thus, rapid fallout of tropospheric radioactive debris tends to produce larger doses of ionizing radiation than does the slower fallout of radioactive particles from the stratosphere.

Nuclear explosions of more than one megaton yield generate a radiant fireball that rises through the troposphere fully into the stratosphere. The fireballs from weapons with yields between 100 and 1,000 kilotons (1,000 kilotons = 1 megaton) will partially extend into the stratosphere. The high temperatures in the fireball chemically ignite some of the nitrogen in the air, producing oxides of nitrogen, which in turn chemically attack and destroy the gas ozone in the middle stratosphere. But ozone absorbs the biologically dangerous ultraviolet radiation from the sun. Thus, the partial depletion of the stratospheric ozone layer, or "ozonosphere," by high-yield nuclear explosions will increase the flux of solar ultraviolet radiation at the surface of the Earth (after the soot and dust have settled out). After a nuclear war in which thousands of high-yield weapons are detonated, the increase in biologically dangerous ultraviolet light might be several hundred percent. In the more dangerous shorter wavelengths, larger increases would occur. Nucleic acids and proteins, the fundamental molecules for life on Earth, are especially sensitive to ultraviolet radiation. Thus, an increase in the solar ultraviolet flux at the surface of the Earth is potentially dangerous to life.

These four effects—obscuring smoke in the troposphere, obscuring dust in the stratosphere, the fallout of radioactive debris, and the partial destruction of the ozone layer—constitute the four known principal adverse environmental consequences that would occur after a nuclear war is "over." There may well be others about which we are still ignorant. The dust and, especially, the dark soot absorb ordinary visible light from the sun, heating the atmosphere and cooling the Earth's surface...

* * *

There has been a systematic tendency for the effects of nuclear weapons and nuclear war to be underestimated. The yield of the first nuclear explosion near Alamogordo, New Mexico, on July 16, 1945, was underestimated by almost all those who designed and constructed the weapon. The extent of fallout from early thermonuclear weapons tests was underestimated; the impairment or destruction of satellites by nuclear weapons explosions in space was a surprise; the depletion of the ozonosphere by high-yield bursts was unanticipated; and nuclear winter was for many—ourselves included—an astonishment. What else have we overlooked?

One, possibly serious, additional effect is the production of toxic gases by city fires. It is now a commonplace that in the burning of modern tall buildings, more people succumb to toxic gases than to fire. Ignition of many varieties of building materials, insulation, and fabrics generates large amounts of such pyrotoxins, including carbon monoxide, cyanides, vinyl chloride, oxides of nitrogen, ozone, dioxins, and furans. Because of differing practices in the use of such synthetics, the burning of cities in North America and Western Europe would probably generate more pyrotoxins than cities in the Soviet Union, and cities with substantial recent construction more than older unreconstructed cities. In nuclear war scenarios in which a great many cities are burning, a significant pyrotoxin smog might persist for months. The magnitude of this danger is unknown.

Another probably very significant and almost unevaluated consequence of nuclear war is what are called synergisms. A very simple example follows from the compromise of the human immune system by both prompt ionizing radiation and ionizing radiation from fallout, as well as from the enhanced post-nuclear winter ultraviolet flux. At the same time that survivors will be much more vulnerable to disease, medical services will have collapsed; insect predators such as birds will have been preferentially killed by the cold, the dark, and the radiation; insects will have proliferated enormously because they can resist these environmental assaults and because the predators that keep them in check will have been greatly reduced in numbers; the radiation may produce particularly virulent forms of microorganisms carried by the insect vectors; and hundreds of millions or billions of corpses will be beginning to thaw. There are many other cases where the interaction of several...environmental assaults...will result in a net adverse consequence much more severe than the simple sum of the component effects. Almost all synergisms are of unknown magnitude; however, almost all of them will have an incremental adverse consequence.

So if the weight of historical evidence and the nature of synergisms imply that the consequences of nuclear war would be even more severe than the present nuclear winter analysis indicates, where does conservatism lie? Is it a proper posture, considering the unprecedented stakes in the answer, to assume that the effects of nuclear war will be less severe than is currently estimated, or more?

It is no longer true that the really serious effects of nuclear war would be restricted to the combatant nations. The biology in equatorial latitudes, for example, is much more vulnerable to even small temperature declines than the biology in more northerly or more southerly latitudes. Agriculture—at least in the Northern Hemisphere, which produces the bulk of the export grain on the planet—would be devastated even by a "small" nuclear war. The propagating ecological consequences all over the Earth are likely to be severe and if, as our and many other studies now show, the cold and the dark move to the Southern Hemisphere, nuclear war implies an unprecedented global catastrophe. It is no longer possible to imagine that nations far from the conflict could merely sit the war out, and inherit a postwar environment freed of the annoyances of big power politics. Instead it seems much more likely that there are no sanctuaries from nuclear war anywhere on Earth. This is one of many implications of the new studies for doctrine, policy, and international politics. A discussion of these subjects is beyond the scope of this meeting and these Conference proceedings, but I have made a preliminary discussion of such implications elsewhere...

* * *

And so we return to Halloween. This meeting on the "World after Nuclear War" is being held, because of circumstances as mundane as the availability of Washington hotels, on October 31. Halloween is celebrated today as a festival of ghosts and goblins and things we know are not real. The horrors of nuclear war, on the other hand, are not fantasies, not projections of our unconscious, but realities that we must deal with in the world of personal emotions and practical politics. Nuclear war is very much worth worrying about and not just on October 31.

Still, if you had to hold such a meeting on a date with some symbolic significance, Halloween seems to be an appropriate choice. It was originally, in pre-Christian times, a Celtic festival called Samhain. It marked the beginning of winter. It was celebrated by the lighting of vast bonfires. And it was named after and consecrated to the Lord of the Dead...

Nuclear weapons are made by human beings. The global strategic confrontation of the United States and the Soviet Union has been devised and carried out by human beings. There is nothing inevitable about these matters. If we are sufficiently motivated, we can extricate the human species from this trap that we have foolishly set for ourselves. But time is very short.

*Carl Sagan and Paul R. Ehrlich*

# The Biological Consequences

## Paul R. Ehrlich

It is a privilege, although a rather somber one, to be able to present to you the consensus of a large and distinguished group of biologists on the likely biological effects of a large-scale nuclear war. That consensus was reached in the course of a meeting immediately following the physicists' meeting that Carl Sagan described and in the course of preparing two documents on the impacts of nuclear war. Those of you familiar with the scientific enterprise know that to get more than fifty scientists to agree, with no significant dissent, to a broad set of conclusions is in itself unusual. To get them to agree on conclusions that bear on a problem of great and current public concern is extraordinary.

The reason for our consensus should be clear to you from Professor Sagan's presentation. The environment that will confront most human beings and other organisms after a thermonuclear holocaust will be so altered, and so malign, that extreme and widespread damage to living systems is inevitable. It is, for example, entirely possible that the biological impacts of a war, *apart* from those resulting directly from a blast, fire, and prompt radiation, could result in the end of civilization in the Northern Hemisphere. Biologists can agree to that as easily as we all could agree that accidentally using cyanide instead of salt in the gravy could spoil a dinner party.

My primary task here today is to give you some technical background to explain why numerous biologists—especially ecologists—are convinced that decision-makers in many nations vastly underrate the potential risks of nuclear war.

### Direct Effects

Most of my focus will be on widely ignored indirect consequences for human beings of such a war that would be transmitted through effects on ecological systems. But I do not want to downplay the potential direct effects, well known as they may be, for they will be truly horrifying. Consider what recent studies indicate would happen in a large thermonuclear war, in which somewhere between 5,000 and 10,000 megatons of weapons were detonated—mostly in the Northern Hemisphere. (To put such a war in perspective, consider it *roughly* equal to the explosion of one-half to three-quarters of a million Hiroshima-sized atomic bombs, which amounts to only a portion of the current nuclear arsenals of the USA and USSR.)

The effects are going to depend to some extent on the size of the war, the distribution of the bursts, numbers of ground bursts and air bursts, and other factors. But I want to emphasize again what Dr Sagan has emphasized so well, that the biological results are robust. That means it would be extremely difficult to design a major nuclear war that would not lead to a biological catastrophe of unprecedented dimensions...

Blast alone, according to one estimate, would be expected to cause 750 million deaths. As many people *as existed on the planet when our nation was founded* would be vaporized, disintegrated, mashed, pulped, and smeared over the landscape by the explosive force of the bombs. Another study predicts that 1.1 *billion* people would be killed and a like number injured immediately by blast, heat, and radiation. In other words, almost *half* of the current global population—including most of the residents of the rich nations of the Northern Hemisphere—could become casualties within a few hours.

It is also crystal clear that the very fabric of industrial society would be destroyed by such a war. Virtually all cities—which are the political, industrial, transport, financial, communications, and cultural centers of societies—would simply cease to exist. Much of humanity's know-how would disappear along with them. Medical care and other disaster-relief services would be essentially nonexistent—there would be no place for help to come from. Survivors in the once-rich nations would not only face the crushing psychological burdens of having witnessed the greatest catastrophe in human history, they would also know there was no hope of succor.

Such a situation is so mind-boggling that many take it to be a worst-case estimate of the potential damage to *Homo sapiens* in World War III. Instead, as we shall now see, I have only described the obvious tip of the iceberg. The fates of the 2–3 billion people who were not killed immediately—including those in nations far removed from targets—might in many ways be worse. They, of course, would suffer directly from the freezing temperatures, darkness, and midterm fallout discussed by Professor Sagan. But the most significant long-term effects would be produced indirectly by the impact of these and other factors on the environmental systems of the planet.

### Ecosystems

To understand this, you need to understand a little about ecological systems—*ecosystems* in biological shorthand. An ecosystem is a biological community—all of the plants, animals, and microbes that live in one area—combined with the physical environment in which those organisms exist. The environment includes solar radiation, the gases of the atmosphere, water in streams, rock fragments in the soil, and so on. And the essence of an ecosystem is a web of processes that connect the organisms with one another and their physical environment.

Those processes include a one-way flow of energy through ecosystems and a cyclic movement of materials within them. Many of you are familiar with the process of photosynthesis, by which green plants "capture" energy from sunlight. Some of that energy then moves up "food chains," being used first by the plants to grow and to drive their other life processes, then by herbivores that eat the plants, then by carnivores that eat herbivores and each other, and finally by decomposers that break down waste products and dead organisms.

Energy from the sun powers all significant ecosystems, not just through photosynthesis, but also through purely physical processes, such as evaporating water from ocean and land surfaces so that it can continue to circulate. Thus you can see immediately why any event that blocked sunlight from Earth's surface might have catastrophic effects on ecosystem functioning.

But what difference would that make? You must understand that *all* human beings are embedded in ecosystems and are utterly dependent upon them for agricultural production and an array of other free "public services." These services include regulating climates and maintaining the gaseous composition of the atmosphere; delivering fresh water; disposing of wastes; recycling of nutrients (including those essential to agriculture and forestry); generating and preserving soils; controlling the vast majority of potential pests of crops and carriers of human disease; supplying food from the sea; and maintaining a vast genetic "library" from which humanity has already withdrawn the very basis of civilization—including all crop plants and domestic animals.

Damage to ecosystems means curtailment of those services. And the 2–3 billion people who might survive the immediate effects of a thermonuclear war would need those services even more than people today.

### Assaults on Ecosystems

What kinds of assaults would ecosystems be subjected to in the event of a full-scale nuclear exchange between the United States and the USSR? Professor Sagan has emphasized the two that would probably be the most important—widespread darkness and very cold continental weather. Others that would not be trivial, however, include wildfires; toxic smog (which might engulf the entire Northern Hemisphere); enrichment of sunlight (when it did penetrate) with dangerous wavelengths of ultraviolet light (UV-B) that, among other things, damages the genetic material (DNA); increased levels of nuclear radiation; acid rains; the release of poisonous chemicals into ground, surface, and onshore oceanic waters; the siltation and sewage pollution of lakes, rivers, and ocean margins; and violent storms in coastal areas.

As I describe some of the impacts of these phenomena, you should keep in mind that most of them will be occurring *simultaneously* in many areas. In addition, often the impacts of two or more concurrent assaults are likely to be *synergistic*—that is, greater than just the sum of their individual effects. For instance, background radiation levels from global fallout (that is, radiation exposure not attributable to local fallout from any particular bomb) may be much higher than has been estimated in previous analyses, because fallout from the upper troposphere had generally been ignored.

It is also important for you to understand that the biologists' conclusions about ecosystemic effects are much less dependent on the exact pattern of detonations than are the direct consequences of blast, heat, and initial radiation. Only in the case of a truly limited small-scale nuclear war is it likely that our qualitative analysis would not apply. Such wars are possible, but whether a nuclear war, once started, could be contained is questionable; many analysts find limited nuclear wars highly unlikely. In any case, decision-makers should be fully apprised of the potential consequences of the large-scale nuclear exchanges that are most likely to cause devastating long-term effects.

Our conclusions may well underestimate those consequences, since we are still far too ignorant of the detailed working of global ecosystems to evaluate all the possible synergistic interactions among the insults to which both human beings and ecosystems would be subjected. Indeed, even if climatic effects did not engulf the Northern Hemisphere or the entire globe, the impacts of nuclear war on the planet's ecosystems would be substantial.

### Cold and Darkness

Reduced temperatures would have dramatic direct effects on animal populations, many of which would be wiped out by the unaccustomed cold. Nevertheless, the key to ecosystem effects is the impact of the war on green plants. Their activities provide what is known as *primary production*—the binding of energy (through photosynthesis) and the accumulation of nutrients that are necessary for the functioning of all biological components of natural and agricultural ecosystems. Without the photosynthetic activities of plants, virtually all animals, including human beings, would cease to exist. All flesh is truly "grass." ...

What all this boils down to is that virtually all land plants in the Northern Hemisphere would be damaged or killed in a war that occurred just prior to or during the growing season. Most annual crops would likely be killed outright, and there would also be severe damage to many perennials if the war were to occur when they were growing actively. Damage might, of course, be less if it happened during the season when they were dormant.

Before a fall or winter war, humanity's main food sources—wheat, rice, corn, and other cereal grains—would have been harvested. But the weather would probably remain unusually cold for months afterward, preventing growth during the next spring and summer, even if other conditions were suitable. Also, since winter temperatures would be far below normal minimums, many perennial plants (for example, fruit trees and important components of the natural vegetation) could be killed. The seed stocks of temperate plants, however, generally would not be damaged by the cold, although those of many tropical plants would be.

While a fall or winter war would probably have a less severe impact on plants at northern latitudes than a spring or summer one, it still could have a severe impact in the tropics, where plants grow throughout the year. The only areas in the Northern Hemisphere where terrestrial plants might not be devastated by severe cold would be in coastal zones and on islands where the temperatures would be moderated by the oceans. Coastal areas, however, would experience especially violent weather because of the enormous temperature differential that would develop between the land and the sea.

Cold, remember, is just *one* of the stresses to which green plants would be subjected. The blockage of sunlight that caused the cold would also reduce or terminate photosynthetic activities. This would have innumerable consequences that would cascade through food chains including those supporting human beings. Primary productivity would be reduced roughly in proportion to the amount of light reduction, even if the vegetation were not otherwise damaged. If the light level declined to 5 percent or less of normal levels—which is likely to be the case for months in the middle latitudes of the Northern Hemisphere—most plants would be unable to maintain any net growth. Thus, even if temperatures remained normal, the productivity of crops and natural ecosystems would be enormously reduced by the blocking of sunlight following a war. In combination, the cold and darkness would constitute an unprecedented catastrophe for those systems.

## Ultraviolet Light

As the cold and darkness abated, green plants would be subjected to another serious insult. Nuclear fireballs would inject large amounts of nitrogen oxides into the stratosphere. These would result in large reductions of the stratospheric ozone shield—on the order of 50 percent. Ozone normally screens out UV-B. In the weeks or months immediately following the war, the atmospheric soot and dust would prevent the increased UV-B from reaching ground level. But the ozone depletion would persist longer than the soot and dust, and, as the atmosphere cleared, organisms would be subjected to UV-B radiation levels much higher than those considered dangerous to ecosystems and human beings.

One response of plants to increased UV-B is reduction of photosynthesis. Furthermore, leaves that have developed in dim light are two to three times more sensitive to UV-B than those that have developed in full sunlight. Thus UV-B will compound the damage caused by earlier low levels of fight. The immune systems of *Homo sapiens* and other mammals are known to be suppressed by even low doses of UV-B. Thus mammals that were subjected to increased ionizing radiation (which also suppresses the immune system), diseases, and a host of other stresses in a postwar world might have one of their most important defenses impaired. It has also been suggested that protracted exposure to increased UV-B could lead to widespread loss of sight. Survivors among people and other mammals might again find themselves in darkness soon after the sky cleared.

## Radioactive Fallout

Ecosystems of the Northern Hemisphere would also be subjected to much higher levels of ionizing radiation from radioactive fallout than has been previously thought. One estimate suggests that a total of about 2 million square miles downwind of the detonations would be exposed to 1,000 rems or more of radiation, mostly within 48 hours. Such levels of radiation would be lethal to all exposed people and to many other sensitive animal and plant species.

As much as 30 percent of the midlatitude land area of the Northern Hemisphere might be exposed to more than 500 rems of radiation within a day. Such a dose would result in death for about half of the *healthy* adult human beings exposed. Because of other stresses, however, few of the adults in those areas would be healthy, so radiation might finish off many millions of wounded, sick, cold, hungry, and thirsty survivors. Those that did not die would be ill for weeks and prone to cancer for the remainder of their lives. The total number of people afflicted would certainly exceed one billion and might include everyone in the Northern Hemisphere—depending on the details of the nuclear exchange.

Lower levels of abnormal exposure, still hundreds of times greater than nominal "background" radiation, would occur over half or more of the hemisphere, making survivors more susceptible to disease, entraining the production of cancers, and causing genetic mutations.

Ecosystemic effects of high levels of radiation are more difficult to predict. Nonhuman organisms are differentially susceptible to radiation damage. The most vulnerable include most of the coniferous trees that form extensive forests over the cooler parts of the Northern Hemisphere. Conifers could be killed over an area making up more than 2 percent of the entire land surface of the Northern Hemisphere. This, in turn, would create conditions conducive to the development of extensive fires.

In addition to the conifers, birds and mammals are prominent among the more sensitive groups. In combination with other assaults, fallout in many areas could thus add to the disruption of the normal functioning of ecosystems. In addition, some radioactive isotopes would enter into nutrient cycles, becoming concentrated in the process, thereby posing possible additional hazards to human survivors.

## Fire, Smog, and Synergisms

This recital by no means exhausts the impacts that ecosystems would suffer. Many ecosystems, of course, would be damaged or destroyed by the blast, fires, and radiation from the thousands of nuclear weapons detonations. Oil wells, coal supplies, peat marshes, coal seams, and so on could continue to burn for months or years. Secondary wildfires, possibly covering 5 percent or more of the Northern Hemisphere's land surface, would have devastating direct effects on ecosystems—especially those not adapted to periodic fires. Multiple air bursts over California in the late summer or early fall could burn off much of the state, leading to catastrophic flooding and erosion during the next rainy season. Silting, toxic runoff, and radioactive rainout could kill much of the fauna of fresh and coastal waters. Human survivors seeking nourishment from filter-feeding shellfish such as mussels at the ocean's edge would be likely to find that they were either dead or had concentrated so much radioactivity that they would be lethal to consume.

There is major uncertainty concerning the extent of firestorms, since the conditions of fuel and ignition that create them are poorly understood. These gigantic conflagrations might, in some circumstances, heat the soil sufficiently to kill the dormant seeds they contain—the "seed banks" on which regeneration of the flora depends. The relatively small firestorm that destroyed Hamburg during World War II sent flames 15,000 feet into the sky and smoke 40,000 feet high. The temperature of the fire was sufficient to melt aluminum, and underground shelters were so hot that, when they were opened and oxygen admitted, flammable materials and even corpses burst into flames. That firestorm covered about 6 square miles; the many firestorms generated in a nuclear war could individually be 100 times or more larger.

The fires and firestorms would generate a hemispheric smog of varying thickness, enriched downwind of incinerated cities by a variety of especially toxic chemicals such as vinyl chlorides. A probable consequence of the injection of nitrogen and sulfur oxides into the atmosphere by the fires would be localized, highly acidic rains. And the altered dynamics of the atmosphere might result in severe droughts in other areas. In general, subjecting ecosystems to various combinations of cold, darkness, fire, UV-B, radioactivity, smog, acid rain, and drought would be likely to cause unprecedented outbreaks of plant pests and diseases that could extend in space and time far beyond the direct devastation of the war.

In many cases, as I have already indicated, the impact of two simultaneous stresses would be much greater than the sum of their effects if they occurred separately. Some such synergisms are easily identified. For example, the loss of sunlight is likely to intensify the effects of other stresses on plants because additional energy (and thus sunlight) is required to cope with stresses and repair any damage they cause. We do not begin to have the knowledge to quantify some of the other synergisms that doubtless would occur in highly altered post-attack ecosystems. It seems safe to predict, however, that there would be many of them—and that overall they might prove much more destructive than some of the individual effects.

## The Fate of Vertebrates and Soil Organisms

The disaster that would befall many or most of the plants of the Northern Hemisphere from the effects of a nuclear exchange would contribute to an equal or greater disaster for the higher animals. Wild herbivores and carnivores and domestic animals either would be killed outright by the cold or would starve or die of thirst because surface waters were frozen. Following a fall or winter war, many dormant animals in colder regions might survive, only to face extremely difficult conditions in a cold, dark spring and summer.

Scavengers that could withstand the projected extreme cold would likely flourish in the postwar period because of the billions of unburied human and animal bodies. Their characteristically rapid population growth rates could, after the thaw, quickly make rats, roaches, and flies the most prominent animals shortly after World War III.

Soil organisms are not directly dependent on photosynthesis and can often remain dormant for long periods. They would be relatively unaffected by the cold and the dark. But in many areas the loss of aboveground vegetation would expose the soil to severe erosion by wind and water. Soil organisms may not be terribly susceptible to the atmospheric aftereffects of nuclear war, but entire soil ecosystems are likely to be destroyed anyway.

## Impacts on Agricultural Systems

Agricultural ecosystems would be subject to the same kinds of impacts as natural ecosystems, but they deserve some extra attention because at present they support human populations far above the carrying capacities of natural ecosystems.

There is little storage of staple foods in human population centers, and most meat and produce are supplied by current production. Only cereal grains are stored in any significant quantities, but the storage sites are usually located in relatively remote areas. Thus, after a nuclear war, supplies of food in the Northern Hemisphere would be destroyed or contaminated, located in inaccessible areas, or quickly depleted. People who survived the other effects of the war would soon be starving. Furthermore, countries that now depend on large imports of foods, including those untouched by nuclear detonations, would suffer immediate and complete cessation of incoming food supplies. They would have to fall back on local agricultural and natural ecosystems. For many developing countries, this could mean starvation for large fractions of their populations.

Reestablishment of agriculture after the war would probably be very difficult. Most crops are highly dependent on substantial subsidies of energy and fertilizers. In addition, producing harvestable yields generally depends on the availability of full sunlight, adequate water, suppression of pests, and relative freedom from stresses such as air pollution and UV-B. Few of these requisites would be available in the immediate postwar world.

After environmental conditions returned more or less to "normal" (except for the loss of irreplaceable soils), the ease with which farming could again be carried out on any appreciable scale would depend on whether societal systems could be reorganized (determined by such factors as the availability of energy and the psychological condition of the population), and on the degree to which seeds and animal breeding stocks had survived. Since seeds for most North American, European, and Soviet crops are not harvested and stored on individual farms, the already limited genetic variability of crops would be further reduced by inevitable losses of seed stocks. Furthermore, those strains that did survive would likely be poorly adapted to the postwar environments in which they would be planted.

In the first few seasons, climate might remain more hostile and unpredictable than usual, resulting in uncertain yields and relatively frequent crop failures. Even small climatic changes can have great effects on agriculture. For example, a mere 3°C decrease in average July temperature would push the northern limit of reliable corn production southward several degrees in latitude to southern Iowa and central Illinois.

Finally, it should be noted that agricultural ecosystems inevitably depend on the natural ecosystems in which they are embedded. War-caused alterations in the latter, especially those influencing their ability to deliver fresh water and pest-control and pollination services, could also retard the restoration of agriculture.

*Carl Sagan and Paul R. Ehrlich*

## The Fate of the Tropics

So far, I have concentrated my remarks on effects in the North Temperate Zone, the presumed locus of the war. But what would happen in the tropics and the Southern Hemisphere? Much, of course, would depend on the precise targeting pattern and how many firestorms were generated (for they could inject huge amounts of material into the stratosphere where it could be readily transported from the Northern to the Southern Hemisphere).

Under any war scenario, the spread of cold and darkness to the extensive tropics of the Northern Hemisphere is highly likely, and it is at least possible that they would spread to the tropics of the Southern Hemisphere as well. Even if the darkness and cold were largely confined to the north temperate regions, pulses of cold air could penetrate well into the tropics. It is therefore appropriate to discuss the probable consequences of such a spread.

Many plants in tropical and subtropical regions do not possess dormancy mechanisms enabling them to tolerate cold seasons. In those regions, large-scale injury to plants would be caused by chilling, even if temperatures did not fall all the way to freezing. In addition, vast areas of tropical vegetation are considered to be very near the photosynthetic "compensation point"—their uptake of carbon dioxide is only slightly more than that given off. If light levels dropped, those plants would begin to waste away—even in the absence of cooling. If light remained low for a long time, or if low light levels were combined with low temperature, tropical forests could largely disappear, taking with them most of one of Earth's most precious non-renewable resources: its store of genetic diversity, including the majority of plant and animal species. Tropical animals, including human beings, are also much more likely to die of the cold than their temperate counterparts. In short, where tropical regions are affected by climatic changes, the consequences could be even more severe than those caused by a similar change in a temperate zone.

Furthermore, even in the absence of cold and darkness, the dependence of tropical peoples on imported food and fertilizer would lead to severe problems. Large numbers of people would be forced to leave cities and attempt to cultivate remaining areas of tropical rain forest, accelerating their destruction as the systems were taken far beyond their carrying capacity.

## The Fate of Aquatic Systems

Finally, what would happen to the parts of our planet that are covered with water? Aquatic organisms tend to be protected from dramatic fluctuations in air temperature by the slowness with which water changes its temperature. In general, therefore, aquatic systems should suffer somewhat less disruption than terrestrial ones. Nonetheless, many freshwater systems would freeze to considerable depths (or completely). After a nuclear war in the spring, for instance, three feet or more of ice would form on all bodies of fresh water, at least in the North Temperate Zone. This would even further reduce light levels in lakes, ponds, rivers, and streams in a darkened world. Oxygen would be depleted, and many aquatic organisms would be exterminated. Moreover, the depth of the freezing would make access to surface water by surviving people and other animals extremely difficult.

In the oceans, the darkness would inhibit photosynthesis in the tiny green plants (algae) that form the base of all significant marine food chains. The reproduction of these plants, known collectively as phytoplankton, would be slowed or stopped in many areas, and the surviving phytoplankton would be quickly eaten up by the small floating animals (zooplankton) that prey upon them. Near the ocean's surface, the productivity of phytoplankton is reduced by present levels of UV-B; so after a war, an increase in this sort of radiation would be an additional stress. In the Northern Hemisphere, marine food chains might be disrupted for long enough to cause extinction of many valuable fish species, especially after a spring or summer war.

Of course, not only would marine life be decimated in rich near-shore waters such as the Georges Bank, but those waters also would be extremely stormy. Furthermore, to the degree that they are in port when the war occurs, the fishing fleets and skilled fishermen that now harvest oceanic riches would have been largely converted into particulate matter helping to shade the oceans. Survivors able and willing to fish would have great difficulty finding fuel and suitable port and processing facilities. Overall, there is little reason to believe that, at least in the Northern Hemisphere, the forms of marine life that serve as major food sources for human beings would be accessible to survivors.

## The Fate of the Earth

Plausible nuclear war scenarios can be constructed that would result in the dominant atmospheric effects of darkness and cold spreading over virtually the entire planet. Under those circumstances, human survival would be largely restricted to islands and coastal areas of the Southern Hemisphere, and the human population might be reduced to prehistoric levels.

When many of us read Jonathan Schell's book, *The Fate of the Earth,* we were very much impressed by the moving way in which he presented the case, but I suspect that most biologists, like myself, thought it was a little extreme to imagine that our species might actually disappear from the face of the planet. It did not seem plausible from what we knew then.

Now, the biologists have had to consider the possibility of the spread of darkness and cold over the entire planet and throughout the Southern Hemisphere. It still seemed unlikely to them that that would immediately result in the deaths of all the people in the Southern Hemisphere. We would assume that on islands, for instance, far from sources of radioactivity and where the temperatures would be moderated by the oceans, some people would survive. Indeed, there probably would be survivors scattered throughout the Southern Hemisphere and, perhaps, even in a few places in the Northern Hemisphere.

But one has to ask about the long-term persistence of these small groups of people, or of isolated individuals. Human beings are very social animals. They are very dependent upon the social structures that they have built. They are going to face a very highly modified environment, one not only strange to them but also in some ways much more malign than people have ever faced before. The survivors will be back in a kind of hunter and gatherer stage. But hunters and gatherers in the past have always had an enormous cultural knowledge of their environments; they knew how to live off the land. But after a nuclear holocaust, people without that kind of cultural background will suddenly be trying to live in an environment that has never been experienced by people anywhere. In all likelihood, they will face a completely novel environment, unprecedented weather, and high levels of radiation. If the groups are very small, there is a possibility of inbreeding. And, of course, social and economic systems and value systems will be utterly shattered. The psychological state of the survivors is difficult to imagine.

It was the consensus of our group that, under those conditions, we could not exclude the possibility that the scattered survivors simply would not be able to rebuild their populations, that they would, over a period of decades or even centuries, fade away. In other words, we could not exclude the possibility of a full-scale nuclear war entraining the extinction of *Homo sapiens.*

## Summary

Let me briefly recap. A large-scale nuclear war, as far as we can see, would leave, at most, scattered survivors in the Northern Hemisphere, and those survivors would be facing extreme cold, hunger, water shortages, heavy smog, and so on, and they would be facing it all in twilight or darkness and without the support of an organized society.

The ecosystems upon which they would be extremely dependent would be severely stressed, changing in ways that we can hardly predict. Their functioning would be badly impaired. Ecologists do not know enough about these complicated systems to be able to predict their exact state after they had "recovered." Whether the biosphere would ever be restored to anything resembling that of today is entirely problematical.

---

From Paul R. Ehrlich, Carl Sagan, Donnald Kennedy, and Walter Orr Roberts, *The Cold and the Dark: The World After Nuclear War* (New York: Norton, 1984).

*Carl Sagan and Paul R. Ehrlich*

## Questions and Topics for Discussion and Writing

1. Why does Sagan view a "limited" nuclear war as virtually impossible?
2. What kind of life forms seem most likely to survive a nuclear war? Explain your answer.
3. The Sagan and Ehrlich lectures point out just how delicately interrelated are all the life-supporting elements of the earth and how easily the entire system can be disrupted by destructive actions. Discuss ways other than nuclear war that the conditions necessary for life as we know it can be disrupted—either by natural or human causes and either suddenly or gradually.
4. Put yourself in the position of someone who supports the building and maintaining of a nuclear arsenal, for example, a strategic military commander. Assume your audience is familiar with Sagan's and Ehrlich's positions. Make the strongest case you can in favour of nuclear weapons.

# The A-Bomb Kid

## John Aristotle Phillips and David Michaelis

The vast majority of atomic weapons are in the arsenals of the United States and the Soviet Union. But other countries have learned to make them also. Britain, France, and China are members of the "Atomic Club" and other countries are rumoured to have secretly developed their own nuclear weapons. Given the virtually instantaneous reaction times of today's sophisticated defence systems, it is very likely that *any* atomic bomb set off by anyone could start a world-wide nuclear war, ending life and civilization as we know it. Hence all members of the "Atomic Club" must guard their atomic weapons very carefully against accidental or malicious detonations. Perhaps these countries are responsible enough to do that. But what if other, *less* stable countries develop nuclear capability? What if terrorists or organized crime develop the "bomb"? How hard is it, after all, to figure out how to make an atomic bomb? With these questions in mind, an undergraduate physics student at Princeton University, John Aristotle Phillips, undertook to find out in the late 1970s. Using only publicly available scientific documents, Phillips researched and wrote a thirty-four-page report on how to build an atomic bomb. The following is an excerpt from the subsequent book about his experience written by Phillips and fellow student David Michaelis, *Mushroom: The Story of the A-Bomb Kid.*

When I return to Princeton, I have a short meeting with Freeman Dyson. He asks about my progress. I show him the documents I've brought back from Washington. He takes them from me very casually, the way an oblivious parent might ask a child, "What did you find when you were playing in the woods, dear?" and then suddenly realize he has a live snake in his hands.

"Good Lord," says Freeman. "Where did you get these?"

"At the National Technical Information Service."

He inspects them very closely. His face is a study in disbelief. "I was sure this kind of information was still classified."

"It only cost me twenty-five bucks."

"Good Lord," he says again.

I tell him about the den mother lady at the desk and what she said. He shakes his head, takes off his spectacles, and wipes them clean. "Well, all I can say is you've done some pretty good detective work John. From now on you'll be on your own." He pauses for a moment and pats the documents. "These will help you far more than I can. I'll be expecting your design in five weeks."

"Do you think I have a good chance?"

"Let's just say I'll be expecting your design. Good luck."

When Freeman leaves he is visibly shaken. He has this distant, glazed look in his eyes, somewhere between shock and fright...

Having witnessed the look in Freeman's eyes, I know I'm going to do it. The only question is: how?...

The material necessary to explode my bomb is plutonium 239, a man-made substance. It is the same element that was used in the bomb which destroyed Nagasaki in World War II. The reason why I must use plutonium—as opposed to, say, lead—is because plutonium is a heavy and *unstable* element: Its atomic structure is constantly changing. The best way to understand the instability of plutonium is to imagine a cliff by the side of a highway where the rocks are crumbling and occasionally falling, thereby dislodging other rocks. The constant decay of the cliff in time haphazardly breaks the rock formations apart, causing the cliff to be unstable. If a runaway avalanche were induced on the cliff by the use of explosives, the energy of the rocks falling would represent the action of plutonium atoms being split apart in the runaway chain reaction of an atomic bomb.

Visualize an atomic bomb as a marble inside a grapefruit inside a basketball inside a beachball. At the center of the bomb is the initiator, a marble-sized piece of metal. Around the initiator is a grapefruit-sized ball of plutonium 239. Wrapped around the plutonium is a three-inch reflector shield made of beryllium. High explosives shaped as breastlike lenses are placed in a symmetrical order around the beryllium shield. Wires are attached to each lens and each wire runs to an electrical source. By the time the lenses and wires have been installed, the bomb is about the size of a beachball.

When the electrical current runs through the wires to the lenses, an explosion is triggered. Because of the symmetrical nature of the placement of the explosives, a spherically imploding shock wave is set off, instantly squeezing the beryllium, plutonium, and initiator. The beryllium shield is pushed inward by the explosion, compressing the grapefruit-sized ball of plutonium to the size of a plum. The plutonium has now gone from a subcritical to a supercritical density, and the initiator at the center has been similarly squeezed. At this moment, the process of atoms fissioning—or splitting apart—begins.

Neutrons released from the initiator strike the plutonium atoms at an extremely fast rate. Each time a neutron hits a plutonium atom, the atom splits, creating two more neutrons, which in turn hit two more atoms, which split into four neutrons, which find four new atoms, thus splitting into eight neutrons, sixteen, thirty-two, sixty-four, one hundred and twenty-eight, two hundred and fifty-six, and so on. This tremendously fast splitting process is called a runaway chain reaction. If this does not occur with precise timing, the bomb will not explode with superior force. Each time an atom is split, a terrific amount of energy is released along with a variety of lethal atomic particles. The sum total of all atoms splitting in the chain reaction creates the atomic fireball which rises into the sky as a great mushroom-cloud energy release. This energy is comprised of heat waves, shock waves, and lethal atomic particles that are exploded outward across the countryside.

There are many subtleties involved in the explosion of an atomic bomb. Most of them center around the actual detonation of explosives surrounding the beryllium shield. Normally, only a very small fraction of the plutonium has a chance to fission before the bomb assembly is blown apart by the energy released from the atoms already split. For this reason, the size of the final mushroom cloud explosion depends upon how long the chain reaction of splitting atoms can be sustained. The timing and efficiency of the chain reactions is in turn determined by the specific arrangement of the explosives surrounding the plutonium.

The grouping of these explosives around the plutonium is one of the most highly classified aspects of the atomic bomb and the area around which the most intense postwar espionage was centered. (Julius and Ethel Rosenberg were executed in 1953 allegedly for passing this secret to the Russians.) In the bombs exploded over Hiroshima and Nagasaki, TNT was the explosive used to set off the chain reaction. But since that time, the US Army has designed more sophisticated and efficient explosives to do the job. How to arrange the explosives and which explosives to use pose the biggest problems for me as I begin to design my bomb. The correct arrangement around the plutonium is the critical factor without which my design will be worthless.

* * *

My base of operations is a small room on the second floor of Ivy Club. The large conference table in the center of the room is covered with my books, calculators, design paper, notes... My sleeping bag is rolled out on the floor. The ashtrays are full. As the next three weeks go by, I stop going to classes altogether. I work all day and all night in the room. My progress is slow at first. I pretend not to listen to the noise coming from the other rooms in the club. Sometimes I find myself in a trance, staring out at the first buds on the magnolia trees beyond the leaded windows. I drink coffee from a thermos, and when I feel drowsy I go downstairs to the kitchen to refill the thermos. I stop eating at regular meals, taking only a bologna sandwich when I'm hungry. The other members at Ivy begin referring to me as the Hobo because of my unshaven face and disheveled appearance. I develop a terrible case of bloodshot eyes.

Sleep comes rarely.

I devise several mind-sets that help me solve

some of the problems I come up against. In my work (and even in my dreams now), I place myself in the position of a terrorist. I approach every problem from a terrorist's point of view. The bomb must be inexpensive to construct, simple in design, and small enough to sit unnoticed in the trunk of a car or abandoned U-Haul trailer.

At other times, I put myself in the shoes of the Los Alamos scientists working on the first bombs. By closely following the technical accounts of their progress which I bought in Washington, I design the bomb as they did, working on each component separately, one at a time. But the one terrific advantage I share with the nations which developed the bomb after the United States is that I know it can be done. The scientists at Los Alamos were not sure that such a bomb would be successful until after the first Alamogordo test. Furthermore, according to *The Curve of Binding Energy*, every bomb that has been designed and built since 1945 has worked on the first try.

---

From John Aristotle Phillips and David Michaelis, *Mushroom: The Story of the A-Bomb Kid*, 97–102 (New York: Morrow, 1978).

---

### Questions and Topics for Discussion and Writing

1. Explain how an atomic bomb works.
2. Discuss the significance of Phillips' achievement.

# The New Star Wars

## John C. Polanyi

In 1983, the President of the United States, Ronald Reagan, announced his plan for a new way to defend the world from nuclear weapons, which he called the Strategic Defense Initiative. It involved establishing a protective "shield" of orbiting satellites and other devices that would detect and destroy missiles in flight. The scheme was quickly nicknamed "Star Wars," after the futuristic space fantasy movies of George Lucas. The Star Wars plans have included proposals for a variety of strange contraptions that sound like they belong in a science-fiction movie. The latest proposals are called "brilliant pebbles." The following article, by University of Toronto chemistry professor and Nobel laureate John Polanyi, questions the logic behind brilliant pebbles and the whole gamut of the Strategic Defense Initiative.

It appears there is only one thing harder to destroy than a nuclear missile in flight, and that is Washington's Star Wars dream.

US Secretary of Defence Dick Cheney recently proposed a major new defence system against antiballistic missiles (ABM). It is called "Brilliant Pebbles" and would consist of thousands of small projectiles (the pebbles) mounted on as many sophisticated (brilliant) rockets. The whole shebang is to circle the globe indefinitely, waiting for Armageddon.

Modern physics and chemistry have been pillaged over the past half-dozen years to produce one hypothetical ABM defence after another—infrared beams, ultraviolet beams, X-ray beams, proton beams, neutron beams, rail-guns, space-based rockets and ground-based rockets have all been marshalled at various times. Each has been the subject of criticism by Star Wars skeptics, defended by its designers and subsequently withdrawn to be replaced by next year's model.

All this at the very modest cost of $16-billion, since we are dealing only with proposals. The real cost, which has been much larger, has been the diversion of effort down futile and hazardous paths.

In 1983 President Ronald Reagan invited US scientists and engineers to join him in the Strategic Defence Initiative—a crusade to design a defence that would render nuclear weapons "impotent and obsolete." The SDI proposal suffered from a number of fundamental flaws:

- It is easier to make a sophisticated device malfunction than it is to make it function. Hence, the technology that could cause Star Wars to succeed will, in reality, ensure that it fails.
- While bemoaning the pernicious effects of arms racing, we are in danger of committing ourselves, through SDI, to two arms races in place of one. More and larger shields engender more and larger swords, which in turn require a new generation of shields. This vastly expensive progression puts the world more and more on a war footing at a time when we have less and less to quarrel about.

  We should be reducing levels of weaponry so that we can turn our attention to the real dangers that threaten us all: environmental degradation and the impoverishment of half of the world.
- Star Wars is inimical to arms control and disarmament. In common with offensive systems, defensive systems do not remain static. They require constant updating. This implies the need for a broad range of new military technologies.

  The antithetical relationship between improving missile defences and reducing the number of missiles is a fact of life. It was not a whim that caused the Soviet Union to link progress in the Strategic Arms Reduction Talks to downgrading the SDI program.

- Star Wars is not an alternative to arms control; it requires arms control. This proposition may seem to be at odds with the previous one, and it is, but Star Wars involves a contradiction.

Star Wars requires arms control because it posits a transition from abundant weaponry to weaponry that has been rendered ineffectual through the erection of high-tech shields. If one side's shield appears at any time to be superior to the other's, the weaker party is placed in a threatening situation.

Mr. Reagan warned of this in announcing the SDI in 1983, and called for a balanced, bilateral introduction of anti-missile defences.

It is, of course, easy to view this process of transition with equanimity if one is confident of retaining at all times a superiority in defence. But not even the United States can be assured of that. There is no way this transition can be managed without a step-by-step agreement between the parties involved. Given the difficulties of assessing the efficacy of novel defensive systems, and the climate of arms-racing, this will be staggeringly difficult to achieve.

The crowning absurdity in this difficult and dangerous undertaking is best illustrated by the apocryphal story of a late-night discussion between the two superpower leaders in Reykjavik four years ago. Mr. Reagan, having exerted all of his considerable powers of persuasion in the hope of obtaining Soviet leader Mikhail Gorbachev's agreement to mutual deployment of anti-missile defences, is supposed to have elicited the response: "Ronnie, you really make an excellent case, but would you mind going over the bit about whom we are defending ourselves against?"

* * *

There are two ways of reducing the destructiveness of nuclear weapons, both requiring the acquiescence and cooperation of the parties involved. One, anti-missile defence, involves crippling expenditures. The other, disarmament, comes free of charge. The former raises profound and legitimate fears. The latter—guaranteed by the sort of inspection procedures being instituted today—involves minimal risks.

The moral should be clear. However, the recent recommendation for a multi-billion-dollar research program to develop and test Brilliant Pebbles indicates it is not.

The new proposal has already emerged as the central focus for SDI, and is readily outlined. Brilliant Pebbles are descended from "Smart Rocks," proposed in an earlier scheme. Both are rocket systems that use optical sensors to find an object, and then hurl a weight in its path to destroy it on impact. No explosive charge is required.

The principal novelty is the compactness of the rocket, which is about three feet long and weighs only 100 pounds, of which the "pebble" accounts for perhaps two pounds. Low cost can be traded for large numbers. The figure quoted most often is 20,000 rockets that would circle Earth for years at a time before having to be replaced.

An important element is the inclusion in each rocket of a computer as small as a cigaret pack but with the power of a CRAY-1 supercomputer. This has allowed designers to give a high degree of autonomy to each rocket, making it able to plot its own path and make sophisticated choices. This renders the system less vulnerable to an attack, but it is not yet clear what it does to the degree of human control.

Taken together, these innovations are regarded as important. In a speech to the US Navy League, Vice-President Dan Quayle said of Brilliant Pebbles: "It could revolutionize much of our thinking about strategic defence."

If US early-warning satellites were to detect what appeared to be the initiation of a missile attack, "brilliant" rockets that happened to be in striking range would race toward the opposing missiles as they rose from their silos. At this point there is plenty of exhaust heat on which to lock the guidance system and only one target for each missile to eliminate, instead of the 10 there would be after a missile's multiple war-heads go their separate ways.

In common with its various predecessors, the Brilliant Pebble system is designed to destroy the current generation of slow-burning Soviet SS–18 intercontinental ballistic missiles, whose boost phase lasts several minutes. However, the Soviet Union now has the technology to reduce this phase to a minute, which the SDI office has conceded would "seriously degrade" the new system.

There are other ways in which it could be degraded. If the world tolerates thousands of orbiting rockets in space, it also must tolerate the presence of some anti-rockets—the space equivalent of mines.

With these simple devices tagging along beside the rockets, the attacker would have the option at any time of exploding a sufficient number to make a rent in the defensive screen, through which he could then send his missiles.

Defensive systems are vulnerable because the attacker chooses the time and place for the encounter. His is the easier task, technically and strategically. If he is determined, he will get through. Of the thousands of missiles available to him, precious few are really required. One or two missiles are capable of destroying 10 major cities.

Tragically, SDI threatens arms control not only in the future but today. The keystone in the arms-control structure is the 1972 ABM Treaty, which Canada has endorsed ringingly and repeatedly. It states under Article V that the parties undertake "not to develop, test or deploy ABM systems or components which are sea-based, air-based or space-based."

This treaty is the rock on which we can at last hope to build. It is not to be exchanged for pebbles of whatever superficial brilliance.

*The Globe and Mail* (May 12, 1989): A7

## Questions and Topics for Discussion and Writing

1. What is the purpose of the Strategic Defense Initiative ("Star Wars")?
2. Explain Polanyi's statement "It is easier to make a sophisticated device malfunction than it is to make it function."
3. Why does Polanyi view the proposal for Brilliant Pebbles futile and dangerous?

# Cold Fusion

## Malcolm W. Browne, Louise Brown, Jack Miller, Stephen Strauss, Christopher Reed, and the Associated Press

What is *potentially* the biggest development in science in the latter half of this century—if not the biggest development in centuries—is cold fusion. Nuclear fusion refers to the release of energy caused by the fusing together of two light atoms to form one heavier atom. This is the principle of the hydrogen bomb. Until March 1989, the scientific community was in general agreement that nuclear fusion could take place only at temperatures similar to those found on the sun. Then, researchers in the United States announced that they had found a process that produced energy from fusion at ordinary room temperatures in an inexpensive, small apparatus. If their claims are true, the implications are profound. Cold fusion could provide an almost limitless supply of *clean* energy, doing away with the need to pollute the atmosphere by burning non-renewable fossil fuels or dispose of incredibly dangerous nuclear waste.

But are their claims true? Some scientists claim to have verified the original researchers' results. Others claim that the announced results are all a mirage—an artifact of mishandling the experiment. What's worse, the cold fusion researchers are chemists, engendering a bit of territorial squabbling as they stray out of their usual domain into that of physics.

Then there is the question whether the researchers were premature in their announcement or whether the recent technological phenomena of electronic bulletin boards and fax machines have changed the pace of scientific communication and thereby upset the checks and balances hitherto in place.

It is too early to assess any of these issues, but it is worthwhile following them as they unfold. To that end, the following is a series of newspaper articles on cold fusion that appeared shortly after the first announcements.

## Fusion in a Jar: Announcement by Two Chemists Ignites Uproar

### Malcolm W. Browne

Two chemists who claimed last week to have triggered nuclear fusion in a jar of water have ignited a major uproar in the scientific world.

At a news conference Thursday, Dr B. Stanley Pons of the University of Utah and Dr Martin Fleischmann of the University of Southhampton announced that they had not only achieved hydrogen fusion in a simple electrolytic cell, but had obtained a substantial yield of energy.

If their assertion is verified, scientists agree, the two would almost certainly win a Nobel Prize and would probably become very wealthy through the commercialization of their process. Such an achievement might also give the world a new source of cheap energy.

Unlike fission, in which atoms are split to release energy, fusion is the joining together of the nuclei of hydrogen atoms to produce helium and enormous amounts of energy.

Scientists have long sought to produce controlled fusion reactions, although some engineers believe that even if the means can be found to maintain a fusion reaction, the construction of a safe fusion reactor plant might be impossible.

The two chemists' claim that they achieved room-temperature fusion has not been ruled out by experts in chemical catalysis, who use similar materials and techniques. But scientists at federally financed laboratories who have worked for decades toward achieving controlled hydrogen fusion are profoundly skeptical of the assertion. In dozens of interviews, most said that although they believed that the two are honest researchers, their results are most probably flukes or errors.

Deepening the professional skepticism was the unorthodox way in which the claimed achievement was announced. Although the two said they had been working on their idea for 10 years, they said they had confided in no other scientists. Up to the present, they have not published their claim in any scientific journal, a procedure that would have required them to submit convincing experimental results to a panel of their peers.

Dr Pons said in a telephone interview that his work had, in fact, undergone peer review by a professional journal, which would publish it. But he refused to say which journal, or to disclose other details of publication.

He said he and Dr Fleischmann had paid for their work with their private funds, about $100,000, becuase they had assumed that "mainstream" scientific financing agencies would not take them seriously.

Even scientists who believe the claim to be hypothetically possible deplored the way in which the achievement was disclosed.

### "Appalling" Disclosure Method

Prof. John Bockris of the chemistry department at Texas A&M University called the two chemists "absolutely solid fellows who would never try to put us on," but deplored the way their claim was publicized as "appalling." Dr David W. Goodman of Texas A&M University, a leading expert in catalytic chemistry and a long-time colleague of the investigators, said their claim should be taken seriously, but that the means by which it was disclosed was "outrageous."

Dr John H. Sinfelt of the Exxon Research and Engineering Company, a leader in the field of catalytic research related to the Pons-Fleischman claim, said he and many colleagues were fascinated by what they had seen on television or read in newspapers. "But the trouble is, we don't know what to think, when we're reduced to picking up crumbs of information from the popular press, rather than having a formal paper to evaluate," he said.

### "Our Hand Was Forced"

Dr James Brophy, chief of research at the University of Utah, acknowledged that the disclosure was highly irregular, but said, "Our hand was forced" by a reporter in Utah who was about to publish an account of the research. "We felt an obligation to make sure the story was released accurately," he said.

Meanwhile, scientists around the world eagerly awaited sufficient details of the experiment to try it for themselves, an essential step in the verification of any scientific claim. But they may have some time to wait. Dr Brophy said that because of current "patent problems" details would be withheld for the time being.

He defended the unorthodox disclosure as having had a precedent in early disclosures of advances in high-temperature superconductivity. "Two years ago, when the first of the high-temperature superconductors was announced, no one knew why it worked," he said. "Even now, the science involved is still very poorly understood. It's the same thing in our case. We don't claim to understand the science underlying cold fusion; we claim merely to have observed that it occurs."

The basic apparatus used by Dr Pons and Dr Fleischmann would seem familiar to a ninth-grade student of general science. It consists of a vessel filled with heavy water (in which ordinary hydrogen atoms are repalced with their heavy isotope equivalent, deuterium). Two electrodes are immersed in the heavy water and an electrical current is passed between them.

The negatively-charged cathode, Dr Pons said, is made of fine palladium wire. Palladium is a silvery metal belonging to the platinum group that is often used to catalyze or stimulate a chemical reaction, including the reactions that cleanse automobile emissions as they pass through catalytic converters.

The electrical current in the Utah cell splits molecules of heavy water ($D_2O$), just as it would split ordinary water ($H_2O$), and the positively charged deuterium atoms separated from the water molecules are drawn to the cathode. The cathode supplies negatively charged electrons, thereby making the deuterium atoms electrically neutral. The neutral deuterium atoms, in intimate contact with the palladium cathode, are then absorbed into the palladium metal.

It is at this point, according to Dr Pons, that the magic occurs.

Palladium is made up of a tightly knit lattice of metal atoms, into which hydrogen (or deuterium) can diffuse. The atomic lattice grips the deuterium atoms and holds them in such close proximity to each other, Dr Pons says, that a quantum-mechanical effect known as "tunnelling" becomes possible. Tunnelling is a phenomenon that occasionally permits particles to leap through energy barriers, breaking the rules that govern objects larger than molecules in the everyday world.

Dr Pons says that once sufficient time has passed to alow many deuterium atoms to take up positions within the palladium lattice, the statistical possibility arises that some of them will overcome their huge mutual repulsion to "tunnel" toward each other and fuse, thereby creating tritium, helium, neutrons and heat energy.

The process, he said is extremely slow, especially if a large cathode is used. Using a cathode consisting of palladium wire, the process required 10 hours before fusion was observed, he said. Asked why no other heavy water experiment in the past half century had revealed the fusion effect he claimed, he said: "Earlier investigators weren't using palladium as a cathode. In any case, fusion takes a long time to occur, and other researchers weren't looking for it."

The traditional approach to fusion required applying extremely high tempratures and pressures to force nuclei that would ordinarily repel one another to fuse together instead.

Fusion experts at the Lawrence Livermore Laboratory in California, and Los Alamos National Laboratory, who requested that their names be withheld, continued to express skepticism after they were informed of Dr Pons's explanation for his work.

"I don't know what he and Fleischmann are seeing," one scientist said "but until someone else does the experiment for all to see, I think we must be very wary of the claim."

Dr Pons said he expected such criticism from traditional physicists. "The fact reamins, there is no way we could have obtained the results we did without fusion," he said.

He cited as evidence a calorimetric experiment designed to compare the energy put into the experiment with the energy it yielded, a four-to-one ratio. He said gas given off by the reaction had been collected and analyzed using a mass spectrometer, which had confirmed the presence of tritium, a hydrogen isotope that could only have been created by fusion. Neutrons ejected from the reactions were captured in water, he said, and the water emitted gamma rays of precisely the energy that would be expected for fusion-produced neutrons.

---

*The New York Times* (March 28, 1989): C1.

## Fusion: Is it the Key to the Future?

### Louise Brown

For the moment, let's say the claims by a US and British scientist about a new method of nuclear fusion are true.

What would it mean if the world really did have some cheap, safe new fuel that would make gasoline, oil and electricity obsolete? The good news is obvious.

- World hunger? Gone.

  "We could make the desert bloom through irrigation," says Norm Rubin, Energy Probe's director of nuclear research.

  "We could plant soybeans at Antarctica, because we could dial up our own weather" with fancy high-tech wizardry that is still too expensive even to study, let alone apply.
- Buildings like the Skydome would become superfluous because all cities would have a dome over their entire area.
- Say goodbye to the household meter reader and eliminate heating costs from your budget. It'll be too small to bother with.
- No more heart-breaking scenes of parched children on Metro beaches, towel in hand, staring at a lake they are not allowed to enter for fear of toxic contamination, despite a ruthless heat wave.

Even if there were a heat wave under this new urban umbrella, Rubin says that "if price were no object, we could extract toxins mechanically out of the water," using, one imagines, a sort of high-tech chemical vacuum cleaner.

- Throw out your gasoline credit cards. They'd be sentimental keepsakes from the days when people actually burned gasoline to get around. That's all they would be. Cars would run on something else.
- Feel free to buy a car you'd want to drive all year round. It wouldn't rust, Rubin figures, because "we wouldn't need road salt to melt snow if energy's free. We'd just heat the pavement and not worry about the cost, since it would be so little."
- Don't worry about winter tires. There would be no blizzards to glaze the pavement, because we could just "dial up the weather under the dome," says Rubin.

In short, no more hunger, heating bills, weather reports.

So why isn't Norm Rubin smiling?

It isn't just scientific skepticism.

"I'm convinced these guys aren't just dreaming in technicolor, that something really is happening," says Rubin.

* * *

He is referring to the dramatic revelation in March of chemist Stanley Pons of the University of Utah and chemist Martin Fleischmann of Britain's University of Southampton that they had produced fusion (merging) of the nucleus of two hydrogen atoms, using common, inexpensive lab equipment, rather than the costly, mammoth apparatus scientists traditionally have used when trying to attempt nuclear fusion.

For every watt Pons and Fleischmann put in to this no-frills cocktail of treated sea water ("heavy water") and simple rods of a chemical called palladium, they claim four watts came out.

That's why it could be so cheap to produce power on a widespread basis—because it actually generates more energy than it's fed.

And although many scientists around the world are having trouble duplicating the results of this experiment, some other universities in both Utah and the Soviet Union say their conclusions point in the same direction.

Make no mistake—some people are thrilled at the possibilities. Canadian newspapers from Toronto to Ottawa and Winnipeg, for example, have long breathless stories in the past two weeks, suggesting applications that smack of Disneyland's futuristic World of Tomorrow displays.

Small wonder people are excited about this new discovery.

"It's the first good news anyone has had about energy in nearly two decades," says energy consultant Charles Komanoff of New York City, director of Komanoff Associates.

"This is the first glimmer of hope people have had about energy in ages. For 15 years, people have heard that energy damages the environment and is in limited supply."

So why isn't there wild celebration among the people who have devoted their lives to promoting safe energy alternatives?

* * *

At first glance, groups such as Energy Probe and Friends of the Earth should be throwing parties over cold fusion. The process looks like a perfect alternative to fossil fuels that are being blamed for heating up the atmosphere to a critical degree. But Rubin of Energy Probe is far from happy.

"If this really is fusion that's happening, it's very bad news for the planet," he says.

Why?

Because fusion creates neutrons, and neutrons are one of the key ingredients in making atomic and hydrogen bombs.

Not that people would make bombs out of their basement cold-fusion gizmo. But in the wrong hands, the wrong people might be tempted, critics say.

"Fusion is a source of very fast energy, and high-energy neutrons are bad for you," says Rubin.

"Neutrons are one of the things needed to change non-radioactive materials, including weapons, into radioactive materials."

Similarly, if one somehow added the fairly common chemical lithium to his or her fusion-fired furnace, you'd get tritium—which Rubin warns is the main ingredient for hydrogen bombs.

"Listen—if the thought of this so-called cold

fusion meant just safe cheap energy, I'd would say 'Yippee!'" explains Rubin.

But when something seems so cheap, it looks like a high-tech new version of the "chicken in every pot" policy, it cries out for serious thought, he says.

"I wonder about the wisdom of our species to act responsibly, if we put an energy source in the hands of everybody," says Rubin.

"We worry already about whether or not (United States President) George Bush will use good judgment when considering pushing that button (that would activate a nuclear attack on another country).

"What if everyone had the makings—not the finished product, but the materials—to create his own 'button? ' And what if people threatened to push that button every time they didn't get what they want?"

A dreary thought that far from dominates Rubin's mind these days.

"When you think of the good you could do with unlimited, free energy—like feed the entire world (largely through sweeping irrigation that is now too expensive); when you think of energy that truly is 'too cheap to meter,' it's an amazing dream," says Rubin, "and once you start running your hands together, you can't help but cackle.

"But 'energy too cheap to meter' combined with neutrons is not fun."

---

*The Toronto Star* (April 15, 1989): D1, D6

## Fusion theory might solve Earth's riddle

### Jack Miller

The "cold fusion" euphoria sweeping the world has ignored what could be one of the greatest scientific discoveries ever made about our planet.

If the little-noticed theory is proven correct (and it looks strong), it means:

- The same sort of nuclear fusion reactions that power the sun go on, at a slower rate, deep in the Earth, and always have;
- You could almost call the planet a cool star, because it generates heat of its own this way.

When two scientists announced in Utah last month that they seemed to have normally deadly fusion fires burning harmlessly in a laboratory jar, it seemed they had corralled the power of the sun. It now seems they actually may have corralled the power of the planet.

Almost no one talks about that, because almost no one ever thought the planet had any such thing to corral.

This "cold" fusion, as it was quickly (and probably wrongly) dubbed by an army of instant fans, may never produce energy concentrated enough to generate cheap electricity everywhere.

The cheap electricity dream is the angle that has got the world so excited and if it fizzles out, the world may soon dismiss the whole idea as a quirk and concentrate on the next big news break.

But the discovery of fusion within the Earth could prove correct regardless, and boost our understanding about how the planet functions, and maybe even how we could put that understanding to work for us.

The problem about where excitement should be focused goes back to what now seems a full-blown feud between two teams of researchers in Utah. One team was at the University of Utah in Salt Lake City, the other at Brigham Young University in the city of Provo.

The Salt Lake City team was headed by professors Stanley Pons of Utah and Martin Fleischmann of Britain's University of Southampton. The Provo team was headed by professor Steven Jones, with professor Paul Palmer as an enthusiastic member who joined the work after seeing it might provide answers to giant questions that had bedevilled him for years.

Both groups were studying the phenomenon of low-intensity nuclear fusion. Late last year, it's reported they heard about each other's parallel efforts, got together to talk, and agreed to submit official papers on their work and their findings to *Nature* magazine simultaneously on March 24 this year.

Instead, the day before, on March 23, Pons and Fleischmann announced their findings at a press conference. They justified jumping the gun by saying rumors of their work had been leaking out and they were afraid someone else might hear about their idea and steal it before they could establish a patent on their process.

At Brigham Young the Jones team, presumably aghast, took the quieter and more conventional tack of submitting their paper to *Nature* that day—also one day ahead of the earlier-agreed deadline.

Both papers normally would have gone through the prestigious magazine's process of tough scientific analysis to see whether they made sense and deserved to be published.

But Pons and Fleischmann had sidestepped this and got world attention. As well, Pons and Fleischmann reported more exciting results.

Nuclear fusion happens when two lightweight atoms (usually hydrogen) weld themselves together to make a single atom of some heavier element (helium or tritium). Each such reaction gives off heat energy, and normally also throws out a neutron or a proton, a form of radiation.

Fusing two atoms into one would not produce enough heat or radiation to worry about. But when billions or more pairs of atoms fuse per second, enough heat comes out to be useful, and enough radiation particles come out to be deadly.

At Salt Lake City, Pons and Fleischmann said they had produced what they believed to be fusion reactions in a simple contraption, giving off lots of heat and apparently little or no dangerous neutrons.

They used heavy water, whose molecules link oxygen with a heavy form of hydrogen called deuterium. They put this liquid in a glass, put palladium and platinum electrodes into it, and ran an electric current through the heavy water, using the palladium as the negative terminal. This would cause positively charged nuclei of deuterium atoms to be drawn to the palladium metal.

It was assumed these deuterium nuclei dissolved into the metal (gases will do that) and that in time, enough would be in there to be squeezed tightly enough to fuse, giving off heat.

Pons and Fleischmann said the heat energy they got out of their jar was four times the electrical energy they put in to make the thing work. If scaled up, that could produce enough heat to boil water economically, to drive electricity generators cheaply.

Meanwhile at Brigham Young, Jones, Palmer and friends were claiming they had produced fusion in the lab, too. But they had found relatively little heat produced—NOT enough to make cheap electricity a genuine promise. But they had not staged a dramatic news conference and, as a result, got almost no attention.

Since then, scientists around the world have been trying to duplicate the Pons-Fleischmann procedure, to prove it right. Most have got no results at all, but a few have reported spotting neutrons.

This has generally been reported as evidence Pons and Fleischmann were right, but in fact these reports have found evidence of fusion at the lower, less heroic level, reported by Jones. In the gleeful hysteria of the crusade, this distinction has not been made clear.

Meanwhile, buried on Page 7 of the Jones team's paper, is the reasoning supplied by Paul Palmer about what it means to have found fusion can happen far removed from raging heat-and-pressure conditions that promote it inside stars.

This could explain why the planet's core boils. Most people have assumed that heat in the Earth's centre dates back to the chaotic pressures created when the planet was forming 4.5 billion years ago. In that scenario, Earth's crust has acted like a giant thermos bottle, keeping the heat in.

Now, we question that.

This may be far less intense, with maybe fewer than one atom-pair fusing per second rather than the trillions-per-second found in stars, Palmer points out. But over the multi-billion-year history of Earth, that could account for enough heat to explain why the heart of the planet is so hot.

Palmer never has agreed with conventional "thermos bottle" thinking.

Earth's centre should be a lot cooler by now, he says, unless something is happening down there to keep generating new heat. Also, there is too much of the element helium-three deep in the Earth. And high levels of the element tritium were found repeatedly in gases spewing from erupting volcanoes.

When Palmer heard Jones, at his university, was exploring the idea of fusion occurring in relatively cool, calm, un-star-like conditions, a light clicked on in his head.

Fusion must be going on under the Earth's crust, he assumed, fuelled by atoms of deuterium in sea water that seeped down through cracks in the crust and then was spewed back up through volcanoes.

*Malcolm W. Browne, Louise Brown, Jack Miller, Stephen Strauss, Christopher Reed, and the Associated Press*

The helium-three in the core would be a product of that fusion, he reasoned. So would the tritium in those volcano gases.

---

*The Toronto Star* (April 15, 1989): D6

# Fusion Confusion: Scientific Proof is Proving Elusive

## Stephen Strauss

They sought it with thimbles, they sought it with care; they pursued it with forks and hope.
Lewis Carroll, *The Hunting of the Snark*

Forks and thimbles, hope and care about sum up the experimental snark hunt that surrounds the potentially revolutionary report of so-called cold fusion.

The process, which was initially trumpeted as so simple that it could be done in a slightly upgraded high school laboratory, is now confounding most of the scientists who are trying to reproduce it.

Instead of a little heavy water, two metal rods, car batteries and some garden-variety heat and radioactivity measuring devices, researchers in the United States are resorting to commandeering multi-million-dollar neutron detectors that were originally designed for particle accelerators. In Canada, University of Toronto researchers are wondering out loud whether it makes sense to build a special, low-level neutron detector to get around their failure to produce results.

What has gone wrong? Why haven't the thousands of scientists slaving away night and day around the world been able to tell us incontrovertibly whether a seemingly simple process works? Certainly it cannot be a lack of motivation. If the cores of atoms can be made to fuse together at room temperature, and if the process can be scaled up, fusion promises cheap, non-polluting, almost non-radioactive energy.

"I think this discovery is one of the same order as the discovery of fire," burbled an enthusiastic Carnegie Mellon researcher two days after Stanley Pons of the University of Utah and English colleague Martin Fleischmann of the University of Southampton held a press conference in Utah on March 23.

Maybe scientists already know the answer but, with typical cautiousness, are waiting to reconfirm six different ways from Sunday before going public. On his behalf, Prof. Pons believes proof is already at hand. On Monday, in Salt Lake City, he told reporters that partial confirmations had been relayed to him from 60 laboratories around the world. However, these laboratories, he went on to add, didn't want their names made public.

If true, this shyness is probably more than the usual scientific reticence. Of the two laboratories in North America that have gone public on the issue, Texas A & M University and Georgia Institute of Technology, one (Texas) has strongly hedged its findings and the other (Georgia Tech) later retracted them. More troubling, say other scientists, is why the respected Atomic Energy Research Institute in Harwell, England, with whom Prof. Fleischmann has been personally working, hasn't come forward to verify the sensational, original work.

A second major problem is what exactly is being proven. "Two things are intertwined in a misunderstood way in the scientific community as well as among the general public—neutrons and heat production," said Alastair Miller, a chemical engineer with Atomic Energy of Canada Ltd.

What the physicists want is some clear, radioactive fingerprint that proves fusion has taken place. When the cores of two heavy hydrogen atoms, deuterium and tritium, combine into one, they produce energy, gamma rays, neutrons and radioactive forms of helium.

If these elements are found in higher-than-normal background levels in experiments by many laboratories, the physical phenomenon of room-temperature fusion would be established. "If it is fusion, you have to see nuclear debris," said Dominic Ryan, a physics professor at McGill University who is trying to replicate the Utah work.

Yet all kinds of neutron-counting problems have cropped up. The Georgia Tech researchers found that heat in their reactions made their neutron counters mistakenly report neutrons. There is naturally appearing radiation—cosmic rays and radon in cement among others—that can vary from geographical location to geographical location or room to room in a given laboratory. When tiny amounts of fusion are reported—a supposed corroboration in Hungary found the reaction producing 3 neutrons per second as opposed to 1 normally—the chance of some kind of systematic mistake is high.

Even if these difficulties can be overcome, engineers are not particularly entranced with the usefulness of the neutron proof. They have been more interested in the phenomenal energy production that the Pons-Fleischmann group also reported. For every watt of electrical energy used to produce the reaction, four watts of heat, later bumped to eight, were reportedly produced.

"To be useful, as opposed to scientifically Interesting, one must have high temperature heat production," Mr. Miller said.

It is this a-lot-for-a-little reaction that promises energy salvation, particularly when compared with a similar fusion report by Steven Jones at Brigham Young University. He suggested that only a trillionth of a watt of heat energy had been produced. Yet, here a scientific split appears. Many scientists are not impressed with heat without fusion.

"It is the least convincing evidence," said Michel Potissou of the TRIUMF particle accelerator facility in Vancouver. Workers there have also been trying to replicate the experiment. The problem is that various heat-producing chemical reactions are known to take place when heavy hydrogen atoms dissolve into metals. This suffusion is apparently the basis for the Pons-Fleischmann fusion reaction.

If scientists have been fussing about how to prove fusion, it is a relatively straightforward debate compared with the question of how to disprove it. "When you see nothing, when do you know that you have covered all the possibilities?" asks David Bailey, a University of Toronto physicist who has been fruitlessly working to reproduce the original result.

"The problem with this is that it is much more difficult to disprove than prove. When you go around and say I couldn't do it, someone else can say 'well I can,'" Prof. Ryan said.

Part of the frustration over clearly proving or disproving the discovery lies in misinformation—either deliberate or accidental—from both the initial discoverers and the people who may have duplicated the work. The Pons-Fleischmann paper on their work leaves out key points. How much total energy did they use to make it operate? How long did it take to begin operating? What added chemicals did they put in that may have made a difference?

"We now know from the bush telegrams (unofficial word) that this is crucial," said James Prentice, another physicist at the U of T. He compares the effort to reproduce work without knowing all its parameters to the problem of someone put in a seven-dimensional world and told 'find your way to the lady imprisoned in the forest.'

The feeling is that some of this information has been left out because of the discovery's commercial implications. "Anyone who sees heat immediately heads to the patent office and doesn't want to talk to anyone else," Prof. Bailey said.

With all this doubt, when will the question of whether the fusion report is a breakthrough or a booboo be resolved?

"I suspect this will be one of those things which just peter out. People will find no confirmation and interest will peter away," Prof. Ryan said.

Mr. Miller, who was quite hopeful the week of the announcement that the phenomenon was real, reflected that his own perceptions had turned negative. "Maybe we've been swept away."

However, just when many scientists were thinking the worst, another of the supposed confirmations was reported Tuesday by Italy's state nuclear energy body Enea. "That's the way it has been. We've been going up and down like a yo-yo," said Mr. Miller when told the news.

The Globe and Mail, Saturday, April 22, 1989 D1, D4

# Computers, Faxes Spreading Fusion Fever Among Researchers

## Stephen Strauss

While the public may have become fixated by vistas of limitless energy pouring forth from a test tube, in light of recent news reports, it is the puckish, perceptive spirit of Marshall McLuhan which has hovered over short-term implications to science of room-temperature fusion.

"The mediums were the message" in researchers' feverish, month-long efforts to learn about the possible new fusion process.

For example Stanley Pons, the University of Utah chemist who co-authored the seminal papcr in the field, was asked in Indiana how could he obtain a copy of the report he had read about only in the press.

"I guess you're not on Bitnet," Mr. Pons said jokingly.

Bitnet is one of several computer bulletin boards that lit up like a second sun when word of the supposed fusion discovery first appeared. Large sections of the research paper had been entered into it and electronically zipped around the world by information-hungry scientists.

Four days earlier, Steven Jones, a Brigham Young University fusion researcher, told about 300 people at a seminar at Columbia University that he had sent out about 20 copies of a paper he had done on cold fusion to selected people. How many people have a copy? he asked.

About 20 hands went up.

"I guess every one of them is here," Mr. Jones said jokingly.

The modern facsimile machine seems to have accelerated the spread of the copies. James Prentice, a University of Toronto physicist who has been trying to replicate the initial experiment, says it is clear how the excitement over fusion differed from earlier potential breakthroughs of which he was a part. "It's only six months since everybody has had a fax machine," he said.

While it seems like an inherently good thing—the quick spread of information about a hot and potentially revolutionary topic—how the technologies circumvented the stately, generally slow process of scientific verification and publication is already causing worry in the community.

"It may turn out to be a horrible lesson in how not to do science," said Alastair Miller, a chemical engineer with Atomic Energy of Canada Ltd. "New technologies have appeared that suggest different ways of doing things, but so far they haven't demonstrated much benefit."

In the traditional method of science there are clear stages of prepublication. A researcher submits his or her paper to a journal. If it appears worthy, the journal asks anonymous referees, other scientists, to judge the merits of the work. They offer critiques and pose questions, which the submitters seek to address in resubmissions, if necessary. The resubmission would then be reviewed and more questions might be raised.

Generally none of the drafts is considered an official publication; while referees often will pass around copies of the work to colleagues in other fields for comment, they are restrained from generally distributing them. Researchers can correct, change, or withdraw information in this stage without any loss of face or anyone generally knowing.

This has not been the case with fusion. It is not clear whether the University of Utah researchers surreptitiously released early versions of their paper or whether reviewers, pressured by their colleagues, decided to break the rules and pass it on.

The difference between information, rumor and misprint blurred in the hunger of researchers to get descriptions of the procedures before they had been reviewed and printed. University of Toronto physicist David Bailey has been trying to recreate the experiment from a bootlegged, faxed copy of the paper by Prof. Pons and Martin Fleischmann. He does not know whether he has an early draft or a final copy. "I think they are off by a factor of a million," he said about one result, "but it's the fax quality maybe."

Whatever the outcome of the fusion research, it is clear that its possible revolutionary implications stirred the scientific community's imaginative unconscious. "There was a lot of stream of consciousness going on. James Joyce would have been proud," Prof. Prentice said.

Indeed one of the early literate memo writers on the Usenet bulletin board may have provided the literary reference that sums up the finding. He quoted Chaucer: "Out of olde boxes, in good feyth, cometh all this newe science."

*The Globe and Mail* (April 22, 1989): D4

## Who Needs An H-bomb When You've Got A Rubber Bucket?

### Christopher Reed

Stanley Pons, who has gone from being a little-known chemistry professor in Utah to one of the world's two most controversial scientists, is happy. Now he is "99 per cent-plus" certain of his claimed breakthrough in fusion energy research.

Yesterday Dr Pons left for Britain to join his partner, Dr Martin Fleischmann, at the University of Southampton, where they will finish their eagerly awaited and somewhat overdue complete report on "fusion in a flask," the process they announced March 23, amazing the scientific community.

Before leaving, he said the paper should be ready by early August, and will "fully explain" how the two confounded international physicists by apparently producing nuclear energy in a rubber bucket at room temperature in the basement of the University of Utah's chemistry department.

During the interview, Dr Pons seemed confident, buoyant and eager to demolish his critics. That would be a considerable accomplishment, for they represent respected academic institutions around the world.

Of major US universities, only Stanford has backed the two chemists. Scientists from Princeton, Yale, the California and Massachusetts institutes of technology, and Brookhaven National Laboratory, have dismissed the Pons-Fleischmann claims as impossible.

The harshest critics have denounced their methods as sloppy and the two chemists as incompetent, inexperienced or misled. Dr Pons promises a terrible revenge.

"It couldn't make me more happy to hear those people say that what we have done doesn't work, because they have made their position clear for all of history. Those who said nasty and stupid things, who got up and immediately ridiculed us after we spent five years the project, and who said we were deluded, or incompetent, or not reputable, those people are marked forever. We will show that it is they who are deluded."

In the near future, he predicted, "you will see from us a very definitive paper with a lot of data. You will see further confirmations from major institutions, more and more positive conclusions, and increasing questions about the premature criticisms. For me it has been the biggest challenge in my life, and now it's going to just get more and more fun."

It certainly wasn't much fun in the beginning. Although the pair expected disbelief, they did not anticipate "so many people being so negative without doing their own experiments first." In the early days, he received 100 telephone calls a day "from well-wishers and those who did not wish us well, from scientists, corporations, the media, and from crackpots who had better inventions." Despite changing his unlisted telephone number three times, Dr Pons still receives 25 to 30 calls a day.

An American, Dr Pons studied in Britain under Dr Fleischmann, and the two became friends. The English chemist has been a visiting professor at Utah, where the pair planned the fusion research. Both appeared at the March 23 press conference, an occasion labelled "hype-energy physics" and lambasted as improper by the scientific establishment.

At that time, Dr Pons said, he was 90 per cent sure of his claims, but is "much more happy now." It has long been assumed that nuclear fusion—the amalgamation of atoms rather than their division, as in fission—requires temperatures rivalling the sun or a hydrogen bomb.

Physicists have spent billions using complex devices to produce hot fusion energy, but they still see decades of experiments ahead before any commercial application.

Cold fusion offers the tantalizing possibility of clean, cheap, inexhaustible energy for the world. A principal source could be sea water and, although the Utah experiments used an expensive and rare metal, palladium, others could be substituted, optimists believe.

Dr Pons and his partner continue to receive poor press. The fury of their critics has energized a long-standing division between physicists and chemists. Physicists believe they follow a superior discipline and are said to look down on chemistry as the "stinks" department. Chemists resent the seeming arrogance and claim to follow a harder-working, more practical calling.

According to Dr Pons, physicists rushed to judgment over the Utah findings and failed to apply the patience needed to replicate the Utah experiment, a required scientific procedure before acceptance of new developments. "We got crucified," he said.

Scientists (mainly physicists) also derided an early Pons-Fleischmann report to a Dutch chemistry journal for being scanty and lacking essential details and sufficient verification. Utah University's vice-president of research, Dr James Brophy, acknowledged that the initial report was "a bit thin." "However," added Dr Pons, "physicists only publish once every 10 years after their colleagues spend three years reading it over. They aren't doing the innovative things happening in chemistry. We have so much stuff to announce, we have to publish fast."

Others dismiss Utah University as a second-rate institution in a state dominated by narrow, parochial officials of the Church of the Latter-day Saints. Dr Chase Peterson, the university's principal, is a Mormon and acted as ringmaster for the media circus over the world's first artificial-heart transplant, performed on Barney Clark at Utah in 1982. However, the university directly sponsored by the Mormon church is nearby Brigham Young, where rival research was pre-empted by the cold-fusion publicity.

Neither he nor Dr Fleischmann are Mormons, said Dr Pons, and "the church doesn't really come into this." His university's inexperienced public-relations staff have often mishandled the fusion affair, but Dr Pons' irritation is directed at the "Eastern establishment," a frequent target of US provincial academe.

He feels those who emerge from the fusion affair as discredited won't be the church, the state, the university or himself and Dr Fleischmann, but physicists from Ivy League colleges—a development he doesn't seem to mind at all.

*The Globe and Mail* (June 13, 1989): A7

## Cold Fusion Gets a Hot New Reaction at Los Alamos

### Associated Press

Los Alamos National Laboratory researchers say they have discovered tritium, a by-product of nuclear fusion, in duplicating the University of Utah's purported breakthrough in producing cold fusion.

Edmund Storms, a staff member at the Department of Energy's New Mexico laboratory, said he and another researcher, Carol Talcott, found tritium in "significant amounts," 1,000 counts per millilitre, in two cells, the Deseret News reported yesterday.

On March 23, Utah chemist Stanley Pons and a British colleague, Martin Fleischmann, announced they had achieved sustained nuclear fusion by immersing rods of palladium in platinum coils in deuterium-rich heavy water, and that the experiment produced up to 100 times the energy needed to produce the reaction.

"Tritium can only be produced through a fusion reaction," Mr. Storms said.

"We are trying now to understand really what environment must exist in the electrodes to allow this to happen."

About a dozen labs around the world have reported finding tritium in their fusion experiments, but Los Alamos is the first government laboratory to do so.

"This result is of fantastic importance because tritium absolutely proves fusion. If you get tritium, there is fusion occurring," said John Bockris, a professor of chemistry at Texas A&M University.

Three labs at Texas A&M have duplicated major aspects of the Pons-Fleischmann experiments, including small amounts of excess heat. In one lab, researchers have seen the neutrons and large amounts of tritium.

The Los Alamos result, "which confirms ours precisely, is of immense, world-influencing importance," Mr. Bockris said.

He is convinced the Los Alamos confirmation will quell the controversy that has raged since the Pons-Fleischmann announcement.

*The Globe & Mail* (June 24, 1989) A4.

## Questions and Topics for Discussion and Writing

1. Why is there so much interest in the scientific community in the verification or falsification of the cold fusion claims?
2. Explain the particular problems of verification of cold fusion reactions.
3. Why does cold fusion have a bearing on the temperature of the earth's core?
4. Why is the creation of neutrons in fusion reactions alarming to peace activists?
5. Discuss the dissemination of information to the public on cold fusion research. How has this dissemination differed from the usual practice in the announcement of scientific results? What are the consequences of the differences?

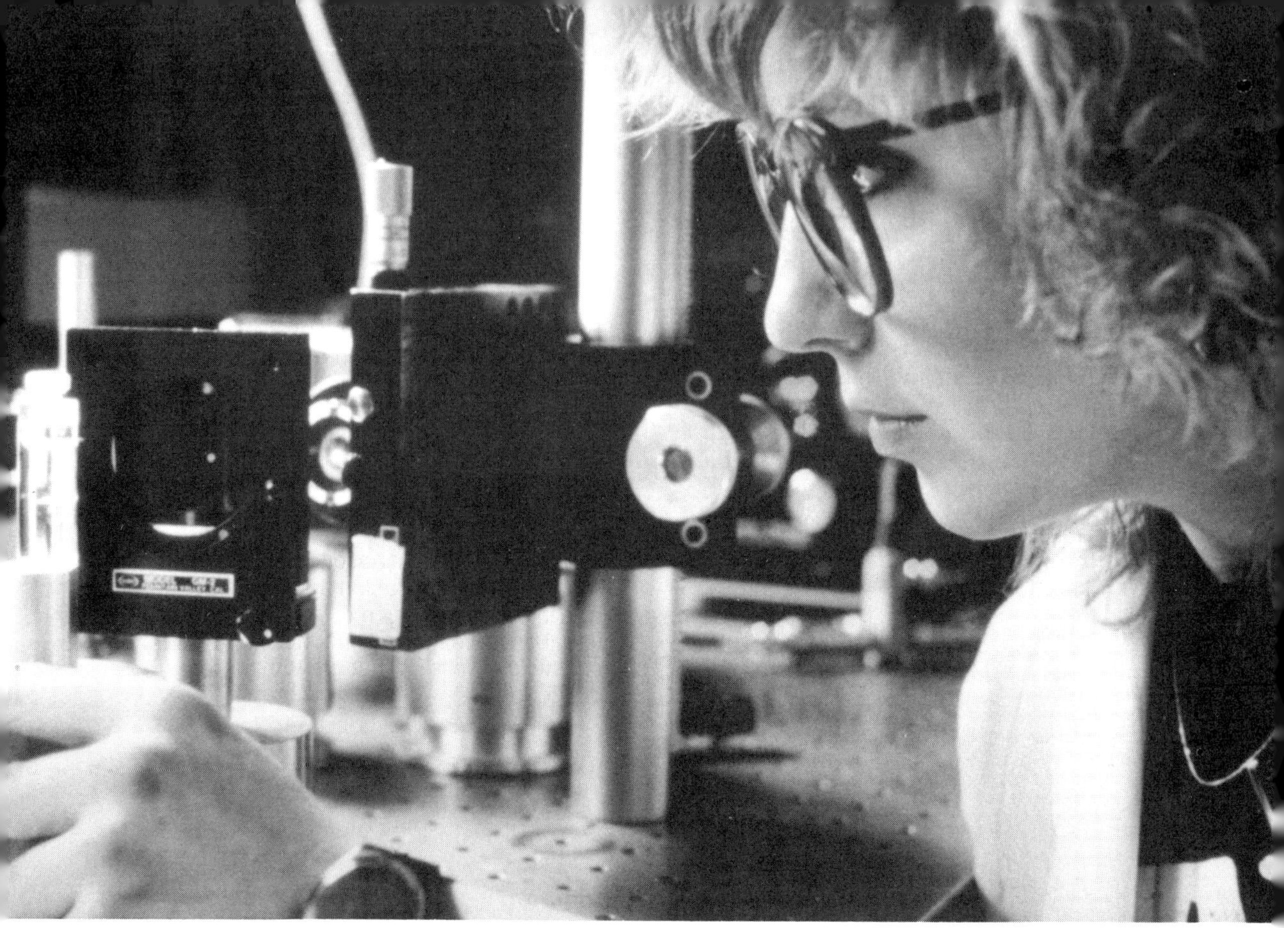

Research for biologically active molecular structures.

*Issue*

# Biotechnology and Related Issues

Here's an outrageously oversimplified history of biology as a science: In 1859, Charles Darwin published *The Origin of Species* and gave the world a theory within which to understand the diversity of life and the relationships of different species to each other. In 1900, Mendel's work on hybrids of pea plants was rediscovered, founding the science of genetics and thereby providing science with a theoretical mechanism through which the characteristics of different life forms are passed from generation to generation. In 1953, Watson and Crick work out the molecular structure of DNA, the carrier of the genetic code, thereby opening the way for a full understanding of the precise biological information that determines the physical characteristics of all life forms.

Though the above "history" is maddeningly incomplete, it highlights the major trends in biological research that have followed each event listed. After 1953, scientists naturally turned their attention to a closer understanding of DNA, how it differs from species to species, how it differs from individual to individual, and how DNA and other complex biochemical building blocks determine the characteristics of the living beings they reside in. It was only a matter of time before scientists found a way to alter the DNA and other biochemical matter in a living being and change the way the animal grew. Thus began biotechnology, or bio-engineering, or gene-splicing as it is sometimes called.

Like every other major new capability that humans have gained to manipulate nature, there are costs and benefits. The benefits are goals that are sought, or at least recognized if they are happened upon accidentally. The costs are often hidden and become obvious only after a period of time. The potential benefits are great, including finding cures for such diseases as cancer and preventing deformities and other abnormalities, as well as turning the power of nature on to industrial and agricultural applications, such as developing organisms to consume and dissolve oil spills and other pollutants on the seas, and developing new breeds of plants for food that are more productive and resistant to disease. As we are but at the beginning of biotechnology, it would be impossible to describe all the possible uses even in outline.

But what about the costs? At this point we can hardly know them specifically, but long before we heard of DNA we have been contemplating the possible costs in nightmares and science fiction stories about mad scientists. The prototype story is Mary Shelley's *Frankenstein, or the Modern Prometheus,* published in 1818. Letting our imaginations run wild may help to conceive the worst scenarios that might come to pass, but it is unlikely to be of much help in making the hard decisions that we must face about the limits, if any, which society should impose on biotechnology. It is more useful to look at the dangers that are already at hand and those that are just around the corner.

The articles in this section are all directed toward raising the question of what should or should not be permitted in genetic engineering and research. The first, "The *Double-Edged* Helix: Genetic Engineering in the Real World" by Liebe F. Cavalieri, is a general discussion of the issues that biotechnology presents, comparing the ethical position of the genetic researchers to that of the nuclear physicists whose work led to the atomic bomb. Thomas Murray's "The Growing Danger from Gene-Spliced Hormones" asks what would be the result if synthetic growth hormones were generally available to any parents who wanted their children to grow a bit taller. Stephen Strauss's "Inside Information: A Quebec Community Must Confront its Unusual Genetic History" concerns a different kind of genetic research—one not based on recombinant-DNA, but on computer-assisted tracing of lines of heredity that make possible the pinpointing of the source and dissemination of genetic disorders. The result is that science may be able to provide individuals with genetic information that would affect such personal decisions as who they should marry (or rather, who they should *not* marry) for the sake of the genes that might be passed on to any offspring.

# The *Double-Edged* Helix: Genetic Engineering in the Real World

## Liebe F. Cavalieri

In the twentieth century it has become more and more difficult to draw a line to distinguish where science ends and technology begins, and therefore more and more difficult for researchers in pure science to free themselves of responsibility for the uses to which their findings can be put. Atomic physicists have come to this realization with the horror of nuclear weapons. But no brand of modern science is immune. In the following selection, taken from the editor's Prologue and Epilogue to a book about genetic engineering, Liebe Cavalieri compares the uses of nuclear science and genetic engineering and the enormous responsibilities that accompany each.

### Twin Perils: Nuclear Science and Genetic Engineering

There is a striking similarity between nuclear science and genetic engineering. Both major scientific accomplishments confer a power on humans for which they are psychologically and morally unprepared. The physicists have already learned this, to their dismay; the biologists, not yet. Molecular biologists are still euphoric over the potential of their achievements. Indeed, Nobel laureate David Baltimore has proclaimed: "We can outdo evolution"—a signal that molecular biologists are about to translate genetic engineering into an instrument of power, much the way the physicists did when they exploited their discoveries at the beginning of the nuclear age.

The early part of this century was a golden era for physics. It was a time when physicists had become deeply submerged in finding the "truth" about the physical universe—a period of reasoned excitement that prevailed for a number of years; but it gave way to an urgent mood with the discoveries of the 1930s. There was no opportunity for the usual lapse of time between scientific discovery and its application. The discovery of nuclear fission defined the bomb. World War II and the threat of Nazi science provided an immediate rationale for physicists to tap one of nature's most powerful forces. "Truth" was converted instantaneously to power. Academic physics was to fuel the field of nuclear technology despite Ernest Rutherford's early disclaimer.

> A lot of nonsense has been talked about transmutation. Our interest in the matter is purely scientific, and the experiments which are being carried out will help us to a better understanding of the structure of matter.

But it was clear from the beginning that industrial and military uses of atomic energy were in the physicists' minds, for Rutherford also felt obliged to say:

> These transformations of the atom are of extraordinary interest to scientists, but we cannot control atomic energy to an extent which would be of any value commercially, and I believe we are not likely to be able to do so.

Leo Szilard and others did not share Rutherford's faulty judgment. The project went forward—the inexorable drive from truth to power was on. After reaching their goal, the physicists finally were left powerless with nothing to do but ponder, in a state of deep dejection, what they had done.

In the spring of 1945, Szilard wondered "what we were working for" when the Germans were close to defeat. "What is the purpose of continuing the development of the bomb, and how would the bomb be used if the war with Japan has not ended by the time we have the first bomb?" Much later, physicist Robert Wilson said:

> I would like to think now that at the time of the German defeat that I would have stopped, taken stock, thought it all over very carefully, and that I would have walked away from Los Alamos at that time. And I—in terms of...everything that I believe in, before and during and after the war—cannot understand why I did not take that, and make that act. On the other hand, it simply was not in the air. I do not know of a single instance of anyone who had made that suggestion or who did leave at the time. There might have been someone that I didn't know, but at the time it just was not something that was part of our lives. Our life was directed to do one thing. It was as though we had been programmed to do that, and we as automatons were doing it.

Frank Oppenheimer, brother of J. Robert Oppenheimer, has said:

> Amazing how the technology tools trap one, they're so powerful. I was impressed because most of the sort of fervor for developing the bomb came as a kind of anti-fascist fervor against Germany. But when VE Day came along, nobody slowed up one little bit. No one said, "Ah well, the main thing—it doesn't matter now." We all kept working. And it wasn't because we understood the significance against Japan. It was because the machinery had caught us in its trap and we were anxious to get this thing to go.

Clearly, physicists kept on going for all the obvious human reasons. Freeman Dyson summed it up accurately and concisely.

> I have felt it myself. The glitter of nuclear weapons. It is irresistible if you come to them as a scientist. To feel it's there in your hands—to release this energy that fuels the stars, to let it do your bidding. To perform these miracles—to lift a million tons of rock into the sky. It is something that gives people an illusion of illimitable power and it is, in some ways, responsible for all our troubles. I would say this—what you might call technical arrogance overcomes people when they see what they can do with their minds.

But the physicists were naive. Scientific prowess faded into insignificance in the face of political power. Szilard and others tried to stop the use of the bomb on Japanese cities, but by then it was too late. The power to create the bomb, originally in the hands of physicists, led insidiously to a new kind of power. The military stepped in, superseding any possibility of decisionmaking by the physicists. Those who had created the technology in the first place were cast aside, left only to reflect on the inhuman consequences of their work. The feeling of the physicists was put most poignantly by J. Robert Oppenheimer.

> We knew the world would not be the same. A few people laughed. A few people cried. Most people were silent. I remembered the line from the Hindu scripture, the *Bhagavad Gita.* Vishnu is trying to persuade the prince that he should do his duty, and to impress him, takes on him multi-armed form and says: "Now I am become death, the destroyer of worlds." I suppose we all thought that, one way or another.

As nuclear physics was approaching its climax, the golden age of molecular biology was just beginning. DNA was shown to be the genetic substance in 1944; in 1953 another milestone was reached, when James Watson and Francis Crick proposed the DNA double helical structure, with its molecular basis for genetic continuity. This was the dawn of molecular genetics; a new paradigm was born.

As in the development of nuclear physics, once the elegance of the new discoveries had been felt and the intrinsic beauty of that corner of nature revealed, the enticement for further investigation and exploitation proved irresistible. Indeed, the science and its applications are so closely coupled that they are virtually one and the same. Like the discoveries in atomic physics, truth and use are hard to separate. Molecular biologists have succumbed to intellectual and emotional forces just as the physicists did, with medical and economic pretexts in place of the Nazi threat. No one heeds the trap that befell the physicists: scientists can choose and control their experiments, but once the science has escaped from the laboratory, they are no longer in control.

Already, an academic-industrial complex has been formed. More and more, the direction of university research is open to influence by commercial considerations. In the normal sequence of events, large corporations will gobble up the smaller, scientist-controlled operations as soon as they prove profitable. Nobel laureate Walter Gilbert says that the genetic engineering venture will be different from other commercial forays—scientists will be in control. Well, perhaps, for a while. It would take far more profound social changes than that to prevent the transfer of power to the corporate structure. The fascination of the endeavor, the "glitter" of genetic engineering and the "illusion of illimitable power" are again in the air; after chemistry and physics, it's now biology's turn. Few have asked with Erwin Chargaff, "Have we the right to counteract, irreversibly, the evolutionary wisdom of millions of years, in order to satisfy the ambition and the curiosity of a few scientists?...My generation, or perhaps the one preceding mine, has been the first to engage, under the leadership of the exact sciences, in a destructive colonial warfare against nature. The future will curse us for it."

There was an attempt not to go all the way with atomic fission. In his initial apprehension, Szilard was clairvoyant when he tried to preserve his idea of a nuclear chain reaction by patenting it so that it could remain secret. And indeed it seemed at first that a similar social action would occur with recombinant DNA technology. In a spirit of humanitarian concern, when molecular biologists first perceived the impact of recombinant DNA technology, they decided to convene a conference in 1975 to seek a modus operandi. The serious and sincere discussion at the outset concerned only the immediate hazards of the research; future effects of the technology did not enter the discussion in any meaningful way. No public health experts, ecologists, ethicists, or philosophers were invited, and the questions discussed were those most intimately connected to the research projects of the molecular biologists in attendance. The discussion was, therefore, not broad-based and it revolved about the possible escape of a pathogen from the laboratory into the environment. No long-range ethical and social consequences were considered.

The ad hoc discussion at the 1975 conference served as a basis for guidelines for recombinant DNA research, first promulgated by the National Institutes of Health in 1976. Since then, the guidelines, under constant pressure from interested parties, have been relaxed to the point where they afford virtually no protection. In extricating themselves from government control, molecular biologists crossed an energy barrier and have become adept at political manipulation. the hope for meaningful public input on the ethical and moral implications of this technology has faded as the momentum of genetic engineering has grown.

With the spectre of Nazism before them, physicists felt it necessary to develop atomic weapons. Molecular biologists, in choosing to push full-speed ahead with genetic engineering, cannot so convincingly rationalize their war on Nature. In this period of unparalleled ecological stress, with so many socially important areas of inquiry unexplored, there is no overriding need to develop a technology that is pregnant with unpredictable consequences. All forms of life are vulnerable to this technology—any DNA can be connected to any other DNA; human DNA can be put into viruses and bacteria and vice versa; cancer virus DNA has already been put into bacteria, and so on. The gene pool of the Earth, the life-determinant of the future, is the experimental subject for genetic engineering. This precious, irreplaceable legacy of natural evolution is in the truest sense a one-time occurrence, and it would be naive to assume that we can manipulate it without harming ourselves. We do not have the infinite wisdom that would be required.

In the face of the infinite complexity of natural systems, the idea that we could improve on the design of nature is not only hubris, it is frightening. In Lewis Thomas' words, we are ignorant "most of all about the enormous, imponderable system of life in which we are embedded as working parts. We do not really understand nature at all." We know that the Earth behaves like an indivisible, delicately tuned mechanism, in which the inanimate environment is strongly conditioned by living things, and vice versa; but we have only begun to decipher the influence of each part on the whole.

For example, we recognize that certain microorganisms convert organic wastes to usable nutrients, and that this recycling process is critical in maintaining the composition of the atmosphere and other conditions favorable to human life and to the web of species that sustain us. But we cannot predict the effects on these vital microorganisms of accelerated evolution, engineered by man, coupled with the accelerated environmental changes now produced by human activities—such as the production of carbon dioxide on a vast scale from fossil fuels, the distribution of novel chemical pollutants around the Earth, the large-scale clearing of forests, displacement of biological diversity by a minuscule number of cultivated species, and so forth.

However, as the result of current efforts to design industrially useful organisms, microorganisms with properties taken from higher forms of life will inevitably escape into the ecosphere; other engineered forms will eventually be released intentionally into the environment for purposes such as the solubilization of trace metals in mining operations or the digestion of oil spills. We are laying the groundwork for unforeseen evolutionary changes that may create an environment inhospitable for present species.

Frequently, one is confronted with specious arguments about how well evolutionary forces have managed thus far and how they will continue to provide viable ecosystems. Certainly, we can find some assurance in nature's resiliency; life has survived environmental upheavals for millions of years. But as conditions have changed, so has the balance of life, with incompatible forms disappearing and new ones arising.

If there were a drastic change in the environment, some forms of life would undoubtedly adapt, but humans, with their many, exacting biological requirements, could not evolve fast enough to become compatible with the new environment.

Genetic engineers have not overlooked the possibility of changing humans themselves. It will not be long before single gene replacement therapy—the correction of a defective gene—will be possible. Although in this case the change will die with the patient, more radical experiments are underway in which eggs or sperm are altered to produce individuals with hereditary alterations. Considerable success along these lines has already been achieved in mice. The rationale for these experiments is that they provide information about mammalian genetics and fetal development. But when the technology for intervention in human evolution has been perfected, will it remain unused? Preliminary experiments with human embryos have been underway in England, for example, for several years. What is more seductive than the power to design human beings?

Although the repair of genetic defects appears laudable, the indistinct boundary between repair and improvement raises serious problems. Who is to decide what qualities define a perfect human? And even if there were a consensus today as to which genes are desirable, it certainly would not ensure that future generations would hold the same views. Clearly, the present cultural milieu is an important determinant of present values. In a changing world, the genetic engineering of perfection would imply a divine intelligence that could peer far into the future.

In "The Fate of the Earth," Jonathan Schell poignantly describes the outcome of the use of nuclear weapons. Even if humanity were somehow to survive, the living conditions on Earth would be radically changed. The gene pool of the Earth would have been largely destroyed; diversity of living organisms, so important to ecological stability, would be minimized if not wiped out. Schell says

> A full scale nuclear attack would devastate the natural environment on a scale unknown since early geological times, when in response to natural catastrophes whose nature has not been determined, sudden mass extinctions of species and whole ecosystems occurred all over the earth. . . . It appears at the outset that the United States would be a republic of insects and grass.

Because of human genetic damage,

> Over the decades not only would the survivors of a limited attack face a contaminated and degraded environment but they themselves—their flesh, bones, and genetic endowment—would be contaminated: the generations that would be trying to rebuild a human life would be sick and possibly deformed generations.

Finally, Schell says

> In weighing the fate of the earth and, with it, our own fate, we stand before a mystery, and in tampering with the earth we tamper with a mystery. We are in deep ignorance. Our ignorance should dispose us to wonder, our wonder should make us humble, our humility should inspire us to reverence and caution should lead us to act without delay to withdraw the threat we now pose to the earth and to ourselves.

In an initially more subtle way, genetic engineering also threatens to create an ecology that is inimical to life as we know it. In the effort to produce more "perfect" plants, animals, or humans we may be headed for a monolithic, undiversified, unstable ecosystem—achieved, however, without a fireball.

The discoveries that energy can be released from

the atomic nucleus and that DNA, the material of the cell nucleus, is the genetic stuff of life are without parallel in human experience. These twin scientific feats demand a new consciousness if human life on this planet is to continue. We have mismanaged the applications of the first discovery. Now, as the second is about to be exploited, we must not permit the biosphere, surpassing as it does our understanding, to become an experimental subject. There is only one Earth, one earthly biosphere, and we are part of it. There is no margin for error...

**Genetic Engineering: A Moral Dilemma**

In the mid–1940s the world witnessed the fruit of nuclear research when the first atomic bomb was exploded. At about the same time we learned that DNA is the genetic substance. Within a matter of months we were presented with two great secrets of nature: one at the core of matter, the other at the core of life. These discoveries, qualitatively different from any we had known previously, thrust us toward new boundaries in human affairs. The advent of genetic engineering—the manipulation of the DNA of living organisms—has accelerated this process. We have before us now a moral dilemma of extreme dimensions.

The problem is that the power of science and technology has grown enormously in recent times, while our psychological and social framework has remained stagnant. Thus, nuclear arms make war obsolete as an instrument for solving political conflicts; yet we act as though this were not so. In the same way, scientists, corporations, and government are treating genetic engineering as though it were an ordinary technology when in fact it has created a new paradigm that demands a new moral consciousness.

Should we proceed with this technology that appears to hold great promise, even though its potential for altering life reaches beyond the scope of our finite wisdom? Some are inclined to answer yes, citing the blessings reaped by humankind from the free technological development of the past. Others observe that there has been a quantum increase in science in the last several decades; it has probed so deeply into the nature of things that we are now capable of altering, often in irreversible fashion, the basic natural order. The applications of science have become increasingly intrusive. Recombinant DNA technology is the tool par excellence for rearranging life's substance. Genes can be split from one another and shuffled about at will, enabling a living organism to acquire characteristics that are normally foreign to it. There are no biological barriers to this maneuvering. Nothing is scientifically forbidden. Accordingly, the temptation to create unique molecular biological structures has increased enormously in recent years. But is it justified to put diphtheria toxin genes...into E. Coli, a bacterium that normally lives virtually everywhere in the environment, including the human gut?

There has also been a change in society's perception of what science can do. Scientific and technological approaches are now more heavily relied upon for solutions to problems that are social or economic in origin. Robots are seen as solutions to manufacturing problems. "Star wars" is the technical solution for peace. A search for root causes is not part of standard operating procedure.

In another new area, human in vitro fertilization and embryo transplants, are being encouraged with no thought given to possible future consequences. These technologies have profound implications. For one thing, man's intervention in human conception opens the possibility of sex choices, which could well lead to change in the sex ratio, with enormous sociological impact. But the short-term view is winning out as usual; hospitals are gearing up for these activities. A commercial firm is already in the business of selling frozen human embryos.

All of this glitter entails heavy risks, which have become part of the bargain. The long-term social and human costs are rising. What price the splitting of the atom? But should we have stopped, back in the 1930s, the Rutherfords, the Bohrs, the Fermis, the Einsteins—the fathers of nuclear science? In the context of those times, surely not. Should we question, regulate, or even try to stop their biologist counterparts today, knowing what we now know? This is a hard question, for we immediately find ourselves confronting the question of freedom of inquiry. Scientists will be among the first to agree that technologies should be monitored and perhaps regulated. But Recombinant DNA technology presents a unique situation because the fundamental science and its applications are in many instances one and the same. The situation is made more

*Liebe F. Cavalieri*

complex by the fact that conflicts of interest have emerged, for many molecular biologists practicing recombinant DNA technology are in university laboratories that are supported by corporate funds. Freedom of technology masquerades as freedom of scientific inquiry.

As early as 1977, a protocol was described for the treatment of blood disorders such as sickle cell anemia or thalassemia using gene therapy. The cells of an afflicted individual would first be removed from the bone marrow and the appropriate genes transplanted. The cells containing the new gene would then be returned to the individual and the inserted gene would supply the normal blood protein. A premature and unsuccessful attempt at this type of therapy was made in 1980. Since then new scientific discoveries have led to new strategies for delivering the normal gene. Before long, one of these procedures may be ready for new trials. Should the treatment prove successful the treated individuals would still remain carriers of the defect, passing it on to their children. Treatments that would produce hereditary cures in which the genetic defect would not be passed on have also been under development for some time.

In 1980 it was shown that foreign DNA could be incorporated into the DNA of a living mammalian cell by microinjection. This experiment created great excitement and it initiated a new round of escalation in genetic engineering experiments. Rabbit globin genes and rat growth hormone genes were put into fertilized mouse eggs in 1982. The rat growth hormone gene, and in later experiments the human growth gene, caused the mouse to grow to at least twice its normal size. These transplanted genes were found to be passed on to the offspring of the treated mice. The possibility of intervention in the germ-line has therefore become a reality. This raises thorny problems, for the same techniques could also be used to "improve" segments of the human race, to select, eliminate and design characteristics according to arbitrary standards. Some scientists say that they are opposed to intervention in the human germ-line to produce hereditary change; others are not. This question requires careful scrutiny, for the line between "cure" and "improvement" is not clear-cut.

The purpose of the original microinjection experiments was to learn how organisms develop, yet simultaneously they perfected the techniques that make possible the engineering of human beings. The science and its application turn out to be inextricably linked. Genetic manipulation is therefore a double-edged sword. Will scientists and the rest of the world community be able to resist the Faustian temptation to push onward?

Television provides a powerful medium for bringing the tragedy of crippling genetic disease dramatically to public attention; consequently the thought of the sick child who potentially could be cured by means of gene therapy tends to dominate our vision, and the more fundamental human issues are obscured. As a society we tend to rush to the aid of single individuals once our imagination is captured. If we learn that 50,000 children, mostly in Third World countries, die daily we are not similarly moved; faceless millions and statistics evoke a minimal reaction. A distortion of values is also evident when comparing the cost in terms of human resources, talent, time, and money of sophisticated medical or surgical procedures that can be available only to relatively few individuals, with much simpler procedures that would benefit large numbers of people suffering from lesser ailments. Of course if resources were unlimited the problem would not exist. For any disease, if a cure exists it ought not to be denied anyone. But the situation is never quite that simple. And in the case of genetic disease if the cure is not yet at hand, how much effort should be devoted to developing one, recognizing the multiple urgent problems that compete for scientific attention and knowing that each genetic disease would require the development of a complex new methodology?

Moreover, how do we deal with the dilemma created by developing genetic engineering knowledge that eventually will place the genetic future of humans and all other species in the hands of a few human beings?

In the context of today's society the pressure to use all available knowledge mounts inexorably and very rapidly. We still call this "progress." But the pervasiveness and irreversibility of genetic intervention, and most importantly our profound ignorance of the interlaced workings of natural systems, make interference in this realm a very risky enterprise.

Some new technologies, for example, will require the deliberate release of genetically engineered organisms into the environment. The first attempt of this kind is now underway, as scientists at the University of California tackle the problem of frost damage to potato plants. It is known that certain bacteria (Pseudomonas Syringae) found on potato plants nucleate the ice crystals that kill the plants in cold weather. The California scientists have genetically engineered the bacteria to remove the ability to cause frost formation. They reason that if this newly-created strain of bacteria is sprayed on potato plants the indigenous ice-forming bacteria may be displaced and the plants would be spared from frost damage. If successful, the technology is likely to be widely adopted, resulting in the distribution of vast numbers of the engineered bacteria around the earth. The natural counterparts of the genetically engineered bacteria, however, are found not only on plants and in soil but in the upper atmosphere, where they cause the formation of water droplets or ice crystals and thereby determine precipitation patterns. If the proposed technology should ultimately be applied over vast areas, as is indeed the intention, massive quantities of the sprayed genetically engineered bacteria would be wafted upward by wind currents—as they normally are—where they might alter normal rain patterns. This would certainly damage all vegetation, in the affected area, and the associated ecosystems involving bacteria, insects, birds, and mammals would surely be disrupted.

In 1983 the scientists asked and received permission from the Recombinant DNA Advisory Committee of the National Institutes of Health to release the genetically engineered bacteria into the environment, i.e., into an open potato field. At the time there were no ecologists or other relevant experts on the NIH [National Institutes of Health] Committee to provide intelligent judgment on possible adverse ecological outcomes, and hence the decision for approval was reached in a scientific vacuum. The case is now under litigation, as critics call for more stringent evaluation, including an environmental impact statement, before release of any engineered organisms.

It is a well-established ecological principle that the introduction of a new organism may cause serious disruption in its new environment. Well-known examples are the gypsy moth, the Japanese beetle, and dutch elm disease. There are countless others. The reason for the disruption is simple: the delicate relationships among bacteria, insects, plants, and animals are thrown out of balance.

Genetic engineering applications of this kind raise difficult questions: Who will bear the responsibility if there is ecological damage in another state; another country? Witness the social and economic problems that are caused by acid rain. In the final analysis, genetic engineering seeks to change the rules of living organisms that have taken not several years but many millions of years to evolve into a coherent, integrated structure that accommodates all. The outcome will change the nature of Nature, irreversibly and with limited predictability. We cannot be sure the ecological disaster may not result in the long run. Do we present humans have the right to satisfy our needs and desires, however important they may seem to us, realizing that we cannot fathom the needs and desires of future humans who will be living under different circumstances? The environment, which in the final analysis is part of the world commons, is not in private ownership but in the public domain. The immediate value of an extended potato crop should not blind us to the grave consequences that could ensue. Only public awareness and a heightened social conscience can solve the dilemma of the commons, threatened as it is by many types of pollution including industrial products and waste chemicals.

The nuclear threat, unimagined before the physics research of 50 years ago, will burden the human race not for years, not for a lifetime, but for the rest of its existence. We are now losing our innocence in another realm, one that has troubled humanity little in the past, as natural evolution proceeded quite successfully on its own path without human help. What will happen when human beings possess the power to interfere with it, for better or worse? Will the struggle between enlightenment and irresponsibility take root there too, as it has in nuclear affairs and in virtually every other domain under human control? Is it appropriate that the nature of life itself should submit to the human will, or is a natural genetic heritage an inalienable birthright? C.S. Lewis pointed out, in a similar context that "what we call Man's power over Nature turns out to be a power exercised by some men over other men with Nature as its instrument." It will not be easy to apply the knowledge of genetic manipulation with wisdom, equity, and moral commitment.

---

From Liebe F. Cavalieri, *The* Double-Edged *Helix: Genetic Engineering in the Real World*, Prologue and Epilogue, 3–10, 158–163 (New York: Praeger, 1985)

## Questions for Discussion and Writing

1. Discuss the comments, made by atomic scientists looking back on the research to develop the atomic bomb, that they found it difficult to stop work even when it was clear that the bomb was not necessary to win World War II.
2. Discuss the possible benefits that genetic engineering can bring.
3. How can genetic engineering lead to disasters?
4. Discuss the moral responsibilities that confront modern scientific researchers today. Where do the responsibilities of scientists begin and end? Where do the responsibilities of the general public begin and end?

# The Growing Danger from Gene-Spliced Hormones

## Thomas Murray

We are merely at the beginning of the age of biotechnology but already we can see some of the ethical issues looming ahead. This article is about the seemingly innocuous and entirely beneficial development of artificial growth hormone. Yet even it presents us with difficult questions about its uses.

My son Peter is twelve, and lately the cuffs of his pants have been racing up his ankles—a sure sign that growth hormone is coursing through his body. He's on about the same growth schedule I was. By my fourteenth birthday I was already just shy of six feet, and hopeful of four or five inches more. I thought that would be enough to give me a shot at playing basketball in college. Alas, I'd reached my limit. Lacking any notable physical talents, I had to rely on guile; if you can't shoot over the guy, get him looking one way and then scurry past on the other side. (As we used to say, "Fake left, go right.") Recent happenings in genetic engineering make me wonder if something similar, albeit unintentional, is going on there: Is our attention being directed one way while what's important is slipping by on the other side?

At least since 1980, worries about using recombinant DNA technology to alter "human nature" have focused on gene therapy—the direct and intentional alteration of genetic material to treat disease. On June 20 of that year the general secretaries of three national organizations, for Protestants, Jews, and Catholics, wrote the President [of the United States] to warn: "History has shown us that there will always be those who believe it appropriate to 'correct' our mental and social structures by genetic means, so as to fit their vision of humanity. This becomes more dangerous when the basic tools to do so are finally at hand. Those who would play God will be tempted as never before."

Nothing garners attention as quickly as a nice little scandal; within a month of the letter, Dr. Martin Cline of UCLA provided one by experimenting on two patients with beta-zero thalassemia, a genetic condition that causes severe anemia. Cline removed some of their bone marrow, treated it with recombinant DNA containing normal hemoglobin genes, then reinserted it into the bone (after making room by killing some of the remaining marrow cells with radiation). The hope was that the treated cells would multiply and produce normal hemoglobin.

They didn't. Worse, Cline didn't have the approval of the UCLA committee that oversees research with human subjects. When the affair became public, the National Institutes of Health (NIH) stripped Cline of $162,000 in grant money and demanded strict supervision of his research. For those suspicious of human gene therapy, the case was proof that scientists couldn't be trusted to regulate themselves.

Since then, such work has proceeded very cautiously. A presidential commission gave its tentative blessing to gene therapy with somatic cells—those that don't pass the altered genes on to future generations. The NIH's watchdog recombinant DNA advisory committee set up a "human gene therapy subcommittee," which has suggested "points to consider" for doctors who propose to tinker with genes. Among them: whether the benefits of the treatment outweigh the risks, how to choose patients fairly, and how to publicize the results of the research. Thanks to the brouhaha over gene therapy, no great threats to humanity are likely to slip by in the near future—at least not on that side.

But there's another side to genetic engineering that has the power to alter us physically and socially. Rather than directly altering our genes, it can modify our bodies by supplementing the natural supply of important regulatory hormones with genetically engineered ones. A prime example is biosynthetically manufactured human growth hormone—hGH. Except for one additional amino acid—methionine (which appears to have no effect on its action)—biosynthetic hGH is identical to the hormone that promotes normal growth.

Produced in the pituitary gland, hGH plays a key role in determining how tall we'll become. So-called pituitary dwarfs usually lack an adequate supply of bioactive hGH. To treat them, for more than twenty years we've been harvesting pituitaries from human cadavers, each of which yields a minute quantity of hGH. Until recently the supply was barely adequate. In 1979 genetic engineers cloned the gene carrying instructions for making hGH, inserted it into a microorganism, and coaxed the bug to produce the human hormone. Just in time, it appears, because some hGH recovered from human pituitaries seems to have been contaminated with the slow virus that causes Creutzfeldt-Jakob disease (CJD), a degenerative infection of the brain. In April 1985 the Food and Drug Administration (FDA) halted the sale of natural growth hormone, and shortly thereafter approved Genentech's biosynthetic version. Since no human tissue is used in producing it, there's no danger of contamination with the CJD virus. Also, we're no longer limited by the scarcity of cadaver pituitaries, and other uses for hGH can be explored.

Biotechnology came to the rescue of kids deficient in growth hormone. But if hGH injections can make extremely short children a bit taller, what can it do for those who aren't dwarfs, but just shorter than average? What about the youngster who would have been only of average height? And what about the basketball player for whom a couple of inches more might mean the difference between the schoolyard and the NBA? In short (no pun intended), why not use hGH to give your child the advantages that come with being tall?

Years ago, at an FDA hearing, I speculated that once biosynthetic hGH was approved, people would want to use it for all sorts of non-therapeutic purposes. One member of the FDA committee told me that several parents had already asked her if they could get the drug for their kids, who weren't hGH-deficient. All, she recalled, were physicians. Rebecca Kirkland, who does clinical trials of biosynthetic hGH at the Baylor College of Medicine, recently said she's had inquiries from five parents wanting to get hGH for their normal children.

Why would parents want to go to such expense (treatment with biosynthetic hGH costs roughly $10,000 a year), cause their children pain (the shots hurt a bit), and risk unknown long-term side effects? Quite simply, because it's advantageous to be tall—within limits. A modest body of scientific evidence supports the commonsense observation that taller people often get the nod over their shorter counterparts, because they're perceived as more intelligent, good-looking, likable, extroverted, and attractive. Being very much taller than average is a mixed blessing, to be sure. But being a few inches above average seems to help.

A survey at the University of Pittsburgh in 1968 found that starting salaries for graduates varied with height: roughly $300 an inch up to six feet two inches. In a study of men whose heights had been recorded twenty-five years earlier, a graduate student at Washington University in St. Louis demonstrated a "height bonus" of approximately $400 per inch.

When a researcher at Eastern Michigan University presented two hypothetical job candidates to recruiters, one eight inches taller than the other, 72 per cent preferred the taller one, 27 per cent said there was no difference, and only one chose the shorter applicant. And much has been made of the fact that the taller candidate for President usually wins. Only two presidents—Madison and Benjamin Harrison—were shorter than the average American male of their eras.

If some parents want to give their child the edge that height seems to confer, what's wrong with that? If it's O.K. to spend $2,500 on orthodontics, to buy your kid private tennis and music lessons, or to spend $10,000 a year and up for prep school and private college, what's a few thousand bucks more to buy a couple of inches? The kid could turn out to be a klutz at tennis, have a tin ear, and major in Michelob, but taller is taller.

In this century, we make a strong presumption in favor of liberty. Before we interfere with the right of parents to bring up their children as they judge best, we demand strong reasons for doing so. Can we find them in the case of hGH?

Let me ask a skeptical question: What's the disease for which human growth hormone is the cure? Philosophers have a difficult time agreeing on the definition of disease, but most would recognize a physiological deficiency in hGH as a genuine disease, and hGH injections as a reasonable treatment. There are some kids, though, who aren't measurably deficient in hGH but who are very short. Their shortness can be the consequence of any one of numerous medical problems, or they may fall into the category of "familial short stature"—that is, short like mom or pop. Either way, can their shortness ever be a disability? A disability is a condition that interferes with the tasks of daily living. If people are so short that they qualify for the elevator riddle,[1] their shortness may well be a disability. Disabilities usually justify medical intervention. But what if a person isn't suffering from a disease, and isn't so short that the lack of height becomes a disability? What if it's merely a disadvantage?

* * *

Even if hGH turns out to be physiologically harmless—some experts have warned of possible effects on glucose regulation, as well as an increased risk of atherosclerosis and high blood pressure—there may be psychological consequences to treating children with the hormone. The unmistakable message given to a child is that shortness is a grave enough problem to justify the considerable expense, inconvenience, and discomfort of hGH treatment. It's likely to increase the child's self-consciousness about height. And since children rarely grow as much with the hormone as they or their parents hope, disappointment is likely.

A study of hGH-deficient children and their families found that the most psychologically mature kids weren't those who grew the most but those whom parents and other adults had treated appropriately for their age rather than their size. Kids who were encouraged to pursue interests where their height wasn't a disadvantage were much happier with themselves.

People differ in so many ways: in intelligence, charm, quickness of hand and foot, facility with words, wit, etc. But when we put children through hGH treatments, we focus almost entirely on their height (where they don't "measure up") and ignore their other talents and abilities. Understandably, short kids receiving hGH may come to feel that they're inadequate and inferior.

* * *

All other things being equal, taller basketball players are more effective than shorter ones. Height is an advantage in basketball and some other sports. In almost all other realms of human endeavor, though, height bears no relationship to the ability to do a job well. But in a culture that regularly imputes desirable characteristics to tall people and undesirable ones to short ones, shortness is surely a disadvantage. (Even our language is laden with "heightisms": we look up to people we admire, look down on those we don't.)

Like other "isms," such as racism and sexism, heightism involves making unwarranted judgments about people based on irrelevant criteria. Does anyone believe the solution to racism is to find a drug that lightens black skin? The mind boggles at the possible biotechnological remedies for sexism. And yet those who want to give their kids hGH are proposing just this sort of technological end-run around heightism.

If we choose to allow hGH to be used for non-disease, non-disabling shortness, then we must make a choice. Either we let those who can afford it buy it for their children, or we make it available at public expense to all children whose parents want it.

Suppose we let it be sold. The children of rich parents will have one more leg up, so to speak, on their peers. You could ask, what's one more advantage in light of all the others available to people with means? But the prospect of two classes—one tall and monied, the other short and poor—is ugly and disquieting. It would allow injustice to be piled upon injustice.

Suppose we take the other route and provide hGH to anyone who wants it. If all parents (short and tall

[1] Why did the man always take the elevator to the sixth floor, then walk up four more flights to his apartment? Because he could only reach the button for the sixth floor.

alike) rushed out to get hGH shots for their kids, the average height of the entire population might increase. But in all likelihood the distribution of height in the population wouldn't change much, if at all. There would still be the taller and the shorter, and since we're doing nothing to diminish heightism, discrimination against the shorter would continue. Some people would benefit, of course—for example, the stockholders in Genentech (which holds a patent for making biosynthetic hGH) and those who produce fabrics (since everyone will be wearing bigger sizes). Meanwhile, at considerable social expense, kids would get their three shots a week with a little pain and, we hope, minimal side effects.

Inevitably, a few eager parents would want to regain the edge for their kids and try to get bigger doses of hGH, like those athletes who take increasing amounts of anabolic steroids in the hope of obtaining an advantage over their rivals. In both cases, individuals pursue their own interests, only to make everyone worse off.

Whether hGH is available just to those who can pay for it or to everyone, the results would be unfortunate. In one instance, we use biotechnology to reinforce the advantages of wealth; in the other, we incur enormous expense and unknown risks without making anyone better off. The wisest course is to restrict hGH to cases of disease and disability.

Although hGH may be the first biosynthetic hormone to tempt us to improve on human nature, it won't be the last. Imagine what we might do with a hormone that prompted damaged nerves to regenerate. Someone would wonder whether it would also stimulate growth in the brain. And soon we'd be trying to enlarge our brains, however misguided that might be, scientifically or morally.

It also occurs to me that simply by writing this article I may spur some parents to seek out growth hormone for their child of normal height. I fervently hope not. But the temptation posed by hGH, and by other fruits of biotechnology as yet unripened, will be great. And it will require all our collective common sense to use them wisely.

*Discover* (February, 1987): 88–92.

## Questions and Topics for Discussion and Writing

1. Is tallness a real advantage for human beings? If so for whom is it an advantage? Is tallness a health matter or a social concern?
2. If growth hormones should be prohibited from use by "normal" individuals, should hair dyes, elevator shoes, and other ways of changing our appearance also be prohibited?
3. Compare the questions raised by artificial growth hormones to those raised by the use of anabolic steroids by athletes.

# Inside Information: A Quebec Community Must Confront Its Unusual Genetic History

Stephen Strauss

Recent research in genetics is not all focused on bio-engineering—the altering of genetic codes and creation of new organisms and substances. Some profound new developments in genetics concern learning much more about existing genetic structures and patterns of inheritance. This is based not so much on new knowledge of biological processes as it is on the application of the methodology of epidemiology and the power of computers to trace and correlate genetic lines through family trees. While this all seems benign and useful, this news article reports on some recent research in a genetic disorder that raises some interesting new ethical issues of the relationship of scientifically determined genetic profiles of individuals to the personal decisions of those individuals to marry and have children.

Researchers studying an unusually high incidence of a genetic disorder in an isolated area of Quebec believe they have traced the origins of the disease back to a couple who were married in what was then New France in 1661.

The accuracy of this genetic pinpointing is a product of the diligence of generations of parish priests, the deep memory of computers and the increasing sophistication of modern scientists. But the study also raises profound questions about genetic research and its implications for society.

The disorder, myotonic dystrophy, is 40 to 50 times more common among inhabitants of the Saguenay-Lac-St-Jean area 200 kilometres north of Quebec City than in the general population. The condition is characterized by muscle weakness, cataracts, shrunken testicles, premature balding, heart disease and mental deterioration.

Epidemiologist Marc De Braekeleer of the University of Quebec at Chicoutimi announced his findings at a recent meeting of French-speaking scientists in Montreal. He said he and his co-workers know the 17th century couple had 14 children, half of whom were carriers of the gene for the disease.

However, the study's findings have much wider implications. Forcing an out-of-the-way corner of Quebec to confront its disturbing genetic history has opened up ethical dilemmas.

The scientists involved are torn about what to do with damning genetic information, specifically whether or not to inform a carrier who is ignorant of his or her genetic risks.

"I am damned if I do, and damned if I don't," said Dr. Charles Scriver, a McGill University geneticist and member of the research team, about whether to inform a carrier who has not asked for information.

The dilemma is compounded because this kind of genetic information could influence the decision to marry or have children.

"They (the Saguenaians) are the metaphor for you and me and everyone elsewhere in the world ... it may be that in 10 or 15 years, most young couples do some (genetic) checking before they hit the bed," said Denis Gauvreau, a researcher with Quebec's Institut National de Recherche Scientifique, who is studying the area's incidence of Alzheimer's disease.

The scientific detective work in the rugged Saguenay region was fuelled by the voluminous personal records local priests have kept since the region was first settled in the 1830s.

The records (of nearly 300,000 people) provided the raw material to construct geneologial trees stretching back almost 150 years. Not only are most of the residents direct descendants of a few thousand founding families, but many came from the Charlevoix region near Quebec City, which has parish records of their ancestries dating back to an initial French settlement in the mid-seventeenth century.

The mountain of data was accessible through the efforts of [University of Quebec at Chicoutimi] historian Gerard Bouchard, who in 1971 began computerizing births, deaths and marriages in the area. Since then, the data bank has grown to include 50 other socio-economic factors, such as income levels.

A second factor driving the genetic research is that the Saguenay, like other isolated regions of Canada, developed a unique and often distressing genetic profile from in-breeding.

The area's pattern of extensive intermarriage was accentuated by the large sizes of its families. Into the 1940s, the average family had 12 people and families of 22 or 23 were not unknown. The social and religious pressure to marry and reproduce was so great that women continued to get pregnant even when they bore children with genetic diseases.

The inbreeding has created a unique genetic profile in the area—and it is being explored by the joint efforts of demographers, historians, epidemiologists and molecular geneticists at the University of Laval, McGill University and the University of Quebec at Chicoutimi.

The research team, known as SOREP, is studying the four or five genetic diseases—including several neuromuscular diseases such as the Charlevoix ataxia and a hereditary form of rickets—now known to be unique to the area and to those descended from the original settlers but who no longer live in the Saguenay.

For scientists, the fortunate marriage of computers, geneological records and genetic conditions has spun off a number of unique research projects. One of the most interesting is a study of the cause and development of Alzheimer's disease, a form of senility.

One of the challenges facing doctors is that Alzheimer's probably has several triggers, including environmental factors and a genetic predisposition.

In such a jumble, it becomes necessary to study large-sized, geographically-fixed families, and to able to compare genetic inter-relations spanning several generations—and this is how meticulous parish records provide an incomparable help. "The computer tells us almost where to go to find distant cousins," Dr. Gauvreau said.

All this has given rise to what Dr. Braekeleer refers to as "natural laboratory." To start, it is good news for families where genetic diseases have been passed down, especially since the area's birth rate has dramatically fallen to an average 1.8 children per family.

Many genetically inherited conditions occur only in one out of every four births. With 10 or more children in a family, the chances were significantly greater some children would be born with genetic problems. But today, with couples having less than two children, it is unlikely the affected gene would be passed on at all.

"With small families you don't have that many throws of the (genetic) dice," Dr. Scriver said.

But Prof. De Braekeleer cautions that if the birth rate shot up again—something the Quebec government has recently tried to encourage with large baby bonuses—"that would be a disaster from a genetic point of view, a catastrophe for the region."

He also suggests Saguenay natives who marry people from outside the Saguenay significantly lower their risk of having children with the genetic conditions common to the region.

But more importantly, the scientists and people of the Saguenay are wrestling with how to pass on the new information and [how they] will act on it. A genetic counsellor at Chicoutimi Hospital said little of the new knowledge has yet been passed on to the general population.

To make the information more available, a genetics seminar has been started at the junior college (CEGEP) in Chicoutimi and has been taught since 1987 in the area's high schools. It teaches students about their genetic histories and the risks at least some face if the diseases run in their family.

The SOREP group, with the help of two lawyers, has adopted a code of ethics in which no specific information about their research findings are released unless requested by an individual or his or her doctor.

But Prof. Michel Parrent, a sociologist who helped organize the genetics seminar, is trying to determine if there is an ethical way of informing people—specifically carriers who may pass on [the] gene but who themselves are not affected—about their genetic risks if they have not asked for the information.

The dilemma of who is to know raises the hackles of the SOREP geneticists, who fear it will involve them in issues of eugenics - active genetic improvement—associated with a befouled Nazi science.

Dr. Laberge is emphatic. "We should do nothing ... this is a personal and a private matter," he said.

It is a thorny problem for ethicists too.

"The act of approaching a person to say, 'would you like to know about this' gives them some reason to worry," said Margaret Somerville, director of the McGill Centre for Medicine, Ethics and Law.

She cautions that the morality of beginning the SOREP type of research cannot easily be disassociated from the problems of communicating its results.

*The Globe and Mail* (June 17, 1989): D2

## Questions and Topics for Discussion and Writing

1. Why is the epidemiological work reported in this article a new development in science? Could this have been done, say, 20 years ago?
2. Is there any particular reason that this research is centered on a small community in Quebec (other than the reason that the research scientist is located at a nearby university)? Can this methodology be applied equally well in other communities?
3. Why does this work raise ethical issues? Compare the ethical issues raised here with other instances wherein scientific information can influence people in matters of personal and private choice.

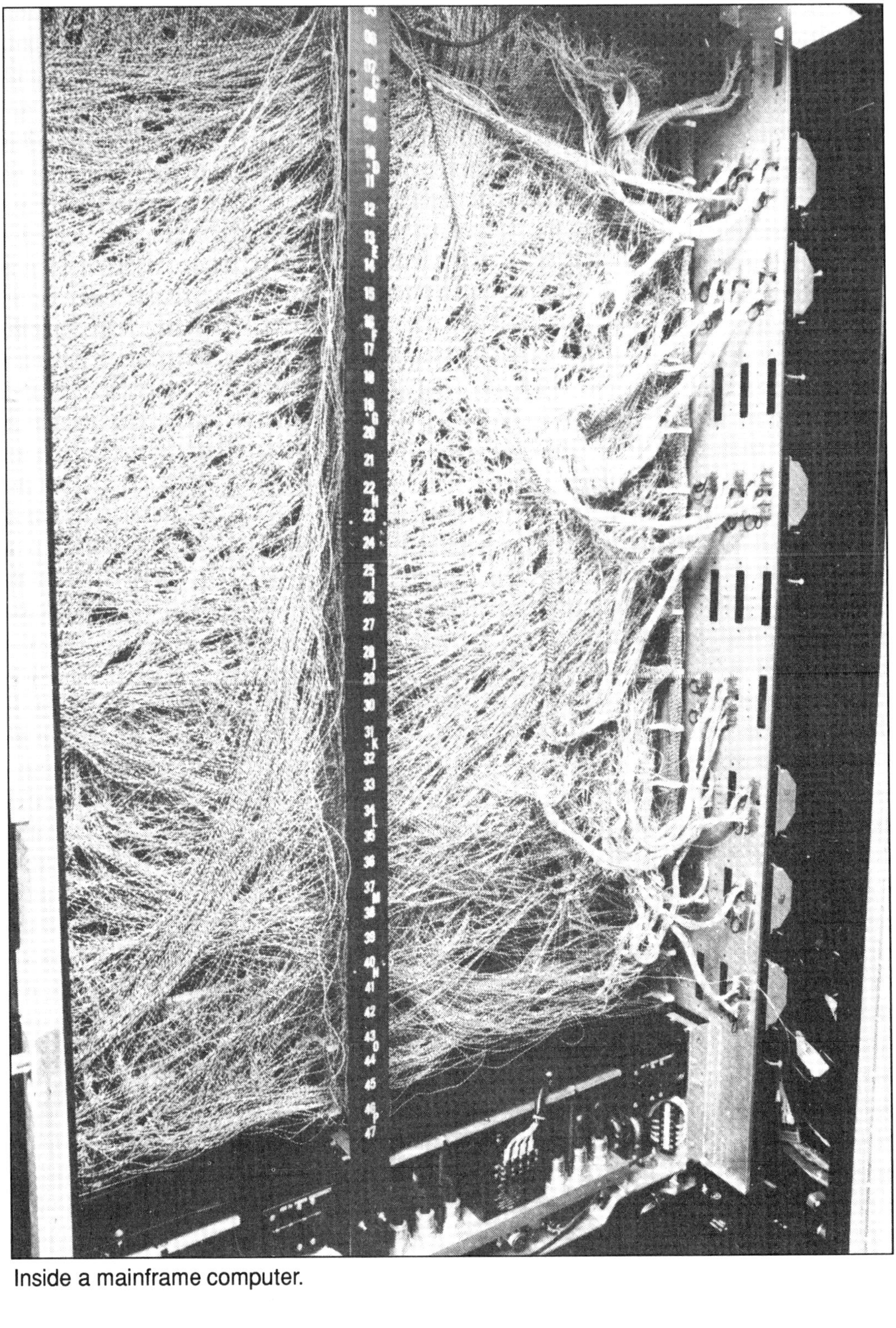

Inside a mainframe computer.

*Issue*

# Computers and Automation in Society

Some people go so far as to place the invention of the computer on a par with the invention of printing by movable type, or even the invention of written language itself. The computer certainly is a major invention which is changing the shape of the modern industrialized world. The speed and complexity of modern computers is so great that problems can be tackled that were unbelievably difficult before, and other tasks that were slow and tedious to perform are made quick and painless.

As with every other technological advance, there are costs that we may not yet be aware of—the most obvious being the dehumanizing of activities that are now better performed by computers. Another, less obvious, cost is the growing tendency for all of us to begin to perceive the world in the same ways that a computer "perceives" the world.

The articles chosen for this section are intended to cast some small light on some of the issues of how the computer has entered, indeed invaded, our culture, and what that might be doing to us. The first two selections are by J. David Bolter, a professor of classics at the University of North Carolina. In "The Computer As a Defining Technology," Bolter argues that the computer is symbol of our age, as clocks and steam engines were of previous eras. In "Turing's Man," Bolter sees the computer being used as the model for understanding everything—in particular, for understanding the minds of human beings. Those who try to characterize the human mind as an "information processor" are what he calls "Turing's men." The next article is a classic statement by one of the greatest of "Turing's men," Norbert Wiener, founder of cybernetics. Wiener's article, first published in 1948, titled simply "Cybernetics," is one of the clearest and most thoughtful statements of the virtues of considering the human mind as analogous to a computer.

The last two articles are discussions of computers and automation in the workplace. Heather Menzies' "The Automated Office" is a quick "day in the life" of person working in the not very distant future in a highly automated, electronic office. In the few short years since Menzies first published this sketch (in 1982), the office she describes as futuristic seems now perfectly feasible, even if still rather unusual. Michael A. Arbib's "Automation in Industry and the Office" reviews the history of automation in industry and shows how computers form a natural extension and improvement upon older assembly line automation in industry, and how automation only reached the office environment with the arrival of the computer.

# The Computer as a Defining Technology

## J. David Bolter

In the late twentieth century, the ultimate technology is embodied in the computer. Computers are both strange and familiar, held in awe and taken for granted. In this article, J. David Bolter, who is both a computer scientist and a professor of classics, portrays the computer as the "defining technology" of our age, the symbolic representation of all technology and a metaphor for the state of our civilization itself.

Computers perform no work themselves; they direct work. The technology of "command and control," as Norbert Wiener has aptly named it, is of little value without something to control, generally other machines whose function is to perform work. For example, the essence of the American space shuttle is the computers that control almost every phase of its operation. But unless the powerful rocket engines provide the expected thrust, there is no mission for the computers to control. The computer leaves intact many older technologies, particularly the technologies of power, and yet it puts them in a new perspective. With the appearance of a truly subtle machine like the computer, the old power machines (steam, gas, or rocket engines) lose something of their prestige. Power machines are no longer agents on their own, subject only to direct human intervention; now they must submit to the hegemony of the computer that coordinates their effects.

As a calculating engine, a machine that controls machines, the computer does occupy a special place in our cultural landscape. It is the technology that more than any other defines our age. Our generation perfected the computer, and we are intrigued by possibilities as yet only half-realized. Ruthlessly practical and efficient, the computer remains something fantastic. Its performance astonishes even the engineers who build it, just as the clock must have astonished craftsmen in the fourteenth century and the power of the steam engine even the rugged entrepreneurs of the nineteenth century. For us today, the computer constantly threatens to break out of the tiny corner of human affairs (scientific measurement and business accounting) that it was built to occupy, to contribute instead to a general redefinition of certain basic relationships: the relationship of science to technology, of knowledge to technical power, and, in the broadest sense, of mankind to the world of nature.

This process of redefinition is not new. Technology has always exercised such an influence; it has always served both as a bridge and a barrier between men and their natural environment. The ability to make and use tools and the subtle capacity to communicate through language have allowed men to live more comfortably in the world, but these achievements have also impressed upon them their separation from nature.

Men and women throughout history have asked how it is that they and their culture (their technology in the largest sense) transcend nature, what makes them characteristically human and not merely animal. For the Greeks, a cardinal human quality was the ability to establish the political and social order embodied in a city-state: men at their best could set collective goals, make laws, and obey them, and none of these could be achieved by animals in a state of nature. Their city-state was a feat of social technology. In the Middle Ages, the accomplishments of technology were perhaps more physical than social, but the use of inanimate sources of power, wind and water, fostered a new view of mankind versus the forces of nature. The discoveries of the Renaissance and the Industrial

Revolution moved men closer to nature in some respects and separated them even more radically in others. Continued emphasis on exploring and manipulating the physical world led to a deeper appreciation of the world's resources. Yet the desire to master nature—to harness her more efficient sources of power in steam and fossil fuels and to mine her metals for synthetic purposes—grew steadily throughout this period. When Darwin showed convincingly that man was an animal like any other, he shattered once and for all the barrier that separated men from the rest of nature in the Greek and medieval chains of being. Yet nineteenth-century engineers with their railroads and still more twentieth-century physicists with their atomic bombs seemed less natural than ever before, less under the control of either nature or a personal deity and more responsible for their own misjudgments.

Continually redrawing the line that divides nature and culture, men have always been inclined to explain the former in terms of the latter, to examine the world of nature through the lens of their own created human environment. So Greek philosophers used analogies from the crafts of pottery and woodworking to explain the creation of the universe: the stars, the planets, the earth, and its living inhabitants. In the same way, the weight-driven clock invented in the Middle Ages provided a new metaphor for both the regular movements of heavenly bodies and the beautifully intricate bodies of animals, whereas the widespread use of the steam engine in the nineteenth century brought to mind a different, more brutal aspect of the natural world. It is certainly not true that changing technology is solely responsible for mankind's changing views of nature, but clearly the technology of any age provides an attractive window through which thinkers can view both their physical and metaphysical worlds.

Technology has had this influence even upon philosophers, like Plato, who generally disdain human craftsmanship and see it as a poor reflection of a greater nonhuman reality. And even in Christian theology and poetry, the pleasures of heaven could only be described as a grand version of the tainted pleasures men know on earth, and the tortures of hell were earthly tortures intensified. Almost every sort of philosopher, theologian, or poet has needed an analogy on the human scale to clarify his or her ideas. Speaking of creation as the imposition of order upon the natural world, he or she generally assumes a creator as well, and this creator is a craftsman or technologist.

It is in this context that I propose to examine electronic technology. The computer is the contemporary analog of the clocks and steam engines of the previous six centuries, it is as important to us as the potter's wheel was to the ancient world. It is not that we cannot live without computers, but that we will be different people because we live with them. All techniques and devices have the potential to become defining technologies because all to some degree redefine our relationship to nature. In fact, only a few devices or crafts in any age deserve to be called defining technologies. In the ancient world, carpentry and masonry were about as important as spinning and pottery, and yet poets and philosophers found the latter two far more suggestive. In medieval Europe, crop rotation and the moldboard plough had a greater economic and social impact than the early clockwork mechanisms. Yet not many philosophers and theologians compared the world to a lentil bean. Certain skills and inventions have moved easily out of the agora into the Academy, out of the textile mill into the salon, or out of the industrial research park into the university classroom.

The vision of particular philosophers and poets is important to such a transference. Descartes and his followers helped to make the clock a defining technology in Western Europe. Certainly the first poet to elaborate the myth of the Fates who spin the thread of life helped to make textiles a defining technology for ancient Greece. But there must be something in the nature of the technology itself, so that its shape, its materials, its modes of operation appeal to the mind as well as to the hand of their age—for example, the pleasing rotary motion of the spindle or the autonomy and intricacy of the pendulum clock.

Such qualities combine with the social and economic importance of the device to make people think. Very often a device will take on a metaphoric significance and be compared in art and philosophy to some part of the animate or inanimate world. Plato compared the created universe to a spindle. Descartes thought of animals as clockwork mechanisms, and scientists in the nineteenth century and early twentieth centuries have regularly compared the universe to a heat engine that is slowly squandering its fuel. Today the computer is constantly serving as a metaphor for the human mind or brain: psychologists speak of the input and output, sometimes even the hardware and software, of the brain; linguists treat human language as if it were a programming code; and everyone speaks of making computers "think."

A defining technology develops links, metaphorical or otherwise, with a culture's science, philosophy, or literature; it is always available to serve as a metaphor, example, model, or symbol. A defining technology resembles a magnifying glass, which collects and focuses seemingly disparate ideas in a culture into one bright, sometimes piercing ray. Technology does not call forth major cultural changes by itself, but it does bring ideas into a new focus by explaining or exemplifying them in new ways to larger audiences. Descartes's notion of a mechanistic world that obeyed the laws of mathematics was clear, accessible, and therefore powerful because his contemporaries lived with clocks and gears. So today electronic technology gives a more catholic appeal to a number of trends in twentieth-century thought, particularly the notions of mathematical logic, structural linguistics, and behaviourial psychology. Separately these trends were minor upheavals in the history of ideas; taken together, they become a major revision in our thinking.

---

From J. David Bolter, *Turing's Man: Western Culture in the Computer Age* (Chapel Hill, NC: University of North Carolina Press, 1984).

---

## Questions and Topics for Discussion and Writing

1. What "defining technologies" does Bolter name for different periods in history? Discuss how well these defining technologies "fit" their ages and how well the computer fits ours.
2. How can a defining technology help make people think?
3. What is the value of identifying a concept such as a "defining technology" and writing about it as Bolter has done?

# Turing's Man

## J. David Bolter

Continuing from the previous selection, in this article J. David Bolter gives a poignant example of the defining technology in action, in the tendency of those thinking and writing today to view all culture, and indeed the human mind itself as a sort of computer.

In the development of the computer, theory preceded practice. The manifesto of the new electronic order of things was a paper ("On Computable Numbers") published by the mathematician and logician A.M. Turing in 1936. Turing set out the nature and theoretical limitations of logic machines before a single fully programmable computer had been built. What Turing provided was a symbolic description, revealing only the logical structure and saying nothing about the realization of that structure (in relays, vacuum tubes, or transistors). A Turing machine, as his description came to be called, exists only on paper as a set of specifications, but no computer built in the intervening half century has surpassed these specifications; all have at most the computing power of Turing machines. Turing is equally well known for a very different kind of paper; in 1950 he published "Computing Machinery and Intelligence." His 1936 work was a forbidding forest of symbols and theorems, accessible only to specialist. This later paper was a popular polemic, in which Turing stated his conviction that computers were capable of imitating human intelligence perfectly and that indeed they would do so by the year 2000. This paper too has served as a manifesto for a group of computer specialists dedicated to realizing Turing's claim by creating what they call "artificial intelligence," a computer that thinks.

Put aside for the moment the question of whether the computer can ever rival human intelligence. The important point is that Turing, a brilliant logician and a sober contributor to the advance of electronic technology, believed it would and that many have followed him in his belief. The explanation is partly enthusiasm for a new invention. In 1950 the computer was just beginning to bring vast areas of science and business under its technological aegis. These machines were clearly taking up the duties of command and control that had always been assumed by human operators. Who could say then where the applications of electronic command and control might end? Was it not natural to believe that the machine would in time eliminate the human operator altogether? Inventors, like explorers, have a right to extravagant claims. Edison had said that the record player would revolutionize education; the same claim was made for radio and, of course, television.

I think, however, that Turing's claim has had a greater significance. Turing was not simply exaggerating the service his machine could perform. (Does a machine that imitates human beings perform any useful service at all? We are not running short of human beings.) He was instead explaining the meaning of the computer for our age. A defining technology defines or redefines man's role in relation to nature. By promising (or threatening) to replace man, the computer is giving us a new definition of man, as an "information processor," and of nature, as "information to be processed."

I call those who accept this view of man and nature Turing's men. I include in this group many who reject Turing's extreme prediction of an artificial intelligence by the year 2000. We are all liable to become Turing's men, if our work with the computer is intimate and prolonged and we come to think and speak in terms suggested by the machine. When the cognitive psychologist begins to study the mind's "algorithm for searching long-term memory," he has become Turing's man. So has the economist who draws up input-output diagrams of the nation's business, the sociologist who engages in "quantitative history," and the humanist who prepares a "key-word-in-context" concordance.

Turing's man is the most complete integration of humanity and technology, of artificer and artifact, in the history of the Western cultures. With him the tendency, implicit in all eras, to think "through" one's contemporary technology is carried to an extreme; for him the computer reflects, indeed imitates, the crucial human capacity of rational thinking. Here is the essence of Turing's belief in artificial intelligence. By making a machine think as a man, man recreates himself, defines himself as a machine. The scheme of making a human being through technology belongs to thousands of years of mythology and alchemy, but Turing and his followers have given it a new twist. In Greek mythology, in the story of Pygmalion and Galatea, the artifact, the perfect ivory statue, came to life to join its human creator. In the seventeenth and eighteenth centuries, some followers of Descartes first suggested crossing in the other direction, arguing, with La Mettrie, that men were no more than clockwork mechanisms. Men and women of the electronic age, with their desire to sweep along in the direction of technical change, are more sanguine than ever about becoming one with their electronic homunculus. They are indeed remaking themselves in the image of their technology, and it is their very zeal, their headlong rush, and their refusal to admit any reservation that calls forth such a violent reaction from their detractors. Why, the critics ask, are technologists so eager to throw away their freedom, dignity, and humanity for the sake of innovation?

Should we be repelled by the notion of man as computer? Not until we better understand what it means for man to be a computer. Why on the face of it should we be more upset by this notion than by the Cartesian view that man is a clock or the ancient view that he is a clay vessel animated by a divine breath? We need to know how Turing's man differs from that of Descartes or Plato, how the computer differs conceptually and symbolically from a clock or a clay pot. And to do this, we must isolate the precise qualities of computers and programming, hardware and software, that have the magnifying effect mentioned earlier—bringing ideas from philosophy and science into a new focus.

---

From J. David Bolter, *Turing's Man: Western Culture in the Computer Age* (Chapel Hill, NC: University of North Carolina Press, 1984).

---

### Questions and Topics for Discussion and Writing

1. What would a computer need to be able to do to imitate human intelligence perfectly?
2. Explain what Bolter means by "Turing's men".
3. Discuss the benefits and dangers of seeing the world as "Turing's men" see it.

# * Cybernetics

## Norbert Wiener

The computer is a machine, designed and built by human beings. Its history can be traced back through calculators and adding machines and automatic devices such as the Jacquard loom. But computers are also much more than any of its predecessors, because the sheer level of complexity and sophistication of a modern computer provides a reasonable model for understanding some of the complex and sophisticated functioning of the human brain. The original formulation of the science that would study the similarities of brains and computers was made by the mathematician Norbert Wiener in 1948 in his book *Cybernetics.* The article that follows is a brief and non-technical exposition by Wiener of some of the main ideas of his book. It was published in the same year, 1948, in *Scientific American.*

Cybernetics is a word invented to define a new field in science. It combines under one heading the study of what in a human context is sometimes loosely described as thinking and in engineering is known as control and communication. In other words, cybernetics attempts to find the common elements in the functioning of automatic machines and of the human nervous system, and to develop a theory which will cover the entire field of control and communication in machines and in living organisms.

It is well known that between the most complex activities of the human brain and the operations of a simple adding machine there is a wide area where brain and machine overlap. In their more elaborate forms, modern computing machines are capable of memory, association, choice and many other brain functions. Indeed, the experts have gone so far in the elaboration of such machines that we can say the human brain behaves very much like the machines. The construction of more and more complex mechanisms actually is bringing us closer to an understanding of how the brain itself operates.

The word cybernetics is taken from the Greek *kybernetes*, meaning steersman. From the same Greek word, through the Latin corruption *gubernator*, came the term governor, which has been used for a long time to designate a certain type of control mechanism, and was the title of a brilliant study written by the Scottish physicist James Clerk Maxwell 80 years ago. The basic concept which both Maxwell and the investigators of cybernetics mean to describe by the choice of this term is that of a feedback mechanism, which is especially well represented by the steering engine of a ship. Its meaning is made clear by the following example.

Suppose that I pick up a pencil. To do this I have to move certain muscles. Only an expert anatomist knows what all these muscles are, and even an anatomist could hardly perform the act by a conscious exertion of the will to contract each muscle concerned in succession. Actually what we will is not to move individual muscles but to pick up the pencil. Once we have determined on this, the motion of the arm and hand proceeds in such a way that we may say that the amount by which the pencil is not yet picked up is decreased at each stage. This part of the action is not in full consciousness.

To perform an action in such a manner, there must be a report to the nervous system, conscious or unconscious, of the amount by which we have failed to pick up the pencil at each instant. The report may be visual, at least in part, but it is more generally kinesthetic, or to use a term now in vogue, proprioceptive. If the proprioceptive sensations are wanting, and we do not replace them by a visual or other substitute, we are unable to perform the act of picking up the pencil, and find ourselves in a state known as ataxia. On the other hand, an excessive feedback is likely to be just as serious a handicap. In the latter case the muscles overshoot the mark and go into an uncontrollable oscillation. This condition, often associated with injury to the cerebellum, is known as purpose tremor.

Here, then, is a significant parallel between the workings of the nervous system and of certain machines. The feedback principle introduces an important new idea in nerve physiology. The central nervous system no longer appears to be a self-contained organ receiving signals from the senses and discharging into the muscles. On the contrary, some of its most characteristic activities are explainable only as circular processes, traveling from the nervous system into the muscles and re-entering the nervous system through the sense organs. This finding seems to mark a step forward in the study of the nervous system as an integrated whole.

The new approach represented by cybernetics—an integration of studies which is not strictly biological or strictly physical, but a combination of the two—has already given evidence that it may help to solve many problems in engineering, in physiology and very likely in psychiatry.

This work represents the outcome of a program undertaken jointly several years ago by the writer and Arturo Rosenblueth, then of the Harvard Medical School and now of the National Institute of Cardiology of Mexico. Dr. Rosenblueth is a physiologist; I am a mathematician. For many years Dr. Rosenblueth and I had shared the conviction that the most fruitful areas for the growth of the sciences were those which had been neglected as no-man's lands between the various established fields. Dr. Rosenblueth always insisted that a proper exploration of these blank spaces on the map of science could be made only by a team of scientists, each a specialist but each possessing a thoroughly sound acquaintance with the fields of his fellows.

* * *

Our collaboration began as the result of a wartime project. I had been assigned, with a partner, Julian H. Bigelow, to the problem of working out a fire-control apparatus for anti-aircraft artillery which would be capable of tracking the curving course of a plane and predicting its future position. We soon came to the conclusion that any solution of the problem must depend heavily on the feedback principle, as it operated not only in the apparatus but in the human operators of the gun and of the plane. We approached Dr. Rosenblueth with a specific question concerning oscillations in the nervous system, and his reply, which cited the phenomenon of purpose tremor, confirmed our hypothesis about the importance of feedback in voluntary activity.

The ideas suggested by this discussion led to several joint experiments, one of which was a study of feedback in the muscles of cats. The scope of our investigations steadily widened, and as it did so scientists from widely diverse fields joined our group. Among them were the mathematicians John von Neumann of the Institute for Advanced Study and Walter Pitts of Massachusetts Institute of Technology; the physiologists Warren McCulloch of the University of Pennsylvania and Lorente de No of the Rockefeller Institute; the late Kurt Lewin, psychologist, of M.I.T.; the anthropologists Gregory Bateson and Margaret Mead; the economist Oskar Morgenstern of the Institute for Advanced Study; and others in psychology, sociology, engineering, anatomy, neurophysiology, physics, and so on.

The study of cybernetics is likely to have fruitful applications in many fields, from the design of control mechanisms for artificial limbs to the almost complete mechanization of industry. But in our view it encompasses much wider horizons. If the 17th and early 18th centuries were the age of clocks, and the latter 18th and 19th centuries the age of steam engines, the present time is the age of communication and control. There is in electrical engineering a division which is known as the split between the technique of strong currents and the technique of weak currents; it is this split which separates the age just passed from that in which we are living. What distinguishes communication engineering from power engineering is that the main interest of the former is not the economy of energy but the accurate reproduction of a signal.

At every stage of technique since Daedalus, the ability of the artificer to produce a working simulacrum of a living organism has always intrigued people. In the days of magic, there was the bizarre and sinister concept of the Golem, that figure of clay into which the rabbi of Prague breathed life. In Isaac Newton's time the automaton became the clockwork music box. In the 19th century, the automaton was a glorified heat engine, burning a combustible fuel instead of the glycogen of human muscles. The automaton of our day opens doors by means of photocells, or points guns to the place at which a radar beam picks up a hostile airplane, or computes the solution of a differential equation.

Under the influence of the prevailing view in the science of the 19th century, the engineering of the body was naturally considered to be a branch of power engineering. Even today this is the predominant point of view among classically minded, conservative physiologists. But we are now coming to realize that the body is very far from a conservative system, and that the power available to it is much less limited than was formerly believed. We are beginning to see that such important elements as the neurones—the units of the nervous complex of our bodies—do their work under much the same conditions as vacuum tubes, their relatively small power being supplied from outside by the body's circulation, and that the bookkeeping which is most essential to describe their function is not one of energy.

In short, the newer study of automata, whether in the metal or in the flesh, is a branch of communications engineering, and its cardinal ideas are those of the message, of the amount of disturbance or "noise" (a term taken from the telephone engineer), of the quantity of information to be transmitted, of coding technique, and so on.

This view obviously has implications which affect many branches of science. Let us consider here the application of cybernetics to the problem of mental disorders. The realization that the brain and computing machines have much in common may suggest new and valid approaches to psychopathology, and even to psychiatry.

These begin with perhaps the simplest question of all: how the brain avoids gross blunders or gross miscarriages of activity due to the malfunction of individual parts. Similar questions referring to the computing machine are of great practical importance, for here a chain of operations, each of which covers only a fraction of a millionth of a second, may last a matter of hours or days. It is quite possible for a chain of computational operations to involve a billion separate steps. Under these circumstances, the chance that at least one operation will go amiss is far from negligible, even though the reliability of modern electronic apparatus has exceeded the most sanguine expectations.

* * *

In ordinary computational practice by hand or by desk machines, it is the custom to check every step of the computation and, when an error is found, to localize it by a backward process starting from the first point where the error is noted. To do this with a high-speed machine, the check must proceed at the pace of the original machine, or the whole effective order of speed of the machine will conform to that of the slower process of checking.

A much better method of checking, and in fact the one generally used in practice, is to refer every operation simultaneously to two or three separate mechanisms. When two such mechanisms are used, their answers are automatically collated against each other; and if there is a discrepancy, all data are transferred to permanent storage, the machine stops and a signal is sent to the operator that something is wrong. The operator then compares the results, and is guided by them in his search for the malfunctioning part, perhaps a tube which has burned out and needs replacement. If three separate mechanisms are used for each stage, there will practically always be agreement between two of the three mechanisms, and this agreement will give the required result. In this case the collation mechanism accepts the majority report, and the machine need not stop. There is a signal, however, indicating where and how the minority report differs from the majority report, If this occurs at the first moment of discrepancy, the indication of the position of the error may be very precise.

It is conceivable, and not implausible, that at least two of the elements of this process are also represented in the nervous system. It is hardly to be expected that any important message is entrusted for transmission to a single neurone, or that an important operation is entrusted to a single neuronal mechanism. Like the computing machine, the brain probably works on a variant of the famous principle expounded by Lewis Carroll in *The Hunting of the Snark*: "What I tell you three times is true."

It is also improbable that the various channels available for the transfer of information generally go from one end of their course to the other without connecting with one another. It is much more probable that when a message reaches a certain level of the nervous system, it may leave that point and

*Norbert Wiener*

proceed to the next by one or more alternative routes. There may be parts of the nervous system, especially in the cortex, where this interchangeability is much limited or abolished. Still, the principle holds, and it is probably holds most clearly for the relatively unspecialized cortical areas which serve the purpose of association and of what we call the higher mental functions.

So far we have been considering errors in performance that are normal, and pathological only in an extended sense. Let us now turn to those that are much more clearly pathological. Psychopathology has been rather a disappointment to the instinctive materialism of the doctors, who have taken the view that every disorder must be accompanied by actual lesions of some specific tissue involved. It is true that specific brain lesions, such as injuries, tumors, clots and the like, may be accompanied by psychic symptoms, and that certain mental diseases, such as paresis, are the sequelae of general bodily disease and show a pathological condition of the brain tissue. But there is no way of identifying the brain of a schizophrenic of one of the strict Kraepelin types, nor of a manic-depressive patient, nor of a paranoic. These we call functional disorders.

This distinction between functional and organic disorders is illuminated by the consideration of the computing machine. It is not the empty physical structure of the computing machine that corresponds to the brain—to the adult brain, at least—but the combination of this structure with the instructions given it at the beginning of a chain of operations and with all the additional information stored and gained from outside in the course of its operation. This information is stored in some physical form—in the form of memory. But part of it is in the form of circulating memories, with a physical basis that vanishes when the machine is shut down or the brain dies, and part is in the form of long-time memories, which are stored in a way at which we can only guess, but probably also in a form with a physical basis that vanishes at death.

There is therefore nothing surprising in considering the functional mental disorders fundamentally as diseases of memory, of the circulating information kept by the brain in active state and of the long-time permeability of synapses. Even the grosser disorders such as paresis may produce a large part of their effects not so much by the destruction of tissue which they involve and the alteration of synaptic thresholds as by the secondary disturbances of traffic, the overload of what remains of the nervous system and the rerouting of messages which must follow such primary injuries.

In a system containing a large number of neurones, circular processes can hardly be stable for long periods of time. Either they run their course, dissipate themselves and die out, as in the case of memories belonging to the specious present, or they embrace more and more neurones in their system, until they occupy an inordinate part of the neurone pool. This is what we should expect to be the case in the malignant worry that accompanies anxiety neuroses. In such a case, it is possible that the patient simply does not have the room—*i.e.*, a sufficient number of neurones—to carry out his normal processes of thought. Under such conditions, there may be less going on in the brain to occupy the neurones not yet affected, so that they are all the more readily involved in the expanding process. Furthermore, the permanent memory becomes more and more deeply involved, and the pathological process which began at the level of the circulating memories may repeat itself in a more intractable form at the level of the permanent memories. Thus what started as a relatively trivial and accidental disturbance of stability may build itself up into a process totally destructive to the normal mental life.

Pathological processes of a somewhat similar nature are not unknown in the case of mechanical or electrical computing machines. A tooth of a wheel may slip under such conditions that no tooth with which it engages can pull it back into its normal relations, or a high-speed electrical computing machine may go into a circular process that seems impossible to stop.

* * *

How do we deal with these accidents in the case of the machine? We first try to clear the machine of all information, in the hope that when it starts again with different data the difficulty will not recur. If this fails and the difficulty is inaccessible to the clearing mechanism, we shake the machine, or, if it is electrical, subject it to an abnormally large electrical impulse in the hope that we may jolt the inaccessible part into a position where the false cycle of its activities will be interrupted. If even this fails, we may disconnect an erring part of the apparatus, for it is possible that what remains may be adequate for our purpose.

In the case of the brain, there is no normal process, except death, that can clear it of all past impressions. Of the normal non-fatal processes, sleep comes closest to clearing the brain. How often we find that the best way to handle a complicated worry or an intellectual muddle is to sleep on it! Sleep, however, does not clear away the deeper memories, nor indeed is a malignant state of worry compatible with adequate sleep.

Thus we are often forced to resort to more violent types of intervention in the memory cycle. The most violent of these involve surgery on the brain, leaving behind permanent damage, mutilation and the abridgement of the powers of the victim, for the mammalian central nervous system seems to possess no power of regeneration. The principal type of surgical intervention that has been practiced is known as prefrontal lobotomy, or leucotomy. It consists in the removal or isolation of a portion of the prefrontal lobe of the cortex. It is currently having a certain vogue, probably not unconnected with the fact that it makes the custodial care of many patients earlier. (Let me remark in passing that killing them makes their custodial care still easier.) Prefrontal lobotomy does seem to have a genuine effect on malignant worry, not by bringing the patient nearer to a solution of his problem, but by damaging or destroying the capacity for maintained worry, known in the terminology of another profession as the conscience. It appears to impair the circulating memory, *i.e.*, the ability to keep in mind a situation not actually presented.

The various forms of shock treatment—electric, insulin, metrazol—are less drastic methods of doing a very similar thing. They do not destroy brain tissue, or at least are not intended to destroy it, but they do have a decidedly damaging effect on the memory. In so far as the shock treatment affects recent disordered memories, which are probably scarcely worth preserving anyhow, it has something to recommend it as against lobotomy, but it is sometimes followed by deleterious effects on the permanent memory and the personality. As it is used at present, it is another violent, imperfectly understood, imperfectly controlled method to interrupt a mental vicious circle.

In long-established cases of mental disorder, the permanent memory is as badly deranged as the circulating memory. We do not seem to possess any purely pharmaceutical or surgical weapon for intervening selectively in the permanent memory. This is where psychoanalysis and the other psychotherapeutic measures come in.

Whether psychoanalysis is taken in the orthodox Freudian sense or in the modified senses of Jung and of Adler, or whether the psychotherapy is not strictly psychoanalytic at all, the treatment is clearly based on the concept that the stored information of the mind lies on many levels of accessibility. The effect and accessibility of this stored information are vitally conditioned by affective experiences that we cannot always uncover by introspection. The technique of the psychoanalyst consists in a series of means to discover and interpret these hidden memories, to make the patient accept them for what they are, and thus to modify, if not their content, at least the affective tone they carry, and make them less harmful.

All this is perfectly consistent with the cybernetic point of view. Our theory perhaps explains, too, why there are circumstances in which a joint use of shock treatment and psychotherapy is indicated, combining a physical or pharmacological therapy for the malignant reverberations in the nervous system and a psychological therapy for the damaging long-time memories which might re-establish the vicious circle broken up by the shock treatments.

We have already mentioned the traffic problem of the nervous system. It has been noted by many writers that each form of organization has an upper limit of size beyond which it will not function. Thus insect organization is limited by the length of tubing over which the spiracle method of bringing air by diffusion directly to the breathing tissues will function; a land animal cannot be so big that the legs or other portions in contact with the ground will be crushed by its weight, and so on. The same sort of thing is observed in engineering structures. Skyscrapers are limited in size by the fact that when they exceed a certain height, the elevator space needed for the upper stories consumes an excessive part of the cross section of the lower floors. Beyond a certain span, the best possible suspension bridge will collapse under its own weight. Similarly, the size of a single telephone exchange is limited.

In a telephone system, the important limiting factor is the fraction of the time during which a subscriber will find it impossible to put a call through. A 90 per cent chance of completing calls is probably good enough to permit business to be carried on with reasonable facility. A success of 75 per cent is annoying but will permit business to be carried on after a fashion; if half the calls are not completed, subscribers will begin to ask to have their telephones taken out. Now, these represent all-over figures. If the calls go through a number of distinct stages of switching, and the probability of failure is independent and equal for each stage, in order to get a high probability of final success the probability of success at each stage must be higher than the final one. Thus to obtain a 75 per cent chance for the completion of the call after five stages, we must have about 95 per cent chance of success at each stage. The more stages there are, the more rapidly the service becomes extremely bad when a critical level of failure for the individual call is exceeded, and extremely good when this critical level of failure is not quite reached. Thus a switching service involving many stages and designed for a certain level of failure shows no obvious signs of failure until the traffic comes up to the edge of the critical point, when it goes completely to pieces and we have a catastrophic traffic jam.

So man, with the best developed nervous system of all the animals, probably involving the longest chains of effectively operated neurones, is likely to perform a complicated type of behavior efficiently very close to the edge of an overload, when he will give way in a serious and catastrophic manner. This overload may take place in several ways: by an excess in the amount of traffic to be carried; by a physical removal of channels for the carrying of traffic; or by the excessive occupation of such channels by undesirable systems of traffic, such as circulating memories that have accumulated to the extent of becoming pathological worries. In all these cases, a point is reached—quite suddenly—when the normal traffic does not have space enough allotted to it, and we have a form of mental breakdown, very possibly amounting to insanity.

This will first affect the faculties or operations involving the longest chains of neurones. There is appreciable evidence, of various kinds, that these are precisely the processes recognized as the highest in our ordinary scale of valuation.

If we compare the human brain with that of a lower mammal, we find that it is much more convoluted. The relative thickness of the gray matter is much the same, but it is spread over a far more involved system of grooves and ridges. The effect of this is to increase the amount of gray matter at the expense of the amount of white matter. Within a ridge, this decrease of the white matter is largely a decrease in length rather than in number of fibers, as the opposing folds are nearer together than the same areas would be on a smooth-surfaced brain of the same size. On the other hand, when it comes to the connectors between different ridges, the distance they have to run is increased by the convolution of the brain.

Thus the human brain would seem to be fairly efficient in the matter of the short-distance connectors, but defective in the matter of long-distance trunk lines. This means that in the case of a traffic jam, the processes involving parts of the brain quite remote from one another should suffer first. That is, processes involving several centers, a number of different motor processes and a considerable number of association areas should be among the least stable in cases of insanity. These are precisely the processes which we should normally class as higher, thereby confirming our theory, as experience does also, that the higher processes deteriorate first in insanity.

* * *

The phenomena of handedness and of hemispheric dominance suggest other interesting speculations. Right-handedness, as is well known, is generally associated with left-brainedness, and left-handedness with right-brainedness. The dominant hemisphere has the lion's share of the higher cerebral functions. In the adult, the effect of an extensive injury in the secondary hemisphere is far less serious than the effect of a similar injury in the dominant hemisphere. At a relatively early stage in his career, Louis Pasteur suffered a cerebral hemorrhage on the right side which left him with a moderate degree of one-sided paralysis. When he died, his brain was examined and the damage to its right side was found to be so extensive that it has been said that after his injury "he had only half a brain." Nevertheless, after his injury he did some of his best work. A similar injury to the left side of the brain in a right-handed adult would almost certainly have been fatal; at the least it would have reduced the patient to an animal condition.

In the first six months of life, an extensive injury to the dominant hemisphere may compel the normally secondary hemisphere to take its place, so that the patient appears far more nearly normal than he would have been had the injury occurred at a later stage. This is quite in accordance with the great flexibility shown by the nervous system in the early weeks of life. It is possible that, short of very serious injuries handedness is reasonably flexible in the very young child. Long before the child is of school age, however, the natural handedness and cerebral dominance are established for life. Many people have changed the handedness of their children by education, though of course they could not change its physiological basis in hemispheric dominance. These hemispheric changelings often become stutterers and develop other defects of speech, reading and writing.

We now see at least one possible explanation for this phenomenon. With the education of the secondary hand, there has been a partial education of that part of the secondary hemisphere which deals with skilled motions such as writing. Since these motions are carried out in the closest possible association with reading, and with speech and other activities which are inseparably connected with the dominant hemisphere, the neurone chains involved in these processes must cross over from hemisphere to hemisphere, and in any complex activity they must do this again and again. But the direct connectors between the hemispheres in a brain as large as that of man are so few in number that they are of very little help. Consequently the interhemispheric traffic must go by roundabout routes through the brain stem. We know little about these routes, but they are certainly long, scanty and subject to interruption. As a consequence, the processes associated with speech and writing are very likely to be involved in a traffic jam, and stuttering is the most natural thing in the world.

The human brain is probably too large already to use in an efficient manner all the facilities which seem to be present. In a cat, the destruction of the dominant hemisphere seems to produce relatively less damage than in man, while the destruction of the secondary hemisphere probably produces more damage. At any rate, the apportionment of function in the two hemispheres is more nearly equal. In man, the gain achieved by the increase in the size and complexity of the brain is partly nullified by the fact that less of the organ can be used effectively at one time.

It is interesting to reflect that we may be facing one of those limitations of nature in which highly specialized organs reach a level of declining efficiency and ultimately lead to the extinction of the species. The human brain may be as far along on its road to destructive specialization as the great nose horns of the last of the titanotheres.

*Scientific American* 179, No. 5 (November, 1948): 14–19.

## Questions and Topics for Discussion and Writing

1. Explain the concept of feedback in humans and machines.
2. Explain Wiener's analogy which compares mental illnesses to computer malfunctions.
3. Why is a computer such a fruitful model for studying and understanding the human mind?

# The Automated Office

## Heather Menzies

Virtually every resident of North America has some regular experience of computers, even if only the experience of being passively recorded and monitored in the vast data banks of governments, banks, schools, and credit card companies. For many, their contact with computers is much greater through their homes, workplaces, or schools. Relatively few, however, will have yet become so much a part of the computer age as the hypothetical person described by Heather Menzies in this article.

As other companies run to match productivity improvements or perish under competitive pressure, microchip technology is spreading as surely as the typewriter and telephone before it. Here's a brief account of a day in an intermediate job in the office of the future, based in part on the automated system already in place in Canadian Pacific's headquarters in downtown Montreal and in its offices across Canada.

It's 9 a.m. and you've just sat down at your workstation, which differs from your old desk only by being plugged in. Through a desktop (or built-in) unit, which might combine a television-like screen with a retractable keyboard connected to a vast telecommunications network, you're all set to work. First, you "log" onto the system, by typing in your personal password, the equivalent of unlocking your office or desk, for an office is really an information system. Having entered the electronic information system, you begin your day by calling up your electronic calendar and those of other people in the company to which you have access—your subordinates, for instance. After checking what's on for the day, you call up your electronic "in" basket, containing documents sent from other departments or the word processing centre, plus telephone messages and electronic memos. Noticing one from your boss, who's noted a spare hour in your calendar and asked if you could meet with her, you deal with that first. On reading an electronic letter from someone in Germany who, along with you, is participating in a computer conference, you key in instructions to have it tucked into your electronic "computer conference" file on that subject.

With an hour to spare before your first meeting you turn your television screen (a video display terminal or VDT) into an electronic workpad, keying in the code to extract from your electronic filing cabinet the report you've been preparing for an upcoming policy meeting. It's instantly displayed on the VDT screen. Through the communications systems at your fingertips, you gather up-to-the-minute sales data from regional offices, documented electronically when sales are made. You run these through some standard computer programs for statistical analysis to yield regional comparisons, combining the results with existing records to compare with market targets and trends. Then you devise the specifications for a graph that, drawn automatically, will highlight any trends or deviations most effectively. For further background information, you key or dial into the company's information services, first checking company records for any precedents on the type of suggestion you're about to make (more decentralized marketing, for instance), then dial outside the company into industry data banks your firm subscribes to and electronic newspaper files provided by InfoGlobe (*The Globe and Mail*) and *The New York Times*. Finally, you do a bit of computer modelling (simulated implementation of different variations on the policy you're suggesting) to demonstrate the possible effects of such a policy.

Beep-beep, your built-in electronic clock signals that your hour is up. You get up, grumbling about the archaic custom of going to meetings, but still going because there's no substituting for face-to-face brainstorming on the important stuff; besides, you're ready for a cup of coffee.

Such an office of the future is a giant step up from just automating individual job functions. Now these automated functions are integrated and set within a telecommunication network that lifts you, as on a winged chariot, from the print-bound information world of the past into the dynamic realm of electronic information. It's a world as vast as the communications networks (comprising telephones, telex, two-way television sets, cable and even the exotic satellites and optical fibres) and as powerful as the computer tools (word processors and other microcomputers, plus the larger mainframes), with automated clerical functions at your disposal.

The wonder of this new technology won't in itself cause companies to install it. They'll be pushed forward by the greater competitiveness the technology confers on the corporate opposition, forcing all companies to become more productive.

From Heather Menzies, *Computers on the Job* (Toronto: James Lorimer, 1982).

## Questions and Topics for Discussion and Writing

1. Describe Menzies' attitude toward the "Automated Office." Does she welcome it enthusiastically, or does she have reservations about it?
2. What skills are required to work in such an office?
3. Menzies describes the automated office as an inevitable course of events which will make companies more productive. Does this mean that technology is now dictating how work (and perhaps leisure also) will be conducted in the future. Is there no choice left for people?

# Automation in Industry and the Office

Michael R. Arbib

Automation in the workplace did not arise from computer technology alone. Indeed in a sense, the entire history of technology is in some way a history of progressive automation and increased productivity. But automation has a price—the reduction of personal human involvement with intimate details of work can also result in a reduction of personal satisfaction and fulfillment from work. Automation is placing more control and responsibility in the hands of the managers, and less variety and involvement in the hands of workers. In consequence, some formerly satisfying occupations have become dull and stupefying while others are eliminated altogether. Moreover, the rapid pace of technological development means that skilled workers may find their skills obsolete more than once in their working careers. This places a strain upon the community as a whole. On the other hand, automation has relieved workers of some of the duller jobs and opened up new, potentially more interesting occupations. In the following excerpt from his book, *Computers and the Cybernetic Society*, Michael A. Arbib reviews the pros and cons of automation in the workplace.

If we are to understand the impact of automation, we must first place it in historical perspective. The first steps were taken toward automation when one of our distant ancestors, perhaps not even human yet, first picked up a stone to use as a weapon for hunting or as a tool for preparing food or clothing. The first tools provided the first technologies. The change from a hunting to an agricultural society that occurred approximately 8000 years ago (although agriculture began much earlier) marked a dramatic upheaval in the way in which people lived. This transformation was the result of another radical alteration in human society—the move from a relatively scattered tribal existence to one of settled town life (a move that perhaps first took place in Jericho some 9000 years ago). However, the change that offers the richest parallels with the changes now being brought about by automation was the first *industrial revolution*. This "revolution" consisted of a drastic change in the means of production, with increasing use of heavy machinery and the concentration of workers in the factories in which the machines were housed.

A revolution of this kind has no sharp beginning or end, but one may say that it started around 1750 in England and had completed its first major phase within 100 years. During this period, for example, spinning and weaving ceased to be cottage industries; instead of farmers' wives and daughters spinning the wool (gathered from their own sheep) and weaving it into rough cloth to be taken to the market, the wool was gathered and taken to the factories. Similarly, the production of many small items that previously had been handled by craftsmen or guild members working with a few apprentices was now turned over to factories where hundreds of people worked long days in the company of machines.

The first industrial revolution not only brought about a great increase in production, but also a great loss of self-esteem among the workers and much human misery. People who had before lived on their own farms now found that they had the choice of either starving or going to live in a factory town. Frequently, the life of the peasant had been one of misery—there was no age of bliss, but a time of low income and often miserable living standards. But, at first, the new industry provided no improvement and a great loss of traditional life-styles, such as they were. Workers in factories found, all too often, that they were forced to live in the substandard houses provided by the factory owner, work twelve to fourteen hours a day in hideous conditions, and accept whatever small wage they were given—using it to buy food at inflated prices from the company store and spending whatever was left on alcohol to drown their sorrows. It was not only adults who had to work in these conditions but young children, too—children who were often horribly mangled when at the end of a long day they dozed off and got caught in the machinery.

It was on the basis of such conditions that Karl Marx wrote his famous critique *Capital* and tried to assess the likely outcome of these conditions. He expected that capitalism would be crushed as the workers of the world united and cast off their chains. Marx's analysis of the situation was in fact an extremely important one, and much of our understanding of the impact of economic processes on society rests on his insights. Nonetheless, the history of the world in the last 100 years has departed quite strongly from his forecasts. Although I am not an expert in the field of economic history and cannot spell out what really has happened, nonetheless, there are several factors to which I can point. One is the fact that the owners of the factories eventually realized that it was not in their interests to exploit the workers to the utmost limits. While it seemed at first that the less the workers were paid, the greater would be the profits that would go to the factory workers, it soon became clear that to earn as much money as possible, a factory owner had to produce as many goods as possible. To produce more goods, there had to be larger markets. This led to the development of the various colonies of European countries during the nineteenth century as markets for European manufactured goods—and not just as simply sources of raw materials for the European factories. It also led to the raising of the living standards of the workers; for if the workers could become consumers, then the level of production could rise, and so the level of profits could be the greater. Henry Ford doubled his workers' wages so that they could afford his cars! The second factor in the failure of Marx's vision was that the interest of workers in mutual support to improve their living standards was not able to overcome the call of nationalism. Many socialists and communists had hoped, before World War I, that the ties between the workers in different countries in Europe would be such that they would never take arms against each other. Yet, in the hysteria of World War I, even many of the most ardent socialists among the German workers were eager to enlist and go to war against their fellow workers in other countries, all in the name of nationalism.

We could continue this sketchy economic history further, but since our main purpose in discussing the first industrial revolution and its consequences is to provide a perspective for our discussion of the impact of automation, we will discuss just one further social change that was an outgrowth of that revolution. We have seen that as factories came to provide cheaper goods, workers in cottages and guilds could no longer compete in the marketplace and so were forced to go to the factory or to starve. In the factory, most workers were badly underpaid and terribly overworked. The eventual response to this was *unionization*—the workers in a factory finally realized that the only chance they had for shorter working hours and higher wages was to band together. If workers, on an individual basis, were to go on strike, they would simply lose their jobs, and someone else would take their place; but if all of them could go on strike at the same time, then the loss of production would pose such a serious threat to the industrialists that they would be forced to take steps to improve their conditions so they could get their employees back to work, and get their profits flowing once again. (It may shock many readers that as recently as the 1930s, United States steel companies could call upon armed "goons," or strikebreakers, to come in and shoot down striking workers, sometimes with the assistance of state and local police. It is only since the 1930s that the unions have had a strong voice in the conditions of labor throughout the United States.) One of the questions that we discuss below is the extent to which the union can still protect the interests of the worker in the age of automation.

### From Assembly Line to Automation

With this background we can move from the first industrial revolution to the invention of the *assembly line*: the discovery that the productivity of humans and machines in a factory can be raised dramatically if each job is analyzed and broken down into simple parts, with each human being required to specialize in a single subtask.... The various parts that had to be assembled into a finished product could then be arrayed along an assembly line that would move steadily past individual workers, with each worker carrying out a few simple tasks as a part moved by. Because the tasks were simple and repetitive, a worker could become very skilled at them and carry them out very quickly. Because the parts were moving along on an assembly line, rather than being reached for by the worker when he or she was ready to handle them, the pace of work could be stepped up. While the introduction of an assembly line into a factory required the installation of new equipment, and could thus be very expensive, these costs could be more than made up for by the increased productivity that resulted. Increased productivity could mean, in turn, increased wages for the workers—as long as they had a strong enough union to help them share in the new profits. The price they paid for this increased economic well-being was having to carry out the same monotonous task day in and day out. While for some people the chance to get paid for a job that required no mental effort on their part was satisfying, for all too many other people the job on the assembly line was intensely boring and frustrating. Thus the first industrial revolution and the subsequent increased use of the assembly line have provided jobs that allow the worker to earn a reasonable amount of money and have made a large contribution to productivity. However, for many workers there has been a great loss of personal involvement and satisfaction in their work.

With this, we can move on to look at *automation*. We may characterize automation as a continuation of the assembly line idea. Once a job is divided into simple subtasks, it becomes easier to turn those tasks over to a machine; consequently, we find that, with automation, more and more of the jobs of the assembly line are being done by machines. The computer can be used to make these machines a little more "intelligent" than might otherwise be possible. Instead of doing a repetitive task the same way every time, a computer-controlled machine can follow a program that allows it to test for various conditions on the assembly line and modify its actions accordingly. The other face of automation is that many tasks that are not of the assembly-line type can also be turned over to the machine—for the computer is an information-processing device. Thus, a whole range of clerical and even higher-level tasks can now be done by the machine.

What are the advantages of automation? From the point of view of the worker, it means that some of the most boring tasks of the assembly line need no longer be done by humans. Much human drudgery can be done away with. Similarly, many jobs that expose the worker to dangerous gases or dangerous conditions, such as coal mining, can be increasingly done by the machine, thus decreasing the number of workers who must put their lives in danger. Furthermore, according to hearings conducted by the US Congress in the early 1980s, automation is the key to increased productivity. Congress concluded that the declining position of the United States relative to Japan is in part due to the increased use of "robots" in Japanese factories.

Of course, these very advantages bring disadvantages. If there are fewer workers exposed to boredom or danger, there are also fewer workers who have jobs. To some cynics, it may seem that this is one of the major reasons why managers should decide to automate their plant, but the savings in payroll may only be comparable to the cost of putting in the new machines. Here, the even more cynical observer would suggest that the reason for automating is that machines "don't go on strike, whereas people do"—by automating, factory owners are "getting the union off their back." This leads to the suggestion that automation is automatically a bad thing, and that one should always design a factory to provide as many jobs as possible for people. This is a somewhat naive viewpoint, however, as we can see by looking at a specific example: the newspaper industry.

Until recently, a newspaper was prepared in roughly the following way. A news story was either typed by the journalist who had covered it or was pulled off the teletype if it had come "over the wire" from some news service. The various typed and teletyped pages were then passed on to an editor who would mark up the pages in various ways, carefully count the number of words, and on this basis lay the material out ready for typesetting. The marked-up pages would then go to a typesetter, whose job it was to sit at a special typewriter device and type the material in such a way as to set type. A print could be taken from this so that a proofreader could check to see whether all the type had been set correctly. The type would then be locked in place and sent off to be used in making the masters that would go on the printing press. Finally, the paper would be printed.

The computer is changing this process. Journalists no longer have to compose their stories at conventional typewriters. Instead, they type at a *graphics terminal* that displays their work on a screen, thereby allowing them to read, edit, and correct their material. When they are satisfied with their stories, they can transmit them to their editors over the office information network. This process can also be used for stories that come in over the wire. A computer system, rather than a teletype, receives the wire stories and codes the information into the computer memory. Editors can then handle the stories in exactly the same manner that they handle the stories of their in-house journalists.

By inserting the stories—in magnetic form—into their machines, editors are able to use consoles to edit the material, removing information that they do not want published and adding new information. They do not have to count words and worry about space, since their computers can show them how much space will be taken by the remaining material. Consequently, it is very easy to adjust the length of articles. Once the editors have finished their jobs, their computers can instruct the appropriate machines to set the type. Thus, it is no longer necessary for humans to transfer the material from manuscripts to the typesetting machines. This, in turn, eliminates the need for human proofreaders to check the accuracy of the typesetting. In all, automation eliminates the need for three job positions: teletype operators, typesetters, and proofreaders.

The monetary benefit resulting from the elimination of jobs is not the only reason that newspaper automation is an attractive approach. Newspaper publishers also find it appealing because of its *flexibility*. Since it takes several hours to set type for a paper, it is almost impossible to get a hot story into the paper if it requires any major revision of material—unless the story is known several hours before the paper goes to press. With a computer typesetting system, however, the changes can be made almost at the last minute. Again, because all the information is stored in the computer…it is very easy to rearrange sections of the paper simply by directing one portion of the tape to be read into the typesetting machine before another. In other words, the advent of automation in the newspaper business is important, not only because of the economies it affords, but also because of the immense flexibility it offers. For further considerations of this kind, see the discussion of office automation below.

### Automation and the Unions

Many newspaper publishers have already converted their operation to the computer mode. This has only been done, however, through the loss of a great many jobs. At first, the *unions* that represented these workers were not willing to see them laid off. In one very bitter strike at the Washington Post during 1975, not only did the printers go on strike, but fires were set and some people took crowbars to the printing presses to ensure that the paper could not be printed on those presses during the typesetters' absence. But this battle has now been lost.

The outcome of labor action of this kind is often quite strange—the union agrees that the managers may install the new equipment, while the managers agree that they will not fire anyone. One of the most dramatic examples of this "featherbedding" was the decision to keep a fireman on every diesel train on American railroads. When the trains were pulled by steam engines, a fireman was required—in addition to the engineer—to shovel coal into the steam engine's furnace. When the diesel engines came along, there was no longer any need for the fireman. However, the unions were strong enough to get the railroad to agree to have a fireman in each engine cab—even though there was no work for him to do—rather than run the risk of a long and costly fight in which the engineers would join the firemen in a walkout. Of course, when firemen retire, no new firemen are hired—and so this attrition slowly cuts down the featherbedding. The job of the union is to protect their members—and this is the key to the survival of any union leader. In this case, we see that the unions succeeded in preserving only one of the conditions of the job—that it was a source of money for the worker—and had left the job with no socially productive role and no aspects of self-fulfillment for the individual worker.

It is true that the unions continue to play a very important role for many workers in ensuring that they are properly paid and that they have tolerable working conditions. For many other workers, however—such as the firemen and the typesetters—it seems that the unions cannot properly respond to changing conditions. A century or two ago, farmers would expect to work the land in the same way as their fathers or their fathers' fathers before them. Artisans would expect to carry out their craft in exactly the same way as the masters to whom they were apprenticed and their masters' masters before them. However, we now live in an age that is marked by dramatic change. Since 1940, we have seen television go from a novelty to a standard medium of communication; we have seen the first computer built and seen millions more following; we have seen airplanes become a standard means of transportation and the jet engine come into its own; we have seen space travel with men walking on the moon and with spacecraft sending back pictures from many of the planets of the solar system; we have become used to satellites relaying our messages around the globe and providing us with surveys of our own resources; and we have seen incredible advances in medical science. With all these changes, many old means of employment have disappeared forever, while totally new opportunities have arisen. The service industries, of which the provision of computer services is the one closest to the theme of this book, have experienced especially large growth.

Even with the growth of new professions, it is still not clear that new technology can create more jobs than automation destroys. It is likely that, if major unemployment is to be reduced, we will see a trend toward earlier retirement and a shorter work week. However, automation will tend to make fewer and fewer jobs available for unskilled workers. Because of this, most countries will need major improvements in the education of the poor, the disadvantaged, and the minorities if these people are not to be permanently frozen out of the marketplace.

The young person starting his or her career today cannot expect to work in the same job for 40 or 50 years. Rather, a quite drastic change in professional orientation—and a continual process of reeducation—must be expected. This puts many unions in a very difficult position. While such unions as a roofers' union can probably count on continuing demand for the services of their members, other unions must face the fact that in a few years the profession that they serve will be, if not obsolete, at least in very little demand. These unions, then, find themselves to be too job-tied for an adaptive society. Their emphasis on the size of the paycheck and the number of hours worked no longer meets the real emotional needs of the workers in finding satisfying employment. Yet, if a union tries to place its members in other professions, other unions will accuse it of poaching. Thus, in the future, unions will have to develop far larger structures of interunion cooperation to enable them to predict patterns of change in employment and work together to ensure that, wherever possible, workers are retrained and given union cards that will enable them to continue to play useful roles in society. However, even such an agreement will not make life easy for the individual who has held a senior position in one profession and can—even after retraining—only qualify for half the pay in another job.

The point being made is that when we consider automation...the real focus is not on the use of computers but rather on the broad social problems of which the computer is but one aspect. We see that there has been a massive change in the conditions of the workplace over the last two centuries, starting with the industrial revolution. Automation continues that process, as it puts some people out of work and creates new employment opportunities for others. It removes certain boring and repetitive tasks from the marketplace, it increases productivity, and it decreases cost through mass production. What remains is to develop a sufficiently flexible social structure to take advantage of these changes by ensuring that each member of society has a fair share of the wealth that is created in this way and has a chance to make meaningful use of his or her talents.

During the 1980s, the major trend in union negotiations has been toward demands for job security and profitsharing—on the part of workers—and toward increased productivity—on the part of management. Another important, but not yet sufficiently widespread, concern is for *worker democracy*. Worker democracy is the approach whereby workers are allowed to join in discussion with union leaders and management on such matters as improvements in working conditions, incentives, and production methods.

*Michael R. Arbib*

### Cars and Computers

We now leave questions of the social impact of automation and devote some pages to a number of examples of the way in which computers are being used in the automobile industry.

First, let us look at some of the ways in which the computer can be used to help with the mental work involved in car production—namely, in the design process, itself. Car parts—fenders, hoods, and so on—are stamped out of sheet metal using large dies. In the past, these dies were hand tooled by craftsmen, working from elaborate blueprints supplied by the designers. Now, however, the blueprint comes, not in the form of a few charts on paper, but in numerical form stored inside a computer. These computer specifications can be used to drive the milling machines, instructing them to carve out pieces of metal in the desired shapes so that they can be used to stamp out the bodies of individual cars. Moreover, the use of the computer does not simply come in to replace the craftsmen and the milling machine. It is also used to help the designer develop the specifications of the car. It is still the practice in such companies as General Motors to build a full-size clay model of half the car (after all, the other half is just the mirror image!) and shape this to form an aesthetically pleasing design. However, once the overall "look" of the car has been agreed on, the remainder of the redesign process will take place inside the computer. An elaborate machine is used to move back and forth over the full-size clay model and feed the measurements of each point into the computer. Engineers can then sit at a graphics console and look at pictures on the screen showing how the car looks from various angles. They can then carry out computations that will "sweeten" the measurements to make sure that all the curves on the surface are smooth.

Not only does this process reduce a great deal of the tedium of the original manner of drawing the blueprint by hand, but it also puts the measurements in a form where they can be easily modified. For example, a windshield has to be designed to fit a specific opening in the car and must meet government regulations about the sweep of the wipers. In addition, it must be shaped in such a way that when molten glass is poured into the mold for the windshield, it will sag into the mold without any distortion or stretching. An engineer might take two weeks to develop a design satisfying all of these requirements if it had to be done by hand—yet it could be done in one afternoon using computer graphics. The cost of the engineer and the machine together for one afternoon may be no less than the cost of the engineer alone for two weeks. What is important, though, is the *turn around time*. It is difficult enough to meet the basic production schedule for a car—but any sudden design changes occurring late in the schedule would be absolutely catastrophic without the capacity that computers give for rapid recomputation of specifications.

...The computer can also be used...in the process of car design [to predict] how a particular type of structure will respond to different types of crash. In the past, the testing of a car body's response to a crash was an elaborate and costly business. to set up a crash in such a way as to ensure that the impact would be of a particular kind might take two weeks, and the total cost would be $30,000. Now, with increased understanding of the effects of different types of impact on different types of metal structure, mathematical models can be used to analyze the resistance to crashing of different regions of a car. However, to successfully model these effects, we need to know various parameters about the strength of different pieces of metal.... These parameters can be obtained by means of a device called a static crusher, which applies force to one portion of the car at a time. Thus, many different parameters can be obtained in a relatively short period of time using a single car. All these numbers, together with a program representing a general mathematical model, can be fed into a computer. the engineer can then simulate different types of crashes and receive as output from the graphics system a sequence of drawings that can be used to make a film—an "animated cartoon"—of the crash. The engineer can then look at this film and discover what seemed to be the most disturbing types of damage to the car in various types of crashes and then develop various modifications for those portions of the car. After a sequence of experiments in the static crusher, followed by a number of runs on the computer, a series of design changes can be made in a short time at relatively little expense. Then, the model can be checked out by building a car that meets the new specifications and running it through a real crash.

*Michael R. Arbib*

Another use of computers is for automatic warehousing—keeping a complete data base...of all parts, including their part number, description, and location in the warehouse. This allows an automatic updating of inventory—perhaps even to the extent of automatically replacing orders at the appropriate factory when a given stock gets below a certain level. The computer can also be used to keep track of the extent to which warranties are used. If a distributor seems to be charging the company for an unnecessary number of repairs under warranty,—as measured by a comparison of his or her replacement rate with the national average—then the company can have a spot check made...

In addition to the use of computers to control particular pieces of machinery, they can also be used to monitor the assembly line process. For example, sensors can be placed at different stages in an automated production line to ring an alarm if a jam-up occurs. These computerized counters are also important for maintenance scheduling. Since the counters are also used to maintain a fast schedule of production, the workers often try to trip mechanical counters. In this way, they can increase their apparent production rate and thus slow down. Computerized counters are less easy to fool in this way.

One of the greatest dissatisfactions of assembly-line workers is with having to carry out a task at a fixed speed set by the assembly line. Volvo set about reducing this problem by introducing *buffers* into many of their existing assembly lines. A buffer is a "waiting room" or "stockpile" between two successive operations in a production chain. Instead of having to carry out a task every fifteen seconds as things go by on the assembly line, a worker can work fast to build up a number of processed units in his or her buffer and then take a break while the next worker catches up. Without buffers, a holdup or breakdown at any place on the assembly line can stop production all the way down the line. With buffers, the workers can enjoy their work more, and there is better productivity because the buffers can absorb the effects of most disturbances.

Volvo has used computer simulation in deciding where to put the buffers on their existing production lines. The programs estimate the maximum annual production of the line, the distribution of each worker's time, the time workers spend waiting due to breakdowns (both of people and of machines), the opportunities workers have to take breaks, and can thus determine the relative merit of different sizes and placements of buffers.

With proper technical planning and worker participation, technology can be used to *humanize* the workplace...

### Office Automation

The office, that traditional "work station" of white-collar workers, remained almost the same for many decades. In the 1920s a "modern," office was supported by a telephone, a typewriter, a calculator, and some kind of a duplicator. By the 1970s, the only major invention was the photocopier machine. Neither the technological support systems nor the work procedures in offices had changed much over a fifty-year period. The past few years, however, have ushered in the era of the automated or technology-enhanced office, the backbone of which is the microprocessor used in microcomputers, in massive networking, and in smart appliances of all kinds (including new forms of telephones and typewriters). There is every indication that in the very near future, the contemporary office will be as different from the turn-of-the-century office as the factory is from a cottage operation.

Why is office automation just arriving? It certainly has not been for lack of need. The most obvious explanation is that the office is primarily a place for dealing with data and information and that the kinds of machines that dominated the earlier eras, with their gasoline engines and steam engines, could not have helped with the main functions of the office. The computer, on the other hand, is an information machine and can thus facilitate office functions. It has only been in the last few years, however, the microprocessor costs have plunged to the point where they favor widespread adoption.

Even though the technology has arrived, are there enough economic factors to cause its widespread adoption? There seem to be. Consider productivity. Its vast increase in prior decades enabled the enormous upsurge in the material standard of living. During recent years, however, improvements in productivity have been small. In combination with other factors, the result has been a stagnant economy saddled with inflation. While there are many explanations, one really stands out: Since the 1940s, more than half the labor force has been employed in service industries, working essentially in offices. While industrial productivity has increased by 90 percent over the past ten years (in great part due to advances in factory equipment and in automation—the investment in equipment in US factories averages $50,000 per worker), office productivity has increased by only 4 percent. This low productivity rate has served to offset the higher rates for industry. As a result, the overall increase in productivity for the economy has been meager and far below the wage and salary increases. This meager growth rate has been one important reason for inflation. Understandably, interest in increasing productivity in the office has been intense—with the attention given to office automation being one manifestation of this.

A by-product of office automation appears to be an increase in home computing. This is not surprising given the fact that so many office workers take home a portion of their work—with or without the computer. As a result, the American home may evolve into a partial extension of the office. If this trend continues, the move to suburbia caused by the automotive revolution may be greatly intensified as the need to come to a centrally situated office decreases. The possibility of using homes as places for office work could well be another stimulus for office automation. The mounting energy prices that have been rapidly pushing up the costs of transportation, heating, and lighting, coupled with inner-city problems, would make it very attractive for businesses to allow their office workers to work right out of their homes. In this way they would be acquiring valuable space at no cost to the businesses. For the workers, the freedom to stay home, or to go wherever their terminals can travel, could mean a most welcomed change in life-style—although it might still be pleasant to visit the office from time-to-time to meet people.

Unfortunately, the word automation gives the erroneous impression that advanced technology will require less judgment, and even less education, on the part of the manager. The opposite seems to be the case. The more technology, the more the need for judgment and training. For instance, secretaries who use computer-based word processors need to be much more sophisticated in their use and understanding of technology—even to the point of having a knowledge of the concepts of programming—than do secretaries who use manual typewriters.

Office automation has many levels. The limited viewpoint is that it is an extension of data processing. As such, it includes technological innovations that reduce the task of typing—for example, word processing (that is, microprocessor-controlled typing, editing, and reproduction of text) and source-data capture through such means as optical character recognition and the like. A broader view of office automation is that *it is a technology that is leading the way to a paperless society through the mechanization of paper-dependent tasks.* Automatic bank tellers, electronic credit cards, and airline-reservation systems would fall under this definition.

It is doubtful that such approaches to office automation will make a significant impact on productivity. Over two-thirds of the white-collar wage bill goes to managers and professionals. In contrast, secretaries and typists receive only 6 percent of the pie. If every typewriter were transformed into a word processor, the impact might be negligible. In fact, the productivity might well go down initially because of the need for considerable retraining. In addition, word processing seems to transfer such traditional tasks as proofreading and sorting from secretaries to managers who are paid many times more than the secretaries. With such displacements taken into account, overall productivity might remain down even after retraining.

A broader view of office automation is that it focuses on providing knowledgeable workers with a better working environment by augmenting their capabilities through powerful, intelligent tools. An even broader perspective is that office automation includes decision-support systems. In fact, Decision-Support Systems is emerging as a field in its own right, combining data processing with decision modeling.... An example of office automation in this context is the computer-based portfolio-management system for use by a bank. It consists of specialized data bases and modeling tools for the construction of forecasts and for answering "what-if" questions. Whereas the narrower concepts of office automation focus on the automated execution of structured tasks, the decision-support view deals with the formalization of semistructured decision processes.

The broadest concept of office automation is that it is the utilization of technology to improve the realization of office functions in terms of its objectives. This viewpoint is a derivative of the so-called systems analysis and design. Thus, one can describe office automation as the computer-based application of systems analysis to the office. Within this framework, office automation is not a matter of technology but of how technology is employed.

If office automation is to be achieved within a systems-analysis framework, the office itself must be well understood in terms of its *functions*. Unfortunately, people all too often describe the office in terms of activities that can be readily detected—for example, in terms of typing, filing, and memo writing. A more appropriate way is to view the office as having activities that are distributed both in time and space and that involve the continuous coordination of parallel activities in many different locations. Office activities are interactive and often involve groups. The information supporting these activities comes in many forms (text, voice, gestures, motions, and pictures) and from many sources. In short, what meets the eye in an office probably constitutes a small and insignificant portion of the function of the office. The problem is that, while managers partake in the real functions of an office, they find it difficult to describe what happens in sufficiently operational terms to permit the introduction of office automation within the framework of systems analysis...

To be truly effective, each office should have its own "tailored" computer-based *expert system* to represent the knowledge of the *particular office* in which the system is to operate. The expert system should combine powerful modeling techniques with techniques of acquiring and representing knowledge in ways that are intriguingly similar to how people acquire and organize knowledge.

The automation of payroll and accounting procedures probably represents the first successful commercial applications of office automation in terms of office activities. Airline-reservation systems might be viewed as the second generation. Then came the construction and subsequent networking of a variety of data banks. Presently, word processing is rapidly spreading across offices, as well as some homes. Electronic mail is also catching on. So is teleprocessing in its various forms, from the teleprocessing of checks to the teleprocessing of medical symptoms.

In these applications, one can discern two important technological innovations in hardware. The first innovation is what we have so often mentioned, namely the microprocessor. This ubiquitous device has resulted in a variety of intelligent appliances. The word processor is one such device. So are the reproduction devices that can handle text as well as graphics. Smart telephones, dictating units, and memo pads are also some of the vast number of smart appliances to arrive in the office.

The second important technological innovation in the area of hardware for office automation is networking. In networking the appliances, as well as data bases, communicate with each other. In the long run, networking is likely to prove to be the more important innovation. One reason for this is that managers spend about 80 percent of their time just communicating. And it is networking that gets to the heart of communication.

Networking plus microprocessor-based appliances have brought about a variety of trends. The following are some examples.

*Personal or Desk-Top Computers.* At less than $2,000, the personal computer can do reports, text editing, and graphics.

*Proliferation of Intelligent Appliances.* Many of the maintenance functions of an office—such as detecting malfunctioning lights and reporting them, turning lights on and off, detecting harmful substances, and so on—will be in the domain of intelligent appliances.

*People-Compatible Input/Output.* Presently, there is a trend to make input/output compatible with people. For instance, attempts are being made to design desk-top computers that can understand spoken commands and talk to their users. This latter capability is already available, the former is still in the design stages. More and more appliances are acquiring people-friendly input/output capabilities and this trend is bound to continue.

*Local-Area Networking.* A good deal of the communication in a company takes place within its own boundaries. Therefore, it is perfectly reasonable to have the need for local-area networks that will conveniently connect a company's personal computers, powerful peripherals, and various appliances together. Several such networks are being marketed already, a primary example being Ethernet developed by Xerox. It is estimated that by the end of the decade local-area network sales will reach many billions of dollars.

*Data-Base Management Systems.* The trends mentioned above all directly relate to hardware technology. However, the real kind of office automation, in terms of the functions of the office, will most likely depend on software technology. The most notable continuing trend in this domain is that of data-base management systems. These systems are rather like a very efficient librarian who can listen to your information needs and then go and fetch the books that best match your description. Some of the commercial products available are ADABAS, System 2000, Mark IV, and System 1022.

*Decision-Support System Banks.* These system banks (financial, not data) are among the more eager consumers of decision-support systems. These systems seek to combine data bases with decision models.

*Knowledge-Based Systems.* In many ways, knowledge-based systems—or expert systems—combine data bases, data-base management systems, and modeling techniques in a very sophisticated way. They depend on recent breakthroughs in representing data in terms of language so that the users can interact with the systems in the same way they might interact with a knowledgeable colleague. Such systems have already reached the application stage, especially in medicine and oil exploration.

*Distributed Processing.* It is reasonable safe to safe to say that American prosperity is in part due to a remarkable innovation in the design of industrial organizations: decentralization as pioneered by General Motors, among others. As an organization grows, its housekeeping tasks grow exponentially, very quickly swam,ping the managers who are supposed to provide guidance. It is because of decentralization that the managers can find the time to manage. The same is true of information management. When centralized, it can quickly grow into an unmanageable and unmanaged activity. Distributed processing can be the salvation. In the past, lack of networking made distributed processing difficult but not impossible (tapes could still be mailed). Now, however, the trend is toward distributed processing with local data bases being developed, maintained, and interconnected.

*Personal Modeling.* The really significant aspect of office automation is provided by managers plugging themselves into the automated office through personal modeling; that is, managers modeling their decision- and information-processing activities and functions and using the technology to enhance and augment themselves as office workers. It is in this manner that the real productivity gain can be registered.

From Michael A. Arbib, *Computers and the Cybernetic Society*, 2nd ed., 338–353 (Orlando, FL: Academic Press, 1984).

## Questions and Topics for Discussion and Writing

1. What is "automation"? Give some examples *not* involving computers.
2. Why does the trend to automation present social problems?
3. Give some examples of new technologies and services that would be totally impossible without automation.
4. In a society tending towards greater automation of all kinds of work, what kinds of work will be the most interesting? What will be the least interesting?

Albert Einstein and J. Robert Oppenheimer.

# Scientists in Society

How is a scientist different from another person, and how does that difference affect the scientist in relation to the rest of society? One difference is that a scientist, especially a research scientist, has an expert knowledge of a particular field that is probably not shared by very many others. Because of that expert knowledge, the scientist may have a responsibility to interpret his or her field of specialization to the community, and may have the responsibility to protect the community's interests from adverse consequences that he or she could foresee, but which non-specialists could not. The other side of the coin is that highly specialized scientists may be so specialized that they are quite ignorant of most everything else outside their field.

Scientists share the human failings found in the population at large, including avarice and envy. Scientists whose careers are built on the quality and quantity of their research output, which few people will be able to understand, are faced with constant temptation to fudge a few results here and there to help an experiment along. Though there is probably very little yielding to that temptation, there is no doubt that there is some.

The articles in this section explore the three themes mentioned above. Robert M. Hutchins' "Science, Scientists, and Politics" deplores the narrow, overspecialized education of scientists, which he sees as fitting them for very little outside of their narrow subdiscipline, and not even very well for that. J. Robert Oppenheimer, the "father" of the atomic bomb, muses in his "Farewell Address to the Los Alamos Scientists" on the thoughts, beliefs, and feelings, shared by himself and his colleagues who worked on the bomb, and what they thought they were doing for humanity. Finally, Beth Savan's "The Academic Hustle" analyzes several notable cases of scientific fraud, how they happened, why they were not detected right away, what motivated the perpetrators, and why science remains vulnerable to such abuses.

# Science, Scientists, and Politics

## Robert M. Hutchins

Robert M. Hutchins was one of the United States' most vigorous and eloquent supporters of a well-rounded liberal education. As President of the University of Chicago he revamped the undergraduate curriculum so that all students, no matter what their eventual field of specialization, were given a comprehensive grounding in the arts *and* the sciences. Later he was appointed to the Center for the Study of Democratic Institutions in Santa Barbara, California, where he continued his advocacy for the integration of all learning. In this essay he takes aim at the excessive specialization of science and argues forcefully that it produces neither a good education nor, ultimately, good science.

It is clear that the behavior of professors is questionable at best. Scientists are worse than other professors because they have special problems. One of these is that their productive lives often end at thirty-five. I knew an astronomer who was contributing to the international journals at the age of eleven. Compare that with the difficulty of contributing at a similar age to an international journal on, let us say, Greek law. A scientist has a limited education. He labors on the topic of his dissertation, wins that Nobel prize by the time he is thirty-five, and suddenly has nothing to do. He has no general ideas, and while he was pursuing his specialization science has gone past him. He has no alternative but to spend the rest of his life making a nuisance of himself.

Scientists are the victims of an education and a way of academic life created by their misinterpreters and propagandists. These misinterpreters have propagandized an entirely inconsecutive chain of consecutive propositions: The pursuit of truth, they say, is the collection of facts. Facts can be experimentally verified. Thus, the only method of seeking truth is the scientific method. The only knowledge is scientific knowledge, and anything else is guesswork or superstition. So Lord Rutherford could say to Samuel Alexander, the great English philosopher, "What is it that you have been saying all your life, Alexander? Hot air. Nothing but hot air."

A recollection I shall always cherish of one of our leading mathematicians, now a professor at Chicago, affords a stunning example of the frame of mind the propagandists have created. He came to Chicago as a graduate student. Toward the close of his first year I asked the chairman of the mathematics department how the boy was doing. "Oh, Mr. Hutchins," he said, "he's a fine mathematician, but I'm sorry to have to tell you, he's crazy." I said, "What do you mean 'crazy'? How does he evidence this unfortunate condition?" And the professor responded, "He's interested in philosophy!"

The misinterpreters' and propagandists' doctrine has paralyzing educational repercussions. According to its tenets, education consists in cramming the student with facts. There is not enough time to stuff in all the facts. Therefore, facts outside a narrow area of specialization must be excluded. One of our Consultants to the Center has described the education in science in the state university from which he graduated as two years of German, two years of military training, and all the rest mathematics, physics, and chemistry.

Seduced by the fact formula, the medical school at the University of Chicago set out on a perfectly sincere, although somewhat misguided, campaign against liberal education. There are countless facts in medicine. A medical school must fill its students with these facts or they will fall behind. This meant that there was no time to teach anything else. The medical school strongly recommended that the whole freshman and sophomore years be abolished—the junior and senior years had already gone—and that the entire curriculum be devoted to science and medicine. I can conscientiously say that any senior in the University of Chicago medical school knew more facts about medicine than any professor in a German university.

The consequences of this line of educational endeavor are clear enough. Everybody specializes. There can be no academic community because scientists cannot talk to one another. The chairman of the anatomy department of the University of Chicago brought this home to me once when we were discussing the great biological symposium that had been held to celebrate the University's fiftieth anniversary. I said, "Tell me, how was it?" He said, "I didn't go." When I asked why not, he replied, "Well, there weren't any papers in my field." Scientists cannot talk to anyone else because there isn't anyone else worth talking to. Hence, university life offers no remedy for the defects of their education.

The propagandists and misinterpreters of science have set the tone for the whole learned world in the United States. Their slogan is, "If you can't count it, it doesn't count." The influence of this slogan is felt in literature, philosophy, languages, and of course in the social sciences. The most striking feature of social science today is the total absence of theory. Its greatest modern achievement is the public opinion poll. Social scientists can count, but cannot comprehend.

Those who live their lives without theory are technicians, or mechanics. As a result there is no significant contemporary social science. Politics is viewed as power because power can be observed and measured. Power is something real. Therefore, using the misinterpreters logic, it is *all* that is real about politics or political science. The most characteristic book title in social science in the past thirty years is *Politics: Who Gets What, When, How.*

In spite of the misinterpreters' nonsense, science contains elements of sense. Serious scientists know that science is just one very important way of looking at the world. When scientists are actually doing science they are caught in a great tradition. They know they are not simply collecting facts or conducting random experiments. No serious scientist believes that if a million monkeys were put down at a million typewriters one of them would eventually turn out *Hamlet.* Nor does he think the scientific method is the only method. Scientists do not use the scientific method outside of science.

How the propagandists and misinterpreters of science have managed to take over all the academic virtues and label them "scientific" escapes me. I ran across a fascinating study of the scientific attitude by a professor of education. This learned gentleman had written to sixteen eminent scientists and asked them what characterized the scientific attitude. These were the replies:

- Chemist: *Openmindedness*
- Physiologist: *Intellectual honesty*
- Botanist: *Openmindedness*
- Zoologist: *Observation, inquisitiveness, perseverance and industry, objectivity and critical independent reflection*
- Physicist: *Objectivity*
- Sociologist: *Objectivity*
- Microbiologist: *Respect observation*
- Mathematician: *Openmindedness*
- Anthropologist: *Openmindedness*
- Chemist: *Practiced willingness to label conclusions tentative until supported by reproducible or confirmed data*
- Agriculturist: *Desire to tolerantly explore ideas*
- Mathematician: *An open mind*
- Physicist: *A will to know the truth*
- Chemist: *Insistence on critical examination*
- Director of Educational Research: *Intellectual curiosity*
- Psychologist: *An inquiring mind.*

Obviously this study shows that science has a corner on all the rational processes of thought.

But there is not an honest scholar in any field who would not insist on being openminded, honest, and objective, and on considering all the evidence before he reached a conclusion. You can hear Thomas Aquinas laughing.

The propagandists of science say, "Sure, but fellows like Thomas Aquinas had commitments. They all had philosophies and principles that distorted their thinking. Scientists haven't any." The answer to this is that everybody has a metaphysics. Every scientists, for example, has a commitment to the reality of the external world. The distortion comes when the metaphysics is denied instead of being recognized and made as rational as possible.

Understanding science is an indispensable part of a liberal education. To demonstrate my sincerity, I point out that at the University of Chicago one whole half of the first two years of every student's education was natural science. St. John's College, with which I also had something to do, is the only college in the United States that requires four years in the laboratory for every student. An education without science is no education at all. The limitations and possibilities of science cannot be understood without scientific training, and our very existence depends on comprehending these limits and possibilities.

We do not know what science is, and partly as a result we do not know what politics is. Mr. C.P. Snow is wrong about the two cultures. There is only one, and it is pseudo-scientific.

The leading phenomena of our time exhibit a curiously ambiguous character. Technology may blow us up, or it may usher in the paradise of which man has been dreaming ever since Adam and Eve got kicked out of the first one. Bureaucracy may stifle democracy or be the backbone of democratic government. Nationalism may disrupt the world or prove to be the necessary precondition of a world community.

Unfortunately these ambiguities do not lend themselves to scientific procedure. Our essential problem is what kind of people we want to be and what kind of world we want to have. Such questions cannot be solved by experiment and observation. But if we know what justice is, which is not a scientific matter, science and many other disciplines may help us get it.

The problems resulting from these ambiguities are not going to be solved by men of fractional or pseudo-culture. The solution depends on moral and intellectual virtues rather than on specialized knowledge. It is a humbling thought to recall that 25 per cent of the SS guards in Nazi Germany were holders of the doctor's degree...

From the Occasional Paper *Science, Scientists, and Politics*, a publication of the Center for the Study of Democratic Institutions in Santa Barbara, California, 1963

## Questions and Topics for Discussion and Writing

1. Why are scientists usually unable to understand research papers outside of their own specialties?
2. Discuss the quoted comment by the chairman of the mathematics department that his graduate student was a fine mathematician but he was "crazy" because he was interested in philosophy.
3. Discuss the comment made by Hutchins that "the leading phenomena of our time exhibit a curiously ambiguous character." What is the relevance of this comment to science education?

# Farewell Address to the Los Alamos Scientists

## J. Robert Oppenheimer

There is no better example of the ethical problems that can beset scientists than that which confronted the nuclear physicists who designed and built the first atomic bombs. The leader of the American team that developed the bomb at Los Alamos, New Mexico, was J. Robert Oppenheimer. In this, his farewell address to his colleagues, Oppenheimer expresses the mixed feelings and ambivalent convictions shared by all of them about why they did what they did, whether they really had any choice in the matter, and whether the secrecy that was necessary in wartime weapons research could ever be reconciled with the traditional openness of science.

In considering what the situation of science is, it may be helpful to think a little of what people said and felt of their motives in coming into this job. One always has to worry that what people say of their motives is not adequate. Many people said different things, and most of them, I think, had some validity. There was in the first place the great concern that our enemy might develop these weapons before we did, and the feeling—at least, in the early days, the very strong feeling—that without atomic weapons it might be very difficult, it might be an impossible, it might be an incredibly long thing to win the war. These things wore off a little as it became clear that the war would be won in any case. Some people, I think, were motivated by curiosity, and rightly so; and some by a sense of adventure, and rightly so. Others had more political arguments and said, "Well, we know that atomic weapons are in principle possible, and it is not right that the threat of their unrealized possibility should hang over the world. [It is] right that the world should know what can be done in their field and deal with it." And the people added to that that it was a time when all over the world men would be particularly ripe and open for dealing with this problem because of the immediacy of the evils of war, because of the universal cry from everyone that one could not go through this thing again, even a war without atomic bombs. And there was finally, and I think rightly, the feeling that there was probably no place in the world where the development of atomic weapons would have a better chance of leading to a reasonable solution, and a smaller chance of leading to disaster, than within the United States. I believe all these things that people said are true, and I think I said them all myself at one time or another.

But when you come right down to it the reason that we did this job is because it was an organic necessity. If you are a scientist you cannot stop such a thing. If you are a scientist you believe that it is good to find out how the world works; that it is good to find out what the realities are; that it is good to turn over to mankind at large the greatest possible power to control the world and to deal with it according to its lights and its values.

There has been a lot of talk about the evil of secrecy, of concealment, of control, of security. Some of that talk has been on a rather low plane, limited really to saying that it is difficult or inconvenient to work in a world where you are not free to do what you want. I think that the talk has been justified, and that the almost unanimous resistance of scientists to the imposition of control and secrecy is a justified position, but I think that the reason for it may lie a little deeper. I think that it comes from the fact that secrecy strikes at the very root of what science is, and what it is for. It is not possible to be a scientist unless you believe that it is good to learn. It is not good to be a scientist, and it is not possible, unless you think that it is of the highest value to share your knowledge, to share it with anyone who is interested. It is not possible to be a scientist unless you believe that the knowledge of the world, and the power which this gives, is a thing which is of intrinsic value to humanity, and that you are using it to help in the spread of knowledge, and are willing to take the consequences. And, therefore, I think that this resistance which we feel and see all around us to anything which is an attempt to treat science of the future as though it were rather a dangerous thing, a thing that must be watched and managed, is resisted not because of its inconvenience—I think we are in a position where we must be willing to take any inconvenience—but resisted because it is based on a philosophy incompatible with that by which we live, and have learned to live in the past.

There are many people who try to wiggle out of this. They say the real importance of atomic energy does not lie in the weapons that have been made; the real importance lies in all the great benefits which atomic energy, which the various radiations, will bring to mankind. There may be some truth in this. I am sure that there is truth in it, because there has never in the past been a new field opened up where the real fruits of it have not been invisible at the beginning. I have a very high confidence that the fruits—the so-called peacetime applications—of atomic energy will have in them all that we think, and more. There are others who try to escape the immediacy of this situation by saying that, after all, war has always been very terrible; after all, weapons have always gotten worse and worse; that this is just another weapon and it doesn't create a great change; that they are not so bad; bombings have been bad in this war and this is not a change in that—it just adds a little to the effectiveness of bombing; that some sort of protection will be found. I think that these efforts to diffuse and weaken the nature of the crisis make it only more dangerous. I think it is for us to accept it as a very grave crisis, to realize that these atomic weapons which we have started to make are very terrible, that they involve a change, that they are not just a slight modification: to accept this, and to accept with it the necessity for those transformations in the world which will make it possible to integrate these developments into human life.

As scientists I think we have perhaps a little greater ability to accept change, and accept radical change, because of our experiences in the pursuit of science. And that may help us—that, and the fact that we have lived with it—to be of some use in understanding these problems...

There are a few things which scientists perhaps should remember, that I don't think I need to remind us of; but I will, anyway. One is that they are very often called upon to give technical information in one way or another, and I think one cannot be too careful to be honest. And it is very difficult, not because one tells lies, but because so often questions are put in a form which makes it very hard to give an answer which is not misleading. I think we will be in a very weak position unless we maintain at its highest the scrupulousness which is traditional for us in sticking to the truth, and in distinguishing between what we know to be true from what we hope may be true.

The second thing I think it right to speak of is this: it is everywhere felt that the fraternity between us and scientists in other countries may be one of the most helpful things for the future; yet it is apparent that even in this country not all of us who are scientists are in agreement. There is no harm in that; such disagreement is healthy. But we must not lose the sense of fraternity because of it; we must not lose our fundamental confidence in our fellow scientists.

I think that we have no hope at all if we yield in our belief in the value of science, in the good that it can be to the world to know about reality, about nature, to attain a gradually greater and greater control of nature, to learn, to teach, to understand. I think that if we lose our faith in this we stop being scientists, we sell out our heritage, we lose what we have most of value for this time of crisis.

But there is another thing: we are not only scientists; we are men, too. We cannot forget our dependence on our fellow men. I mean not only our material dependence, without which no science would be possible, and without which we could not work; I mean also our deep moral dependence, in that the value of science must lie in the world of men, that all our roots lie there. These are the strongest bonds in the world, stronger than those even that bind us to one another, these are the deepest bonds—that bind us to our fellow men.

*J. Robert Oppenheimer*

## Questions and Topics for Discussion and Writing

1. What does Oppenheimer mean when he says that the scientists at Los Alamos built the bomb"because it was an organic necessity"? Are all possible scientific and technological developments organic necessities?
2. Discuss the view that science should not be secret, but should be open and available to all.
3. Oppenheimer lists many seemingly conflicting accounts of why the scientists worked on the bomb and what its true significance is, and then says that he believes them all to be true. What does he mean by this? Why does he say these things? How can they all be true?

# * The Academic Hustle

## Beth Savan

A research scientist investigates a problem, performs experiments, analyses data, and, upon obtaining worthwhile results, publishes them in a scientific journal. Other scientists can then read those journal articles and build upon those results in further work. Thus the scientific community around the world shares its findings and works together to increase the sum total of scientific knowledge in the world. This cooperative enterprise relies on a great deal of trust that what is said in reputable scientific journals is true—that experiments were in fact done the way it is said that they were done and the results that were said to be obtained were obtained. Scientists need to share their work with each other freely and doing so is encouraged and respected, and is the route by which individual scientists advance their own careers while assisting in the advancement of science in general.

But there is one serious flaw in the otherwise highly commendable process through which individual aspirations are reconciled with societal good: cheating. A scientist looking to leap into the limelight will dearly want to publish outstanding results and to get credit for being the first to discover some new and important fact. Careers are built on such things. However at odds with the spirit of science it may be, and however devastating the effect upon one's career it is for those who are caught, there is still a strong temptation to fudge a few numbers here and there, or hurry up an experiment that would just take too long or require too much material.

This article by Beth Savan reviews some recent cases of scientific fraud—why the fraud was perpetrated, why it went undetected for so long, and why the scientific community remains vulnerable to such abuses.

If you are a successful scientist, you start off and do experiments with your own hands. You then progress to a point where you have a small group, three or four people working with you, all on related projects. You then probably progress up the ladder to a point where you have a large group, say 15 to 20 people even, doing experiments, and you have varying degrees of involvements in the individual experiments....People will bring to you their data books and they will say: We got this result, we got that result—and you will discuss with them the interpretation, what experiment they should do next, and so on....If someone in your laboratory says to you this is the experiment I did, these are the reagents I used, and this is the result I obtained, you are not likely to doubt it. Not unless there is something that looks very awry about it or the person has an aberrant personality.

*B. Lewin,* personal communication, 1982.

This is the way Dr. Benjamin Lewin, editor of the respected journal *Cell*, explains how a senior scientist can be unaware of research irregularities in his own laboratory. Fortunately, deliberate falsification of research data is exceptional; for fraud is anathema to scientific endeavour. In every aspect of their work, scientists rely heavily on the techniques, methodology, knowledge, and theories developed by their colleagues. If they hope to advance their field, they must build on previous work—and must assume that their predecessors did the experiments they claimed to have done, and reported their results accurately.

Although this assumption has not been seriously disputed, there have always been cases of scientific deception. Analysis of these occasional and reprehensible acts of fabricating research evidence reveals several basic flaws in the way academic research is organized. Current and traditional scientific practices don't often prevent, and can sometimes even encourage tampering with data for personal gain. As long ago as 1830, Charles Babbage defined three kinds of fraud in scientific research: forging, cooking, and trimming. Over the years, examples of each kind of malpractice have been exposed in many scientific disciplines, and several recent cases of scientific finagling have become media sensations. Whether the proportion of dishonest scientists has increased, we can't know; but the public is certainly becoming much more aware of those who do tamper with their results.

A case of alleged deception involved a prolific post-doctoral researcher in the laboratory of Dr. Louis Siminovitch, at the University of Toronto in Canada. The post-doctoral fellow, Dr. Demetrios Spandidos, was working in the exciting new field of gene transfer. Simply put, his experiments were designed to transfer one piece of genetic material, DNA, from a certain type of cancer cell to a recipient, non-cancer cell, to determine whether cancer has a simple genetic basis. (This gene-transfer methodology, now called "DNA transfection," has since been perfected and is used routinely to study genetic characteristics.) Ultimately, the implications of this technique are enormous. Eventually, defective genes might be replaced by fit ones and malignant cells might be rendered benign by genetic manipulation.

This field was just breaking in 1977, when Spandidos reported that he had developed a new technique for transferring genes from one strain of animal cells to another. His new method of gene transfer seemed more successful than any other process.

At the Brookhaven Symposium on Biology in 1977, O.W. McBride and R.S. Athwal, two well-known cancer researchers at the National Institutes of Health in the United States, welcomed Spandidos and Siminovitch's results, and described the dramatic impact their work might have on the young field: "Spandidos and Siminovitch...have made significant modifications in the procedure for chromosome uptake and concurrently greatly increased the frequencies for CMGT [Chromosome Mediated Gene-Transfer] (by about 100-fold)... This method will become extremely important if it can provide similar transfer frequencies in interspecies transfer systems." It was, therefore, especially disappointing when Louis Siminovitch alleged that Demetrios Spandidos had never done some of their reported work. To understand how this might have happened without the knowledge of the senior scientist, it is important to remember the tremendous excitement and anticipation that surrounded this research in the mid–1970s. Spandidos was one of the pioneers of chromosome-mediated gene transfer, working at the leading edge of a brand-new field—the manipulation of purified genetic material.

Spandidos came to Louis Siminovitch's laboratory highly recommended by a respected scientist in the field who was a very good friend of Siminovitch's. And he worked very hard: "seven days a week, nine to twelve." Before long, he was churning out volumes of data. In only a year and a half, he had produced six papers, which were co-authored with Siminovitch and published in top journals. His impressive record prompted Siminovitch to recommend him for the Canadian Medical Research Council's prestigious Centennial Fellowship, which he received.

Spandidos continued to be enormously prolific, but Siminovitch began to wonder what all this productivity was based on: "I had a technician working with him and she felt he just didn't know how to do experiments. And that got me a little nervous because you would have to have real golden fingers to do as much as he had done so accurately and so well..." Nevertheless, Spandidos continued to produce the data and the publications to add to his and Siminovitch's growing list.

Then, in early 1978, a post-doctoral researcher, Bill Lewis, spent several months in Siminovitch's lab trying to apply Spandidos's techniques exactly as they had been reported. Despite repeated efforts, he failed to transfer genes from one batch of cells to another. Finally, he worked with Spandidos on one experiment. While Spandidos claimed it was a success, Lewis charged that sloppy technique had allowed the genes in fragments of the original cancer cells to contaminate the culture containing the recipient cells, i.e., that the supposedly transferred genes had never been actually incorporated into the recipient cells. (Subsequent efforts to repeat Spandidos's specific experiments on the part of Siminovitch and Lewis also failed, but the general gene transfer technology has proved successful.)

Another incident further eroded Siminovitch's confidence in Spandidos, and finally resulted in Spandidos's expulsion from the laboratory:

> One experiment just required an enormous amount of material, and I checked for the amount of mate-

> rial he was using and it wasn't there. And then the referee's report came and it said, "I'm amazed at how much this man has done, because it would have taken this and this many plates," etc. That was the [last] straw. So I called him in and...he says, "I haven't made anything up," and I said, "Well you would have needed six hundred plates for this experiment"—I don't remember the exact number and I didn't say 600 plates—he says, "Well they were there," and I said, "Well I looked at the thing and they weren't there," and he says, "Well they were there," and the next day he came in to see me and he says, "It wasn't 600 plates because I used bigger plates, it was 300 plates," so I said, "Well, I didn't see 300 plates," he said, "Well, I often used bottles too." And [he continued with] that sort of rationalization...
>
> I told him I did't want him to do any more experiments but I gave him two or three months to clean up his affairs...
>
> What no one's ever really understood is that I never let him go because we couldn't repeat the experiments. I let him go because he wasn't doing some of the experiments. I never did know how much was true and how much wasn't true.

In June 1978, Demetrios Spandidos returned to Greece, claiming that "pressure from Siminovitch (perhaps through unofficial conversations with colleagues) had made it impossible for him to secure another position in North America; but he continued to plead his case. He sent out a form letter to the Medical Research Council, as well as to the editors of several journals in the field. In it, he asserted his innocence and asked for their support in gaining "the fundamental right to defend myself in front of an independent scientific committee to present in more detail my side of this extremely important issue." Spandidos never got that opportunity.

The lack of official recognition of these allegations ultimately allowed Spandidos's career to prosper. At first Spandidos was forced to leave Canada, and was deprived of his MRC scholarship as a result. He lost the glowing job prospects that Siminovitch's influence and praise might have secured him. Likely job offers at renowned American biomedical research establishments failed to materialize after Siminovitch asked Spandidos to leave his laboratory. Yet less than a year later, Spandidos's MRC fellowship was restored to him when he secured a position in the laboratory of Dr. John Paul, at the Beatson Institute for Cancer research in Glasgow. He has published prolifically since then, authoring seven articles in 1984 alone, one of them in the prestigious journal *Nature*.

Unfortunately, the Spandidos-Siminovitch case is not an isolated example of research misconduct. A rash of scandals has erupted recently in prestigious biomedical research laboratories in the United States and Canada. Within the past several months alone a team of scientists who had published a major research article in *Science* admitted that "those biological data...are incorrect, and we wish, therefore, to retract the data and the conclusions based on them": the US National Institutes of Health (NIH) severely censured Dr. Charles J. Glueck, a medical scientist they had funded, whose published work they concluded incorporated "extensive inconsistencies and errors...[which] represent substantial departures from generally accepted standards of scholarship"; and a panel of the US National Institute of Mental Health found that Dr. Stephen E. Breuning, a well-known researcher of mental retardation, "knowingly, willfully, and repeatedly engaged in misleading and deceptive practices in reporting results of research." (Several institutions changed the treatment of patients to conform to Dr. Breuning's results, which have now been judged to be fraudulent.)

Between 1980 and 1982, the US National Institutes of Health was notified of forty-five allegations of research misconduct, and in 1983 similar charges were made to them a couple of times each month. In 1976, *New Scientist* surveyed its readers, who are, predominantly, engaged in scientific work, on the subject of cheating in science. More than 50 per cent of the respondents reported personal knowledge of intentional bias in research; a further 17 per cent learned of it from colleagues who had direct contact. Dr. Robert G. Petersdorf, Vice Chancellor for Health Sciences at the University of California at San Diego, blames the "highly competitive pressures in modern science." It is, he says, "similar to cheating on exams, where we keep finding scandals wherever we look for them."

Despite Daniel E. Koshland Jr.'s confident proclamation in an editorial in *Science* in January 1987 that "99.9999 percent of reports are accurate and truthful," there has been no comprehensive survey establishing what proportion of published scientific reports are either honest or dishonest.

In one of the most shocking of these cases, Mark Spector, a brilliant graduate student at Cornell University, fabricated a whole series of experiments. Spector had worked in the laboratory of the eminent biochemist Efraim Racker, who, like Siminovitch, was a senior investigator heading a large research team; as well, he had many other scientific and administrative responsibilities. During the year and a half Spector worked at Cornell, his colleagues were impressed by his intelligence, technical expertise, and devotion to his work.

Spector's results were spectacular—they promised a unifying framework for the cause of cancer. They were presented in *Science*, under the ambitious title, "Warburg Effect Revisited: Merger of Biochemistry and Molecular Biology." The article began with a quote from G.K. Chesterton: "There are no rules of architecture for a castle in the clouds." Unfortunately, Spector's promised panacea turned out to be just such a castle. His work was deception on a massive scale. After a colleague discovered in some of Spector's research material a radioactive isotope that had no business being there, Dr. Racker retracted their co-authored papers. He worked long days in the laboratory, repeating all of Spector's work, to find out whether any of it could be reliably reproduced. Spector left the laboratory and withdrew from Cornell in mid–1981.

Nor is the problem confined to North America. The eminent Swiss biologist Dr. Karl Illmensee of the University of Geneva was accused by his own lab colleagues of not adhering to standard protocol in his experimental work. (American researchers have also had difficulty repeating his spectacular success in manipulating mouse embryos.) Illmensee claimed to have cloned mice, to have incorporated human or rat genetic material into mice, and to have reared mice having genes from only their mothers. After one of his collaborators publicly discredited some of Illmensee's results in 1983, several commissions were set up to investigate Illmensee and his alleged achievements. The University of Geneva's external review committee drew no formal conclusion regarding Illmensee's motivation. They did, however, find that "the numerous errors and discrepancies found in the experimental protocols...throw grave doubts on the scientific validity of the conclusions." The committee appointed by the Jacson Laboratories at Bar Harbor, Maine, where Illmensee did some of his work in conjunction with Dr. Peter Hoppe, also found no evidence of fraud, but advised Illmensee and Hoppe to repeat their experiments using a more conclusive experimental design.

These incidents raise a number of crucial questions, concerning their causes as well as their consequences. What is the long-term impact of alleged research fraud on progress in the field? What is the ultimate effect of such scandals on the careers of those involved? How can the scientific community find out what really happened in such cases, and what can be done to ensure a just outcome? Most important, what can be done to prevent similar incidents from occurring in the future?

The advancement of any field of science can be traced by the publications in it. While there is often a great deal of lively debate among specialists at conferences and seminars, science is now too international and vast a venture to be communicated primarily by word of mouth: scientists must rely on publications to keep up to date. Consequently, the possibility that published work is ill-founded can have widespread repercussions.

There are few established rules for dealing with the allegation that some published papers are based on incorrect or inadequate data. Dr. Benjamin Lewin, editor of *Cell*, believes that a formal retraction is the wisest course:

> If we learn that there is some degree of doubt about some series of experiments which we have published, we usually say to the senior author that it would be proper to publish some sort of statement in the journal to clarify the position....there's a responsibility to a much wider range of people who read the article, who don't keep abreast of everything in the filed and who can be easily misled—and there's a responsibility to set the record straight. We do encourage authors in those circumstances to publish a retraction.

This is exactly what Efraim Racker did. He retracted his co-authored paper with Spector as soon as he suspected any wrongdoing, even before he know how much of their published work might stand up to his attempts to repeat it. In a formal letter of retraction to *Science*, where one of the suspect articles had been printed, he declared:

> ...an article was recently published in *Science* (17 July, p. 303) that I coauthored with Mark Spector, a graduate student in my laboratory. I feel compelled to withdraw some of the claims that we made in that article....We are now checking all published data, and it will take us many months before we know what is correct. We suspect that some of the data dealing with cells transformed by various tumor viruses are incorrect. We did not deal with these experiments in the *Science* article, but they were subjects of a paper in press which we are withdrawing. They were also presented by me and others in seminars, and I wish to withdraw these claims until we can verify them.

He sent a similar letter to *Cell*, which had also published a paper based on Spector's work.

Louis Siminovitch adopted a strikingly different course. In the two years that Demetrios Spandidos had worked in Siminovitch's laboratory, they had co-authored five papers in such journals as *Cell, The Proceedings of the National Academy of Sciences of the USA, Brookhaven Symposium on Biology,* and *Nature*. Two more articles had been submitted to journals and two were in press in April 1978 when Demetrios Spandidos was asked to leave Siminovitch's laboratory.

These four papers were withdrawn prior to publication, but the published articles were not formally retracted in the usual way: a letter of retraction in the journal of initial publication. Instead, Dr. Siminovitch initially relied on word of mouth: he phoned several of his friends to let them know what had happened; and in a public lecture in the summer of 1978, he voiced his suspicions regarding the gene-transfer research. Two years later, in his next publication on the subject, he included the following disclaimer: "One of us (L.S) was involved as a co-author in a series of papers in which the transfer of a variety of other genes by means of metaphase chromosomes was described. We have not been able to repeat those experiments, and we have reason to believe that the technologies described in those papers do not result in successful gene transfer."

*Somatic Cell Genetics*, which published the article quoted above, was not one of the journals that published the six suspect articles. Dr. Siminovitch gave two reasons for his chosen course of action: "I just didn't know how much was true and how much wasn't true: where do I stop and where do I begin, saying that they're wrong? ...but the second reason of course is that I was concerned that if I said that, I would have a lawsuit....How do you prove that you hadn't done an experiment? I mean, he could always say that the conditions have changed and the salts have changed, and you could go on for years that way." One thing is certain: if the scientific community at large is informed of the concern promptly and effectively, the damage will be minimized. But this happens all too rarely. In the case of Spandidos and Siminovitch, such action was not taken and, incredibly, the several articles co-authored by Spandidos and Siminovitch are still cited in publications on related matters. In 1980, for example, they were referenced very favourably in a widely read review of chromosome-mediated gene transfer. If Dr. Siminovitch's suspicion is correct, that some of the experiments were not performed, many researchers may be on expensive and time-consuming wild goose chases, trying to follow up on Spandidos's apparently exciting results.

The procedures followed in several similar scandals in the United States have ensured that most allegations of scientific fraud in that country are eventually investigated and judged, and that some disciplinary action is taken when cheating has been established. Thorough audits by external review teams have determined the extent of the deception in various recent American incidents. The US National Institutes of Health, a major research funding agency, has been particularly rigorous in acting on allegations of research misconduct. They have investigated and publicly reported on a long series of cases of research malpractices, most recently, in May 1987, by excluding Dr. Charles Glueck from NIH funding and peer-review service and by notifying the publishers of his discredited work. The National Institute of Mental Health has been even more aggressive in its review of Dr. Stephen Breuning's work by distributing its report to his employer and to professional boards and associations and by initiating action to recover grant funds he had misused. Such action has not always been immediate, however; investigations generally begin several months after the concerns have been raised.

Almost a whole year elapsed in the case of Vijay Soman, an assistant professor at Yale Medical

School who was accused of plagiarizing a manuscript he had reviewed. Despite the assurances of Soman's supervisor and then vice-chairman of the Department of Medicine, Dr. Philip Felig, that Soman's data were credible, a professor at Harvard Medical School, Dr. Flier, was called in to thoroughly audit the data that had been published in an article Soman and Felig had co-authored. This article was found to misrepresent the experimental results; Soman was asked to resign immediately, which he did. His supervisor, Felig, was forced to give up his recently assumed and very prestigious past as chairman of medicine at the Columbia College of Physicians and Surgeons in New York. All the other raw data that Soman had produced at Yale were subsequently audited and found to be seriously inadequate.

Eventually, eleven papers based on these experiments were retracted.

A similar procedure was followed in the case of John Roland Darsee, who was caught concocting data in May 1981 at Harvard Medical School's Cardiac Research Laboratory. When it was learned that Darsee's work on a major National Institutes of Health (NIH) study was also suspect, the university set up a committee of senior academics to investigate the affair. The NIH also appointed four university-based cardiologists to do an independent audit. As a result of these investigations, Darsee was barred from the National Institutes of Health and disqualified from funding for ten years.

Sadly, Canadian official behaviour is in stark contrast to the relatively vigorous US investigations of allegations and subsequent disciplinary actions. Spandidos and Siminovitch seem to have been protected by the lack of any conclusive investigation of the suspicions concerning their co-authored work. Spandidos has always asserted that the work he did in Siminovitch's lab was valid, and has since cited it favourably. When he was asked to leave the laboratory by Siminovitch, one of the first things Spandidos did was write to the Medical Research Council of Canada (MRC), pleading his innocence and requesting that the council set up a committee to investigate the allegations against him. Dr. James M. Roxburgh, then director of the grants program at the MRC, which had paid for the questionable research effort, replied to Spandidos, refusing his request. Roxburgh later defended this action, saying: "It was just not considered necessary or desirably to set up such a committee....These allegations were simply hearsay to us. They were reported to us by Spandidos. They were never made publicly by his supervisor or to us formally by his supervisor." Dr. Roxburgh did admit that, although there had been no formal allegations made by Louis Siminovitch against Demetrios Spandidos, he had "known Lou Siminovitch for many many years, and naturally...there were informal conversations off the record." The MRC feels that it's up to the supervisor to establish the truth in these cases. But Louis Siminovitch was clearly reluctant to carry out his own investigation, partly because "it might have taken even more than two years," and partly because it may, in the end, have been inconclusive.

Many scientists argue that the consequences of incidents such as the one involving Spandidos and Siminovitch are ultimately insignificant. Dr. Robert Weinberg, an associate professor at the Center for Cancer Research at MIT and a pioneer in the development of gene-transfer techniques, doesn't view the Spandidos and Spector affairs as a "disaster for science... They're amusing, and they have an interesting human element to them, but in a field in which there are many scientists interested, these two incidents represent small tempests in a teapot and are soon glossed over. One soon finds out what the real answers are and forgets these minor impediments which happened along the way."

Other scientists try to rationalize such controversies, assuring themselves and others that actual perpetrators are incredibly rare, pathologically abnormal individuals. No sane scientist, they say, would risk the inevitable censure of the academic community by engaging in such a risky venture. The consequences of Soman's large-scale distortion of data are undoubtedly grave. But for a young scientist working largely independently, with the prospect of a glittering career, the risks may seem meagre when compared with the potential rewards.

Strong similarities between the cases that continue to come to light point to a problem inherent in the way scientific research is organized. The system of academic-career advancement, coupled with the laboratory hierarchies prevalent in most research establishments, creates both the motive and the mechanism for scientific malpractice. A detailed analysis of these structures is essential to understanding how to prevent future scandals of this kind.

Many of these cases have occurred in exciting new areas of biomedical research, where the potential achievements are enormous; so too is the pressure to be the first to prove its initial promise. Competition for the prestige, positions, and generous salaries awaiting successful scientists in these fields is, therefore, very stiff. After Vijay Soman was caught having doctored his data, he was reported in *Science* to have blamed his transgressions on "the cut-throat rate of research." "My actions," he was quoted as saying, "were done in the midst of significant pressure to publish these data as fast as possible so as to obtain priority." Indeed, competition to enter and secure a position in academic science has been increasing; across-the-board demand for this kind of employment has remained high, despite cuts to research-funding agencies. This competitive atmosphere is heightened for those young scientists working in other scientists' laboratories who are seeking their crucial first positions as independent researchers.

Dr. Louis Siminovitch admits that this can pose difficulties for many young researchers:

> Post-doc years are probably the best years of a scientist's life, or should be. They're the most traumatic for them, because they've always got to think, there's a job problem at the end.... As a post-doc you're given money, you're given space, you're given all of biology there to be discovered, and you're young and what else? ...essentially it is an idyllic situation. except for the problem that at the end of the road you've got a job to find, and of course, you're deciding whether you're going to make it. Not everybody responds to...that situation.

It is no mere coincidence that all the cases I have referred to involved young researchers, in their late twenties or early thirties, on the brink of emerging from the laboratories of eminent senior scientists. While studying with such people certainly enhanced the young researchers' credibility, it also meant they were forced to work much more independently than they would have in the more modest labs of less eminent scientists. For as they become better known, senior scientists usually spend less and less time in the laboratory milieu.

A successful scientist's growing reputation is almost always accompanied by mounting administrative commitments. To provide a large laboratory with the necessary space, equipment, and financing demands time: time to apply for grants, to hire staff, to screen new students, and to maintain a favourable position within the institution. The director of a large research laboratory also has responsibilities: to the university department, to the agencies providing funding, and to the community at large. Senior scientists participate in academic and professional committees to keep abreast of the latest developments in the field and to maintain the political influence that can ease their and those of their students' future careers.

Grant-awarding bodies and scientific journals need well-established scientists for their peer-review committees, as they naturally want the best advice on what work to fund or publish. Well-known scientists who maintain some research involvement are also sought after on political or educational advisory committees and on journal editorial boards. The external commitments of established research scientists mount with their reputations and, inevitably, these investigators spend progressively less time doing hands-on research. This trend is supported by an investigation of the science faculty of US graduate schools, who are generally relatively well-established in their fields: the National Science Foundation in the United States reported that "scientists and engineers on the faculty of graduate schools in the United States spend *an average* of only 16 hours a week on research."

Louis Siminovitch is a good example of the successful senior scientist. In 1981 he was awarded the $25,000 Gairdner Foundation Wrightman Award for outstanding leadership in medicine and medical science in Canada. The citation included the following statements: "His awareness of the impact of science on society is as important a part of his scientific makeup as is his enthusiasm for new ideas and his insistence on high standards in his experimental work." In the same year, he was also one of the first recipients of the newly established Izaak Walton Killam Memorial prizes, in recognition of his "distinguished lifetime achievement and an outstanding contribution to the advancement of knowledge."

His list of external advisory positions is impressive. In 1982, he was vice-president of the National cancer Institute of Canada (NCIC); he was on the executive of the Medical Research Council of Canada; on the Ontario Council of Health; on the executive of the Ontario Council of Health; on the board of the Ontario Cancer Treatment Research Foundation; on the Advisory Board for Cancer Treatment and Prevention of the National Cancer Institute in the United States; chair of the Canadian Broadcasting Corporation Advisory Committee on Science and Technology; chair of the special Terry Fox Committee of the NCIC; an editor of about seven journals; on the board of the Mount Sinai Research Foundation; and on the advisory board of the Connaught Medical Research Laboratories. He subsequently became director of the new research unit of Toronto's Mount Sinai Hospital.

As Siminovitch himself admits, "I've got my fingers in quite a few pies." This administrative work doesn't itself produce an iota of data. What it does is ensure that Siminovitch has the power to choose the direction of research, the supply of money and equipment required to produce the results, and the access to publications that he needs to perpetuate his laboratory empire.

Siminovitch's preoccupation with professional and administrative activities may well have prevented him from having a very detailed knowledge of the work in progress in his lab: "So in fact it's very common, all over science, that the professors don't work in the lab, or very little, and that the work is done mostly by students, post-docs and technicians....I'm a bit unusual in the sense that I have quite a bit of extraneous activity, compared to most people. But you know I work pretty hard, I'm here all the time, when I'm not out of town, at a meeting or something like that, but I don't think I'm overly unusual." The political nature of senior scientists' work brings with it definite benefits. They maintain the knowledge and the friendships to get their own and their associates' work published and funded. One of their attractions for young scientists is that their connections with journal editors can greatly ease the often lengthy and frustrating business of getting research reports published. Dr. Lewin, editor of *Cell*, describes the trust and high expectations which await papers co-authored by established scientists:

> Look at it this way. You work in some area of molecular biology. You publish very distinguished papers for a series of years, all your papers are well received, you make what are acknowledged to be significant contributions to the field. When you write another paper, that paper has a certain degree of credibility. And that's undeniable. By virtue of the reputation of your laboratory in the past, when you write a new paper it brings to that an enhanced degree of credibility. Perhaps an enhanced degree of credibility is too strong—it brings to it a belief that what you say you have done in this paper is exactly what you *have* done.

In a kind of Catch–22 situation, this borrowed or vicarious credibility may allow the work of a relatively junior scientist to escape serious critical scrutiny altogether: the senior scientist co-authoring resulting papers may not have the time or inclination to go over the research procedures, analysis, and interpretations in detail, and journal reviewers may also neglect to investigate these matters as intensively as they should, assuming the respected senior scientists has already done so. This seems to have been the case for many of the 109 papers published by Dr. John Darsee and his 47 co-authors. A detailed analysis of these papers, by Walter Stewart and Ned Feder at the National Institutes of Health, published in *Nature* in 1987, found that all but two of the papers contained from one to thirty-nine errors and discrepancies per paper (an average of about twelve). All of these "lapses from generally accepted standards of research," which included unacknowledged republication, misleading statements, discrepancies in figures, and impossible data (e.g. parents aged eight or nine), were presumably missed by co-authors, journal reviewers, and editors alike. Dr. Darsee, now judged to have repeatedly published fraudulent work, may have ben responsible for many of these "lapses," but the presence of reputable co-authors likely meant that these papers did not receive a very critical reception.

There is a kind of unwritten contract between the junior and senior scientist: the senior scientist secures the research funds, using his or her reputation to ensure generous grants to cover the cost of equipment, chemicals, and even the junior researcher's salary. In exchange, the junior investigators churn out data and draft the papers, which they and the senior scientists co-author, to provide the senior researcher with proof that his or her research funds are well spent.

A hierarchical lab structure is established: the junior researchers work in the laboratory, largely unsupervised, and merely meet with the senior scientist to discuss experimental results and revisions to draft papers. It is easy to see how intense competition for rich scientific rewards in a hierarchical laboratory setting could conspire to tempt ambitious young scientists to concoct especially interesting data. These serious flaws in the structure of current scientific research make an internal checking system for science all the more necessary.

Scientists place great faith in what is called "experimental replication": if research work cannot be reproduced exactly by an independent investigator, it becomes suspect. But this ultimate arbiter of scientific "truth" cannot always be relied upon. If the experiments are tedious to perform and their results not surprising, they are likely to be assumed to be correct without follow-up. It's clearly more exciting and prestigious to author original work than merely to report on the reliability of someone else's data, and for this reason, exact replications of experiments are not popular. In an analysis of publication decisions, T. Sterling, professor of computer science at Simon Fraser University in Vancouver, noted that "significant results published...are seldom verified by independent replication." This means that fraud not discovered in the perpetrator's laboratory may remain undetected for years, and scientists may be building on false results. As Sterling notes: "The possibility thus arises that the literature...consists in substantial part of false conclusions.

Of course, it is impossible to know how often data are fabricated and whether the cases of proven fraud are the tip of an iceberg of scientific subterfuge or merely rare, atypical incidents. Nevertheless, the similar circumstances of these cases indicate that there may be something in the lab situation that spawns them. The combination of pressure, competition, large potential gains, and a lack of close supervision may convince some junior scientists that the risk of getting caught is minimal compared with the possible rewards of deception. All research institutions should, therefore, develop consistent and enforceable procedures for preventing and dealing with fraud. Several American institutions have established mechanisms for ascertaining the validity of questionable results and for censuring guilty parties when detected. (For projects funded by the US National Institutes of Health, such procedures are now required by law.)

Dr. Eugene Braunwald, who worked with Dr. John Darsee, later found to have engaged in fraudulent research practices, goes so far as to urge investigation and criminal prosecution in alleged cases of dishonest research reporting. At the very least, clear, comprehensive laboratory protocols, requiring, for example, open data books and the involvement of the laboratory directors in experimental design and analysis, would help to define appropriate behaviour and identify individual responsibility. But independent institutional initiatives will not change the system of scientific advancement and rewards that gives cheating its appeal; they can only make it more difficult. Therefore what is much more important is to develop a style of science that provides less incentive for deception. For this to be effective, universities and other research institutions must collaborate in a concerted effort to make researchers more accountable, labs more egalitarian, and to make the rewards of scientific success, publications and grants more equitably available to all those conducting scientific enquiries. Research dollars will always be scarce, relative to the demand for spending them, and some open, fair, and responsive procedures should ensure that they go to those scientists with the best research ideas.

---

From Beth Savan, *Science Under Siege*, 89–108 (Toronto: CBC Enterprises, 1988).

---

### Questions and Topics for Discussion and Writing

1. Why do senior scientists find themselves in the embarrassing position of having to retract articles which they have co-authored from major journals?
2. Compare the motivation for a scientist to report false data from experiments with the motivation for high level amateur athletes to take illegal performance enhancing drugs.
3. How can science best rid itself of the problem of falsified research? (Or at least minimize it.)

*Beth Savan*